ANNUAL REVIEW OF ENTOMOLOGY

EDITORIAL COMMITTEE (1971)

ANNUAL REVIEW OF ENTOMOLOGY

RAY F. SMITH, *Editor*
University of California

THOMAS E. MITTLER, *Editor*
University of California

1971

PUBLISHED BY
ANNUAL REVIEWS, INC.
IN CO-OPERATION WITH THE
ENTOMOLOGICAL SOCIETY OF AMERICA

———

ANNUAL REVIEWS, INC.
4139 EL CAMINO WAY, PALO ALTO, CALIFORNIA

94306

ANNUAL REVIEWS, INC.
PALO ALTO, CALIFORNIA, U.S.A.

Standard Book Number 8243-0116-1

© 1971 BY ANNUAL REVIEWS, INC.

Library of Congress Catalog Card Number: A56–5750

FOREIGN AGENCY

Maruzen Company, Limited
6, Tori-Nichome Nihonbashi

Tokyo

PRINTED AND BOUND IN THE UNITED STATES OF AMERICA
BY GEORGE BANTA COMPANY, INC.

CONTENTS

REPRINTS

The conspicuous number (6000 to 6015) aligned in the margin with the title of each chapter in this volume is a key for use in the ordering of reprints.

Beginning with July 1970, reprints will be available from all future *Annual Reviews* volumes. Reprints of most articles published in the *Annual Reviews of Biochemistry and Psychology* from 1961 to 1970 and the *Annual Reviews of Microbiology and Physiology* from 1968 to 1970 are now maintained in inventory.

Available reprints are priced at the uniform rate of $1 each, postpaid. Payment must accompany orders less than $10. The following discounts will be given for large orders: $5–9, 10%; $10–24, 20%; $25 and over, 30%. All remittances are to be made payable to Annual Reviews, Inc. in U. S. dollars. California orders are subject to a sales tax. One-day service is given on items in stock.

For orders of 100 or more, any *Annual Review* article will be specially printed and shipped within 6 weeks. Reprints which are out of stock may also be purchased from the Institute for Scientific Information, 325 Chestnut Street, Philadelphia, Pa. 19106. Direct inquiries to the Reprint Department, Annual Reviews, Inc.

The sale of reprints of articles published in the *Reviews* has been expanded in the belief that reprints as individual copies, as sets covering stated topics, and in quantity for classroom use will have a special appeal for students and teachers.

EDWARD ARTHUR STEINHAUS
1914–1969

Publication of the *Annual Review of Entomology* was first announced in Members Newsletter, Number 2, of the Entomological Society of America, dated August, 1954. It reported briefly on the first meeting of the A.R.E. Editorial Committee held May 28, 1954, in Urbana, Illinois. A photograph of the participants in this meeting appeared in Volume I, No. 1, of the Bulletin of the Entomological Society of America. The next issue of the Bulletin had an article by E. A. Steinhaus (although the article is unsigned) giving additional details on the *Review* and topics and authors for the first two volumes. The preface to the first volume, which appeared in January 1956, discussed briefly those early events that led to its publication. The objectives as conceived by the Editorial Committee were also mentioned. Six more volumes followed with E. A. Steinhaus as editor or co-editor with Ray F. Smith. In 1963, Steinhaus resigned as editor because of the pressure of other responsibilities.

Nothing in the documents mentioned above does justice to the key role of Edward Arthur Steinhaus in the establishment and founding of the *Annual Review of Entomology*. First of all, it was his idea. His completely selfless dedication and unbounded enthusiasm sold others on it. He explored the possibilities for publication. He convinced the Governing Board of E.S.A. of its importance and, at a time when Society finances were especially tight, persuaded the Society to support it financially. He negotiated the contract between Annual Reviews, Inc. and the Entomological Society of America. Then, with enthusiasm, skill and hard work, he developed the *Review* into the outstanding and valuable series it is today.

When the word was passed that Steinhaus had died in his home at Newport Beach, California, on October 20, 1969, friends and fellow scientists all over the world were shocked and saddened. We who worked closely with him and knew of his many contributions to the *Annual Review of Entomology* feel that we have lost a distinguished scientist, a true friend, and a driving spirit whose influence will always be felt in the pages of these volumes. A fuller treatment of his life and accomplishments may be found in the *Journal of Invertebrate Pathology*, **14**, iii–v (1969) and the *Journal of Entomology*, **63**, 689–691 (1970).

BLOOD-SUCKING BEHAVIOR OF TERRESTRIAL ARTHROPODS

6000

BRIAN HOCKING

Department of Entomology, University of Alberta,
Edmonton, Alberta, Canada

This review covers the blood-feeding behavior and associated physiology of terrestrial arthropods from the stage where the host search begins, up to and including the successful completion of a blood meal. Sufficient morphology is included to make this comprehensible and the outcome of the blood meal is referred to only where it illuminates behavior. Some parts of this field have been reviewed more recently than others so that starting dates of the literature covered range backwards from about 1960; no complete coverage beyond the end of 1969 has been attempted. Tatchell (207) has reviewed the subject from a parasitological viewpoint, and there are several relevant *Annual Review* papers (53, 70, 117, 146, 190, 197).

More than half the literature on blood-sucking arthropods relates to mosquitoes and nearly half of this relates to *Aedes aegypti* or the *Culex pipiens* complex because of their domesticity and resulting ready colonization. Mosquitoes are atypical blood-sucking arthropods and these two species are far from representative of mosquitoes. To offset this bias I have given special attention to feral and less familiar blood-suckers. Medical and economic interests have led to a bias in the literature toward mammals as hosts; I have done what I can to redress this.

THE BLOOD-SUCKING GROUPS AND THEIR HOSTS

General.—The insects and the arachnid order of Acarina include nearly all the blood-sucking forms, although the feeding behavior of most arachnids, and even indeed of the Onychophora, is carnivorous and not far removed from blood-sucking, and there are records of other arachnids perhaps feeding on blood. It is necessary here to make a distinction between blood-sucking and feeding by injection of enzymes into the body of the victim, usually arthropod, followed by sucking back the liquified contents, including, of course, blood. This is a distinct feeding method, which has, however, not always been distinguished. Among the arachnids the Acarina is a derived rather than a primitive group. Within it, the Ixodoidea and a few families of mites in the suborder Mesostigmata are blood-suckers.

1

In the hemimetabolous insects three orders, the Anoplura, the Thysanoptera, and the Hemiptera all include blood-sucking forms. These are all derived orders, the Hemiptera being the most specialized. All groups with sucking mouthparts have some representatives which suck blood.

In the Holometabola, blood-sucking groups are also to be found in three orders, the Lepidoptera (7, 8), the Diptera, and the Siphonaptera. Again, these are derived groups, all of them members of the Panorpoid complex (96) and among the more specialized holometabolous orders. Of all blood-sucking groups the flies, by our present knowledge, are the most interesting (54). The Hymenoptera have sucking mouthparts but are not known to use them to feed on blood. The beetles, alone among the major orders of insects, have made but two moves towards fluid feeding, the flower-feeding genus *Nemognatha* (Meloidae) with mouthparts resembling those of moths, and the very peculiar ectoparasite of beavers, *Platypsyllus castoris*, with reduced mandibles and of uncertain feeding habits.

In each of these three major groups then, a trend from the primitive manipulative mouthparts adapted for the ingestion of solid food, toward a cibarium and pharynx adapted for suction is apparent. Along with this, usually, the mouthparts and adjoining structures are adapted for piercing or penetrative cutting and piping in fluid food. In each group, some smaller groups have used this equipment to exploit the blood of the other animals as a source of food.

Hosts.—We know what hosts these blood-sucking arthropods feed on by direct observation (3, 193, 218), by inference from availability (32, 74), or by the presence of a specific parasite in a blood-fed animal, and by histological or biochemical (precipitin or haemagglutination) tests on stomach contents (22, 37, 94, 223, 225). Known hosts (Table 1) include annelids (123), insects (89), arachnids (229), all the major classes of terrestrial vertebrates, and even fish (192). Marsupials (102, 133) and monotremes (55) are recorded, and most other species of mammals and birds. The reptiles include crocodiles (30, 204), turtles (47, 90, 161, 195), lizards (63, 93), and snakes (91), and the amphibia (168) frogs (26, 145), toads (91), and salamanders (91). While only the mammals are known to be fed on by all blood-sucking groups, this is probably only because as mammals we have the best opportunity to observe them. Similarly, the tastes of mosquitoes (lower Diptera) may be only apparently catholic; many more blood-sucking species are found in this group than in others. Much of our recent knowledge of hosts has come from the search for reservoirs and alternate hosts of the neurotropic viruses and other agents of human disease. We have much to learn, but the general picture is not one of highly specific relationships, although such do occur, for example, *Simulium euryadminiculum* on the common loon (138).

TABLE 1. THE HOST RELATIONSHIPS OF BLOOD-SUCKING TERRESTRIAL ARTHROPODS, AND SOME REFERENCES (NUMBERS)

Hosts	Invertebrates			Vertebrates					
				Poikilotherms				Homoiotherms	
Blood-Suckers	Annelida	Arachnida	Insecta	Teleosts	Dipnoi	Amphibia	Reptiles	Birds	Mammals
Arachnida									
Chelonethi									**
Acarina, mites	**	**	**			**	**	39	39
Acarina, ticks							102	102	102
Hemimetabola									
Anoplura								228	222
Thysanoptera			**						5
Hemiptera			157			216*	216*	216	216
Holometabola									
Lepidoptera									7
Diptera, lower	123	229	54	192	**	91	91	91	91
Diptera, higher						30*	30	30	13
Aphaniptera							63*	12	12

* Under laboratory conditions only.
** Considered likely, but no conclusive evidence that blood is sucked.

Notes on the blood-sucking groups.—Since there is little review literature outside of the mosquitoes, we may comment briefly on the other groups. In the Acarina, the hard ticks, Ixodidae, and the soft ticks, Argasidae, are exclusively blood-sucking, frequently on a succession of hosts, and members of a few families in the mesostigmatid mites share this habit with them.

In the Anoplura, all the sucking lice, Siphunculata, and a few species of Mallophaga (228) have been recorded sucking blood. In the Thysanoptera a dozen or more species in several families have been recorded (5, 6) "biting" man, several of them clearly ingesting blood. This parallels rather closely the situation in the Hemiptera, in which members of four families, the Cimicidae, Reduviidae (Triatominae), Anthocoridae, and Polyctenidae are primarily vertebrate blood feeders, and occasional sucking of vertebrate blood is indulged in by several dozen species (157, 215) representing seven families of normally plant-feeding bugs. Some 20 species representing four families of predators on insects, normally feed on the haemolymph of other arthropods (Bequaert in 215).

Recent reports of blood feeding by Lepidoptera (7, 8) involve the Pyralidae, Noctuidae, and Geometridae feeding on some ten species representing the four mammalian orders Artiodactyla, Perissodactyla, Proboscidea, and Primates. The habit appears to have been derived from that of feeding on lachrymal secretions shown by at least 20 species representing five families: the interest taken by Lepidoptera in salt is well known. Downes (54) has reviewed the blood-sucking habit in Diptera in the context of parasitology; the difference between the "lower" blood-sucking Diptera, including the Blephariceridae, phlebotomine Psychodids, Culicidae, Simuliidae, Ceratopogonidae, Tabanidae, and Rhagionidae, and the "higher" groups including a section of the Muscidae, the Hippoboscidae, Streblidae, and Nycteribidae is rather marked and centers around the loss of the mandibles. These structures have recently been reported in the Tanyderidae, and Chironomidae, and blood meals have been found in the Chaoboridae (54). The Siphonaptera is an isolated group, probably derived from forerunners of both the Diptera and the Lepidoptera, the Mecoptera. Fleas in some respects parallel the position of the lice in the Exopterygota, and are exclusively blood feeders with accompanying loss of wings and reduced sensory equipment.

SENSORY EQUIPMENT AND RESPONSES TO STIMULI

Visual and radiant heat receptors.—In the blood-sucking mites, ticks, lice, higher flies except Muscidae, and the fleas, eyes and ocelli are reduced or lacking. This does not mean, however, that these animals are not sensitive both to light (42, 95) and to photoperiod (176). Other groups have well-developed compound eyes, concerning the function of which we have learned little; Sato, Kato & Toriumi (183) have followed up earlier morphological work by Sato and found remarkable sensitivity in the dark-adapted eyes of *Culex pipiens;* Pucat, Reddy, and I have provided dimensional data for eyes of three species of simuliids, two tabanids, and *Aedes aegypti* and indicated resolving power in *Hybomitra affinis* nearly equal to that of *Apis* (99). We have as yet no proof of visual color discrimination in any blood-sucking insect, but data on differential landing on colors continue to accumulate (43, 114). Wood & Wright (232) have shown responses by *A. aegypti* to movement in the visual field. This species has also been shown to rest preferentially on a surface highly reflective to infrared (143). Earlier claims of this for other species have not been substantiated; indeed, the only insect species for which infrared sensitivity has been clearly established and characterized is the buprestid *Melanophila acuminata* (59). Sippell & Brown (188) and Wenk (224) have also demonstrated the importance of visual stimuli, and Eldridge (58) has shown sensitivity to photoperiod in *Culex pipiens.*

Chemical, humidity, and temperature senses.—The recognition and description of cuticular sense organs of the antennae (150), palps (151), tarsi (2), labellae (2), labrum (23, 71), and cibarium (23, 164, 208) has at last been accomplished for some species and the advent of stereoscanning electron microscopes should facilitate this for many others. Many of these organs on *A. aegypti* and *Culex tarsalis* are clearly olfactory, others probably humidity and temperature receptors. Von Gernet & Buerger (71) and Buerger (23) have shown the four apical sensilla on the labra of females of many genera of mosquitoes to be lacking from males and nonblood-sucking females. Bässler (9) has found behavioral evidence for carbon dioxide receptors on the maxillary palpi of male mosquitoes; Kellogg (113) has recorded the responses of humidity receptors on the antennae and carbon dioxide receptors on the maxillary palpi of female *A. aegypti* electrophysiologically. Lacher (125), recording from single olfactory receptors on female *A. aegypti* antennae, found A1 cells to be excited by fatty acids and depressed by essential oils, while A2 cells were depressed by lower fatty acids and excited by higher. The electroantennogram approach (200) has not been used successfully with blood-sucking arthropods.

Halcrow (87) has described tarsal sensilla on a number of species of mosquitoes; Hudson (106) reports "sensory-like" structures at the tip of the hypopharynx of *Wyeomyia smithii, Aedes atropalpus,* and *Aedes stimulans.* Haller's organ (185) in the Acarina should repay study with newer techniques. Boeckh, Kaissling & Schneider (14) have reported on the function of insect olfactory receptors in general, and a bibliography with abstracts covering 2265 entries up to 1959 on all aspects of smell in insects is available (98).

Edwards (57) found the tibiae and antennae of *Rhinocoris carmelita* sensitive only to water, and gustatory discrimination to lie in the tip of the rostrum and an unspecified internal site.

Feir, Lengy & Owen (60) have measured acceptance thresholds of *Culiseta inornata* tarsal and labellar receptors for sucrose and glucose. The values they give are well above the levels normal in human blood, as are those of the tarsal receptors of tabanids examined by Lall & Davies (127). Salama (180) has measured the thresholds of acceptance for 39 sugars and of rejection for 25 electrolytes in *A. aegypti* and shown that cibarial receptors alone control the ingestion of liquids and are stimulated by blood, sugars, and unacceptable compounds. He has also (181) found differences in the sugar responses of *C. inornata.*

Hosoi's (103, 104) original demonstration that adenosine 5-phosphate induced gorging in *C. pipiens* has been followed up by many workers with other species: *Rhodnius* (179), tabanids (126), *A. aegypti,* and *Ornithodoros* (66, 67). Galun (66) has suggested that it is the chelating properties of nucleotides which are responsible for their withdrawing divalent zinc leading to membrane depolarisation.

Yelizarov (234) recorded potentials from chemosensory sensilla in two species of ixodid ticks and found increases in impulse frequency in response to increasing concentrations of amino acids and sodium chloride. Owen (164) was able to evoke feeding responses by applying water and sugar, but not blood, to the chemoreceptor hairs on the labella and ligula of *Culiseta inornata*.

Mechanoreceptors.—Only minor contributions have been made concerning sense organs of this type, despite the fact that they have been known for many years to be important in ticks (134). Stretch receptors consisting of three multiterminal neurones on each side of the anterior wall of the cibarial pump (169) have been described in *Glossina austeni* and *Calliphora erythrocephala*. It is suggested that these monitor, and perhaps control the rate of cibarial pumping.

Grzybowska (84) described the innervation and an unusual distribution of campaniform sensilla on the wings of *Ornithomyia biloba*. Day (46) has suggested a pressure sensitivity for a sense organ described in the head of mosquitoes and other Nematocera.

Host Finding and Host Selection

The problems.—There are many hundreds of papers on the physiological responses of blood-sucking arthropods, especially mosquitoes, in the laboratory to the various stimuli supposedly provided by their hosts, and hundreds more ecological studies on the mechanisms of host finding and selection in the field. Rarely has so much work yielded so little consensus of opinion; results which are apparently contradictory abound, even in the same paper (227) and more so among different workers (19, 211, 186, 213). There appear to be three interrelated reasons for this. First, the variation in behavior among taxa at all levels, even the subspecific (86, 120, 137, 140). This is well known but often lost sight of. Autogenous strains naturally behave differently from their anautogenous counterparts (197). Second, variation in physiological state; facultative autogeny often appears as a simple reflection of nutritional state (36, 221); the internal and the external environments influence thresholds of response, and host-seeking behavior seems to show a special plasticity to stimulus history (124). In particular, an endogenous rhythm of avidity with a period of four or five days has been demonstrated in several species (25, 128, 160) and may be unrelated to the ovarian cycle. There is also daily rhythm (35, 73, 140, 154, 162, 198, 217, 220). Third, behavioral work is always more difficult in the absence of an adequate knowledge of sensory physiology; even for mosquitoes we are only now acquiring a working knowledge of the structure and distribution of receptors (150, 165, 199) and we are still a long way from a full understanding of their functions. For other groups we are still more ignorant. Rudolfs' (177) 1922 assessment of the host-finding procedure in mosquitoes is remarkably similar to that of Clements' (34) in 1963.

TABLE 2. A Summary of the Principal Stages in Securing a Blood Meal and the Stimuli Controlling Them in Winged Terrestrial Arthropods

Stage	Stimuli		
	Facilitating/ Initiating	Orienting	Terminating or inhibiting
Rest	Cold, darkness bright light/ Engorgement, exhaustion, tarsal contact	Gravity, wind optomotor, odor	Mechanical, visible movement, no tarsal contact
Flight & searching	Moderate wind/ mechanical, visible movement, no tarsal contact	Wind, odor, optomotor	Cold, bright light, tarsal contact, exhaustion
Settling on host	Exhaustion/ Odor, convection current	Convection current, temp. gradient, gravity	Surface texture & temp., repellent odor
Probing & biting	Hunger/ Surface temp. sugars, ATP	Sometimes gravity	Abdominal tension. All thresholds elevated.
Take-off	Nourishment/ Abdominal tension, voiding of fluid	Darkness?	Excessive engorgement?

Kalmus & Hocking (111) divided the process of host-finding and feeding by *Aedes* mosquitoes into eight phases and studied the conditions favoring them and the stimuli required for initiation, inhibition, and termination of each phase. Some of the findings which have been supported, or extended to at least other groups of the lower flies by subsequent work, are summarized as modified by such work in Table 2. This sequence of events can be somewhat simplified for flightless arthropods but it remains applicable though the stimuli differ. It is clear that at least some of the steps in the host-finding process are links in a chain reflex (233). An experimental study which starts in the middle of a chain will yield very different results from one which starts at the beginning.

Generalizations.—In the light of these problems, it is hazardous but none the less important to generalize on host-finding behavior. The favorable climatic conditions for host-seeking are mostly moderate: wind speed and turbulence, however, are low, and at least for flying insects, barometer low, falling, or both, with or without electrical activity; light intensity changing

(118, 139, 231). Somewhat different factors are important for flightless forms (39, 40, 122). Some at least of the effect of environmental conditions arises from the masking or emphasizing of host stimuli such as convection currents.

The stimulus which evokes approach in daylight seems to be usually the visual one of host movement, perhaps reinforced by host form, especially for simuliids (64, 219, 224). Clearly, there is little specificity at this stage, as witnessed by the readiness of tabanids to pursue vehicles of many kinds; all that seems to be needed is an appropriate combination of size, distance, and speed. The mechanism may operate at night, but only for insects with good vision, such as mosquitoes and tsetse flies. A positive anemotaxis triggered and maintained by an olfactory stimulus (69, 111, 116, 147, 156, 187) seems to be more important at night. This, too, calls for good but less specific vision—recognition of relative movement of the immediate environment, rather than of a host at a distance. Vision, then, is of great importance in host-finding by Diptera; it is of much less importance in bugs (226), although it is well developed, and of almost no importance in most fleas, lice, mites, and ticks. Glasgow (74) considers that with *Glossina* olfactory stimuli are of primary importance at the beginning of the host search, visual stimuli take over when the host is sighted, and thermal stimuli are dominant in initiating probing.

Giglioli (72) has shown that searching flight by *Anopheles* is channelled along visible lines such as the junctions between woodland and grassland. I have data on this for other genera and also on behavior which may conflict with it, namely, cross-wind searching flight which is theoretically advantageous since it provides the best prospect of picking up an odor cone (101, 115). Both types of behavior lead into upwind flight.

Kairomones?—We have seen from the distribution of blood-sucking groups among host groups that the phenomenon of blood-sucking is not in general a specific one. A single blood-sucking species may feed on several species of reptiles, birds, mammals (91), and occasionally amphibia and other insects; but at a given time in a given habitat there is commonly, as Ardö (4) has said, a single obvious host. Through techniques such as the precipitin test (21, 148), this gives an illusion of specificity, and since only the olfactory sense (at least in arthropods) has the versatility to account for specificity, much effort has been devoted to finding a specific attractant emitted by the animal we are most interested in—man—whereby he attracts mosquitoes. Every imaginable part of the human body and its products must by now have been fractionated in every possible way in the search for a specific chemical attractant (16, 121, 144, 189, and other papers by these authors and their associates). While certain attractive materials have been found: lysine and alanine (18), sex hormones (172–174), carbamino compounds (17), L-lactic acid (1), methionine (108, 109), unspecified (184), nothing proves to be as attractive as the intact man. This seems to suggest that blood-sucking arthropods recognize man for what he is, by his complex of effluvia and not by any specific indicator. Only this can explain the evident discrimination by mosquitoes between men, women, and children (171, 194, 196, 210, 212), between one race and another (171), and even between one individual and another (15, 111, 230). This simply means that the kairomone (20)—a useful term—is a complex mixture of which sex hormones are, incidentally, a component.

One of the most controversial materials in the whole story is carbon dioxide, which sometimes attracts (16), sometimes repels (227), and sometimes is inactive (12). Results have often been complicated by using it in the form of "dry ice," thereby introducing a severe temperature gradient. When this form increases general catches in a field trap the increase would seem to be attributable to chilling and CO_2 anaesthesia rather than to any specific attractiveness. Low concentrations at ambient temperature certainly seem to activate mosquitoes rather than orient them, and the same seems to apply to fleas (12) and mites (182) but it has been used to collect unengorged ticks (24), and has also been shown attractive to *Culicoides* (48, 158). The facts that the same olfactory stimuli which evoke attraction usually evoke repulsion at higher concentrations, and that adaptation follows continued stimulation have not always had due attention.

Thermal stimuli.—As we have seen there is little to support the idea of direct sensitivity to the infrared radiated by warm-blooded hosts as an important stimulus. This is not to be expected since cold-blooded hosts are common and the blood-sucking habit probably antedates warm blood. There is even evidence of specialization on cold blood in *Aedes canadensis* on turtles (38) and in *Deinocerites* (61). But this is not to say that thermal stimuli are not important: temperature gradients (which may be set up by thermal radiation) have been shown to be significant for ticks (134), bed bugs (170), lice (222), and tsetse flies (74).

The convection currents set up by warm-blooded animals under temperate conditions are well within the range of sensitivity of stationary insects and of insects on the wing if these have good visual reference points. It is important to remember that when, in the tropics, air temperature exceeds skin temperature, convection currents are cool and downward, rather than warm and upward. The importance of warm upward currents for host-finding has been demonstrated for ticks (135), for blackflies (224), for fleas (187), and for mosquitoes (114). I know of no work on cool downward currents, though I have observed these; they may account, in part, for ankle feeding by tropical species such as *Simulium damnosum* (100, 136, 167). Higher hydrostatic pressure of the blood may also have favored this habit.

It has recently been shown (119), however, that an important contribution to the effectiveness of convection currents is made by the odors (note the plural) they carry. In fact, one may look upon convection currents as simply up-ended versions (either way) of the cones of odor which are so important in the field. Their smallness tends to limit their effectiveness to indoors and sheltered situations.

There is some evidence of thermonegative orthokinesis as the mechanism which leads to settling down on the host and quiescence during the meal. This may be reinforced by various other "comfort" stimuli such as, for example, in *Glossina,* an easy grip with the tarsi or spines (49); the host's comfortable temperature is also that for its visitor.

Mobility.—The differences in host-seeking behavior between the Acarina, Hemimetabola, and Holometabola are mainly those due to their different mobilities. The mites and ticks, lacking wings at all stages, have to depend on the mobility of their hosts and rely largely on alerting stimuli. Hence, as an insurance against failure, enormous meals are ingested which nourish the maturation of as many as 8000 eggs. The hemimetabolous insects, even if winged as adults, affect mostly hosts with definitive dwellings, necessarily rather specific, from which they stray for the most part, only on their hosts. This is true also of fleas. Only the Lepidoptera and the flies have full mobility in the adult, blood-sucking stage, and are hence both free to, and indeed obliged to, seek out a suitable host. They alone have and need some freedom of choice; they are correspondingly less host-specific.

Evidence of this opportunism in feeding is to be seen in the many reports in the literature, usually under the term cannibalism, of unfed insects robbing host blood from fed insects of the same or different species, often when caged together in the laboratory (13, 27, 149, 170, 191). Further evidence is seen in species, not normally haematophagous, feeding at wounds made by blood feeders (68, 206). Seasonal changes have quantitative and qualitative effects on the availability of hosts; *Culex nigripalpis* feeds to a maximum on mammals from July to October, shifting back to birds in winter and spring (56).

THE FEEDING APPARATUS AND METHOD OF FEEDING

One-way and two-way.—There are two basically different types of blood-feeding apparatus. The first has a single and usually broad channel extending distally from the definitive mouth and serving to convey both the salivary secretion outward and the blood meal inward, usually serving these two purposes alternately. This type is found in the mites and also in the ticks (81, 82) in which it is formed from parts of the pedipalpal coxae (= hypostome) below and the chelicerae above; the salivarium, carrying the opening of the salivary glands, is withdrawn into the body. This type is also found in the Lepidoptera in which it is formed by the juxtaposition of the hollowed medial faces of the elongate galeae of the maxillae. Nearly all members of this order as adults feed exclusively on nectar which does not ordinarily call for the outward passage of saliva. The second type of feeding apparatus, found in all other groups, has two channels extending forward (or rotated to downward or even backward) from the mouth, which is a small dorsal one and through which blood is ingested, and a ventral one smaller still, usually within the hypopharynx, through which saliva is passed out into the host. In most insect groups the way is prepared for the sucking tube by the mandibles, with or without the aid of the maxillae. The loss of mandibles by the Lepidoptera and the higher flies has led to this function being taken over by the galeae and the prestomal teeth of the labium, respectively. Among the blood-sucking lower flies only, only the female takes blood, and both sexes usually feed on nectar.

Feeding procedures.—Two other dichotomies in feeding procedures are found. First, there are those insects which insert the tip of the food canal with some precision into a blood vessel of the host and are known as vessel feeders or solenophages (76, 129), and those which cut and stab hither and yon and feed on the extravasated blood, and are known as pool feeders or telmophages. Despite translation difficulties, the saving of two characters for each term seems unwarranted now that so few know Greek. Vessel feeders include lice (130), bugs (51, 132), fleas (131), most mosquitoes (77), and *Melophagus* (159). Pool feeders include ticks (81), *Leptoconops* (Ceratopogonidae) (129), tabanids (52), and some of the higher flies (129). The second dichotomy, not evidently related to the first, is in the rate of feeding: females of most of the hard ticks and some ceratopogonid flies that feed on other insects require several days (up to 14) to complete a blood meal, and larvae and nymphs of *Otobius megnini* require perhaps months (80); members of the other groups usually accomplish this in a similar number of minutes. O'Rourke (163), however, has found a bi-modal distribution of time taken to feed in a population of *Aedes aegypti* and suggests that the slow feeders are, in fact, pool feeders.

The end of the structure carrying the food canal is necessarily small in vessel feeders, so that the food canal itself is also small. The pressure difference required for a given blood flow rate varies inversely as the fourth power of the radius of the tube; rapid feeding has survival value; pressure is clearly a problem (10, 208), and since extravasation means a pressure drop, other things being equal, vessel feeding is to be preferred since it allows the full capillary blood pressure of the host to be exploited by the parasite. Also influencing the duration of feeding is the size of the meal, which reaches a maximum in ticks up to 30 times the basic weight of the animal (110) as compared with from one to five times in most other groups (13, 41, 62, 201).

Blood-sucking insects may have a single or double channel inside the mouth as well as outside. In the Lepidoptera and flies, but not in fleas or Mecoptera (166), a channel branching ventrally from the posterior end of the esophagus leads to a diverticulum which, in the lower flies, generally serves as a holding organ for sugar meals while blood meals go direct to the mid-gut (45, 141, 153). Sugars, through cibarial receptors, appear to provide the stimulus which evokes the shunt, but all is not yet clear since there is wide variation within and between species (214). In the mosquitoes two further diverticula run dorsally from a similar position into the anterior angles between the tergo-sternal and longitudinal median dorsal muscles, the antagonistic muscles of flight. My suggestion (97) that the dorsal diverticula may provide these muscles direct access to sugars and hence explain the remarkable ability of many mosquitoes to fly at low temperatures has been questioned (33) on the grounds of the equally remarkable impermeability of the ventral diverticulum to water. Romoser & Venard (175), however, have shown that the dorsal diverticula have a different origin from the ventral, so that they could differ also in permeability. In the hippoboscid *Stilbometopa,* both gut and ventral diverticulum receive blood (205).

Salivation and termination.—Although it is theoretically possible for arthropods with two-way mouthparts to eject saliva and ingest blood at the same time, this apparently does not always happen. Kashin (112) has developed for mosquitoes an electrical method similar to one used for aphids (152) to record these events. A direct current electrical circuit through the host and the blood-sucker is completed by the fluid columns in the mouthparts when these are continuous. A decrease in electrical resistance marks salivation, an increase marks swallowing; both occurred in *A. aegypti* at a frequency of about 5 cycles per second. This technique has been successfully adapted for ticks (83). Friend & Smith (65) have combined it with ATP and split-screen television in a study of the feeding of *Rhodnius* in which they found continuous salivation during probing. A radioactive technique (50) has been used to measure the quantities of saliva left in the host by *A. aegypti,* and cutting the main salivary duct (107) of *Aedes stimulans* has shown that saliva is the source of the reactions to bites in man and rabbits; this does not prevent feeding. The reactions to bites from the point of view of the biter have had little attention except in the ticks which seem to ingest more than blood (82). Most hard ticks probably are cemented to their hosts by a secretion presumed to be of salivary origin (81); this has a cortex of carbohydrate-containing protein, stabilized by quinone tanning and disulphide linkages, and an internum of lipoprotein (155).

Gwadz (85) has shown, for six species of mosquito, that cutting the ventral nerve cord in the second abdominal segment results in blood meals up to four times normal size and postulates that stretch receptors posterior to this mediate the end of engorgement. Bennet-Clark (11) indicates that in *Rhodnius* this results at a critical level of back pressure from the abdomen, but Maddrell (142) suggests that stretch receptors in the abdomen are involved here too.

Blood Meals in the Life Cycle

The story of blood-sucking behavior is not complete with the withdrawal of the mouthparts and the take-off or departure of the insect, for most arthropods take more than one blood meal. It is this, in fact, that enables them to transfer the members of a third group of organisms, the parasites and pathogens which are more or less equally at home in the host and the blood sucker and which constitute a major reason for our interest in blood-sucking behavior. No arthropod takes a blood meal immediately after completing one. Time, from a few hours in lice to a year or more in ticks, is needed to restore blood hunger. Usually something else is needed too: the previous meal must be converted either into tissue for growth, which calls for moulting, or into eggs or larvae of the next generation.

The mites, ticks, and hemimetabolous insects which feed regularly on blood do so at all stages of the life cycle, the holometabolous insects only as adults, and except for the fleas and the higher flies, only in the female sex. Apart from reports of larval tabanids (202) and rhagionids (123) feeding on blood which may be more properly regarded as external digestion, the only holometabola which can feed exclusively on blood are some fleas and the higher flies, in each group through a highly peculiar adaptation. In some fleas, the adults feed, as it were, on behalf of the next generation, excess blood is voided in the feces which accumulate in the nest of the host where they are fed upon by the larvae. Presumably digestive enzymes of the adult carry out at least some predigestion on behalf of the larvae. Clearly, such species are associated only with hosts which have permanent or semipermanent sleeping quarters or nests.

In the higher flies and the Polyctenidae, adults do even more for the offspring, which are nourished to maturity inside the body of the female, on a secretion originating from the blood of the host. Incidentally, the remarkable convergence between these two groups explains the taxonomic commuting of the polyctenids between the Diptera, the Anoplura, and the Hemiptera, where they now seem likely to remain.

The ecdysotrophic cycle.—By analogy with the well-established expression "gonotrophic cycle" used to describe the alternation of blood feeding and egg development and laying in, for example, mosquitoes, I have coined the term ecdysotrophic cycle to describe a parallel alternation of blood feeding and moulting in mites, ticks, and hemimetabolous insects.

In respect of the ecdysotrophic cycle, the argasid ticks are more akin to lice than to bugs. There may be more than one nymphal stage, all spent on the host, and more than one blood meal between each moult; but each moult is preceded and followed by a period without feeding. The adults also may

California Baptist College
Riverside, California

feed and oviposit more than once and remain in close association with the host. Larval feeding may take from a few minutes to a few days (44). In the ixodids the usual life cycle involves a single meal as a larva, followed by a moult, a single meal as a nymph, followed by a further moult, and a single meal as an adult followed by a single egg laying. In *Rhipicephalus sanguineus* a direct relationship is found between the weight of the engorged female and the number of eggs laid (203). The three meals may be separated by intervals as long as a year, and each may be from a different host species, the first two commonly from rodents, the last commonly from an ungulate. The readiness of both male and female adult *Rhipicephalus appendiculatus* to feed increases for at least five weeks after the final moult (110).

Gooding (unpublished findings) has shown that human lice will feed again 3 hr after a meal in all except the first instar which may require an interval up to 6 hr. It is normal for lice to feed several times between moults, but the interval between meals is longer before and after moulting. Gooding (75) has also found that young adult male *Pediculus* are unable to inseminate females until at least 12 hr have passed after they have had a blood meal; that is, they are anautogenous. Gregson (79) found that adult male and female *Dermacentor andersoni* need to feed for five days and one day, respectively, before they will mate, but that they feed only slowly until mating has taken place (78).

Among the bugs, the cimicids can be induced to moult after a succession of small meals (209), but in nature a single large meal between moults appears more usual, after the manner of ixodid ticks (216). Females of some species have been reported to lay a very small number of eggs when neither sex has had a blood meal after the last moult (178). In *Triatoma sanguisuga,* Hays (92) has found a poorly defined gonotrophic cycle in the adult females as well as the ecdysotrophic cycle in both sexes. In bugs, as in ticks and lice, there is a direct relationship between blood ingested and eggs laid. Reproduction is quite strongly influenced in both *Cimex* (216) and *Rhodnius* (29) by the blood meal of the male though not to the extent seen in *Pediculus*. The number of eggs laid by a female mated with an unfed male may be reduced to 25 percent, but the proportion of fertile eggs may be unchanged.

The gonotrophic cycle.—The gonotrophic cycle in mosquitoes has been well reviewed by Clements (34) and more recently Downes (54) has extended this to the rest of the blood-sucking Diptera and further. Here, as in *Rhodnius,* the hormonal control of the cycle initiated by the stretching of the abdomen has been worked out.

Nothing appears to be known of the relationship between blood feeding and reproduction in the Lepidoptera. Although most records of this strange habit do not give the sex of the moth, both sexes apparently take blood, and it appears that while lachrymal secretions pass to the ventral diverticulum, blood goes to "the digestive system"—presumably the stomach. This is a strange parallel to the separation of sugars from blood in the Diptera. Perhaps this is related in some way to the special problems in digestion of blood, not the least of which is the presence of enzyme inhibitors (105).

Phase relationships.—In investigating the responses of blood-sucking arthropods to their hosts and their reactions to host stimuli, a knowledge of the precise nature of the ecdysotrophic and gonotrophic cycles of the species under study is essential for there is great variation in detail. Furthermore, the variations arising from these cyclic events are superimposed in many species upon three other largely independent cycles of change: a general endogenous rhythm (25, 128); a circadian rhythm, both intrinsic and in climatic factors; longer term lunar, seasonal six-monthly (in the tropics), or annual (higher latitudes) cycles in climatic factors, especially photoperiod (176), with the consequent induction of aestivation or hibernation. Not only this, but there is also demonstrable variation in the attractiveness of hosts related to diet, age, sex, and color, influenced in their turn by environmental changes. Some of these are also cyclic: as might be expected from Roessler's work, the attractiveness of women to mosquitoes appears to be influenced by the menstrual cycle. The phase relationships between the four cycles in the blood-sucker and a similar number in the host, decide whether the hunger drive of the blood-sucker will be undetectable or unstoppable, or where in the spectrum between these limits it will lie. It is small wonder that even in the laboratory it is easier to get discordant than consistent results, or that even consistent results in the laboratory may not be reproducible in the field (Fig. 1).

CONCLUSIONS

The main orders of arthropods which include blood-sucking forms, with the probable exception of fleas, were in existence in the Permian. Some, at least, of the families of blood-sucking species go back to the Jurassic period. As for their hosts, the amphibia arose in the upper Devonian, the reptiles in the Permian, the birds and mammals in the Cretaceous. Hoogstraal (102) considers the ticks to have arisen in association with reptiles in the late Palaeozoic or early Mesozoic period, partly on account of the preference of structurally primitive forms for lizard hosts.

In the Acarina, and more or less contemporaneously in the Hemimetabola and the Holometabola, trends toward fluid feeding seem to have developed. It may be supposed in each instance that the initial adaptation was to feeding on exposed fluids, but that as plants and animals generally became better adapted to life on really dry land, by developing substantially impermeable skins and cuticles, a taste for nutrient fluids demanded a combination of suction with cutting or penetration. The same drying conditions of the Mesozoic period which led to these changes, emphasized more strongly the moisture conservation problem of the terrestrial arthropods, especially acute by virtue of their small size and large ratio of surface to volume. This was met by reducing water loss with a remarkable epicuticle, by a reduced mouth opening, and by increasing water intake by specializing on fluid food. Initially this was probably plant sap, somewhat deficient in appropriate amino acids; plants were more abundant, so that the nearest neighbour distance was smaller and the move from host to host less hazardous. Digestion was easier.

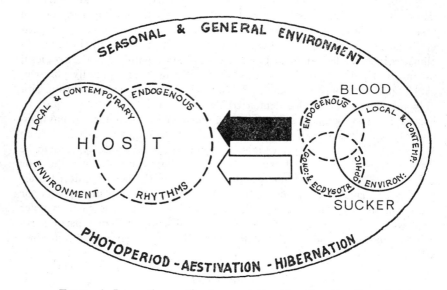

FIGURE 1. Interacting cyclic changes influencing the attraction of blood-suckers to their hosts.

The most readily available blood initially was probably arthropod blood; that many different groups today feed on this suggests that it may have been an early development. On the other hand, the poor performance of mosquitoes on the blood of other insects seems to deny this (88), though the data are scanty. Ticks, with their very large meals, must have had to turn to something larger, and once an example was set, other groups were probably quick to follow (68); there is some evidence of aggregation in blood-feeding, both within and among species (28) as in many other insect activities.

The readiness with which plant-feeding Hemiptera will attempt a blood meal and the switch of Lepidoptera from a saline secretion to blood indicate the facility with which this change can be made. The restricted occurrence of blood-feeding in the bugs reflects the difficulty imposed by the increased nearest neighbour distance of vertebrates and the delaying of the mobility of flight to the adult stage of the insects. At the same time, the greatest need for the extra protein of blood is in the immature stages.

In the Holometabola the possibility of divergent adaptation between the immature and the adult stages has been used by the lower flies to develop highly mobile adult females specializing on blood for egg production. The aquatic immature stages tie the adult females at both ends of their existence to the home of the amphibia; water is also a periodic forum frequented by all vertebrates which also have a water problem. Preferences for cold-blooded hosts persist among the lower flies (31, 61). Only in fleas and the higher flies has the ability to run the whole life cycle on blood, as in the Acarina and Hemimetabola, been achieved. Most of these insects have paid the price in that they have had to stick so close to their hosts that they have become louse-like. *Glossina,* however, has retained full mobility and sensory equipment—and needs both.

ACKNOWLEDGMENTS

I am glad to acknowledge the aid of colleagues in sending reprints and allowing me to use unpublished material.

LITERATURE CITED[1]

1. Acree, F., Sr., Turner, R. B., Gouck, H. K., Beroza, M., Smith, N. 1968. L-Lactic acid: a mosquito attractant isolated from humans. *Science* 161:1346–47

2. Adams, J. R., Holbert, P. E., Forgash, A. J. 1965. Electron microscopy of the contact chemoreceptors of the stable fly, *Stomoxys calcitrans. Ann. Entomol. Soc. Am.* 58(6):909–17

3. Anderson, J. R., DeFoliart, G. R. 1961. Feeding behavior and host preferences of some black flies in Wisconsin. *Ann. Entomol. Soc. Am.* 54(5):716–29

4. Ardö, P. 1958. On the feeding habits of the Scandinavian mosquitoes. *Opusc. Entomol.* 3:171–91

5. Arnaud, P. H., Jr. 1970. Thrips "biting" man. *Pan-Pac. Entomol.* 46(1):76

6. Bailey, S. 1936. Thrips attacking man. *Can. Entomol.* 68(5):95–98

7. Bänziger, H. 1968. Preliminary observations on a skin-piercing bloodsucking moth (*Calyptra eustrigata* (Hmps.)) in Malaya. *Bull. Entomol. Res.* 58:159–63

8. Bänziger, H., Büttiker, W. 1969. Records of eye-frequenting Lepidoptera from man. *J. Med. Entomol.* 6(1):53–58

9. Bässler, U. 1958. Versuche zur Orientierung der Stechmücken: die Schwarmbildung und die Bedeutung des Johnstonschen Organs. *Z. vergl. Physiol.* 41:300–30

10. Bennet-Clark, H. C. 1963. Negative pressures produced in the pharyngeal pump of the blood-sucking bug, *Rhodnius prolixus. J. Exp. Biol.* 40:223–29

11. Bennet-Clark, H. C. 1963. The control of meal size in the blood-sucking bug, *Rhodnius prolixus. J. Exp. Biol.* 40:741–50

12. Benton, A. H., Lee, S. Y. 1965. Host finding reactions of some Siphonaptera. *Proc. Int. Congr. Entomol., 12th, London, 1964,* 792

13. Bequaert, J. 1953. The Hippoboscidae or louse-flies of mammals and birds. I Structure, physiology and natural history. *Entomol. Americana* 32:97–116

14. Boeckh, J., Kaissling, K. E., Schneider, D. 1965. Insect olfactory receptors. *Cold Spring Harbor Symp. Quant. Biol.* 30:263–80

15. Brouwer, R. 1960. Variations in human body odour as a cause of individual differences of attraction for malaria mosquitoes. *Trop. Geogr. Med.* 12:62–66

16. Brown, A. W. A. 1958. Factors which attract *Aedes* mosquitoes to humans. *Proc. Int. Congr. Entomol., 10th, Montreal, 1956* 3:757–63

17. Brown, A. W. A. 1966. The attraction of mosquitoes to hosts. *J. Am. Med. Assoc.* 196:249–52

18. Brown, A. W. A., Carmichael, A. G. 1961. Lysine and alanine as mosquito attractants. *J. Econ. Entomol.* 54:317–24

19. Brown, A. W. A., Sarkaria, D. S., Thompson, R. P. 1961. Studies on the responses of the female *Aedes* mosquito. Part I. The search for attractant vapours. *Bull. Entomol. Res.* 42(1):105–14

20. Brown, W. L., Jr., Eisner, T., Whittaker, R. H. 1970. Allomones and kairomones: transspecific chemical messengers. *BioScience* 20(1):21–22

21. Bruce-Chwatt, L. J., Göckel, C. W. 1960. A study of the blood-feeding patterns of *Anopheles* mosquitoes through precipitin tests. *Bull. WHO* 22:685–720

22. Bruce-Chwatt, L. J., Garrett-Jones, C., Weitz, B. 1966. Ten years' study of host selection by Anopheline mosquitoes (results for the period 1955–1964). Unpublished document. *WHO Vector Control 66.202.* 59 pp.

23. Buerger, G. 1967. Sense organs on the labra of some blood-feeding Diptera. *Quaest. Entomol.* 3:283–90

24. Burgdorfer, W. 1969. Ecology of tick vectors of American spotted fever. *Bull. WHO* 40:375–81

25. Burgess, L. 1959. Probing behaviour of *Aedes aegypti* (L.) in response to heat and moisture. *Nature* 184:1968–69

[1] For supplementary bibliographic material (56 pages), order NAPS Document 01091 from ASIS National Auxiliary Publications Service, c/o CCM Information Corporation, 909 Third Avenue, New York, N.Y. 10022.

26. Burgess, L., Hammond, G. H. 1961. Mosquitoes feeding on a frog. *Can. Entomol.* 93:670–71

27. Burton, G. J. 1963. Feeding of *Mansonia tiltillans* (Walker) on other mosquitoes. *Mosquito News,* 28: 164

28. Büttiker, W. 1962. Notes on two species of Westermanniinae from Cambodia. *Proc. Roy. Entomol. Soc. London, Ser. B* 31(5–6):73–76 (1962)

29. Buxton, P. A. 1930. The biology of a blood-sucking bug, *Rhodnius prolixus. Trans. Entomol. Soc. London* 78:227–36

30. Buxton, P. A. 1955. *The natural history of tsetse flies.* Mem. London Sch. Hyg. Trop. Med. 10, xviii + 816. London: K. Lewis Co.

31. Chaniotis, B. N. 1967. The biology of California *Phlebotomus* under laboratory conditions. *J. Med. Entomol.* 4:221–33

32. Chorley, T. W. 1948. *Glossina pallidipes* Austen attracted by the scent of cattle-dung and urine. *Proc. Roy. Entomol. Soc. London* 23:9–11

33. Christophers, Sir S. R. 1960. *Aedes aegypti (L.) The Yellow Fever mosquito. Its Life History, Bionomics, and Structure.* Cambridge Univ. Press, xii + 739 pp.

34. Clements, A. N. 1963. *The Physiology of Mosquitoes.* Oxford: Pergamon Press, ix + 393 pp.

35. Corbet, P. S. 1966. Diel patterns of mosquito activity in a high arctic locality: Hazen Camp. Ellesmere Island, N.W.T. *Can. Entomol.* 98(12):1238–52

36. Corbet, P. S. 1967. Facultative autogeny in Arctic mosquitoes. *Nature* 215(5101):662–63

37. Crans, W. J. 1969. An agar gel diffusion method for the identification of mosquito blood-meals. *Mosquito News,* 29(40):563–66

38. Crans, W. J., Rockel, E. G. 1968. The mosquitoes attracted to turtles. *Mosquito News* 28:332–37

39. Cross, H. F., Wharton, G. W. 1964. A comparison of the number of tropical rat mites and tropical fowl mites that fed at different temperatures. *J. Econ. Entomol.* 57: 439–43

40. Cross, H. F., Wharton, G. W. 1964. A comparison of the number of tropical rat mites and tropical fowl mites that fed under varying conditions of humidity. *J. Econ. Entomol.* 57:443–45

41. Crosskey, R. W. 1958. The body weight in unfed *Simulium damnosum* Theobald. and its relation to the time of biting, the fat body and age. *Ann. Trop. Med. Parasitol.* 52:149–57

42. Darling, P. G. E. 1969. Observations on the relation of light to the dropping of the tick *Ixodes texanus* Banks. *J. Entomol. Soc. Brit. Columbia* 66:26–28

43. Davies, D. M. 1961. Colour affects the landing of blood-sucking black flies on their hosts. *Proc. Entomol. Soc. Ont.* 91:267–68

44. Davies, G. E., Mavros, A. J. 1956. An atypical *Ornithodoros hermsi* from Utah. *J. Parasitol.* 42(3): 293–96

45. Day, M. F. 1954. The mechanism of food distribution to midgut or diverticula in the mosquito. *Austr. J. Biol. Sci.* 7:515–24

46. Day, M. F. 1955. A new sense organ in the head of mosquitoes and other nematocerous flies. *Austr. J. Zool.* 3:331–35

47. DeFoliart, G. R. 1967. *Aedes canadensis* (Theobald) feeding on Blanding's turtle. *J. Med. Entomol.* 4(1):31

48. DeFoliart, G. R., Morris, C. D. 1967. A dry ice-baited trap for the collection and field storage of hematophagous Diptera. *J. Med. Entomol.* 4(3):360–62

49. Dethier, V. G. 1954. Notes on the biting responses of tsetse flies. *Am. J. Trop. Med. Hyg.* 3:160–71

50. Devine, T. L., Venard, C. E., Myser, W. C. 1965. Measurement of salivation by *Aedes aegypti* (L). feeding on a living host. *J. Insect Physiol.* 11:347–53

51. Dickerson, G., Lavoipierre, M. M. J. 1959. Studies on the methods of feeding of blood-sucking arthropods II. The method of feeding adopted by the bed-bug (*Cimex lectularius*) when obtaining a blood-meal from the mammalian host. *Ann. Trop. Med. Parasitol.* 53(3):347–57

52. Dickerson, G., Lavoipierre, M. M. J. 1959. Studies on the methods of feeding of blood-sucking arthropods III. The method by which

Haematopota pluvialis obtains its blood-meal from the mammalian host. *Ann. Trop. Med. Parasitol.* 53 :465–72

53. Downes, J. A. 1958. The feeding habits of biting flies and their significance in classification. *Ann. Rev. Entomol.* 3 :249–66

54. Downes, J. A. 1970. The ecology of blood-sucking Diptera—an evolutionary perspective. In : *Symposium on Ecology and Physiology of Parasites.* Toronto : Toronto Univ. Press

55. Dyce, A. L. 1963. An observation of mosquitoes biting an *Echidna. J. Entomol. Soc. Queensland* 3 :83

56. Edman, J. D., Taylor, D. J. 1968. *Culex nigripalpus:* Seasonal shift in the bird-mammal feeding ratio in a mosquito vector of human encephalitis. *Science* 161 :67–68

57. Edwards, J. S. 1962. A note on water uptake and gustatory discrimination in a predatory Reduviid. *J. Insect Physiol.* 8 :113–15

58. Eldridge, B. F. 1968. The effect of temperature and photoperiod on blood-feeding and ovarian development in mosquitoes of the *Culex pipiens* complex. *Am. J. Trop. Med. Hyg.* 17(1) :133–40

59. Evans, W. G. 1966. Perception of infrared radiation from forest fires by *Melanophila acuminata* De Geer. *Ecology* 47(6) :1061–66

60. Feir, D., Lengy, J. J., Owen, W. B. 1961. Contact chemoreception in the mosquito *Culiseta inornata* (Williston) ; sensitivity of the tarsi and labella to sucrose and glucose. *J. Insect Physiol.* 6(11) : 13–20

61. Fisk, F. W. 1941. *Deinocerites spanius* at Brownsville, Texas, with notes on its biology and a description of the larva. *Ann. Entomol. Soc. Am.* 34 :543–50

62. Foulk, J. D. 1967. Blood meal size of *Leptoconops kerteszi. Mosquito News* 27(3) :424

63. Fox, I., Fox, R. I., Bayona, I. G. 1966. Fleas feed on lizards in the laboratory in Puerto Rico. *J. Med. Entomol.* 2(4) :395–96

64. Fredeen, F. J. H. 1961. A trap for studying the attacking behaviour of black flies, *Simulium arcticum* Mall. *Can. Entomol.* 93(1) :73–78

65. Friend, W. G., Smith, J. J. B. Feeding in *Rhodnius prolixus:* Visual

observations of mouthpart activity and salivation, and their correlation with changes of electrical resistance. *J. Insect Physiol.* In press

66. Galun, R. 1967. Feeding stimuli and artificial feeding. In Seminar on the ecology, biology, control and eradication of *Aedes aegypti,* 16–20, Geneva, Switzerland. *Bull. WHO* 36(4) :509–93

67. Galun, R., Kindler, S. H. 1968. Chemical basis of feeding in the tick *Ornithodoros tholozani. J. Insect Physiol.* 14(10) :1409–22

68. Garcia, R., Radovsky, F. J. 1962. Haematophagy by two non-biting muscid flies and its relationship to tabanid feeding. *Can. Entomol.* 94 : 1110–16

69. Gatehouse, A. G. 1970. Host finding by the stable fly, *Stomoxys calcitrans* (L.) *Proc. Roy. Entomol. Soc. London, Ser. C.* 34(8) :37

70. Gelperin, A. 1971. Neuronal control of feeding and nutrition. *Ann. Rev. Entomol.* 16 :365–78

71. Gernet, G. von, Buerger, G. 1966. Labral and cibarial sense organs of some mosquitoes. *Quaest. Entomol.* 2 :259–70

72. Giglioli, M. E. C. 1965. The influence of irregularities in the bush perimeter of the cleared agricultural belt around a Gambian village on the flight range and direction of approach of a population of *Anopheles gambiae melas.* Proc. Int. Congr. Entomol., 12th, London, 1964 757–58

73. Gillett, J. D., Teesdale, C., Trpis, M., Rao, T. R. 1969. Diurnal activity cycle of *Aedes aegypti* as assessed by hourly landing rates on man. Unpublished document. *WHO/ VBC/69.158.* 3 pp.

74. Glasgow, J. P. 1967. Recent fundamental work on tsetse flies. *Ann. Rev. Entomol.* 12 :421–38

75. Gooding, R. H. 1968. A note on the relationship between feeding and insemination in *Pediculus humanus. J. Med. Entomol.* 5(2) :265–66

76. Gordon, R. M., Crewe, W. 1948. The mechanisms by which mosquitoes and tsetse-flies obtain their blood-meal, the histology of the lesions produced, and the subsequent reactions of the mammalian host ; together with some observations on the feeding of *Chrysops* and *Cimex.*

Ann. Trop. Med. Parasitol. 42 : 334–56

77. Gordon, R. M., Lumsden, W. H. R. 1939. A study of the behaviour of the mouthparts of mosquitoes when taking up blood from living tissue; together with some observations of the ingestion of microfilariae. *Ann. Trop. Med. Parasitol.* 33 :259–78

78. Gregson, J. D. 1944. The influence of fertility on the feeding rate of the female of the wood tick. *Dermacentor andersoni,* Stiles. *74th Ann. Rep. Entomol. Soc., Ontario* (Contrib. No. 2278) 46–47

79. Gregson, J. D. 1946. Feeding periods prerequisite to the mating of *Dermacentor andersoni. Entomol. Soc. Brit. Columbia,* 43 :3–6

80. Gregson, J. D. 1956. The Ixodoidea of Canada. *Can. Dep. Agr.,* Ottawa 92 pp.

81. Gregson, J. D. 1960. Morphology and functioning of the mouthparts of *Dermacentor andersoni* Stiles. *Acta Trop.* 17(1) :48–49

82. Gregson, J. D. 1967. Observations on the movement of fluids in the vicinity of the mouthparts of naturally feeding *Dermacentor andersoni* Stiles. *Parasitology,* 57(1) :1–8

83. Gregson, J. D. 1969. Electrical observations of tick feeding in relation to disease transmission. *Proc. Int. Congr. Acarol., 2nd, Sutton Bonnington 1967,* 329–39. Budapest : Akademiai Kiado

84. Grzybowska, B. 1956. The innervation and sense organs in the wings of *Ornithomyia biloba,* Dufour. *Pol. Pismo Entomol.* 26 :175–90

85. Gwadz, R. W. 1969. Regulation of blood meal size in the mosquito. *J. Insect Physiol.* 15 :2039–44

86. Haddow, A. J., Gillett, J. D., Mahaffy, A. F., Highton, R. B. 1950. Observations on the biting-habits of some Tabanidae in Uganda, with special reference to arboreal and nocturnal activity. *Bull. Entomol. Res.* 41 :209–20

87. Halcrow, J. G. 1953. Preliminary observations on some possible tarsal sensillae of English mosquitoes. *Entomologist* 86 :104–5

88. Harris, P., Cooke, D. 1969. Survival and fecundity of mosquitoes fed on insect haemolymph. *Nature* 222 :1264–65

89. Harris, P., Riordan, D. F., Cooke, D. 1969. Mosquitoes feeding on insect larvae. *Science* 164 :184–85

90. Hayes, J. 1965. New host record for *Aedes canadensis. Mosquito News* 25(3) :344

91. Hayes, R. O. 1961. Host preferences of *Culiseta melanura* and allied mosquitoes. *Mosquito News* 21(3) : 179–87

92. Hays, K. L. 1965. Longevity, fecundity, and food intake of adult *Triatoma sanguisuga* (Leconte). *J. Med. Entomol.* 2(2) :200–2

93. Henderson, B. E., Senior, L. 1961. Attack rate of *Culex tarsalis* on reptiles, amphibians and small mammals. *Mosquito News* 21 :29–32

94. Herndon, B. L., Ringle, D. A. 1967. Identification of host and host antibodies from mosquito blood meals. *Nature* 213 :624–25

95. Hindle, E., Merriman, G. 1912. Sensory perception of *Argas persicus* (Oken). *Parasitology* 5 : 203–16

96. Hinton, H. E. 1958. The phylogeny of the panorpoid orders. *Ann. Rev. Entomol.* 3 :181–206

97. Hocking, B. 1953. The intrinsic range and speed of flight of insects. *Trans. Roy. Entomol. Soc. London* 104 :223–345

98. Hocking, B. 1960. Smell in insects— a bibliography with abstracts. Defence Research Board, Ottawa. vi + 266 pp.

99. Hocking, B. 1964. Aspects of insect vision. *Can. Entomol.* 96(1–2) : 320–34

100. Hocking, B., Hocking, J. M. 1962. Entomological aspects of African onchocerciasis and observations on *Simulium* in the Sudan. *Bull. WHO* 27 :465–72

101. Hocking, B., Khan, A. A. 1966. The mode of action of repellent chemicals against blood-sucking flies. *Can. Entomol.* 98(8) :821–31

102. Hoosgtraal, H. 1965. Phylogeny of *Haemaphysalis* Ticks. *Proc. Int. Congr. Entomol., 12th, London, 1964.* 760–61

103. Hosoi, T. 1958. Adenosine-5'-phosphates as the stimulating agent in blood for inducing gorging of the mosquito. *Nature* 181 :1664–65

104. Hosoi, T. 1959. Identification of blood components which induce

gorging of the mosquito. *J. Insect Physiol.* 3 :191–218

105. Huang, C.-T. 1970. Vertebrate serum inhibitors of *Aedes aegypti* trypsin. PhD thesis, University of Alberta, Canada

106. Hudson, A. 1970. Notes on the piercing mouthparts of three species of mosquitoes viewed with the scanning electron microscope. *Can. Entomol.* 102 :501–9

107. Hudson, A., Bowman, L., Orr, C. W. M. 1960. Effects of absence of saliva on blood feeding by mosquitoes. *Science* 131 :1730–31

108. Ikeshoji, T. 1967. Enhancement of the attractiveness of mice as mosquito bait by injection of methionine and its metabolites. *Jap. J. Sanit. Zool.* 18(2/8) :101–7

109. Ikeshoji, T., Umino, T., Sazuki, T. 1963. On the attractiveness of some amino acids and their decomposed products for the mosquito *Culex pipiens pallens. Jap. J. Sanit. Zool.* 14(3) :152–56. English summary

110. Joyner, L. P., Purnell, R. E. 1968. The feeding behaviour on rabbits and *in vitro* of the Ixodid tick *Rhipicephalus appendiculatus* Neumann, 1901. *Parasitology* 58 :715–23

111. Kalmus, H., Hocking, B. 1960. Behaviour of *Aedes* mosquitos in relation to blood-feeding and repellents. *Entomol. Exp. Appl.* 3 : 1–26

112. Kashin, P. 1966. Electronic recording of the mosquito bite. *J. Insect Physiol.* 12 :281–86

113. Kellogg, F. E. Water vapour and carbon dioxide receptors in *Aedes aegypti. J. Insect Physiol.* 16 :99–108

114. Kellogg, F. E., Wright, R. H. 1962. The guidance of flying insects. V. Mosquito attraction. *Can. Entomol.* 94(10) :1009–16

115. Kellogg, F. E., Frizel, D. E., Wright, R. H. 1962. The olfactory guidance of flying insects. IV. *Drosophila. Can. Entomol.* 94(8) :884–88

116. Kennedy, J. S. 1939. The visual responses of flying mosquitoes. *Proc. Zool. Soc. London* 109 :221–42

117. Kettle, D. S. 1962. The bionomics and control of *Culicoides* and *Leptoconops. Ann. Rev. Entomol.* 7 :401–18

118. Kettle, D. S. 1969. The biting habits of *Culicoides furens* (Poey) and *C. barbosai* Wirth & Blanton. II. Effect of meteorological conditions. *Bull. Entomol. Res.* 59 :241–58

119. Khan, A. A., Maibach, H. I., Strauss, W. G. 1968. The role of convection currents in mosquito attraction to human skin. *Mosquito News* 28(3) :462–64

120. Khan, A. A., Maibach, H. I., Strauss, W. G. 1969. Gross variations in the response to man among yellow-fever mosquito populations in the laboratory. *J. Econ. Entomol.* 62(1) :96–98

121. Khan, A. A., Maibach, H. I., Strauss, W. G., Fenley, W. R. 1965. Screening humans for degrees of attractiveness to mosquitoes. *J. Econ. Entomol.* 58(4) :694–97

122. Knülle, W. 1967. Significance of fluctuating humidities and frequency of blood meals on the survival of the spiny rat mite, *Echinolaelaps echidninus* (Berlese). *J. Med. Entomol.* 4(3) :322–25

123. Krivosheina, N. P. 1961. *Rhagio thingarius* larvae sucking earthworm blood. *Zool. Zh.* 40 :715–18

124. Laarman, J. J. 1965. The plasticity of response patterns in host-seeking mosquitoes. *Acta Leidensia* 33/34 : 136–38

125. Lacher, V. 1967. Elektrophysiologische Untersuchungen an einzelnen Geruchsrezeptoren auf den Antennen weiblicher Moskitos (*Aedes aegypti* L.). *J. Insect Physiol.* 13 :1461–70

126. Lall, S. B. 1969. Phagostimulants of haematophagous tabanids. *Entomol. Exp. Appl.* 12 :325–36

127. Lall, S. B., Davies, D. M. 1967. Tarsal sensitivity of female tabanid flies to sucrose and sodium chloride. *Can. J. Zool.* 45 :461–64

128. Lavoipierre, M. M. J. 1958. Biting behaviour of mated and unmated females of an African strain of *Aedes aegypti. Nature* 181 :1781–82

129. Lavoipierre, M. M. J. 1965. Feeding mechanism of blood-sucking arthropods. *Nature* 208 :302–3

130. Lavoipierre, M. M. J. 1967. Feeding mechanism of *Haematopinus suis,* on the transilluminated mouse ear. *Exp. Parasitol.* 20 :303–11

131. Lavoipierre, M. M. J., Hamachi, M. 1961. An apparatus for observations on the feeding mechanism of

the flea. *Nature* 192:998–99

132. Lavoipierre, M. M. J., Dickerson, G., Gordon, R. M. 1959. Studies on the methods of feeding of blood-sucking arthropods I. The manner in which Triatomine bugs obtain their blood-meal, observed in the tissues of a living rodent, with some remarks on the effects of the bite on human volunteers. *Ann. Trop. Med. Parasitol.* 53(2):235–50

133. Lee, D. J., Clinton, K. J., O'Gower, A. K. 1954. The blood sources of some Australian mosquitoes. *Austr. J. Biol. Sci.* 7:282–301

134. Lees, A. D. 1948. The sensory physiology of the sheep tick, *Ixodes ricinus* L. *J. Exp. Biol.* 25:145–207

135. Lees, A. D., Milne, A. 1951. The seasonal and diurnal activities of individual sheep ticks (*Ixodes ricinus* L.) *Parasitology* 41:189–208

136. Lewis, D. J. 1953. *Simulium damnosum* and its relation to onchocerciasis in the Anglo-Egyptian Sudan. *Bull. Entomol. Res.* 43:597–644

137. Lewis, D. J. 1957. Simuliidae and their relation to onchocerciasis in the Sudan. *Bull. WHO* 16:671–74

138. Lowther, J. K., Wood, D. M. 1964. Specificity of a black fly *Simulium euryadminiculum* Davies, towards its host, the common loon. *Can. Entomol.* 96:911–13

139. Lumsden, W. H. R. 1947. Observations on the effect of micro-climate on the biting of *Aedes aegypti* (L.). *J. Exp. Biol.* 24:361–73

140. Lumsden, W. H. R. 1958. Periodicity of biting behaviour of some African mosquitoes. *Proc. Int. Congr. Entomol., 10th, Montreal, 1956.* 3:785–90

141. MacGregor, M. E., Lee, C. U. 1929. Preliminary note on the artificial feeding of mosquitoes. *Trans. Roy. Soc. Trop. Med. Hyg.* 23(2):203–4

142. Maddrell, S. H. P. 1963. Control of ingestion in *Rhodnius prolixus* Stal. *Nature* 198:210

143. Magnum, C. I., Callahan, P. S. 1968. Attraction of near infra-red radiation to *Aedes aegypti*. *J. Econ. Entomol.* 61:36–37

144. Maibach, H. I., Skinner, W. A.,

Strauss, W. G., Khan, A. A. 1966. Factors that attract and repel mosquitoes in human skin. (*Aedes aegypti*) *J. Am. Med. Assoc.* 196:263–66

145. Marks, E. N. 1960. *Uranotaenia albescens* biting *Hyla caerulea* (Amph.) *Austr. J. Sci.* 23:89

146. Mattingly, P. F. 1962. Mosquito behaviour in relation to disease eradication programmes. *Ann. Rev. Entomol.* 7:419–36

147. Mayer, M. S., James, J. D. 1969. Attraction of *Aedes aegypti* (L.): Responses to human arms, carbon dioxide and air currents in a new type of olfactometer. *Bull. Entomol. Res.* 58:629–42

148. McClelland, G. A. H., Weitz, B. 1963. Serological identification of the natural hosts of *Aedes aegypti* (L.) and some other mosquitoes caught resting in vegetation in Kenya and Uganda. *Ann. Trop. Med. Parasitol.* 57(2):214–24

149. McCrae, A. W. R. 1969. Midges biting mosquitoes. *East Afr. Virus Res. Inst. Rep. 1968.* 102. Entebbe, Uganda

150. McIver, S. B. 1969. Antennal sense organs of female *Culex tarsalis*. *Ann. Entomol. Soc. Am.* 62(6):1455–61

151. McIver, S., Charlton, C. 1970. Studies on the sense organs on the palps of selected culicine mosquitoes. *Can. J. Zool.* 48:293–95

152. McLean, D. L., Kinsey, M. G. 1964. A technique for electronically recording aphid feeding and salivation. *Nature* 202:1358

153. Megahed, M. M. 1958. The distribution of blood, water, and sugar solutions in the mid-gut and oesophageal diverticulum of female *Culicoides nubeculosus* Meigen. *Bull. Soc. Entomol. Egypte* 42:339–55

154. Monchadskii, A. S. 1956. Blood-sucking flies in the U.S.S.R. and some regularities of their attack on man. *Entomol. Obozr.* 35(3):547–49 In Russian, English summary

155. Moorhouse, D. E., Tatchell, R. J. 1966. The feeding processes of the cattle tick *Boophilus microplus* (Canestrini): A study in host-parasite relations. *Parasitology* 56:623–32

156. Müller, W. 1968. Die Distanz- und

Kontakt-Orientierung der Stechmücken (*Aedes aegypti*) (Wirtsfindung, Stechverhalten und Blutemahlzeit). *Z. vergl. Physiol.* 58: 241–303

157. Myers, J. G. 1929. Facultative bloodsucking in phytophagous Hemiptera. *Parasitology* 21(4):472–80

158. Nelson, R. L. 1965. Carbon dioxide as an attractant for *Culicoides*. *J. Med. Entomol.* 2(1):56–57

159. Nelson, W. A., Petrunia, D. M. 1969. *Melophagus ovinus:* Feeding mechanism on transilluminated mouse ear. *Exp. Parasitol.* 26(3): 308–13

160. Nielsen, E. T., Nielsen, A. T. 1953. Field observations on the habits of *Aedes taeniorhynchus*. *Ecology* 34:141–56

161. Nolan, M. P., Jr., Moussa, M. A., Hayes, D. E. 1965. *Aedes* mosquitoes feeding on turtles in nature. *Mosquito News* 25(2): 218–19

162. Ogata, K. 1954. Studies on the diurnal rhythm of biting activity in *Simulium venustum* Say (Studies on black fly. 3). *Jap. J. Appl. Zool.* 19:136–41

163. O'Rourke, F. J. 1956. Observations on pool and capillary feeding in *Aedes aegypti* (L.) *Nature* 177: 1087–88

164. Owen, W. B. 1963. The contact chemoreceptor organs of the mosquito and their function in feeding behaviour. *J. Insect Physiol.* 9: 73–87

165. Owen, W. B. 1965. Structure and function of the gustatory organ of the mosquito. *Proc. Congr. Entomol., 12th, London, 1964* 793

166. Potter, E. 1938. The internal anatomy of the order Mecoptera. *Trans. Roy. Entomol. Soc. London* 87: 467–510

167. Raybould, J. N. 1967. A study of anthropophilic female Simuliidae at Amani, Tanzania: the feeding behaviour of *Simulium woodi* and the transmission of onchocerciasis. *Ann. Trop. Med. Parasitol.* 61: 76–88

168. Remington, C. L. 1945. The feeding habits of *Uranotaenia lowii* Theobald. *Entomol. News* 56:32–37, 64–68

169. Rice, M. J. 1970. Cibarial stretch receptors in the tsetse fly (*Glossina austeni*) and the blowfly

(*Calliphora erythrocephala*). *Insect Physiol.* 16:277–89

170. Rivnay, E. 1932. Studies in tropisms of the bed bug. (*Cimex lectularius* L.) *Parasitology* 24:121–36

171. Roberts, F. H. S., O'Sullivan, P. J. 1948. Studies on the behavior of adult Australasian anophelines. *Bull. Entomol. Res.* 39:159–78

172. Roessler, P. 1960. Anlockung weiblicher Stechmücken (*Aedes aegypti* L.) mit Duftstoffen. *Naturwissenschaften* 47:549–50

173. Roessler, P. H. 1961. Versuche zur geruchlichen Anlockung weiblicher Stechmücken (*Aedes aegypti* L.). *Z. vergl. Physiol.* 44:184–231

174. Roessler, P. H., Brown, A. W. A. 1964. Studies on the responses of the *Aedes* mosquito; cross comparison of oestrogen and aminoacids as attractants. *Bull. Entomol. Res.* 55:395–403

175. Romoser, W. S., Venard, C. E. 1967. Development of the dorsal oesophageal diverticula in *Aedes triseriatus*. *Ann. Entomol. Soc. Am.* 30(3):617–23

176. Rowan, W., Gregson, J. D. 1935. Winter feeding of the tick, *Dermacentor andersoni*, Stiles. *Nature* 135:652

177. Rudolfs, W. 1922. Chemotropism of mosquitoes. *New Jersey Agr. Exp. Sta. Bull.* 367:1–23

178. Ryckman, R. E. 1958. Description and biology of *Hesperocimex sonorensis*, new species, an ectoparasite of the purple martin. *Ann. Entomol. Soc. Am.* 51(1):33–47

179. Salama, H. S. 1966. Taste sensitivity to some chemicals in *Rhodnius prolixus* Stal and *Aedes aegypti* L. *J. Insect Physiol.* 12:583–89

180. Salama, H. S. 1966. The function of mosquito taste receptors. (*Aedes aegypti*). *J. Insect Physiol.* 12: 1051–60

181. Salama, H. S. 1967. Nutritive values and taste sensitivity to carbohydrates for mosquitoes (*Aedes aegypti*). *Mosquito News* 27(1): 32–35

182. Sasa, M., Wakasugi, M. 1957. Studies on the effect of carbon dioxide as the stimulant on the tropical rat mite *Bdellonyssus bacoti* (Huist, 1913). *Jap. J. Exp. Med.* 27:207–15

183. Sato, S., Kato, M., Toriumi, M. 1957. Structural changes of the

compound eye of *Culex pipiens* var *pallens* Coquillet in the process to dark adaptation. *Sci. Rep. Tohoku Univ.* Fourth Series: Biology 23(3), 91–100

184. Schaerffenberg, B., Kupka, E. 1953. Orientierungsversuche an *Stomoxys calcitrans* und *Culex pipiens* mit einem Blutduftstoff. *Trans. Int. Congr. Entomol., 9th, Amsterdam, 1951* 1:359–61

185. Schulze, P. 1941. Das Geruchsorgan der Zecken. Untersuchungen über die Abwandlung eines Sinnesorgans und seine stammesgeschichtliche Bedeutung. *Z. Morphol. Oekol. Tiere* 37:491–564

186. Schulze, P., Schröder, E. 1949. Zur Sinnesphysiologie der Lederzecken. *Biol. Zentralbl.* 68:321

187. Sgonina, K. 1939. Wirtsfindung und Wirtsspezifität von Flohen. *Trans. Int. Congr. Entomol., 7th, Berlin, 1938* 3:1663–68

188. Sippell, W. L., Brown, A. W. A. 1953. Studies of the responses of the female *Aedes* mosquito. Part V. The role of visual factors. *Bull. Entomol. Res.* 43(4):567–74

189. Skinner, W. A., Tong, H., Pearson, T., Strauss, W., Maibach, H. 1965. Human sweat components attractive to mosquitoes. *Nature* 207 (4997):661–62

190. Slifer, E. H. 1970. The structure of arthropod chemoreceptors. *Ann. Rev. Entomol.* 15:121–42

191. Slooff, R. 1964. *Culicoides (Trithecoides) culiciphagus* on blood or body fluids of *Anopheles koliensis*, 1st record. *Entomol. Ber.* 24:39–40

192. Slooff, R., Marks, E. N. 1965. Mosquitoes (Culicidae) biting a fish (Periophthalmidae). *J. Med. Entomol.* 2:16

193. Smith, A. 1955. The transmission of Bancroftial filariasis on Ukara Island, Tanganyika. IV. Host-preferences of mosquitoes and the incrimination of *Anopheles gambiae* Giles and *A. funestus* Giles as vectors of Bancroftial filariasis. *Bull. Entomol. Res.* 46:505–15

194. Smith, A. 1956. The attractiveness of an adult and child to *A. gambiae*. *East Afr. Med. J.* 33:409–10

195. Smith, S. M. 1969. The black fly, *Simulium venustum*, attracted to the turtle, *Chelydra serpentina*. *Entomol. News* 80:107–8

196. Spencer, M. 1967. Anopheline attack on mother and infant pairs Ferguson Island. *Papua New Guinea Med. J.* 10(3):75

197. Spielman, A. 1971. Bionomics of autogenous mosquitoes. *Ann. Rev. Entomol.* 16:231–48

198. Standfast, H. A. 1967. Biting times of nine species of New Guinea Culicidae. *J. Med. Entomol.* 4(2): 192–96

199. Steward, C. C., Atwood, C. E. 1962. The sensory organs of the mosquito antenna. *Can. J. Zool.* 41: 578–93

200. Stürckow, B. 1965. The electroantennogram (EAG) as an assay for the reception of odours by the gypsy moth. *J. Insect Physiol.* 11: 1573–84

201. Suenaga, O. 1965. A rearing method of stable fly and quantity of blood taken up by a fly. *Endem. Dis. Bull. Nagasaki Univ.* 7(4):296–301

202. Surcouf, J. M. R., Fischer, E. 1924. Notes sur la vie larvaire et nymphale du *Tabanus bromius* L. *Bull. Soc. Entomol. France* 29: 232–37

203. Sweatman, G. K. 1967. Physical and biological factors affecting the longevity and oviposition of engorged *Rhipicephalus sanguineus* female ticks. *J. Parasitol.* 53:432–45

204. Symes, C. B., McMahon, J. P. 1937. The food of the tsetse-flies (*Glossina swynnertoni* and *G. palpalis*) as determined by the precipitin test. *Bull. Entomol. Res.* 28:31–42

205. Tarshis, I. B. 1957. The diverticulum of *Stilbometopa impressa* (Bigot) as a functioning organ. *Ann. Entomol. Soc. Am.* 50:519–21

206. Tashiro, H., Schwardt, H. H. 1953. Biological studies of horse flies in New York. *J. Econ. Entomol.* 46(5):813–22

207. Tatchell, R. J. 1969. Host parasite inter-actions and the feeding of blood-sucking arthropods. *Parasitology* 59:93–104

208. Tawfik, M. S. 1968. Feeding mechanisms and the forces involved in some blood-sucking insects. *Quaest. Entomol.* 4:92–111

209. Tawfik, M. S. 1968. Effects of the size and frequency of blood meals

on *Cimex lectularius* L. *Quaest. Entomol.* 4 :225–56

210. Thomas, T. C. E. 1951. Biting activity of *Anopheles gambiae*. *Brit. Med. J.* 2 :1402

211. Thompson, R. P., Brown, A. W. A. 1955. The attractiveness of human sweat to mosquitoes and the role of carbon dioxide. *Mosquito News* 15 :80–84

212. Thompson, R. C. M. 1951. The distribution of anopheline mosquito bites among different age groups. *Brit. Med. J.* 1 :1114–16

213. Totze, R. 1933. Beiträge zur Sinnesphysiologie der Zecken. *Z. vergl. Physiol.* 19 :110–61

214. Trembley, H. L. 1952. The distribution of certain liquids in the esophageal diverticula and stomach of mosquitoes. *Am. J. Trop. Med. Hyg.* 1 :693–710

215. Usinger, R. L. 1934. Blood-sucking among phytophagous Hemiptera. *Can. Entomol.* 66 :97–100

216. Usinger, R. L. 1966. Monograph of Cimicidae. Thomas Say Found., College Park, Maryland. xi + 585 pp.

217. Usova, Z. V., Kulikova, Z. P. 1958. The activity of black flies in Karelia. *Entomol. Rev.* 37 :869–82

218. Vanderplank, F. L. 1944. Studies of the behaviour of the tsetse-fly (*Glossina pallidipes*) in the field ; the attractiveness of various baits. *J. Anim. Ecol.* 13 :39–44

219. Vladimirova, V. V., Potapov, A. A. 1963. The new type of traps for horse flies and black flies. *Med. Parasitol., Moscow* 1 :83–87

220. Wada, Y. 1969. Ecological studies of *Culex tritaeniorhynchus summorosus* I. Biting rhythm of the mosquito. *Jap. J. Sanit. Zool.* 20(1) : 21–26

221. Walton, G. A. 1957. Observations on biological variation in *Ornithodoros moubata* (Murr.) in East Africa. *Bull. Entomol. Res.* 48 :669–710

222. Weber, H. 1929. Biologische Untersuchungen an der Schweinelaus (*Haematopinus suis* L.) unter besonderer Berücksichtigung der Sinnesphysiologie. *Z. vergl. Physiol.* 9 :564–612

223. Weitz, B. 1960. Feeding habits of blood-sucking arthropods. *Exp. Parasitol.* 9 :63–82

224. Wenk, P. 1962. Zur Wirtsorientierung mammalophiler, einheimischer Simuliiden. *Naturwissenschaften* 49 :165–66

225. West, A. S. 1950. The precipitin test as an entomological tool. *Can. Entomol.* 82 :241–44

226. Wigglesworth, V. B., Gillett, J. D. 1934. The function of the antennae in *Rhodnius prolixus* and the mechanism of orientation to the host. *J. Exp. Biol.* 11 :120–29, 408

227. Willis, E. R., Roth, L. M. 1952. Reactions of *Aedes aegypti* (L.) to carbon dioxide. *J. Exp. Zool.* 121(1) :149–79

228. Wilson, F. H. 1933. A louse feeding on the blood of its host. *Science* 77 :490

229. With, W. W. 1956. New species and records of biting midges ectoparasitic on insects. *Ann. Entomol. Soc. Am.* 49 :356–64

230. Woke, P. A. 1962. Observations on differential attraction among humans for *Mansonia nigricans* and *Anopheles albimanus*. *Proc. 49th Ann. Meeting N.J. Mosquito Extermination Assoc.*

231. Wolfe, L. S., Peterson, D. G. 1960. Diurnal behavior and biting habits of black flies in the forests of Quebec. *Can. J. Zool.* 38 :489–97

232. Wood, P. W., Wright, R. H. 1968. Some responses of flying *Aedes aegypti* to visual stimuli. *Can. Entomol.* 100 :504–13

233. Wright, R. H., Daykin, P. N., Kellogg, F. E. 1965. Reaction of flying mosquitoes to various stimuli. *Proc. Int. Congr. Entomol., 12th, London, 1964.* 281–82

234. Yelizarov, Y. A. 1965. Study of the chemoreception of insects and ticks. Physiology of contact chemoreception of insects and ticks. *Zool. Zh., Moskva* 44(10) :1461–72

INSECT CELL AND TISSUE CULTURE[1,2]   6001

MARION A. BROOKS AND T. J. KURTTI

*Department of Entomology, Fisheries, & Wildlife,
University of Minnesota, St. Paul, Minnesota*

INTRODUCTION

We shall attempt to appraise the current status of cell and tissue culture in the field of entomology, and make some comparisons with vertebrate culture. In doing this, we tend to emphasize problems which appear to have been neglected. These problems, if diligently pursued, may richly reward us.

In 1959, when Day & Grace (13) reviewed insect tissue culture for the *Annual Review of Entomology,* it was already an old problem in the sense that attempts had been made since 1915, yet there was not a single established line of cells capable of indefinitely replicating under regular subculture. Grace (20) soon reported the prolonged survival and growth of cells, although he was as yet unable to subculture them successfully. The first successful establishment was Grace's (21) work with pupae of *Antheraea eucalypti,* a large saturniid silkmoth. The ability of insect cells to replicate indefinitely in vitro was finally demonstrated. Soon after this there occurred the First International Colloquium on Invertebrate Tissue Culture, held in Montpellier, France, in October 1962, which was a great stimulus to this work; this was followed by the Second International Colloquium on Invertebrate Tissue Culture held in Tremezzo and Como, Italy, in September 1967. Since the first colloquium, the production in insect cell culture has grown at an increasing rate. The last general review of the subject was published in 1966 (46).

In large part, the interest in insect cell culture is mission-oriented; the overriding aspiration is to use the cultured cells for propagation of viruses associated with arthropods. For instance, lepidopteran cell culture is designed for the propagation of viruses suggested as control agents; leafhop-

[1] Paper Number 7256, Scientific Journal Series, Minnesota Agricultural Experiment Station. Original research was supported by Research Grant No. AI 00961, from the National Institute of Allergy and Infectious Diseases of the National Institutes of Health, U. S. Public Health Service.

[2] The commonly used culture media and animal additives are referred to in the text by the following abbreviations: GMA (Grace's medium); AEH (*Antheraea eucalypti* hemolymph [heat-treated]); APH (*Antheraea perneyi* hemolymph [heat-treated]); FBS (fetal bovine serum); BPA (bovine plasma albumin [fraction V]); CEE (chick embryo extract); WEU whole egg ultrafiltrate).

per cells, for the study of viruses that cause plant diseases; and dipteran cells (mostly mosquitoes), for the cultivation of arboviruses. These aspects have been thoroughly reviewed by Grace (23, 25, 28), Suitor (102), and Vaughn (113) so we will not go over them again here. Most recently, an international symposium, The Earl Suitor Symposium on Arthropod Cell Cultures and their Application to the Study of Viruses, was held in Bethesda, Maryland, in March 1970, and enthusiastically supported.

The slow beginning in insect cell culture can in general be attributed to the unavailability of a culture medium which would permit the long-term growth and survival of the cells. There was not a single, sweeping innovation or radical departure from methods used for mammalian cultures to which we can ascribe the recent progress. Evidently it is one of those things which can be done once the experimenter is convinced it is possible. The pioneering work of G. R. Wyatt and S. S. Wyatt (118, 119) did much to supply the basic framework for designing an insect cell culture medium which Grace (21) modified and applied to the successful culture of *Antheraea* cells. Information on hemolymph can be used in media design for other insects (85). At present, several cell lines have been established using hemolymph-simulated media (21, 24, 26, 36–38, 56, 95). But since lines have also been established on media derived by somewhat of an empirical approach, it remains to be determined how closely the composition of the hemolymph should be followed.

The terms applicable to cell culture have been defined by Fedoroff (17) and have been accepted by the Tissue Culture Association. We take the liberty here of changing authors' terminology to accord with the TCA definitions. For example, Grace (21), Mitsuhashi (73) and others referred to the establishment of "strains" which should be understood as "lines."

CHARACTERIZATION OF THE CELL LINES

Cell lines of 12 species have been established from the orders Lepidoptera, Diptera, Hemiptera, and Blattariae. The lines are listed in Table 1. They are derived cells from hemocytes, ovaries, whole embryos, and neonate, whole, first instar larvae. When more than one line is maintained from any given species, these lines may or may not be replications of particular cell types. Cultures are grown as monolayers, or as suspensions, and some are mixtures of both growth forms (24, 27).

Insect cell culturists have not yet decided upon a standard form of designating lines by a meaningful code. Only a few recent lines so far bear codes, and some of these are too brief (27, 36–38, 75). The code should indicate the laboratory, the strain (which would refer to the species), and the clone (17). Furthermore, to conform to other practices, a minimum number of transfers, or subcultures, without diminution in cell density or dramatic change in morphology, should be adhered to before a culture is called a line. This number has not been determined for vertebrate cells, but as Fedoroff indicates, in experience with human fibroblastlike cells, 70 passages seem to be a minimum.

TABLE 1. Sources of Insect Cell Lines

Order	Species	Stage of development	Inoculum	Reference
Blattariae	*Periplaneta americana*	embryos post-dorsal closure	whole embryos	(56)
Hemiptera	*Agallia constricta*	blastokinetic embryos	whole embryos	(10)
Lepidoptera	*Antheraea eucalypti*	diapausing pupae	ovaries	(21)
Lepidoptera	*Bombyx mori*	pupae	ovaries	(26)
Lepidoptera	*Carpocapsa pomonella*	embryos	whole embryos	(37)
Lepidoptera	*Chilo suppressalis*	diapausing larvae	ovaries	(73)
Lepidoptera	*Heliothis zea*	adult	ovaries	(38)
Lepidoptera	*Trichoplusia ni*	adult	ovaries	(36)
Diptera	*Aedes aegypti*	prepupae	eviscerated larval bodies	(24)
Diptera	*Aedes aegypti*	neonate larvae	whole larvae	(97)
Diptera	*Aedes albopictus*	neonate larvae	whole larvae	(97)
Diptera	*Anopheles stephensi*	young first instar	whole larvae	(95)
Diptera	*Drosophila melanogaster*	8-hr embryos	whole embryos	(43)
Diptera	*Drosophila melanogaster*	?	?	(14)

We find no indication that cells of insects have been placed in the American Type Culture Collection Cell Repository.

When one tries to compare the characteristics of the cell lines it is a surprisingly difficult thing to do. Some of the lines closely resemble each other in cultural properties and cell morphology; as a result there is an absence of reliable criteria to distinguish them. Degree of ploidy and karyotypes are sometimes used. In general, the lepidopteran and cockroach lines are polyploid while the dipteran lines are diploid, except for Grace's line of *Aedes aegypti*. In many ways this line cannot be distinguished from the *Antheraea eucalypti* line (chromosome morphology, ploidy, cultural characteristics), hence, other means are needed to characterize these two. Perhaps the best way is by the use of antigenic markers.

All together—cultural similarities, lack of good karyotypes, difficulty in identifying the tissue sources (discussed below)—these factors make the characterization of cell lines perplexing.

Selection of Donor Tissue

In most cases the investigator has an interest in, or a need to propagate, a particular cell type; or at least he desires that the cells should be from a particular organ. In the early stages serious efforts were made to establish cell lines from specific types of tissues. These efforts usually ended in aborted primary cultures, with the notable exception of the lines obtained from pupal or adult ovaries and larval hemocytes.

A current practice is to use whole embryos and neonate larvae as inocula. While this gives the desired effect it also introduces several unidentified factors to the culture; and we are still faced with the problem of how to get cells from selected tissues and types.

Sometimes much difficulty is encountered in inducing cell migration or outgrowth from certain tissue explants. Usually those cells which do grow out are the fibroblastlike cells and the lining epitheliocytes accompanied by small spherical cells sometimes thought to be hemocytes from the bathing hemolymph (13, 49). Although people have speculated that cells in lines originate from hemocytes, there is no proof of this since shape and free-floating behavior are not unique characteristics.

Cells derived by outgrowth from untreated tissue explants are refractory to establishment in continuous culture; this seems to be a universal difficulty with cells which are organized into specialized tissues. There is a greater likelihood of establishing lines following dissociation of some kind.

The problem is, that after any method of tissue preparation and regardless of the organ originally selected, the apparent same two or three morphological cell types survive. Are the spindle- and round-shaped cells derived from *A. eucalypti* ovaries the same as those derived from whole *A. aegypti* larvae (21, 24)? What are the similarities between the morphologically unlike lines obtained for *A. aegypti* by Grace (24) and by Singh (97)? It is difficult to know how cells are related to one another if we know nothing of their tissue of origin, much less if they are derived from the functional cells of the organs rather than from epithelial or connective tissue coverings. If only basically undifferentiated cell types grow, they may not be valid cells for analyzing some relationships which depend on tissue specificity in situ. The propagation of viruses and symbiotes, for example, may require the specialized cells which they normally infect. Studies in cell physiology may not have much meaning if they are done on cells which are incapable of specialized function.

The use of embryos or neonate larvae which were obtained in an aseptic state by surface-sterilizing the eggs has been a very productive method. The technique was perfected by Maramorosch (62) and by Mitsuhashi & Maramorosch (76) for working with leafhoppers to study insect-transmitted plant viruses. Boorman (5) described the technique for rearing aseptic mosquitoes. The embryonic tissues are removed from the crushed eggs or the larvae are allowed to hatch in axenic conditions. This is much simpler and less prone to contamination than surface-disinfecting a hirsute larva and dissecting it aseptically. The aseptic egg technique has led to the development of many of the lines from embryos (or early post-embryos), giving the impression that embryonic tissues have the greatest potential for growth in vitro. This may be true, but critical experiments on different ages have not been done on many species.

Certain ages of embryos, or pupae, give rise to cell outgrowths or replications from dissociated cells much more readily than other ages do. Hirumi & Maramorosch (39–41), in experimenting with leafhopper embryos (*Macrosteles fascifrons*), soon found that not all undifferentiated cells were suitable for cultivation. By making a systematic study of the 11-day incubation period, they determined that the 7th - 8th day was the only satisfactory period. This is the time of blastokinesis, a rotation of the embryo in the egg during which time mitotic growth is arrested. Chiu & Black (10) used only 7-day old embryos of their leafhoppers (*Agallia constricta*), which have a total embryonic developmental time of 11–12 days.

Horikawa & Fox (42) found that *Drosophila* eggs taken during the first 2 hr following fertilization did not yield cells capable of multiplying in vitro, but 8-hr eggs yielded a type of cell which entered logarithmic growth. Reflecting on *Drosophila* embryogenesis, we recall that the first 5 or 6 hr is the period of rapid mitotic activity with the blocking out of the presumptive tissues. By the 8th hr (at 25°C) the embryo is in a less active stage mitotically, while the emphasis is on differentiation. Echalier et al (14) found *Drosophila* eggs between 8–14 hr most suitable, and Lesseps (61) used *Drosophila* embryos about 11 hr old. Eide & Chang (15) reported that 6 hr was the best age to select from a total incubation period of 10 hr in *Musca domestica*. Similarly, Landureau (54) found that the only satisfactory stage of embryos of *Blabera fusca* were those which already had a functional dorsal vessel—a relatively late stage in embryogenesis, past active mitoses.

Imaginal wing discs have been a fruitful source of cells for primary cultures (51). The discs grow at a uniform rate, being less subject to mitotic fluctuations compared to embryos and larval tissues. In pupae also there are periods during which cell multiplication is more likely to succeed. The ovaries of pupae, which are good sources of cells (21, 101), may be an example of a resting stage of a tissue with an innate capacity for further growth and differentiation. Stanley & Vaughn (101) found that ovary explants which give the best cell growth have well developed but immature follicle cells and undifferentiated ovariole sheath muscle incapable of contraction.

Primary cultures have been obtained from adults of *M. domestica* and *M. sorbens* (32) and from adults of *Culex pipiens molestus* (34). In the latter case, ovaries and midguts were dissected from recently emerged adult females which were internally aseptic, and cultured as explants. Monolayers of mixed cell types developed rapidly from the ovaries but slowly from the midguts. This is the first report of any extensive growth from midgut tissue of adults. Judy (47) obtained cell outgrowth and aggregates from hindgut fragments of diapausing hornworm pupae *Manduca sexta*.

Statistically significant differences in rate of growth of cells taken from genotypically different stocks of *Drosophila* embryos (89) was another variable which affected establishment of cultures.

In mammalian blood cell cultures, only the amoebocyte survives (117); but insect hemocytes of apparently more than one kind have been cultured by a number of people (53, 66, 68, 72, 73).

The "tumors" often described in insects are not a source of readily proliferating cells as in vertebrates. Harshbarger & Taylor (35) describe insect tumors as inflammatory reactions rather than true neoplasms demonstrating uncurbed growth.

A tissue or organ large enough to give a high seeding density of homotypic cells is often sought, because this would contribute greatly to the identification of the cell line. However, this purity of cell type may be a disadvantage in the establishment of the line, particularly if the medium is inadequate in some respect. Perhaps cell interactions between heterotypic cells is a necessary function at least during the early stages of adaptation.

Many papers explain that in the first few hours or days following inoculation several types of cells grow out from an organ explant or a mixed suspension, only to be followed by a predominance of two or three types (24). This is a common occurrence in mammalian culture also, where it is characterized as consisting of three phases: rapid proliferation, followed by nonspecific degeneration, and finally (but not invariably), resumed proliferation of the same or altered cells (105). The almost universal ease with which primary cultures of short duration can be established in mixed populations is thought to be due to symbiotic reactions between the cells. Vertebrate culturists explain this phenomenon of cell interaction as due to the "leaking" of products from the cells into the medium, and presumably what one cell leaks is a growth factor for another type of cell. The failure of the cells to achieve the proper genetic combination leads to the failure of sustained proliferaton (105). Perhaps then we should not concentrate so much on finding a source of pure, unmixed cell types, but rather on identifying the cells which do grow.

The generalizations discussed above on selecting donor tissue which have been drawn from the several life stages of Lepidoptera, Diptera, Hemiptera, and Blattariae may apply universally, but there are no records known to us of studies on the other orders. Hymenoptera and Coleoptera should be fruitful material; probably the lack of culture studies for those groups is because of the absence of applied work in virology.

ADAPTATION

Grace's success in establishing the first line of insect cells was heralded as a triumph after years of unsuccessful attempts. A most striking feature of his work was the long period of time required for adaptation of the cells to the culture conditions. After 10 months of apparent quiescence, the cells finally began to proliferate. Some, but not all, of the lines established subsequent to that have shown similarly long periods of adaptation. Whether or not insect cells need a longer time than do cells from other animals is debatable. Swim (105) cites a case of certain cloned cells which took one year to adapt. Although a requirement for long adaptation is depressing, recent reports in insect work of consistent and continued growth between the primary culture and subsequent subcultures is an encouraging development.

The processes operating during adaptation are far from understood. Willmer (117) suggests three possibilities: (a) an alteration in the nature of surfaces, organelles, or enzymes of the individual cells; (b) a selection of most favored cell types originally present and elimination of cells less suited to the new environment; and (c) genetic mutation in the cells given impetus by the conditions of culture, followed by selection of the most favored mutants. In cultures which do not adapt for several months, certainly all three possibilities should be seriously considered. Perhaps some information on the first could be obtained by electron microscopy of adapting primary cultures and established cell lines. But it is difficult to draw rational inferences as to what extent each of the three possibilities actually occurs until experiments are done comparing the primary and adapted cells throughout.

Cell lines of *A. aegypti* (97), *A. albopictus* (97), *Anopheles stephensi* (95), and *Drosophila melanogaster* (43) are basically diploid, whereas the remainder of the lines that have been characterized are highly polyploid. Schneider (95) pointed out that the polyploid lines showed a short initial period of growth followed by a period of slowing down or disappearance of mitosis, whereas the diploid lines seemed to have shown a smooth transition from the primary culture to the subsequent subcultures. Thus, the emergence of polyploidy may have been responsible for the adaptation period; but the level of ploidy of the cells in the original inoculum was not reported.

A recent paper by Mitsuhashi & Grace (75) offers some insight as to what might be happening during adaptation. In these experiments, the *A. aegypti* and the *A. eucalypti* cell lines which had always been maintained on a medium containing insect hemolymph were adapted to an altogether different medium which contained FBS. The two media were mixed in different proportions so that the ratio of FBS medium was gradually increased until the cells were finally adapted solely to the FBS medium. During this process, some cells died and cell debris was noted in the media, particularly in certain of the combinations, suggesting that those cells which failed to adapt had died. The authors made no mention of viability counts of the cells, hence what portion of the cells was destroyed with each transfer is not known. Nevertheless, this study indicates that there may be selection of cells during the process of adaptation.

Once adapted to the new medium, cells may or may not continue to survive in the original medium in which their predecessors were cultured. The *A. eucalypti* line adapted to WEU, FBS, and BPA medium by Yunker et al (121), when returned to the original medium containing insect hemolymph (APH), showed signs of degeneration for a month. The healthy condition of the cells could be restored by replacing them into the hemolymph-free medium to which they had been adapted. Sohi (98) was able to adapt Grace's *A. aegypti* cells from 5 percent APH + 5 percent FBS to 10 percent FBS in which they eventually grew as well as they had in the hemolymph medium. But there was evidence for loss of the least common cell type after adaptation.

In contrast to the preceding two cases of successful adaptation, Mitsuhashi & Grace (75) found that hemolymph is still beneficial. They adapted *A.*

aegypti and *A. eucalypti* lines from a medium containing 3 percent APH to an altogether different medium containing 10 percent FBS. The cells grew at a reduced rate in the new medium but regained their original growth rate when the adapted sublines were cultivated in the APH medium. Mitsuhashi & Grace apparently did not prolong the period of adaptation as much as Sohi (98) did, who took eight months to adapt his lines. It is also interesting to note that the cells adapted by Yunker et al (121) showed a faster doubling time, which was taken to mean that the new medium was superior for growth; but Mitsuhashi & Grace (75) interpreted it to mean that there had been a change from the original line.

There may be some question as to whether adapted cells can give the experimenter the desired results or even valid results. Suitor (103), for example, found that *Antheraea* cells adapted to GMA + *Philosamia cynthia pryeri* hemolymph could not become infected with Japanese encephalitis virus; it was necessary to use APH to which the cells had been adapted earlier to get virus growth. More information on the process of adaptation is needed and would undoubtedly lead to a greater degree of success in inducing primary cultures to establish themselves as lines; and perhaps enable us to culture specific cell types. It would be good to know how the drosophilan imaginal cells that are cultured in adult flies (18) would adapt to in vitro conditions, since it has become possible lately to grow imaginal tissues.

CELL-TO-CELL INTERACTIONS

Insect cell culture work is still embryonic in the study of cell-to-cell interactions. Adequate and comparative descriptions of the relationships between cells in vitro cannot be done unless many factors are identified. Some of these factors are: the identity of the cells, texture of the vessel and properties of the interface between the medium and substrate, temperature, pH, ions, proteinaceous and fatty constituents as well as lipoproteins in the medium, and cell density (78, 117). The generalized stages in morphology and locomotion assumed by insect cells when they are first released from the confines of tissue organization and liberated into the fluid medium are similar to the described stages in cells of other animals. The movement of the cells by gliding or pseudopodia formation is often mentioned.

Contact inhibition operating in cultures results in dispersion of the cells (1). This comes about from the fact that a cell will not use another cell as a substrate for its locomotion, and when two cells meet on a collision course, they alter their direction of movement. Judy (47) made the interesting observation that ecdysterone (β-ecdysone) induced a marked shift towards cell dispersion. After the hormone was added to the medium for hindgut organ culture of tobacco hornworm pupae, the fat-filled cells of fat body adhering to the gut disaggregated and moved independently. Thus, it appears that ecdysone in vitro had stimulated an increased contact-inhibition type phenomenon.

Contact-dependent phenomena in which the cells are stopped from moving apart, or are drawn toward the point of contact (1), are described for insect cell cultures. In reactions of this kind, the cell membranes recognize cells of the same type (31). Self-aggregation of like cells was observed in hanging drop suspensions of dissociated *Drosophila* embryos (61). Within a short time the cells moved about, piling up into spheres or discs within which various cell types aggregated together and remained segregated from the other cell types. Since the embryos were quite well developed when sacrificed, many tissues and cell types quickly differentiated. Dissociated cells of advanced embryos of *M. domestica* also displayed self-aggregation, followed by differentiation into muscle tissue and nerve cells (15). The spindle-shaped cells growing out from ovary explants of adult *Culex* and *Aedes* mosquitoes aggregated into small colonies which subsequently grew into sheets (49).

In blood taken from developing pupae of saturniid moths, Walters & Williams (114) observed self-aggregating ameboid plasmatocytes carrying with them immotile fat body cells. The fat body cells by themselves were incapable of movement. This was in contrast to Judy's work (47) mentioned above in which fat body cells as well as hemocytes moved independently after the addition of hormone.

We (unpublished observation) have obtained monolayers of pupal fat body cells which were capable of active migrations. It remains to be demonstrated to what extent the fat body cells are dependent upon the plasmatocytes for morphogenetic movement during pupation.

Whenever entire insects are used for inoculum, the mixture of cells is highly heterotypic, so the symbiotic interactions of the cells may be contributing as much to the success of the cultures as is the critical stage of development of the insect (discussed above in "Selection of Donor Tissue"). A demonstration of symbiotic interactions was manifest in cultures of mixed cell types obtained from reorganizing tissues in pupae of the forest tent caterpillar *Malacosoma disstria* and the cabbage looper *Trichoplusia ni.* Myoblasts aggregated and differentiated into extensive networks of myotubes (52). When relatively pure cultures of myoblasts were set, there was no significant growth or development.

IDENTIFICATION OF CULTURED CELLS AND THEIR TISSUE OF ORIGIN

In attempting to establish a culture, the first requisite is a system for microscopic examination of the cells and their behavior. Various manipulations are employed in the first few minutes of introducing the cells into the vessel to encourage them to settle and attach to the floor. There is considerable evidence that those cells which remain spherical and do not attach ultimately perish, or at least are discarded if the culture fluid is replaced (24, 49, 72). Thus, a bias is introduced immediately in favor of those cells which readily attach and grow on a flat surface. In some cases, the floating cells have been recovered by centrifugation and used to inoculate new cultures (87).

The plating efficiency of single cells is thought to be a measure of the fraction of cells which are capable of growth in conventional tissue culture situations and it is very low; this means that from the outset, selection of some sort is operating in favor of only those few cells which can tolerate the conditions.

Cells in primary cultures are most often described in morphological terms, such as epitheliocytes, fibroblastlike cells, spindle-shaped cells, and small and large round cells. The latter are especially difficult to homologize with mammalian types (21, 49, 101).

The cell types in embryo lines are epitheliocytes from leafhoppers (10), fibroblasts from cockroaches (56), and spherical cells from *Drosophila* (43). The cells in lines from larvae are diverse too. For example, from *Anopheles* there are derived epithelial cells which do not form true monolayers but tend to aggregate (95), from *A. albopictus* there are various round and spindle-shaped cells (97), and from *A. aegypti* there are round and spindle-shaped (24) or mainly epithelial type tending to form hollow vesicles (97). Two kinds of spindle-shaped cells make up the line from larval (pre-pupal) ovaries of *Bombyx mori* (26).

What, if any, significance should be attached to the fact that one or the other morphology predominates in the various lines is a mystery. Seemingly, any organ used as cell source provides epitheliocytes, fibroblastlike cells, and round cells; but the relation of each type of cell to the functional cells of the organ is not known. Whether or not the spherical cells are, in fact, a type of hemocyte is speculative. The hemolymph of diapausing larvae of *Chilo suppressalis,* the rice stem borer, gave rise to primary cultures of predominantly prohemocytes and plasmatocytes; the plasmatocytes gradually disappeared so that eventually the line consisted of mostly spherical cells thought to be prohemocytes (73). There were also a few spindle cells which could change into spheres. The cells which persisted in successful lines grew in suspension while those cells which attached to the glass (during rotation of the vessel) degenerated.

Efforts to culture the cuboidal or columnar epithelium of Malpighian tubules, midgut, or salivary glands, or the polygonal, highly differentiated fat body cells have not resulted in cell lines. It is not clear how much of this refractoriness is due to the innate inability of this type of cell to grow in the conditions we provide; or how much is due to their dedifferentiation or alteration to spherical or flattened types of cells which we cannot identify by visual inspection. None of these problems is unique to insect culture. Willmer (117) discusses the simplification undergone by cells to generalized forms and their approximation of each other in appearance and behavior. It is almost impossible to identify a cell type on the basis of morphology alone, yet morphologically indistinguishable cells may differ in biochemical and physiological properties. Aside from the esthetic need to know the type of cell with which one is working, the information is essential for making meaningful comparisons of activities in vitro with the normal processes of development and physiology in situ. That is, the information would lead to making rational decisions as to which of numerous biosynthetic reactions or products should be sought in the cultured cells. Most likely one would not do analyses for the production of urine on cells known to have been derived from salivary glands, or attempt to characterize cholinesterase in cells derived from follicular epithelium.

If the cells of the original inoculum were completely dissociated so that cloning could be used for subculturing, it might be easier to identify the cell type in the resulting homogeneous population. The techniques of selecting individual cells are laborious; but more seriously, there is overwhelming evidence from mammalian culture that single cells almost always fail to grow (92). Single cell suspensions of day-old rat liver have a plating efficiency of approximately 10^{-4}. Only a few mammalian lines have been established by cloning. So it is of great significance that some insect cells have been cloned, demonstrating that it is at least possible (27, 104). Cells isolated directly from the insect and cloned have not been used to establish a cell line; but cloning has been done from cell lines consisting of large populations of cells. Perhaps with improvements in the media, it will become possible to use primary inocula of smaller size for cloning.

Suitor et al (104) cloned *A. aegypti* cells by drawing single cells into capillary tubes. After many attempts, one clone was obtained from a single spindle-shaped cell, and this in turn was recloned but the rate of success was very low—only 1.2 percent of the cells grew. Definite indications were observed that the spindle-shaped cells gave rise to round daughter cells, and it was only after the cells had increased considerably in number that they resumed the spindle morphology. Grace (27) developed clones from lines of *A. eucalypti* cells by serially diluting and subculturing suspended cells, followed by recovery of colonies assumed to have been derived from single cells. This technique yielded a higher cell population in the clones than was obtained by Suitor et al (104). The clones indicated that there are, in fact, distinctly different cell types in the original cultures of *Antheraea,* not merely morphological variants or transient forms. In most of the ten clones successfully established, there was uniformity as to size and shape of the cells. Photomicrographs of clones developing from single cells show no morphological deviation of the daughter cells from the parent cells.

Larsen (59) collected some interesting data on the effects of several kinds of irradiation on cultured heart fragments in the presence of dyes. Since the dyes did not seem to inhibit growth (vesicles formed), one wonders if it might be feasible to utilize markers of this sort to identify cell sources, provided that different tissues have unique affinities for the dyes, that the colors can be visualized, and that they are indeed harmless. Stanley (99) tried using Calco Oil Red as a marker by mixing it in the food of *Galleria mellonella* larvae. The larvae grew normally and became pink, but the differences in intensity of color between most tissues was so slight that the dye could not be used to identify the origin of the cells in culture.

It would seem that the identification of a cell product should be good evidence of the nature of the cell. Ritter & Bray (90) reported that acellular chitinous material was synthesized in cultures of blood cells. They found that a few epidermal cells of unknown origin appeared ultimately, but the timing of the secretion of the chitinous material and the appearance of the epidermal cells was such that the latter seemed unlikely to be the source of the chitin secretions. In this case, positive identification of the secretory cells is mandatory because, accepted at its face value, this finding is contrary to the general principle that only cells of ectodermal origin can synthesize chitin.

It was mentioned above (in the discussion of "Adaptation") that electron microscopy might be applied to detect alterations in the ultrastructure of cells during the period of adaptation. In the present connection, it would seem that electron microscopy could be advantageous also in identifying the types of the cells in established cultures. The degree to which ultrastructure could be useful here is limited by the following two factors: (a) the extent to which cytoarchitecture is peculiar to each type of cell, and (b) the absence of alterations in the characteristic cytoarchitecture. Granados et al (30) have described a technique for preparing blood cells of leafhoppers for electron microscopy following which the cell organelles were well preserved. We have not seen electron microscopy studies of cultured insect cells used specifically to identify unknown types.

Stanley & Vaughn (101) used a correlation between the abundance of healthy cells in situ and the predominance of certain cell types in vitro in attempting to identify the origin of cells which grew from moth ovarian tissues. This method was a good predictor of stages or ages, or both, of ovaries which give good cell cultures, but it seems to us that it did not constitute the final proof as to where the cells originated.

Kitamura (49) devised a technique which was favorable for watching cell growth from mosquito ovaries, and this enabled him to decide that cells of the ovary sheath did not migrate out nor proliferate, which is contrary to what most other workers find regarding moth ovaries. Hemocytelike cells or derivatives of hemocytes adhering to the outer surface of the ovary were the proliferating cells, and they appeared in the form of epitheliocytes and fibroblastlike cells. Kitamura found no correlation between rate of outgrowth of cells and developmental stage or condition of the ovaries used as explants.

On the organismic level, there is always the possibility of accidental contamination of a laboratory culture with cells from another animal culture maintained in the same laboratory. Immunodiffusion techniques may be useful here to establish relationship of the cells (45). Cases of contamination of monkey kidney cells with mouse L cells, and human liver cells with nonprimate (probably mouse) cells are known (50). In the latter case, the mixed line was segregated into its two components by virtue of the nonspecific lytic effect of unheated human serum against the nonprimate cells, or by selective destruction of the human cells by Coe or poliovirus. It is unlikely that insect cell lines could become contaminated with mammalian cells because the latter probably would not grow at the low temperatures used for insect cells, and the media probably would not be suitable. Nagle et al (80) reversed the media for *Antheraea* cells and for HeLa or L cells and found that neither would grow in the other medium.

DIFFERENTIATION AND SPECIALIZED FUNCTION

How much differentiation can occur in cells or organs in vitro? Are proliferation and maintenance of specialized function mutually incompatible? Is product formation a reliable indicator of differentiation? These questions are germane to insect cell culture as well as to cell culture in general (31).

Organs such as salivary glands, eyes, gonads, imaginal discs, fragments of epidermis, or of embryos, are useful for observing differentiation. Some large cells such as salivary gland cells and neuroblasts are so easy to observe acting like osmometers that they can be used for adjusting balanced salt solutions (8, 16). The effect of osmotic stress was distinguished from the effect of particular ions on puff formation in *Drosophila* salivary glands (16). Many processes continue to operate, such as DNA synthesis, chromosome puff formation, and puff condensation in *Sciara* salivary glands in vitro (6, 7) so that chemically defined media seem to be satisfactory for DNA synthesis, at least. But it is not known if salivary secretions continue.

Organ cultures of imaginal discs have served for studying metamorphic developments. The rate of differentiation of *Drosophila* eye-antennal discs was prolonged in vitro compared to in situ (93); a latter paper (94), however, revealed that there were some differences in the relative competence of the tissues to undergo differentiation in this system. Male and female gonads of *G. mellonella* have been maintained in organ culture for seven months, during which time spermatocytes underwent meiosis and transformed to spermatids, and the oocytes entered into previtellogenesis (60). Even meiotic divisions in some oocytes in organ culture have been reported (100, 116).

Recently there has been much activity in using tissue and organ culture systems to bioassay for hormone influence on various biosynthetic processes. Miciarelli et al (69) found that larval epidermis from the desert locust, *Schistocerca gregaria,* is capable of molting in vitro if taken after the epidermal cells had already undergone mitosis. Marks & Leopold (64) demonstrated the induction of chitinous cuticle in *Leucophaea maderae* leg regenerates by ecdysterone. Similar results were obtained by Agui et al (2) with cultured integument from diapausing larvae of the rice stem borer.

Endocrine activity is assayed with lepidopteran wing discs also. Oberlander & Fulco (83) found that ecdysone added to the culture medium will induce the partial metamorphosis of competent wing discs from last-instar greater waxmoth larvae, and Agui et al (3) found that ecdysterone will induce metamorphosis of winglike tissues from wing discs of rice stem borer larvae. Oberlander (81, 82) demonstrated that ecdysone influences DNA synthesis in cultured waxmoth wing discs. Ecdysterone was noted to cause an increase in the rate of development of rice stem borer testes and spermatocysts in vitro (120).

Ecdysterone affected the action of migrating cells in organ cultures of the tobacco hornworm by causing an increase in speed and frequency of the undulating membranes and increased action of pseudopodia, accompanied by formation of pinocytotic vesicles (47).

For a comprehensive review of the action of hormones in insect cell and organ cultures, the reader is referred to a recent treatment by Marks (63).

In organ cultures the cells of that organ eventually die and no cells are derived from it which are capable of reproducing themselves indefinitely in a line. This leaves doubt as to whether something was missing from the culture parameters, or if the advanced stage of differentiation precluded proliferation of the cells. This latter opinion has been expressed often. Sato & Yasumura (91) contrast the conditions of fragment or organ type culture, "where the environment of each cell varies depending on its position in the tissue mass" with "dispersed cell cultures (which) represent homogeneous systems." Since the extent of differentiation and specialized function observed in organ culture is never seen in dispersed cell culture, Sato & Yasumura continue, it is important to relate the properties of the cells in culture to those of the homologous cells in situ; the cells in vitro must retain the differentiated properties which characterized them in vivo. This must hold even for those cells which, in adapting to culture conditions, have undergone morphological alterations so they are no longer identifiable. That such specialized function is indeed retained by rapidly growing adrenal cortex cells in dispersed cell culture was demonstrated. In another experiment, Sato et al (92) were able to demonstrate by serological methods that a type of cell, which in liver constitutes a minor element, rapidly overgrows the population, so that loss of specialized function is a result of the cultures having descended from nonliver cells.

Nothing similar to the preceding work exists in insect cell culture, but there is evidence that certain dispersed insect cells can differentiate, for example, the myoblasts and neuroblasts isolated from *Drosophila* embryos (96). The differentiation of dispersed cells from embryonic house fly tissues has been obtained (15). Lepidopteran pupal myoblasts, if taken from the right developmental stage, possessed the competence to differentiate into functional muscle (52). Marks et al (65) reflect that perhaps it is this type of apparent inability of insect cells to dedifferentiate and enter into logarithmic growth which is responsible for much of our difficulty in obtaining cell lines.

A beginning toward identifying products of insect tissue in vitro has been made, but for the most part these are done on organ fragments—not on self-reproducing dispersed cells—which survive for a few hours or days but are not "cultured" axenically in the tissue culture context. For instance, Berridge (4) has identified substrates used in metabolic production of urine by surviving Malpighian tubules. Reddy & Wyatt (88) incubated fragments of pupal wing epidermis to determine the incorporation of labeled uridine and leucine into RNA and protein, and the effect of ecdysone thereon.

The report of the production of chitin by hemocytes in culture (90) is based on identification of the product with two histochemical tests which may be interpreted alternatively. However, if this sort of reaction can be documented independently, it could be an example of the modulation or transformation of a cell to a function which it normally is incapable of performing in situ.

Clements & Grace (11) and Grace & Brzostowski (29) have analyzed culture fluid for the uptake of amino acids and sugars by *Antheraea* cells. Their findings are directed at improving the design of the culture fluid. Landureau (57) has determined vitamin requirements, and Landureau & Jollès (58) have determined both uptake and production of a number of amino acids in cell cultures of *Periplaneta americana;* these analyses appear to be the most complete and sophisticated work on dispersed cells.

PARAMETERS IN NEED OF INVESTIGATION

Perhaps the greatest single handicap in working with insects is their small size which makes difficult the acquisition of a large seeding density for the inoculum, or the accumulation of enough cells to set up replications of cultures. In addition to the mechanical difficulties of manipulating the insects, there is the problem that only a small proportion of cells in primary cultures adapt to become capable of sustained growth. Unless large organs from large insects are used, many individuals of the smaller organisms must be pooled. Of course, pooling is inimical to setting up a relatively pure culture of identified cell types. The several lines (discussed above), of which the exact identities or origins of the cells are unknown, emerged from pooled insects.

The morphological alterations of insect cells in vitro are comparable to those that occur in other animal cells (105). The difficulty lies in identifying the cell before alteration occurs. Isolated cells are so plastic, and assume such unfamiliar forms when they are released from the tissue by which we recognize them, that their identification is uncertain (117).

"Transformation" in the sense of changing as the result of the introduction of new genetic material affecting the cultural properties is a virgin field in insect cell culture. There are two reports that foreign genetic material (virus) can be introduced into the host cell without causing destruction of the cell unless the virus is induced (19, 22). Perhaps transduction-type phenomena could be used with insect viruses to transfer desired genetic traits from one cultured cell to another; or perhaps somatic hybridization could modify the genotype of cultured cells. Although somatic hybridization between two different species of insect cells is unknown, interphylum heterokaryons between human (HeLa) and mosquito (*Aedes aegypti*) cells was induced with ultraviolet-inactivated Sendai virus (122).

Basic as those problems are to cell culture in general, it is our opinion that the greatest need in insect cell culture at present is a concerted attack on the tedious work of setting up permutated variables in media to determine their effects on growth. Several insect cell culturists could cooperate to amass works comparable to the classical studies of Earle (70). Nutrition provided in the culture media needs to be defined by the step-by-step process of single omissions and deletions. There is an increasing tendency to purchase one of the several media available commercially; this practice must have a selective influence on the cell types which can be proliferated in primary cultures. Kitamura (48) suggested using differential media to select deliberately only certain types of cells out of a heterotypic inoculum.

Entire media have been tested for their effects on morphology and DNA synthesis in *Drosophila* glands in organ culture (107). Yunker et al (121) tested over 60 combinations of additives in their experiments to improve the Grace medium. But a completely defined medium permitting the continuous cultivation of cells has not yet been achieved. The uncharacterized constituents—insect hemolymph, FBS, BPA, CEE, and WEU—should be analyzed to find suitable substitutes. Grace & Brzostowski (29) determined which amino acids and sugars were utilized by the *A. eucalypti* cell line; but since the medium was supplemented with insect hemolymph containing high levels of amino acids, it was impossible to simply vary each amino acid and find the concentration giving optimum growth. Instead, it was necessary to assay for changes in the amino acid levels during growth. Seven amino acids remained the same or increased while 14 others decreased in concentration. Glucose and fructose were utilized 90 percent and 60 percent, respectively, while only 23 percent of the sucrose was used.

Landureau & Jollès (58) deleted amino acids one at a time for the *P. americana* cell line and found that 16 amino acids were essential for maintenance of the line. In addition, there was an indication that the protein supplements were not utilized as a source of amino acids or peptides. Of the amino acids in the basic medium, only alanine appeared not to be required. The amino acid constitutents were patterned after the cockroach hemolymph, indicating that the hemolymph may be a fairly good representation of the amino acid requirements of the cells. Nine B-vitamins plus choline were determined by Landureau (57) to be essential while ascorbic acid and *para*-aminobenzoic acid were not. Landureau stated that vitamins involved in the synthesis of nucleic acids and lipids appear to be important and even limiting factors for cultured cells.

The requirement for lipids and sterols by the established cell lines is unknown. Grace's medium is without lipid or sterol until the insect hemolymph or FBS is added. Landureau's medium, completely defined except for the Cohn Fraction V and α2-macroglobulin, has no added lipid or sterol. In light of the sterol requirement of cockroaches, it is interesting that the cells grow without any in the medium unless it is being supplied by the human protein plasma additives.

The nutritional requirements of an intact organism may not reflect the requirements of cultured cells, as shown by studies of amino acids required by intact chick and man compared to those required by certain cell lines derived from those organisms (105).

A fundamental question is how far to carry our simulation of insect hemolymph in designing media. The fact that Grace (21) was successful in establishing silkmoth cell lines in Wyatt's medium which resembled silkworm hemolymph does not necessarily mean that this is the best of all possible media. Some of the constituents in hemolymph are wastes and must be inhibitory to cell growth. On the other hand, a silkworm-type medium is suitable for non-lepidopterans in some cases. The growth and establishment of several dipteran cell lines in Grace's medium or in media partially like Grace's (24, 95) indicate that either the properties of the silkworm hemolymph are not unique or that the insect cells are somewhat plastic. Another arresting thought is that the cells of three genera of leafhoppers (10) would not survive subculturing in the medium devised for leafhoppers (77) but did so in a medium patterned after *Drosophila* hemolymph (93).

Much of our difficulty may stem from the fact that we know little about the interstitial fluids which actually bathe the cells, and which may bear only slight resemblance to hemolymph. Perhaps the factor we need to duplicate is very subtle.

There is no doubt that inorganic salts must be provided in a balanced saline solution approximating the inorganic composition of the body fluid, i.e., hemolymph. The Na/K ratios have been given more attention than other ions (8, 84, 106). The effect of hypertonicity on phagocytosis by hemocytes has been studied (86). The measurement and effect of total osmolarity has been determined for *Drosophila* glands (16), for cricket embryos (33), for cockroach embryos (56), and for cutworm hemocytes (66), but otherwise this aspect has been almost completely neglected. The pH is usually adjusted to somewhere between 6.4 and 7.4 but is seldom critically evaluated as in reports by Ellgaard & Kessel (16) or Grellet (33).

The environmental factors affecting cells have usually been mentioned only in passing; we need more critical studies to determine the extent of their influence on growth and other characteristics. The use of a matrix for certain types of cells might prove to be worthwhile. For example, Nagle et al (80) considered methyl cellulose to be the single component responsible for permitting the suspended cultivation of *A. eucalypti* cells in a medium devoid of hemolymph. Vago & Chastang (109) and Vago & Flandre (110) found that culturing cells in a clot of plasma was beneficial. This seems to be retrogressive; circumventing the need of a plasma clot was thought to be a great advance by mammalian culturists, since a clot precludes growth in suspension (91, 105).

Occasionally there is a report on the effect of carbon dioxide at different levels (9, 32, 67) but we do not know if these levels were determined to be optimal. Growth of the cells without added carbon dioxide has been reported (10, 34, 42). Mitsuhashi & Maramorosch (77) noted that only slight amounts of air sealed into the small vessels seemed to provide adequate aeration, while Varma & Wallers (112) suggested that the confined system they used for tick cells may have been benefited by virtue of increased carbon dioxide tension.

The only critical paper on temperature effects on the growing cells is that of Mitsuhashi (74) who determined 25°C to be optimal with both 20°C and 30°C causing a decrease in cell number. Preservation of cultured cells can be done in a manner similar to that used for mammalian cells. Stock cultures of Grace's *A. eucalypti* line have retained their viability for more than nine months either at +4°C or at −185°C in nitrogen with 20 percent glycerol in the culture medium (87). Mitsuhashi (74) stored hemocytes of *C. suppressalis* infected with iridescent virus for periods up to 120 days at 5°C with some loss of viable cells, while the rate of growth when the cells were returned to culture conditions was slightly inferior to that of the controls. Grace (27) froze cloned cells at −80°C in medium containing 10 percent glycerol and found the cells still viable after 12 months. Schneider (95) successfully regenerated cells of *Anopheles* frozen at −68°C in culture medium containing either 10 percent glycerol or 10 percent dimethyl sulfoxide.

More work should be done on the action of rolling, shaking, or other forms of agitation on the development of suspensions, as was done by Nagle et al (80).

We see no mention of the effect of different spectra of visible light on cultures. Unpublished communication by I. Schneider indicates that visible light may have an adverse effect on the growing cells. With increasing evidence for endogenous rhythms in biological systems and control of numerous functions by photoperiod, this area seems to be one worthy of investigation in cell culture. For instance, Turner & Acree (108) reported that the principal hydrocarbons in hemolymph of *P. americana* fluctuate in a circadian manner and that constant light depresses the magnitude of the fluctuation. Photoperiod and circadian rhythms in effect at the time of sacrificing the insect have not been examined for possible control of subsequent growth behavior in vitro. The range of rhythmic changes in arthropods is tremendous, including such things as volume of cell, nucleus, and nucleolus (12).

The past history of an insect—viz., disease, starvation, or chemical stress—may cause enormous variations in growth potential. Personally, we have observed that hemocytes of *G. mellonella* taken from pupae which had been held in cold storage (4°C) for several days did not survive long in vitro.

The effects of various methods of preparing the tissues and cells for inoculum into the cultures is seldom studied critically. Following initial adverse results from using trypsin to dissociate the cells (13), there was a general tendency to avoid the use of this enzyme and to prepare cells by mechanical mincing or even just by placing a fragment of an organ into the vessel. Later, Grace (21) found that trypsinizing *Antheraea* ovaries was without ill effect. Schneider (95) and Singh (97) both used trypsin successfully in establishing lines of *Anopheles* and *Aedes* mosquitoes, respectively. All of the lines have been developed from either trypsinized or minced tissues.

Pronase (another protease) was used by Schneider (95) who found that it was more effective in separating the cells, did not affect the rate of growth, and seemed to encourage better formation of monolayer; although there was an indication that it might be altering the morphology. Varma & Pudney (111) said that a 0.1 percent solution of pronase was more satisfactory than 0.25 percent trypsin for separating embryonic cells of a reduviid bug, *Triatoma maculata*. Seemingly, if different concentrations of proteolytic enzymes are carefully checked against cell morphology and subsequent growth for the particular tissue being studied, it will be found that the enzymes can be used to advantage. The action of a dissociating enzyme overcomes the inability of the cells to migrate out from organized tissues. It is often noted that a tissue explant appears healthy and remains contractile for up to 80 days or more but the few cells which leave it fail to become established in a monolayer or suspension at high cell population. Stanley & Vaughn (101) pointed out that contractility of an organ reduces cell production because the explant must be attached to the substrate to favor cell outgrowth. The great number of cells obtainable after enzymatic dissociation increases the ratio of cell number to medium volume, an important factor in overcoming the low plating efficiency. Seeding densities resulting in good cell reproduction are in the neighborhood of 2×10^5–3×10^6 cells/ml (15, 24, 51, 57, 111).

Ethylenediaminetetraacetic acid (Versene®) as a cell dissociator seemingly has been used by few insect cell culturists (61).

Antibiotics are almost routinely incorporated in the media with the exception of those projects concerned with propagation of protozoal or bacterial symbiotes (53, 55). Cells of *A. eucalypti* and *A. aegypti* were found to be markedly suceptible to amphotericin B (Fungizone), an antifungal agent (71). The insect cells were more susceptible to amphotericin B than were the fungi in question. Since this antibiotic is thought to be a competitive inhibitor of sterols, the intolerance of the cultured cells suggests that the sterol requirement was not being satisfied under these conditions. We have no information with which to compare insect and mammalian cells with respect to their responses to antibiotics in general.

To summarize this section, it seems that the greatest need for the immediate future is to identify growth factors and arrive at a medium which is defined as well as TC 199 or other media used for vertebrate cell culture.

Are our present concepts of culture fluid adequate, or is an innovation needed? Should we adhere to the principle of including all available nutrients, or should we consider the possibility that perhaps a "complete" medium possesses "retardants" which, through product accumulation and feedback, inhibit further growth and metabolism?

Does each basic cell type need a special medium? Is there such a thing as "an" insect culture medium? Hirumi & Maramorosch (41) stated that the medium they devised could "not be considered as 'an insect tissue culture medium' since differences between groups of arthropods are often more pronounced than differences between mammals on one side, and birds or reptiles on the other."

SERUM

Insect hemolymph or FBS, or both, are often used to compensate for imperfections in the culture medium. The hemolymph is usually heated to 60°C for 5–10 min to inactivate the phenolases, frozen overnight, and then centrifuged to remove the cells; so, in a way, it is analogous to serum. Results from using hemolymph for different species may be quite varied. We ask if this is caused by fundamental differences in the insect cells themselves, the complexity of interactions of all the culture ingredients, or by subtle variations in the manner of working by the investigators.

Waymouth (115) has stated that the requirement for sera by cultured cells exists because of qualitative and quantitative deficits in the culture medium. Studies are urgently needed to determine what these deficits are in the present group of media employed by insect tissue culturists.

Without the addition of serum, and in some instances even with certain sera, the culture medium may be a hostile environment for the continuous cultivation of insect cells. Grace & Brzostowski (29) showed that if the 3 percent APH was deleted from GMA, *A. eucalypti* cells could not survive in it. An abrupt decrease in the growth rate was seen and the cells died in three to four weeks. A year later it was shown independently by both Nagle et al (80) and Yunker et al (121) that the cells could be adapted to FBS as a replacement for hemolymph. Horikawa & Fox (42) had reported earlier that while limited increase in cell density was possible in *Drosophila* embryonic cells in an unsupplemented medium, cell lines were obtainable only with hemolymph supplement.

Rahman et al (87), in attempting to find heterologous hemolymph that could be substituted for APH in GMA, found instead that the hemolymph from several species of Lepidoptera caused lysis of the *A. eucalypti* cells.

The use of homologous hemolymph is not always feasible, especially for very small species; and actually, homologous hemolymph has never been used again to establish a line following Grace's work (21). It was about that time when Vago & Chastang (109) found that calf serum could be substituted satisfactorily to a great extent for insect hemolymph. Their experiments indicated that a small additional amount of hemolymph was still beneficial, but recent studies have demonstrated that cell lines can be established using FBS in place of any insect hemolymph supplement (10, 36–38, 42, 56, 73, 95, 97).

Insect hemolymph may be most beneficial during the early stages of adaptation. Chiu & Black (10) found that FBS was a suitable supplement to the medium for the initiation of primary leafhopper cultures, but unsatisfactory when subcultures were made. For subcultures a medium containing 2.5 percent insect or lobster hemolymph and 17.5 percent FBS was employed; once the line was established it could be maintained on a medium supplemented solely with FBS. Mitsuhashi (72) found that both FBS and hemolymph were necessary to halt degeneration in *C. suppressalis* hemocyte primary cultures.

In spite of all the adaptations and substitutions of FBS, BPA, WEU, or CEE for hemolymph (44, 80, 98, 121), Mitsuhashi & Grace (75) continue to find that bovine components are not satisfactory substitutes for hemolymph. In Grace's work (24, 26), only 1 percent FBS or BPA was used, whereas other investigators use from ten to twenty times that much, which suggests that the discrepancy may be merely quantitative.

The processes of adaptation and selection probably operate in favor of different cells in AEH primary cultures compared to FBS cultures. Adapting the cells to different media usually requires an extended period of time (several weeks to months), and undoubtedly results in cells with cultural characteristics different from those of their predecessors. Most reports make little or no mention of changes in this respect. Yunker et al (121) found the growth rate of the adapted *A. eucalypti* line was greater in the medium with bovine and chick factors than that of cells in the hemolymph medium. The log phase of growth was extended, resulting in a greater yield of cells. Nagle et al (80) also got a higher yield of cells in the *A. eucalypti* line in medium with 10 percent FBS and methylcellulose. Nagle (79) improved the growth of the *A. aegypti* line in medium with only 5 percent FBS by elevating the choline to 50 mg/l.

As a result of such work, we cannot assume that cell lines that have been adapted are identical to those being propagated in another laboratory, which points out the need to get these lines into the American Type Culture Collection so that a standard is available for comparison.

Conclusion

Insect cell and tissue culture has evolved from a mystifying and uncertain stage to the point at which primary cultures of numerous species can be initiated with assurance. The use of trypsin to dissociate the cells, the availability of commercially prepared media, and the enhancement of the media with serum from noninsectan sources contribute to the success of this process.

The factors operative in the establishment of a continuous line of self-reproducing cells is another matter. This still is approached by poorly defined manipulations involved in the extent of disturbing the cells and the frequency of feeding or transferring them. The reactions occurring between the cells and the selective forces impressed by the media during the early stages of adaptation are completely unknown for insect cells.

There is not a completely defined medium which will support an insect cell line. Current interest is very lively in the areas of defining the nutrients in hemolymph and bovine serum additives.

The roles of endocrines, lipids and sterols, and their interconversions, should be examined for their effects on isolated cells in vitro.

LITERATURE CITED

1. Abercrombie, M. 1965. The locomotory behaviour of cells. In *Cells and Tissue Culture*, ed. E. N. Willmer, Vol. 1, Chap. 5, 177–202. New York: Academic Press, 788 pp.
2. Agui, N., Yagi, S., Fukaya, M. 1969 a. Induction of moulting of cultivated integument taken from a diapausing rice stem borer larva in the presence of ecdysterone. *Appl. Entomol. Zool.* 4:156–57
3. Agui, N., Yagi, S., Fukaya, M. 1969 b. Effects of ecdysterone on the *in vitro* development of wing discs of rice stem borer, *Chilo suppressalis. Appl. Entomol. Zool.* 4:158–59
4. Berridge, M. J. 1966. Metabolic pathways of isolated Malpighian tubules of the blowfly functioning in an artificial medium. *J. Insect Physiol.* 12:1523–38
5. Boorman, J. 1967. Aseptic rearing of *Aedes aegypti* Linn. *Nature* 213:197–98
6. Cannon, G. B. 1964. Cultures of insect salivary glands in a chemically defined medium. *Science* 146:1063
7. Cannon, G. B. 1965. Puff development and DNA synthesis in *Sciara* salivary gland chromosomes in tissue culture. *J. Cell. Comp. Physiol.* 65:163–82
8. Carlson, J. G. 1961. The grasshopper neuroblast culture technique and its value in radiobiological studies. *Ann. N.Y. Acad. Sci.* 95:932–41
9. Chen, J. S., Levi-Montalcini, R. 1969. Axonal outgrowth and cell migration *in vitro* from nervous system of cockroach embryos. *Science* 166:631–32
10. Chiu, R.-J., Black, L. M. 1967. Monolayer cultures of insect cell lines and their inoculation with a plant virus. *Nature* 215:1076–78
11. Clements, A. N., Grace, T. D. C. 1967. The utilization of sugars by insect cells in cultures. *J. Insect Physiol.* 13:1327–32
12. Danilevsky, A. S., Goryshin, N. I., Tyshchenko, V. P. 1970. Biological rhythms in terrestrial arthropods. *Ann. Rev. Entomol.* 15:201–44
13. Day, M. F., Grace, T. D. C. 1959. Culture of insect tissues. *Ann. Rev. Entomol.* 4:17–38
14. Echalier, G., Ohanessian, A. 1968.

Une souche stabilisée de cellules embryonnaires de *Drosophila. Int. Colloq. Tissue Culture Invert.*, 2nd, Tremezzo-Como, Italy, 1967
15. Eide, P. E., Chang, T. H. 1969. Cell cultures from dispersed embryonic house fly tissues. *Exp. Cell Res.* 54:302–8
16. Ellgaard, E. G., Kessel, R. G. 1966. Effects of high salt concentration on salivary gland cells of *Drosophila virilis. Exp. Cell Res.* 42:302–7
17. Fedoroff, S. 1966. Proposed usage of animal tissue culture terms. *In Vitro* 2:155–59
18. Gehring, W. 1968. The stability of the determined state in cultures of imaginal disks in *Drosophila*. In *The Stability of the Differentiated State*, ed. H. Ursprung, 136–54. New York: Springer-Verlag, 154 pp.
19. Grace, T. D. C. 1958. Induction of polyhedral bodies in ovarian tissues of the tussock moth *in vitro. Science* 128:249–50
20. Grace, T. D. C. 1959. Prolonged survival and growth of insect ovarian tissue under *in vitro* conditions. *Ann. N.Y. Acad. Sci.* 77(Art. 2):275–82
21. Grace, T. D. C. 1962 a. Establishment of four strains of cells from insect tissues grown *in vitro. Nature* 195:788–89
22. Grace, T. D. C. 1962 b. The development of a cytoplasmic polyhedrosis in insect cells grown *in vitro. Virology* 18:33–42
23. Grace, T. D. C. 1963. Insect tissue and cell culture. In *A Post-Graduate Course in Cell Culture*, ed. D. O. White, 229–42. Melbourne, Australia: Cell Culture Soc. of Victoria, University of Melbourne, Univ. Bookroom, 332 pp.
24. Grace, T. D. C. 1966. Establishment of a line of mosquito (*Aedes aegypti* L.) cells grown *in vitro. Nature* 211:366–67
25. Grace, T. D. C. 1967 a. Insect cell culture and virus research. *In Vitro* 3:104–17
26. Grace, T. D. C. 1967 b. Establishment of a line of cells from the silkworm *Bombyx mori. Nature* 216:613
27. Grace, T. D. C. 1968. The develop-

ment of clones from lines of *Antheraea eucalypti* cells grown *in vitro*. *Exp. Cell Res.* 52:451–58

28. Grace, T. D. C. 1969. Insect tissue culture and its use in virus research. *Advan. Virus Res.* 14:201–20

29. Grace, T. D. C., Brzostowski, H. W. 1966. Analysis of the amino acids and sugars in an insect cell culture medium during growth. *J. Insect Physiol.* 12:625–33

30. Granados, R. R., Ward, L. S., Maramorosch, K. 1968. Insect viremia caused by a plant-pathogenic virus: electron microscopy of vector hemocytes. *Virology* 34:790–96

31. Green, H., Todaro, G. J. 1967. The mammalian cell as differentiated microorganism. *Ann. Rev. Microbiol.* 21:573–600

32. Greenberg, B., Archetti, I. 1969. *In vitro* cultivation of *Musca domestica* L. and *Musca sorbens* Wiedemann tissues. *Exp. Cell Res.* 54:284–87

33. Grellet, P. 1968. Milieu de culture pour embryons de gryllides. *J. Insect Physiol.* 14:1735–61

34. Gubler, D. J. 1968. A method for the *in vitro* cultivation of ovarian and midgut cells from the adult mosquito. *Am. J. Epidemiol.* 87:502–8

35. Harshbarger, J. C., Taylor, R. L. 1968. Neoplasms of insects. *Ann. Rev. Entomol.* 13:159–90

36. Hink, W. F. 1970. Established insect cell line from the cabbage looper, *Trichoplusia ni. Nature* 226:466–67

37. Hink, W. F., Ellis, B. J. Establishment and characterization of two new cell lines (CP-1268 and CP-169) from the codling moth, *Carpocapsa pomonella. Curr. Top. Microbiol. Immunol.* In press

38. Hink, W. F., Ignoffo, C. M. Establishment of a new cell line (IMC-HZ-1) from ovaries of cotton bollworm moths, *Heliothis zea* (Boddie). *Exp. Cell Res.* 60:307–9

39. Hirumi, H., Maramorosch, K. 1964 a. The *in vitro* cultivation of embryonic leafhopper tissues. *Exp. Cell Res.* 36:625–31

40. Hirumi, H., Maramorosch, K. 1964 b. Insect tissue culture: further studies on the cultivation of embryonic leafhopper tissues *in vitro*. *Contrib. Boyce Thompson Inst.* 22:343–52

41. Hirumi, H., Maramorosch, K. 1964 c. Insect tissue culture: use of blastokinetic stage of leafhopper embryo. *Science* 144:1465–67

42. Horikawa, M., Fox, A. S. 1964. Culture of embryonic cells of *Drosophila melanogaster in vitro*. *Science* 145:1437–39

43. Horikawa, M., Ling, L. N., Fox, A. S. 1966. Long-term culture of embryonic cells of *Drosophila melanogaster. Nature* 210:183–85

44. Hsu, S. H., Liu, H. H., Suitor, E. C., Jr. 1967. Adaptation of Grace's continuous line of mosquito (*Aedes aegypti* L.) cells to hemolymph-free medium. (Abstr.) *J. Formosan Med. Assoc.* 66:83

45. Ibrahim, A. N., Sweet, B. H. Antigenic relationships of mosquito cell lines as determined by immunodiffusion techniques. *Curr. Top. Microbiol. Immunol.* In press

46. Jones, B. M. 1966. Invertebrate tissue and organ culture in cell research. In *Cells and Tissues in Culture*, ed. E. N. Willmer, Vol. 3, Chap. 7, 397–457. New York: Academic, 826 pp.

47. Judy, K. J. 1969. Cellular response to ecdysterone *in vitro*. *Science* 165:1374–75

48. Kitamura, S. 1965. The *in vitro* cultivation of tissues from the mosquito, *Culex pipiens* var. *molstus* (sic) II. An improved culture medium useful for ovarian tissue culture. *Kobe J. Med. Sci.* 11:23–30

49. Kitamura, S. 1966. The *in vitro* cultivation of tissues from the mosquitoes. III. Further studies on the cultivation of ovarian tissues of three mosquito species and the examination of the origin of cells grown *in vitro. Kobe J. Med. Sci.* 12:63–70

50. Kunin, C. M., Emmons, L. R., Jordan, W. S., Jr. 1960. Detection of cells of heterologous origin in tissues and their segregation by the use of differential media. *J. Immunol.* 85:203–19

51. Kurtti, T. J., Brooks, M. A. 1970. Growth of lepidopteran epithelial cells and hemocytes in primary cultures. *J. Invert. Pathol.* 15:341–50

52. Kurtti, T. J., Brooks, M. A. 1970. Growth and differentiation of lepidopteran myoblasts *in vitro. Exp. Cell Res.* 61:407–12

53. Kurtti, T. J., Brooks, M. A. Growth of a microsporidian parasite in cultured cells of tent caterpillars (*Malacosoma*). *Curr. Top. Microbiol. Immunol.* In press

54. Landureau, J. C. 1966 a. Cultures *in vitro* de cellules embryonnaires de Blattes. *Exp. Cell Res.* 41: 545–56

55. Landureau, J. C. 1966 b. Des cultures de cellules embryonnaires de Blattes permettent d'obtenir la multiplication *in vitro* des bactéries symbiotiques. *C. R. Acad. Sci.* 262:1484–87

56. Landureau, J. C. 1968. Cultures *in vitro* de cellules embryonnaires de Blattes II. Obtention de lignées cellulaires à multiplication continue. *Exp. Cell Res.* 50:323–37

57. Landureau, J. C. 1969. Étude des exigences d'une lignée de cellules d'insectes (souche EPa). II. Vitamines hydrosolubles. *Exp. Cell Res.* 54:399–402

58. Landureau, J. C., Jollès, P. 1969. Étude des exigences d'une lignée de cellules d'insectes (souche EPa). I. Acides amines. *Exp. Cell Res.* 54:391–98

59. Larsen, W. 1965. Reactions of an insect organ culture to some vital stains. *Am. Zool.* 5:255 (Abstr. # 238)

60. Lender, T., Duveau-Hagege, J. 1963. La survie et al différenciation en culture *in vitro* des gonades de larves de dernier âge de *Galleria mellonella*. *Develop. Biol.* 6:1–22

61. Lesseps, R. J. 1965. Culture of dissociated *Drosophila* embryos: Aggregated cells differentiate and sort out. *Science* 148:502–3

62. Maramorosch, K. 1965. New applications of tissue culture in the study of leafhopper-borne viruses. In *Proc. Int. Conf. Plant Tissue Culture, 1963.* eds. P. R. White, A. R. Grove, 541–46. Berkeley, California: McCutchan Publ. Corp., 553 pp.

63. Marks, E. P. The action of hormones in insect cell and organ cultures. *Gen. Comp. Endocrinol.* In press

64. Marks, E. P., Leopold, R. A. 1970. Cockroach leg regeneration: Effects of ecdysterone *in vitro*. *Science* 167:61–62

65. Marks, E. P., Reinecke, J. P., Caldwell, J. M. 1967. Cockroach tissue *in vitro*: A system for the study of insect cell biology. *In Vitro* 3: 85–92

66. Martignoni, M. E., Scallion, R. J. 1961 a. Preparation and uses of insect hemocyte monolayers *in vitro*. *Biol. Bull.* 121:507–20

67. Martignoni, M. E., Scallion, R. J. 1961 b. Multiplication *in vitro* of a nuclear polyhedrosis virus in insect amoebocytes. *Nature* 190: 1133–34

68. Mazzone, H. M. Cultivation of gypsy moth hemocytes. *Curr. Top. Microbiol. Immunol.* In press

69. Miciarelli, A., Sbrenna, G., Colombo, G. 1967. Experiments of *in vitro* cultures of larval epiderm of desert locust (*Schistocerca gregaria* Forsk.). *Experientia* 23:64–66

70. Microbiological Associates. 1964. A complete list of publications from the laboratory of Wilton R. Earle. *Tissue Culture Bibliography* 4: 10–18

71. Millam-Stanley, M. S., Vaughn, J. L. 1967. Marked sensitivity of insect cell lines to Fungizone (amphotericin B). *J. Insect Physiol.* 13:1613–17

72. Mitsuhashi, J. 1966. Tissue culture of the rice stem borer, *Chilo suppressalis* Walker. II. Morphology and *in vitro* cultivation of hemocytes. *Appl. Entomol. Zool.* 1:5–20

73. Mitsuhashi, J. 1967. Establishment of an insect cell strain persistently infected with an insect virus. *Nature* 215:863–64

74. Mitsuhashi, J. 1968. Tissue culture of the rice stem borer, *Chilo suppressalis* Walker. III. Effects of temperatures and cold-storage on the multiplication of the cell line from larval hemocytes. *Appl. Entomol. Zool.* 3:1–4

75. Mitsuhashi, J., Grace, T. D. C. 1969. Adaptation of established insect cell lines to a different culture medium. *Appl. Entomol. Zool.* 4: 121–22

76. Mitsuhashi, J., Maramorosch, K. 1963. Aseptic cultivation of four virus transmitting species of leafhoppers. *Contrib. Boyce Thompson Inst.* 22:165–73

77. Mitsuhashi, J., Maramorosch, K. 1964. Leafhopper tissue culture: embryonic, nymphal, and imaginal tissues from aseptic insects. *Contrib. Boyce Thompson Inst.* 22:435–60

78. Moscona, A. A. 1965. Recombination of dissociated cells and the development of cell aggregates. In *Cells and Tissue Culture*, ed. E. N. Willmer, Vol. 1, Chap. 14, 489–529. New York: Academic, 788 pp.

79. Nagle, S. C., Jr. 1969. Improved growth of mammalian and insect cells in media containing increased levels of choline. *Appl. Microbiol.* 17:318–19

80. Nagle, S. C., Jr., Crothers, W. C., Hall, N. L. 1967. Growth of moth cells in suspension in hemolymph-free medium. *Appl. Microbiol.* 15: 1497–98

81. Oberlander, H. 1969 a. Effect of ecdysone, ecdysterone and inokosterone on the *in vitro* initiation of metamorphosis of wing disks of *Galleria mellonella*. *J. Insect Physiol.* 15:297–304

82. Oberlander, H. 1969 b. Ecdysone and DNA synthesis in cultured wing disks of the wax moth, *Galleria mellonella*. *J. Insect Physiol.* 15: 1803–6

83. Oberlander, H., Fulco, L. 1967. Growth and partial metamorphosis of imaginal disks of the greater wax moth, *Galleria mellonella, in vitro*. *Nature* 216:1140–41

84. Peleg, J. 1965. Growth of mosquito tissues *in vitro*. *Nature* 206:427–28

85. Plantevin, G. 1967. Dosage de Na⁺, K⁺, Ca²⁺ et Mg²⁺ de l'hemolymphe de *Galleria mellonella* L. par spectrophotometrie de flamme. *J. Insect Physiol.* 13:1907–20

86. Rabinovitch, M., deStefano, M.-J. 1970. Interactions of red cells with phagocytes of the wax-moth (*Galleria mellonella*, L.) and mouse. *Exp. Cell Res.* 59:272–82

87. Rahman, S. B., Perlman, D., Ristich, S. S. 1966. Growth of insect cells in tissue culture. *Proc. Soc. Exp. Biol. Med.* 123:711

88. Reddy, S. R. R., Wyatt, G. R. 1967. Incorporation of uridine and leucine *in vitro* by Cecropia silkmoth wing epidermis during diapause and development. *J. Insect Physiol.* 13:981–94

89. Rezzonico Raimondi, G., Gottardi, A. 1967. Genotypically controlled behavior of embryonic cells of *Drosophila melanogaster* cultured *in vitro*. *J. Insect Physiol.* 13:523–29

90. Ritter, H., Jr., Bray, M. 1968. Chitin synthesis in cultivated cockroach blood. *J. Insect Physiol.* 14:361–66

91. Sato, G., Yasumura, Y. 1966. Retention of differentiated function in dispersed cell culture. *Trans. N.Y. Acad. Sci., Ser. II*, 28:1063–79

92. Sato, G., Zaroff, L., Mills, S. E. 1960. Tissue culture populations and their relation to the tissue of origin. *Proc. Nat. Acad. Sci. U.S.* 46:963–72

93. Schneider, I. 1964. Differentiation of larval *Drosophila* eye-antennal discs *in vitro*. *J. Exp. Zool.* 156:91–104

94. Schneider, I. 1966. Histology of larval eye-antennal disks and cephalic ganglia of *Drosophila* cultured *in vitro*. *J. Embryol. Exp. Morphol.* 15:271–79

95. Schneider, I. 1969. Establishment of three diploid cell lines of *Anopheles stephensi*. *J. Cell Biol.* 42:603–6

96. Seecof, R. L., Unanue, R. L. 1968. Differentiation of embryonic *Drosophila* cells *in vitro*. *Exp. Cell Res.* 50:654–60

97. Singh, K. R. P. 1967. Cell cultures derived from larvae of *Aedes albopictus* (Skuse) and *Aedes aegypti* (L.). *Curr. Sci.* 36:506–8

98. Sohi, S. S. 1969. Adaptation of an *Aedes aegypti* cell line to hemolymph-free culture medium. *Can. J. Microbiol.* 15:1197–1200

99. Stanley, M. S. M. 1967. *In vivo* staining of *Galleria mellonella* fed Calco Oil Red. *Ann. Entomol. Soc. Am.* 60:1121–22

100. Stanley, M. S. M., Vaughn, J. L. 1968 a. Histologic changes in ovaries of *Bombyx mori* in tissue culture. *Ann. Entomol. Soc. Am.* 61:1064–67

101. Stanley, M. S. M., Vaughn, J. L. 1968 b. Origin of migrating cells in cultures of moth ovarian tissue based on developmental stages producing optimum outgrowth. *Ann. Entomol. Soc. Am.* 61:1067–72

102. Suitor, E. C., Jr. 1966 a. Arthropod tissue culture. A brief outline of its development and descriptions of several of its applications. *Lecture Rev. Ser., Namru-2-Lr-023.* U.S. Naval Med. Res. Unit No. 2, Taipei, Taiwan

103. Suitor, E. C., Jr. 1966 b. Growth of Japanese encephalitis virus in

Grace's continuous line of moth cells. *Virology* 30:143–45

104. Suitor, E. C., Jr., Chang, L. L., Liu, H. H. 1966. Establishment and characterization of a clone from Grace's *in vitro* cultured mosquito (*Aedes aegypti* L.) cells. *Exp. Cell Res.* 44:572–78

105. Swim, H. E. 1959. Microbiological aspects of tissue culture. *Ann. Rev. Microbiol.* 13:141–76

106. Ting, K. Y., Brooks, M. A. 1965. Sodium: potassium ratios in insect cell culture and the growth of cockroach cells. *Ann. Entomol. Soc. Am.* 58:197–202

107. Tulchin, N., Mateyko, G. M., Kopac, M. J. 1967. *Drosophila* salivary glands *in vitro*. *J. Cell Biol.* 34:891–97

108. Turner, R. B., Acree, F., Jr. 1967. The effect of photoperiod on the daily fluctuation of haemolymph hydrocarbons in the American cockroach. *J. Insect Physiol.* 13:519–22

109. Vago, C., Chastang, S. 1962. Cultures de tissues d'insectes a l'aide de serum de mammifères. *Entomophaga* 7:175–79

110. Vago, C., Flandre, O. (1962) 1963. Culture prolongée de tissus d'insectes et de vecteurs de maladies, en coagulum plasmatique. *Ann. Epiphyt.* 14 (N° hors ser. 3):127–39

111. Varma, M. G. R., Pudney, M. 1967. The culture of embryonic cells from the bug *Triatoma maculata* (Erichson). *Exp. Cell Res.* 45:671–75

112. Varma, M. G. R., Wallers, W. 1965. An improved method for obtaining, *in vitro*, uniform cell monolayer sheets from tissues of the tick, *Hyalomma dromedarii*. *Nature* 208:602–3

113. Vaughn, J. L. 1968. A review of the use of insect tissue culture for the study of insect-associated viruses. In *Insect Viruses*, ed. K. Maramorosch. *Curr. Top. Microbiol. Immunol.* 42:108–28

114. Walters, D. R., Williams, C. M. 1966. Reaggregation of insect cells as studied by a new method of tissue and organ culture. *Science* 154:516–17

115. Waymouth, C. 1965. Construction and use of synthetic media. In *Cells and Tissues in Culture*, ed. E. N. Willmer, Vol. 1, Chap. 3, 99–142. New York: Academic, 788 pp.

116. White, J. F., Sastrodihardjo, S. 1966. Meiosis in pupal ovarian cells of *Samia cynthia* cultured *in vitro*. *Nature* 212:314–15

117. Willmer, E. N. 1965. Morphological problems of cell type, shape and identification. In *Cells and Tissues in Culture*, ed. E. N. Willmer, Vol. 1, Chap. 4, 143–76. New York: Academic, 788 pp.

118. Wyatt, G. R. 1961. The biochemistry of insect hemolymph. *Ann. Rev. Entomol.* 6:75–102

119. Wyatt, S. S. 1956. Culture *in vitro* of tissue from the silk worm, *Bombyx mori* L. *J. Gen. Physiol.* 39:841–52

120. Yagi, S., Kondo, E., Fukaya, M. 1969. Hormonal effect on cultivated insect tissues. I. Effect of ecdysterone on cultivated testes of diapausing rice stem borer larvae. *Appl. Entomol. Zool.* 4:70–78

121. Yunker, C. E., Vaughn, J. L., Cory, J. 1967. Adaptation of an insect cell line (Grace's *Antheraea* cells) to medium free of insect hemolymph. *Science* 155:1565–66

122. Zepp, H., Conover, J., Hirschhorn, K., Hodes, H. L. Interphylum heterokaryons (human x mosquito) induced by UV-inactivated Sendai virus. *Curr. Top. Microbiol. Immunol.* In press

STEROID METABOLISM IN INSECTS[1]   **6002**

W. E. Robbins, J. N. Kaplanis, J. A. Svoboda, and M. J. Thompson

*Insect Physiology Laboratory, Entomology Research Division,
Agricultural Research Service, United States Department
of Agriculture, Beltsville, Maryland*

Since the discovery by Hobson (27) in 1935 that the larva of a blow fly required a dietary source of sterol for normal growth and development, the sterol requirement of insects has been investigated by scientists from a number of disciplines. In the past decade, the utilization, metabolism and function of sterols in insects has become an intensely investigated area of insect physiology and biochemistry and studies directed toward the nutritional, physiological, biochemical and chemical aspects of this problem have greatly expanded our knowledge of this area. The sum of knowledge derived from these studies, which have recently been discussed and summarized in three comprehensive reviews (9, 20, 56) and a compendium of nutritional data (1), permits certain conclusions:

(i) Insects require a dietary or exogenous source of sterol for normal growth, metamorphosis and reproduction. The only exceptions are those insects in which a sterol source may be attributed to associated symbionts.

(ii) The dietary requirement for sterols results from a lack of the sterol biosynthetic mechanism in insects. Thus, insects, along with some related arthropods and certain other invertebrates, differ from plants and vertebrates which fulfill their sterol requirement through the endogenous biosynthesis of sterols from simple molecules.

(iii) In insects, as in mammals, the sterols serve a dual role; as structural components of cells and tissues and as precursors for essential steroid metabolites and regulators (e.g. hormones).

(iv) Specific sterol structures are required by insects and not all sterols can be utilized. However, certain insects can modify dietary sterols and these metabolic modifications serve to provide structures

---

[1] The following abbreviations will be used: GLC (gas-liquid chromatography); TLC (thin-layer chromatography); NADPH (nicotinamide adenine dinucleotide phosphate, reduced form).

appropriate to the specific physiological and biochemical functions of sterols in insects.

This review will be concerned with those metabolic modifications that sterols undergo in insects during their utilization and metabolism. Since the last review (56) considerable progress has been made in elucidating the intermediate steps in two important pathways of steroid metabolism in insects; (I) the conversion of C_{28} and C_{29} plant sterols to cholesterol and (II) the biosynthesis and metabolism of the molting hormones (ecdysones). These aspects, which will be reviewed in detail, are the major known metabolic pathways for sterols in insects.

CONVERSION OF C_{28} AND C_{29} PLANT STEROLS TO CHOLESTEROL

Insects lack the capacity for the *de novo* biosynthesis of the steroid nucleus and thus must obtain their essential cholesterol either as cholesterol per se from the diet or from a dietary sterol that can be readily converted to cholesterol. Nutritional studies (1) have indicated that a number of the phytophagous and omnivorous insects use C_{28} and C_{29} plant sterols in lieu of cholesterol and that in certain species the conversion of the phytosterols to cholesterol may be a prerequisite to the utilization of these compounds. Many of the earlier studies, however, dealt with the utilization of the plant sterols only in terms of growth and if the sterol content of the insect was examined the methods of analysis usually lacked the sensitivity and specificity to accurately differentiate cholesterol qualitatively and quantitatively from other sterols. Such studies were often further complicated by the presence of sterol contaminants in the dietary components or the use of impure commercial sterol preparations. The development of instrumentation for more sensitive and specific sterol analyses, the availability of highly purified sterol preparations, particularly radio-labeled sterols and the use of semidefined artificial insect diets, has made it possible to study directly the metabolism of the C_{28} and C_{29} plant sterols in insects.

Clark & Bloch (7) were the first to demonstrate conclusively that phytosterol dealkylation does occur in an insect. These workers showed that the German cockroach *Blattella germanica* converts ^{14}C-ergosterol to [14]C-22-dehydrocholesterol. This transformation involves the dealkylation of the 24-methyl from the ergosterol side chain and the reduction of the Δ^7-bond in the nucleus (Fig. 1). Since this insect cannot reduce the Δ^{22}-bond, 22-dehydrocholesterol is the end product of ergosterol metabolism (7, 8). The German cockroach has also been shown, through the use of ^{3}H-β-sitosterol and GLC, to convert the ubiquitous plant sterol β-sitosterol to cholesterol (59). This conversion involved the removal of the C-24 ethyl group from β-sitosterol, thus this insect can de-ethylate as well as demethylate. The ^{3}H-β-sitosterol was separated from the ^{3}H-cholesterol by the use of GLC, which efficiently separates C_{27}, C_{28}, and C_{29} sterols, and the relative amount of radioactivity associated with each of the sterols was determined from fractions trapped from the GLC effluent. The conversion of β-sitosterol to cholesterol has since been shown to occur in a number of species of insects, including the pine sawfly (*Neodiprion pratti*) (65), the silkworm (*Bombyx*

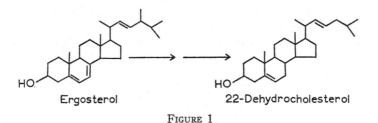

Ergosterol   22-Dehydrocholesterol

FIGURE 1

mori) (36), the boll weevil (*Anthonomus grandis*) (15), the tobacco hornworm (*Manduca sexta*) (78) and a cockroach (*Eurycotis floridana*) (56). Not all insects, however, are capable of converting C_{28} and C_{29} plant sterols to cholesterol. For example, both the larva and adult of the house fly, *Musca domestica,* are unable to convert β-sitosterol to cholesterol (38, 39) and thus must rely entirely on a dietary source of cholesterol to satisfy its "essential" cholesterol requirement.

The biochemical mechanism for the conversion of plant sterols, such as β-sitosterol, to cholesterol is not unique to insects. An insect-parasitic nematode as well as a free-living nematode, both of which require an exogenous or dietary sterol for normal growth, development, and reproduction (11, 14), have been shown to convert β-sitosterol to cholesterol (12, 13). Perhaps this metabolic transformation is widespread in invertebrates, particularly among those that lack the mechanism for the *de novo* biosynthesis of sterols (56).

Biochemical studies on the conversion of phytosterols to cholesterol were initially carried out with the omnivorous German cockroach for a number of reasons; it is a commonly used experimental insect, the modified Nolands diet (53) provides a satisfactory vehicle for administering the sterols under study, and the insects can be reared aseptically. However, two difficulties are encountered with this cockroach—it is cannibalistic and it has a relatively slow nymphal growth rate. Most of the recent research on phytosterol metabolism has been carried out with a phytophagous insect, the oligophagous tobacco hornworm (73–75, 78, 79, 81) and the choice of this insect has indeed proven to be a fortuitous one. Starting with newly-hatched larvae that weigh approximately one milligram, one is able to obtain a greater than 7000-fold increase in the mass of laboratory-reared insects during the normal 15-day larval period. A satisfactory agar-base artificial diet (30) is available, and the combination of rapid growth on an artificial diet and essentially no cannibalism renders this plant-feeding insect particularly well suited for metabolic studies. Although comparative and confirma-

tory studies have been carried out with a number of different insects, representing all three types of metamorphosis, the hornworm has proven to
be the most useful test organism.

The hornworm is also both biochemically efficient and versatile in that it
converts a number of different C_{28} and C_{29} plant sterols (Fig. 2) to cholesterol (75). In addition to removing the 24-R-ethyl from the β-sitosterol or
stigmasterol side chains, this insect can also dealkylate the 24-S-methyl of
campesterol, the 24-S-methyl of brassicasterol or 22,23-dihydrobrassicast-

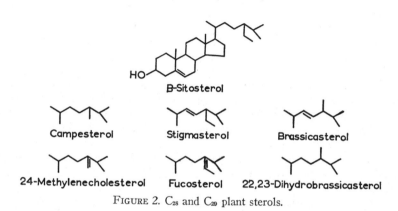

HO

β-Sitosterol

Campesterol   Stigmasterol   Brassicasterol

24-Methylenecholesterol   Fucosterol   22,23-Dihydrobrassicasterol

FIGURE 2. C_{28} and C_{29} plant sterols.

erol, the methylene of 24-methylenecholesterol and the 24-ethylidene of fucosterol. Some degree of specificity was observed in that the 24-S-methyls
of brassicasterol and 22,23-dihydrobrassicasterol did not appear to be as
readily dealkylated as the 24-R-alkyls of β-sitosterol, campesterol, and stigmasterol. Such selectivity could be an adaption to the sterol complex of tobacco, the major host plant of this oligophagous species, in which the latter
three compounds are the predominant sterols. The hornworm also reduces
the Δ^{22}-bond of stigmasterol and brassicasterol and in this respect differs
from the German cockroach (7). This latter conversion may again reflect
the hornworm's adaption to its host plant since stigmasterol is the major
sterol in tobacco.

During the course of studies on the utilization and metabolism of ^3H-β-
sitosterol in the tobacco hornworm ^3H-desmosterol (24-dehydrocholesterol)
(Fig. 3) was isolated and identified as the first known intermediate in the
conversion of β-sitosterol to cholesterol in an insect (78). This sterol was
found to be a constant metabolite of ^3H-β-sitosterol in all stages of the tobacco hornworm. This, taken with the observation that hornworms grow
and develop normally when desmosterol is used as the dietary sterol and
that near complete conversion of ^{14}C-desmosterol to cholesterol is effected
by this insect, further substantiates desmosterol as an intermediate (78).

These results have been verified by the subsequent finding that desmosterol is an intermediate in the conversion of β-sitosterol to cholesterol in a number of insects including the firebrat (*Thermobia domestica*), the German cockroach, the American cockroach (*Periplaneta americana*), the corn earworm (*Heliothis zea*), and the fall armyworm (*Spodoptera frugiperda*) (76). The occurrence of desmosterol as an intermediate in the formation of cholesterol in insects is interesting from a comparative standpoint, since Δ^{24}-sterol intermediates are also involved in sterol biosynthesis in both

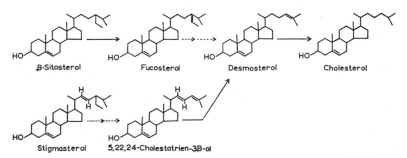

FIGURE 3. Conversion of β-sitosterol and stigmasterol to cholesterol in insects.

plants and vertebrates. Desmosterol is a precursor for cholesterol in the *de novo* biosynthesis of sterols in vertebrates (70, 72), and in plants a Δ^{24}-sterol is involved in the primary step in which supernumerary carbon atoms are added at C-24 on the side chain (alkylation) in the biosynthesis of C_{28} and C_{29} plant sterols (21).

To supplement the in vivo evidence, in vitro studies were also carried out to determine whether desmosterol is indeed converted to cholesterol in the hornworm. In vitro preparations, patterned after the system used for the Δ^{24}-sterol reductase enzyme(s) of vertebrates (71), were made from a number of tobacco hornworm tissues. A preparation of the hornworm midgut efficiently converted ^{14}C-desmosterol to cholesterol, whereas all other tissue preparations were essentially inactive (81). Subcellular fractionation of midgut tissue indicated that the Δ^{24}-sterol reductase activity is associated with the microsomal fraction and NADPH is a necessary cofactor. These results provided the first in vitro system for studying a specific sterol-metabolizing enzyme from an insect. Similar in vitro preparations of American cockroach tissues have a very low level of Δ^{24}-sterol reductase activity compared to the hornworm (77). Thus, there appears to be a significant difference either in the titer of the Δ^{24}-sterol reductase(s) or in the kinetics of the enzyme(s) of these two species.

Recently, another step in β-sitosterol metabolism has been elucidated. ^{3}H-Fucosterol has been isolated and identified as a metabolite of ^{3}H-β-sitosterol and a probable intermediate in the conversion of β-sitosterol to cholesterol in the tobacco hornworm (Fig. 3) (80). Fucosterol was conclusively identified through analysis of the ^{3}H-fucosterol and several of its derivatives by mass spectrometry, GLC and TLC. Fucosterol is a constant minor metabolite of ^{3}H-β-sitosterol in the hornworm, accounting for less than 1 percent of the total ^{3}H-sterols (80), and as previously pointed out it is efficiently converted to cholesterol in this insect (75). The occurrence of fucosterol as a metabolite and possible intermediate had previously been suggested by its tentative identification, solely on the basis of GLC analysis, from both the German cockroach and *E. floridana* (56). Provided that fucosterol and desmosterol are the first and terminal metabolites, respectively, in the major pathway of conversion of β-sitosterol to cholesterol in the hornworm, then dealkylation in insects may well be the reverse of the alkylation process in phytosterol biosynthesis in plants (21).

Certain hypocholesterolemic agents have been shown to exert their effect by inhibiting the Δ^{24}-sterol reductase in vertebrates and thus blocking the conversion of desmosterol to cholesterol (2, 85). Since desmosterol is an intermediate in the conversion of β-sitosterol to cholesterol in the tobacco hornworm, two of these inhibitors, 22,25-diazacholesterol and triparanol (Fig. 4), were tested in the hornworm (74). Both of these compounds, when fed in combination with β-sitosterol, inhibited the Δ^{24}-sterol reductase

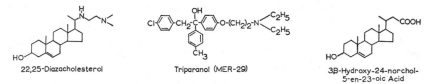

22,25-Diazacholesterol   Triparanol (MER-29)   3β-Hydroxy-24-norchol-5-en-23-oic Acid

FIGURE 4. Inhibitors of sterol metabolism.

of this insect and caused an accumulation of desmosterol and unmetabolized β-sitosterol at the expense of cholesterol formation. In addition to these inhibitive effects on sterol metabolism, larval development was disrupted by these inhibitors, particularly by the diazasterol. The sterol inhibitors have opened up new areas of investigation on sterol utilization and metabolism in insects. Results with these compounds demonstrate, for the first time, that sterol utilization in an insect can be blocked through the inhibition of specific enzymes involved in sterol metabolism and that this block may be accompanied by the disruption of larval development and metamorphosis.

These inhibitors have also proven valuable for studying the utilization and metabolism of plant sterols in insects; the 20,25-diazacholesterol was subsequently used to demonstrate that desmosterol is a common intermediate in the conversion of each of the C_{28} and C_{29} plant sterols in Figure 2 to cholesterol in the hornworm (75). The usefulness of the azasterols in metabolic studies again became apparent when a new sterol, 22-*trans*-5,22,24-

cholestatrien-3β-ol (Fig. 3), was isolated and identified as an intermediate in the conversion of stigmasterol to cholesterol in several insects (35, 73). This sterol, first found to accumulate in tobacco hornworms when a diazasterol was fed in combination with stigmasterol, was later shown, however, to be a normal intermediate that precedes desmosterol in the conversion of stigmasterol to cholesterol (Fig. 3). The sequence of conversion has been determined by feeding the 22-*trans*-5,22,24-cholestatrien-3β-ol either alone or in combination with a diazasterol (73). Cholesterol (77 percent) and desmosterol (15 percent) were the major hornworm sterols when this sterol was fed alone. When the sterol was fed in combination with a diazasterol, cholesterol and desmosterol comprised only about 3 and 9 percent, respectively, of the insect sterols. In neither case was any 22-dehydrocholesterol detected. This indicates that, in the conversion of 22-*trans*-5,22,24-cholestatrien-3β-ol to cholesterol, the Δ^{22}-bond is first reduced and then the Δ^{24}-bond is reduced (73). Interestingly, studies with *trans*-22-dehydrocholesterol as a dietary sterol for the hornworm have shown that there is little if any conversion of this sterol to cholesterol. It would appear then that a $\Delta^{22,24}$-sterol is required as the substrate for reduction of the Δ^{22}-bond by the hornworm sterol reductase. The in vitro preparation of hornworm midgut tissue that reduces desmosterol to cholesterol also effects the conversion of ^{3}H-22-*trans*-5,22,24-cholestatrien-3β-ol to desmosterol and cholesterol (73).

In studies concerned with phytosterol metabolism in the tobacco hornworm it was found that desmosterol accumulated in test insects when a certain commercial sample of β-sitosterol was used in the diet. This observation suggested the presence of a biologically active impurity in the β-sitosterol and led to the isolation, identification, and synthesis of the major active component, 3β-hydroxy-24-norchol-5-en-23-oic acid (Fig. 4), a new Δ^{24}-sterol reductase inhibitor (79). In comparative studies this steroidal acid inhibited the Δ^{24}-sterol reductase of tobacco hornworms and rats, both in vivo and in vitro (81). Although all three types of inhibitors, the azasterols, triparanol and the steroidal acid (Fig. 4), block the conversion of desmosterol to cholesterol, the steroidal acid differed from either the active azasterols or triparanol in that it had no effect on hornworm larval development and metamorphosis. The azasterols, in particular, caused a number of additional inhibitive effects on both sterol metabolism and larval development of the hornworm (76). These include an overall decrease in the total quantity of dealkylated sterols (cholesterol plus desmosterol) which could be due to either a feedback effect or to an additional block(s) in the metabolic pathway (74, 76). The severe reduction in cholesterol formation, caused by an azasterol, is usually accompanied in hornworms with the disruption of larval development and metamorphosis; insects die during the early larval molts, some larvae form prepupae an instar early (4th-instar prepupae) and many insects are unable to pupate normally (76). In the case of the boll weevil, partial inhibition of larval development occurs even when the azasterol is fed in combination with cholesterol (15). The azasterols then apparently become directly or indirectly involved in metabolic pathways other than the conversion of plant sterols to cholesterol. The observed interference with the hormone-mediated processes of molting and morphogenesis in the hornworm strongly suggests that the affected pathways may well involve ecdysone biosynthesis or metabolism.

Little is known of the end products of sterol metabolism in insects except that the major pathways of sterol degradation and excretion that occur in mammals (e.g. coprostanol and bile acids) are absent or of only minor importance in insects, although unknown radioactive "polar steroids" do occur in the intestinal tract and excreta of insects following the ingestion or injection of radioactive sterols (9, 20, 56). Recently, polar sterol metabolites from the meconium of the tobacco hornworm were isolated and identified as the sulfates of cholesterol, β-sitosterol, and campesterol (34). The relatively large percentage of unmetabolized phytosterols in the sterol sulfates from meconium, suggest the selective elimination of these conjugates during pupal-adult development. Based on this information it would appear that the sterol sulfates in insects serve as a means of sterol excretion as has been postulated for the sulfates in mammals (90). Sterol sulfates are also believed to serve as intermediates in the biosynthesis of vertebrate steroid hormones (22, 63) and it will be of interest to determine if they have such a role in insects.

Although we are just beginning to assemble a body of information on phytosterol metabolism in insects, the advances made in the past three years augur well for future progress in this area. The intermediates or metabolites that have been identified should serve to guide future metabolic studies. The availability of a steroid-metabolizing in vitro system hopefully will provide the impetus for detailed studies on the nature of such enzymes. Finally, the discovery of a number of potent inhibitors that block specific enzymes or pathways involved in sterol metabolism in insects, not only makes these compounds available as experimental tools but also suggests the eventual extension of this information to practical application.

BIOSYNTHESIS AND METABOLISM OF THE MOLTING HORMONES (ECDYSONES)

Cholesterol as a probable precursor for physiologically active metabolites in insects was eloquently presented by Clark & Bloch (6) from their studies with the hide beetle *Dermestes maculatus*. This insect, known to possess a stringent requirement for cholesterol (6, 10), was able to grow and develop on a diet containing subminimal amounts of cholesterol in conjunction with a sparing sterol(s) which alone is incapable of supporting growth. Thus, they postulated that the sparing sterol serves in a structural capacity, whereas the essential cholesterol was used for metabolic purposes. That cholesterol is indeed metabolized to physiologically active compound(s) in insects was borne out by the identification of α-ecdysone (Fig. 5), the first molting hormone isolated in crystalline form from an insect (4) and found to be a steroid by chemical methods (49) and X-ray diffraction spectoscopy (33). Shortly after their preliminary identification of α-ecdysone as a steroid (49), members of the same school (48) reported on the transformation of injected ³H-cholesterol to ³H-α-ecdysone by the larvae of the blow fly

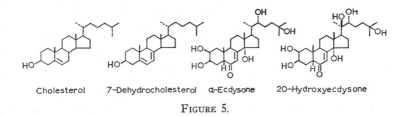

Cholesterol   7–Dehydrocholesterol   α–Ecdysone   20–Hydroxyecdysone

FIGURE 5.

Calliphora vicina. In essence, the study consisted of injecting the larvae with a large dose of high specific activity ³H-cholesterol and examining the extractives, first by paper chromatography and then by dilution and cocrystallization of the eluate with authentic α-ecdysone. These investigators, however, did not indicate whether or not the paper chromatographic system employed was capable of separating α-ecdysone from yet another molting hormone (β-ecdysone) (46) which also has been reported to be present in the blow fly *C. vicina* (46). Though there can be no doubt that cholesterol is a precursor for the molting hormone(s) it is possible that the radioactive fraction from the chromatogram in this study represented a mixture of these two ecdysones.

The structural characterization of α-ecdysone as a steroid was indeed a breakthrough which spearheaded a series of significant events in an already rapidly developing area of insect endocrinology. Shortly after the identification of α-ecdysone, another steroid with molting hormone activity was isolated and identified from a crustacean, the sea-water crayfish *Jasus lalandei* (23). The compound, which was assigned the trivial name of crustecdysone (23) is structurally similar to α-ecdysone but possesses an additional hydroxyl group at the C-20 position. Independently, and almost simultaneously, several laboratories in different parts of the world isolated and identified the same compound from a number of different insects. Thus, crustecdysone isolated from *Antheraea pernyi* (32), β-ecdysone (46), ecdysterone (31), and 20-hydroxyecdysone (29) isolated from *Bombyx mori*, and THE-I (45) isolated from *Manduca sexta* are all the same compound. The nomen, 20-hydroxyecdysone (Fig. 5), immediately relates the compound structurally to α-ecdysone and will be used throughout this discussion for the sake of uniformity and to avoid confusion.

Further biochemical proof that cholesterol is a precursor of the molting hormones in insects was recently provided by Galbraith et al (19). In this report, the tritium-labeled cholesterol after injection was converted to ³H-20-hydroxyecdysone by the larvae of the blow fly *Calliphora stygia*. Interestingly, the incorporation of ³H-cholesterol into the 20-hydroxyecdysone by the larvae of *C. stygia* was 150 times greater than the incorporation of ³H-cholesterol into α-ecdysone by the larvae of *C. vicina* (48).

The molting hormones, α-ecdysone, and 20-hydroxyecdysone are not restricted to insects or related arthropods. These two hormones per se have been found together and have also been isolated in crystalline form and identified from two species of ferns (26, 44). Just as strikingly, a number of related steroids with molting hormone activity have been isolated and identified from a wide variety of plants (55) representing all three major classes of the Pteropsida. The concentration of some of these steroidal compounds in plants is a thousand times greater than that reported for the richest insect source. Although a discussion of this facet is not in the realm of this review, one cannot ignore pointing out that these two widely separated biological systems possess similar biochemical mechanisms for synthesizing these unique and biologically active molecules. In plants as in insects, cholesterol serves as a precursor for these steroidal structures. Heftmann et al (25), and Sauer et al (64) have shown that when 4-¹⁴C-cholesterol is applied to *Podocarpus elata* seedlings, the compound is transformed to ¹⁴C-20-hydroxyecdysone. It is quite apparent then that insects and plants possess the hydroxylases required for the transformation of cholesterol to the ecdysones.

The conversion of cholesterol to the polyhydroxy-keto-steroidal molting hormone represents quite an elaborate transformation. One well wonders what the sequence of biochemical events or intermediate steps are during the conversion of cholesterol to the ecdysones. Unfortunately, a complete story on the biosynthetic scheme from cholesterol to α-ecdysone is not currently available. However, a metabolite of cholesterol, 7-dehydrocholesterol (Fig. 5), occurs in a number of insects and the Δ^7-bond that is common to both 7-dehydrocholesterol and the ecdysones points to this sterol as an intermediate in molting hormone biosynthesis. Metabolic studies with the 4-¹⁴C-cholesterol have shown its conversion to ¹⁴C-7-dehydrocholesterol (37, 40, 51, 57, 60, 62, 65). In the first such report (40) in which 4-¹⁴C-cholesterol was injected into adult female house flies, the adult flies and the eggs oviposited by these flies contained radiolabeled 7-dehydrocholesterol both in the free and in the esterified form. Similarly, when house flies are reared on semidefined diets containing 4-¹⁴C-cholesterol, the eggs from these flies also contained free and esterified ¹⁴C-7-dehydrocholesterol at relatively high concentrations (51, 57). On the other hand, little or no 7-dehydrocholesterol was present in the larvae, pupae, or the newly emerged adults (51). That the insect per se has the metabolic capability for dehydrogenation of cholesterol at the C-7 position, independent of intestinal microorganisms, was clearly shown with the German cockroach (62). When this insect is reared either aseptically or nonaseptically on diets containing 4-^{14}C-cholesterol, the relative percentage of ^{14}C-7-dehydrocholesterol was about the same by either method. As much as 70 percent of the total sterols present in an insect may be in the form of 7-dehydrocholesterol (3). Thus, the relative percentage of 7-dehydrocholesterol present in the total sterols of insects varies from species to species as well as between the different developmental stages within a species.

One species of insect has been shown to require an exogenous source of Δ^7-sterol and thus probably lacks the mechanism for forming the Δ^7-bond (24). However, the absence of 7-dehydrocholesterol in a developmental stage of an insect does not necessarily indicate that either the species or the stage of development lacks the ability to dehydrogenate cholesterol at the C-7 position. Ishii et al (37) have shown that only a fraction of a percent (0.4–0.9) of the 4-^{14}C-cholesterol was converted to ^{14}C-7-dehydrocholesterol in the adult American cockroach. Yet, when a certain specific tissue, the prothoracic gland of the adult American cockroach, was examined by Chen et al (5) as much as 7 percent of the total sterols present in this organ was 7-dehydrocholesterol. It has been shown that the relative percentage of cholesterol and 7-dehydrocholesterol in the prothoracic glands varies according to the stages of development and that 7-dehydrocholesterol may account for as much as 25 percent of the total sterols present in this tissue (5). The fact then that 7-dehydrocholesterol is localized at relatively high concentration in a specific endocrine organ not only points to a possible site for the biosynthesis of 7-dehydrocholesterol but further implicates this compound as a precursor of the molting hormones.

A report timely for inclusion in the discussion of this facet of sterol metabolism is the recent findings of Galbraith et al (19). These investigators have shown that 1-^{3}H-7-dehydrocholesterol, like 1-^{3}H-cholesterol, is metabolized to [3]H-20-hydroxyecdysone in *C. stygia*. Particularly noteworthy was the relative incorporation of each of these precursors into [3]H 20 hy droxyecdysone. The incorporation into [3]H-20-hydroxyecdysone was 0.015 and 0.025 percent for the labeled cholesterol and 7-dehydrocholesterol, respectively. This 1.7-fold difference in the incorporation of the two sterols, points to 7-dehydrocholesterol as a more efficiently utilized precursor than cholesterol.

Thus far, the discussion has dealt with but two of the molting hormones isolated from insects—α-ecdysone and 20-hydroxyecdysone. There is yet a third ecdysone with molting hormone activity that has been isolated and identified from 7-day-old tobacco hornworm pupae at peak ecdysone titer. This compound differs from α-ecdysone in the presence of two additional hydroxyl groups at the C-20 and C-26 positions and has been assigned the trivial name of 20,26-dihydroxyecdysone (Fig. 6) (86). Unlike α-ecdysone and 20-hydroxyecdysone which are about equally active (45), in the house fly assay (42), the 20,26-dihydroxyecdysone is 1/10 to 1/15 as active (86).

Since α-ecdysone, 20-hydroxyecdysone, and 20,26-dihydroxyecdysone have been isolated from a single insect source at peak ecdysone titer, it has been proposed that these compounds represent a metabolic sequence in the bio-synthetic-degradative scheme of the ecdysones (86).

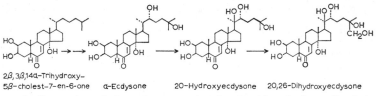

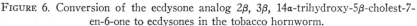

2β,3β,14α-Trihydroxy-   α-Ecdysone   20-Hydroxyecdysone   20,26-Dihydroxyecdysone
5β-cholest-7-en-6-one

FIGURE 6. Conversion of the ecdysone analog 2β, 3β, 14α-trihydroxy-5β-cholest-7-en-6-one to ecdysones in the tobacco hornworm.

That ^{3}H-α-ecdysone can serve as a precursor of [3]H-20-hydroxyecdysone was recently demonstrated in the larvae of the blow fly *C. vicina* as well as in two species of Crustacea (50). In this well-executed study, the injected ^{3}H-α-ecdysone yielded, in addition to unchanged α-ecdysone, a more polar and biologically active ^{3}H-metabolite. This polar metabolite was isolated from all three species in a sufficiently pure form to give an ultraviolet absorption spectrum typical for the molting hormones. Further analyses of the biologically active metabolite and several of its derivatives by TLC indicated the compound to be 20-hydroxyecdysone. Galbraith et al (18) have shown that the blow fly *C. stygia* also converts ^{3}H-α-ecdysone to [3]H-20-hydroxyecdysone.

Of particular interest is the derivation of all three of the insect molting hormones from a synthetic ecdysone analog as the common precursor. This synthetic ecdysone analog, 2β,3β,14α-trihydroxy-5β-cholest-7-en-6-one (Fig. 6) (16, 87), embraces all of the structural features of the ecdysone nucleus but lacks the hydroxyl groups on the side chain. In addition to molt-ing hormone activity (28), this analog has been shown to inhibit larval growth and ovarian maturation in several species of insects (43, 58, 61, 89) and to terminate pupal diapause (41). When it was injected into male dia-pausing tobacco hornworm pupae, the ^{3}H-labeled analog was converted into ^{3}H-α-ecdysone and [3]H-20-hydroxyecdysone (41). Based on the specific ac-tivity of the ^{3}H-ecdysones, approximately one-half of each of the hormones was derived from endogenous sterol precursors and the other half from the ^{3}H-ecdysone analog. In addition to the crystalline ^{3}H-α-ecdysone and [3]H-20-hydroxyecdysone, preliminary evidence also indicated the presence of [3]H-20,26-dihydroxyecdysone. It is possible since the synthetic analog readily enters the steroid pool and is so effectively incorporated into the insect's ecdysones, that this compound is a normal intermediate in the biosynthesis of the molting hormones in the tobacco hornworm (Fig. 6).

Another synthetic ecdysone analog, 25-deoxyecdysone (Fig. 7), has been studied as a possible natural precursor of 20-hydroxyecdysone (88). After injection of 3rd instar larvae of *C. stygia* with the [3]H-25-deoxyecdysone, these workers found that the compound is converted to [3]H-20-hydroxyecdysone as well as to two ecdysone analogs, ^{3}H-ponasterone A and ^{3}H-inokosterone, which are found in plants (52, 84). Since the two phytoecdysones, ponasterone A and inokosterone (Fig. 7) are not normally detected in significant amounts in extracts from this insect, the investigators concluded that this compound is not a major natural precursor for 20-hydroxyecdysone in *C. stygia*. The results do indicate, however, that this insect hydroxylates the C-26 as well as the C-25 position as does the tobacco hornworm (86).

Once these potent molecules have fulfilled their function—how, when, and where are they rendered inactive by the insect? Biological, biochemical, and chemical approaches have been used in an attempt to enhance our understanding of the inactivation of these potent biological regulators in insects. Ohtaki et al (54) have reported that when α-ecdysone is injected into mature larvae of the flesh fly *Sarcophaga peregrina,* it is converted into compounds which no longer have molting hormone activity. The rate of inactivation was found to be dose-dependent—the lower the dose injected the more rapid the inactivation of the hormone. Physical factors, such as low temperature and anaerobic conditions, were found to block inactivation. These investigators also concluded that the inactivating mechanism resides in the tissues and not in the blood. That an exogenous source of ecdysone is rapidly inactivated by an insect following injection and that inactivation is mediated by an enzyme system which resides in the fat body, has been reported from the work with the blow fly *C. vicina* (47, 69). Even though there are some minor disagreements on the length of time required for an exogenous supply of ecdysone to be inactivated possibly because of difference in the age of test insects, radiotracer studies with the labeled molting hormone have in essence corroborated the above cited studies. It has been shown with the blow fly *C. stygia* (17) that if this insect is injected with [3]H-20-hydroxyecdysone 6 hr after puparium formation, most of the hormone is recovered unchanged. However, if this insect is injected with the [3]H-20-hydroxyecdysone 20 hr after puparium formation, the major part of the hormone is converted to more polar water-soluble metabolites. In addition to the above-mentioned studies on rates of disappearance which point to alterations of the ecdysone molecule, the isolation and identification of these modified structures might well provide information on biochemical and physiological processes required for inactivation. The lower biological activity of the 20,26-dihydroxyecdysone from the tobacco hornworm suggests that further hydroxylation may be a step toward inactivation (86). With the limited information currently on hand, however, the possibility that the 20,26-dihydroxyecdysone might also have in its own right a specific function in the tobacco hornworm (86) cannot be ruled out. Other workers

(17) have proposed side chain cleavage and the production of such polar compounds as the C_{19} steroid rubrosterone (83) (Fig. 7). However, based on their experiments with *C. stygia,* it is at most a minor pathway in this insect. Other forms of modification might well be the formation of conjugates such as glycosides and sulfates. A glycoside derivative of a phytoecdysone

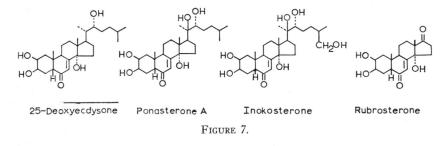

| 25-Deoxyecdysone | Ponasterone A | Inokosterone | Rubrosterone |

FIGURE 7.

has been isolated and identified from a plant (82) and sterols as the sulfate derivatives have already been isolated from the meconium of the tobacco hornworm (34). Possibly the molting hormones are also rendered inactive by insects in this way.

CONCLUDING REMARKS

The conversion of C_{28} and C_{29} plant sterols to cholesterol and the biosynthesis and metabolism of the ecdysones have been treated separately in this review, for the sake of discussion. They should perhaps be more correctly considered as portions of a single overall metabolic scheme for steroids in insects. Such a metabolic scheme for steroids in the tobacco hornworm, based on known intermediates and metabolic conversions, is shown in Figure 8. Exogenous stigmasterol or β-sitosterol, major sterols of the tobacco plant, are converted to cholesterol through a series of intermediates. A portion of the unmetabolized plant sterols or the cholesterol may be excreted by the hornworm as sterol sulfates. In addition, the monohydroxy sterols may occur as sterol esters. Cholesterol, which occupies a central position in the scheme, serves both as a structural component of the insects' tissues and as a precursor for the ecdysones. It is readily apparent that this is a rather complex scheme even though our knowledge of it is far from complete. Certain portions of the scheme in particular require further investigation; the initial steps in the dealkylation of plant sterols and the early and terminal metabolites in the biosynthesis and metabolism of the ecdysones. A more serious deficiency is reflected in our near complete lack of knowledge of the enzyme systems involved in these sterol transformations in insects.

In addition to the above discussed pathways of steroid metabolism in insects there are a number of other important aspects worthy of discussion and additional research effort. One such example concerns the sterol esters of insects; the role of the esters in sterol transport and storage and the nature of the enzymes involved in sterol ester formation and hydrolysis. Of particular interest are the unknown polar steroids found to occur in insects following the administration of radiotracer-labeled sterols (9, 20, 56). It is not currently known whether these compounds are essential steroid metabolites or simply the end products of steroid metabolism. Conceivably there may occur in insects a complex of steroid hormones, which may or may not be structurally related to the ecdysones, that have functions analogous to the steroid hormones of vetebrates. Indeed, Schildknecht and co-workers (66–68) have isolated and identified a number of steroids similar or identical to the steroid hormones of vertebrates from the thoracic bladders of dytiscid water beetles. It will be important to ascertain if these steroids, which apparently act to protect these beetles against poikilothermous vertebrates, also function in the regulation of the biochemical and physiological processes of the insect. Alternately, the polar steroids in insects may well be excretory products, as would appear to be the case for the sterol sulfates in the tobacco hornworm (34). If so, then their isolation and identification

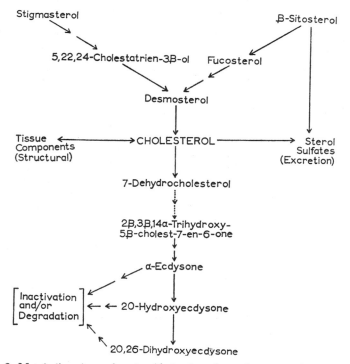

FIGURE 8. Metabolic scheme for steroids in the tobacco hornworm based on known intermediates and metabolic conversions.

could provide us with information both as to the final steps in sterol metabolism and the mechanisms involved in the inactivation or degradation of essential steroid metabolites, such as the ecdysones.

Information concerning the utilization, metabolism, and function of steroids in insects contributes significantly to our overall understanding of insect physiology and biochemistry and provides us with comparative information on steroids in living systems generally, animal and plant. Of particular interest from the comparative standpoint are the intermediates in the dealkylation of phytosterols in insects that have structural features similar or identical to intermediates in the formation of these sterols in plants (21). The biosynthesis, metabolism, and function of the phytoecdysones in plants (55) and the synthesis and the role of the vertebrate steroid hormones in insects (66–68) are two additional areas of study that will certainly provide valuable comparative information. Finally, it should be noted that certain of the azasterol inhibitors (73–76) and the synthetic ecdysone analogs (43, 58, 61) that have been used to elucidate the pathways of steroid metabolism are extremely potent disruptors of insect growth, metamorphosis, and reproduction. Continued research in these areas may lead to the development of safe and selective chemicals for insect control.

LITERATURE CITED

1. Altman, P. L., Dittmer, D. S. 1968. *Metabolism,* 158–61. Bethesda: Fed. Am. Soc. Exp. Biol., 737 pp.
2. Avigan, J., Steinberg, D., Thompson, M. J., Mosettig, E. 1960. Mechanism of action of MER-29, an inhibitor of cholesterol biosynthesis. *Biochem. Biophys. Res. Commun.* 2:63–65
3. Beck, S. D., Kapadia, G. G. 1957. Insect nutrition and metabolism of sterols. *Science* 126:258–59
4. Butenandt, A., Karlson, P. 1954. Uber die Isolierung eines Metamorphose-Hormons der Insekten in kristallisierter Form. *Z. Naturforsch.* 9B:389–91
5. Chen, D., Kaplanis, J. N., Robbins, W. E. Unpublished observations
6. Clark, A. J., Bloch, K. 1959. Function of sterols in *Dermestes vulpinus. J. Biol. Chem.* 234:2583–88
7. Clark, A. J., Bloch, K. 1959. Conversion of ergosterol to 22-dehydrocholesterol in *Blattella germanica. J. Biol. Chem.* 234:2589–93
8. Clayton, R. B. 1960. The role of intestinal symbionts in the sterol metabolism of *Blattella germanica. J. Biol. Chem.* 235:3421–25
9. Clayton, R. B. 1964. The utilization of sterols by insects. *J. Lipid Res.* 5:3–19
10. Clayton, R. B., Bloch, K. 1963. Sterol utilization in the hide beetle *Dermestes vulpinus. J. Biol. Chem.* 238:586–91
11. Cole, R. J., Dutky, S. R. 1969. A sterol requirement in *Turbatrix aceti* and *Panagrellus redivivus. J. Nematol.* 1:72–75
12. Cole, R. J., Krusberg, L. R. 1968. Sterol metabolism in *Turbatrix aceti. Life Sci.* 7:713–24
13. Dutky, S. R. 1967. An appraisal of the DD 136 nematode for the control of insect populations and some biochemical aspects of its host-parasite relationships. *Proc. Joint United States-Japan Seminar Microbial Control Insect Pests, Fukuoka, 1968,* 139–40
14. Dutky, S. R., Robbins, W. E., Thompson, J. V. 1967. The demonstration of sterols as requirements for the growth, development, and reproduction of the DD-136 nematode. *Nematologica* 13:140
15. Earle, N. W., Lambremont, E. N., Burks, M. L., Slatten, B. H., Bennett, A. F. 1967. Conversion of β-sitosterol to cholesterol in the boll weevil and the inhibition of larval development by two aza sterols. *J. Econ. Entomol.* 60:291–93
16. Furlenmeier, A., et al. 1966. Zur Synthese des Ecdysons. Synthesen von 2β,3β,14α-trihydroxy-6-keto-Δ⁷-α/β-cis Steroiden. *Helv. Chim. Acta* 49:1591–1601
17. Galbraith, M. N., et al. 1969. The catabolism of crustecdysone in the blowfly *Calliphora stygia. Chem. Commun.* (19):1134–35
18. Galbraith, M. N., Horn, D. H. S., Middleton, E. J., Hackney, R. J. 1969. Molting hormones of insects and crustaceans: The synthesis of 22-deoxycrustecdysone. *Austr. J. Chem.* 22:1517–24
19. Galbraith, M. N., Horn, D. H. S., Middleton, E. J., Thomson, J. A. 1970. The biosynthesis of crustecdysone in the blowfly *Calliphora stygia. Chem. Commun.* (3):179–80
20. Gilbert, L. I. 1967. Lipid metabolism and function in insects. *Advan. Insect Physiol.* 4:69–211
21. Goad, L. J. 1967. Aspects of phytosterol biosynthesis. *Terpenoids in Plants,* Chap. 10, 159–90. New York: Academic, 255 pp.
22. Gurpide, E., Roberts, K. D., Welch, M. T., Bandy, L., Lieberman, S. 1966. Studies on the metabolism of blood-borne cholesterol sulfate. *Biochemistry* 5:3352–62
23. Hampshire, F., Horn, D. H. S. 1966. Structure of crustecdysone, a crustacean moulting hormone. *Chem. Commun.* (2):37–38
24. Heed, W. B., Kircher, H. W. 1965. Unique sterol in the ecology and nutrition of *Drosophila pachea. Science* 149:758–61
25. Heftmann, E., Sauer, H. H., Bennett, R. D. 1968. Biosynthesis of ecdysterone from cholesterol by a plant. *Naturwissenschaften* 55:37–38
26. Heinrich, G., Hoffmeister, H. 1967. Ecdyson als Begleitsubstanz des Ecdysterons in *Polypodium vulgare* L. *Experientia* 23:995

27. Hobson, R. P. 1935. CCXXXVI. On a fat-soluble growth factor required by blowfly larvae. II. Identity of the growth factor with cholesterol. *Biochem. J.* 29:2023–26

28. Hocks, P., et al. 1966. Synthetische Steroide mit Häutungashormonaktivität. *Angew. Chemi.* 78:680

29. Hocks, P., Wiechert, R. 1966. 20-Hydroxy-Ecdyson, isoliert aus Insekten. *Tetrahedron Lett.* (26): 2989–93

30. Hoffman, J. D., Lawson, F. R., Yamamoto, R. 1966. Tobacco hornworms. In *Insect Colonization and Mass Production*, Chap. 34, 479–86. New York: Academic, 621 pp.

31. Hoffmeister, H., Gruetzmacher, H. F. 1966. Zur Chemie des Ecdysterons. *Tetrahedron Lett.* (33):4017–23

32. Horn, D. H. S., Middleton, E. J., Wunderlich, J. A. 1966. Identity of the moulting hormones of insects and crustaceans. *Chem. Commun.* (11):339-41

33. Huber, R., Hoppe, W. 1965. Zur Chemie des Ecdysons, VII Die Kristall- und Molekulstrukturanalyse des Insektenverpuppungshormons Ecdyson mit der automatisierten Faltmolekülmethode. *Chem. Ber.* 98:2403–24

34. Hutchins, R. F. N., Kaplanis, J. N. 1969. Sterol sulfates in an insect. *Steroids* 13:605–14

35. Hutchins, R. F. N., Thompson, M. J., Svoboda, J. A. 1970. The synthesis and the mass and nuclear magnetic resonance spectra of side chain isomers of cholesta-5,22-dien-3β-ol and cholesta-5,22,24-trien-3β-ol. *Steroids* 15:113–30

36. Ikekawa, N., Suzuki, M., Kobayashi, M., Tsuda, K. 1966. Studies on the sterol of *Bombyx mori* L. IV. Conversion of the sterol in the silkworm. *Chem. Pharm. Bull.* 14: 834–36

37. Ishii, S., Kaplanis, J. N., Robbins, W. E. 1963. Distribution and fate of 4-C[14]-cholesterol in the adult male American cockroach. *Ann. Entomol. Soc. Am.* 56:115–19

38. Kaplanis, J. N., Monroe, R. E., Robbins, W. E., Louloudes, S. J. 1963. The fate of dietary H[3]-β-sitosterol in the adult house fly. *Ann. Entomol. Soc. Am.* 56:198–201

39. Kaplanis, J. N., Robbins, W. E., Monroe, R. E., Shortino, T. J., Thompson, M. J. 1965. The utilization and fate of β-sitosterol in the larva of the housefly, *Musca domestica* L. *J. Insect Physiol.* 11:251–58

40. Kaplanis, J. N., Robbins, W. E., Tabor, L. A. 1960. The utilization and metabolism of 4-C[14]-cholesterol by the adult house fly. *Ann. Entomol. Soc. Am.* 53:260–64

41. Kaplanis, J. N., Robbins, W. E., Thompson, M. J., Baumhover, A. H. 1969. Ecdysone analog: conversion to alpha ecdysone and 20-hydroxyecdysone by an insect. *Science* 166:1540–41

42. Kaplanis, J. N., Tabor, L. A., Thompson, M. J., Robbins, W. E., Shortino, T. J. 1966. Assay for ecdysone (molting hormone) activity using the house fly, *Musca domestica* L. *Steroids* 8:625–31

43. Kaplanis, J. N., Thompson, M. J., Robbins, W. E. 1968. The effects of ecdysones and analogs on ovarian development and reproduction in the house fly, *Musca domestica*. *Int. Congr. Entomol. Proc., 13th, Moscow*

44. Kaplanis, J. N., Thompson, M. J., Robbins, W. E., Bryce, B. M. 1967. Insect hormones: alpha ecdysone and 20-hydroxyecdysone in bracken fern. *Science* 157:1436–38

45. Kaplanis, J. N., Thompson, M. J., Yamamoto, R. T., Robbins, W. E., Louloudes, S. J. 1966. Ecdysones from the pupa of the tobacco hornworm, *Manduca sexta* (Johannson). *Steroids* 8:605–23

46. Karlson, P. 1956. Biochemical studies on insect hormones. *Vitam. Horm.* 14:227–65

47. Karlson, P., Bode, C. 1969. Die Inaktivierung des Ecdysons bei der Schmeissfliege *Calliphora erythrocephala* Meigen. *J. Insect Physiol.* 15:111–18

48. Karlson, P., Hoffmeister, H. 1963. Zur Biogenese des Ecdysons, I Umwandlung von Cholesterin in Ecdyson. *Z. Physiol. Chem.* 331: 298–300

49. Karlson, P., Hoffmeister, H., Hoppe, W., Huber, R. 1963. Zur Chemie des Ecdysons. *Ann. Chem.* 662:1–20

50. King, D. S., Siddall, J. B. 1969. Conversion of α-ecdysone to β-ecdysone by crustaceans and insects. *Nature* 221:955–56

51. Monroe, R. E., Hopkins, T. L., Valder, S. A. 1967. The metabolism and utilization of cholesterol-4-C^{14} for growth and reproduction of aseptically reared house flies, *Musca domestica* L. *J. Insect Physiol.* 13: 219–33

52. Nakanishi, K., Koreeda, M., Sasaki, S., Chang, M. L., Hsu, H. Y. 1966. Insect hormones. The structure of ponasterone A, an insect-moulting hormone from the leaves of *Podocarpus nakaii* Hay. *Chem. Commun.* (24):915–17

53. Noland, J. L., Baumann, C. A. 1949. Requirements of the German cockroach for choline and related compounds. *Proc. Soc. Exptl. Biol. Med.* 70:198–201

54. Ohtaki, T., Milkman, R. D., Williams, C. M. 1968. Dynamics of ecdysone secretion and action in the fleshfly *Sarcophaga peregrina*. *Biol. Bull.* 135:322–24

55. Okauchi, T. 1969. Recent studies on phytoecdysones. *Botyo Kagaku.* 34:140–56

56. Ritter, F. J., Wientjens, W. H. J. M. 1967. Sterol metabolism in insects. *TNO Nieuws* 22:381–92

57. Robbins, W. E. 1963. Studies on the utilization, metabolism and function of sterols in the house fly, *Musca domestica*. *Radiation and Radioisotopes Applied to Insects of Agricultural Importance, Athens, 1963*, 269–80

58. Robbins, W. E., et al. 1968. Ecdysones and analogs: effects on development and reproduction of insects. *Science* 161:1158–60

59. Robbins, W. E., Dutky, R. C., Monroe, R. E., Kaplanis, J. N. 1962. The metabolism of H^3-β-sitosterol by the German cockroach. *Ann. Entomol. Soc. Am.* 55:102–4

60. Robbins, W. E., Kaplanis, J. N., Monroe, R. E., Tabor, L. A. 1961. The utilization of dietary cholesterol by German cockroaches. *Ann. Entomol. Soc. Am.* 54:165–68

61. Robbins, W. E., Kaplanis, J. N., Thompson, M. J., Shortino, T. J., Joyner, S. C. 1970. Ecdysones and synthetic analogs: molting hormone activity and inhibitive effects on insect growth, metamorphosis and reproduction. *Steroids* 16. In press

62. Robbins, W. E., Thompson, M. J., Kaplanis, J. N., Shortino, T. J.

63. Roberts, K. D., Bandy, L., Calvin, H. I., Drucher, W. D., Lieberman, S. 1964. Evidence that steroid sulfates serve as biosynthetic intermediates. IV. Conversion of cholesterol sulfate *in vivo* to urinary C$_{19}$ and C$_{21}$ steroidal sulfates. *Biochemistry* 3:1983–88

64. Sauer, H. H., Bennett, R. D., Heftmann, E. 1968. Ecdysterone biosynthesis in *Podocarpus elata*. *Phytochemistry* 7:2027–30

65. Schaefer, C. H., Kaplanis, J. N., Robbins, W. E. 1965. The relationship of the sterols of the Virginia pine sawfly, *Neodiprion pratti dyar*, to those of two host plants *Pinus virginiana* mill and *Pinus rigida* mill. *J. Insect Physiol.* 11:1013–21

66. Schildknecht, H., Birringer, H., Maschwitz, U. 1967. Testosterone as protective agent of the water beetle *Ilybius*. *Angew. Chem.* 6:558–59

67. Schildknecht, H., Siewerdt, R., Maschwitz, U. 1966. A vertebrate hormone as defensive substance of the water beetle (*Dytiscus marginalis*). *Angew. Chem.* 5:421–22

68. Schildknecht, H., Siewerdt, R., Maschwitz, U. 1967. Cybisteron, ein neues Arthropoden-Steroid. *Ann. Chem.* 703:182–89

69. Shaaya, E. 1969. Studies on the distribution of ecdysone in different tissues of *Calliphora erythrocephala* and its biological half life. *Z. Naturforsch.* 24B:718–21

70. Steinberg, D., Avigan, J. 1960. Studies of cholesterol biosynthesis II. The role of desmosterol in the biosynthesis of cholesterol. *J. Biol. Chem.* 235:3127–29

71. Steinberg, D., Avigan, J. 1969. Rat liver sterol Δ^{24}-reductase. In *Methods Enzymol.*, Chap. 21, 15:514–22. New York: Academic, 903 pp.

72. Stokes, W. M., Fish, W. A. 1960. Sterol metabolism II. The occurrence of desmosterol (24-dehydrocholesterol) in rat liver. *J. Biol. Chem.* 235:2604–7

73. Svoboda, J. A., Hutchins, R. F. N., Thompson, M. J., Robbins, W. E. 1969. 22-*Trans*-cholesta-5,22,24-trien-3β-ol—An intermediate in

the conversion of stigmasterol to cholesterol in the tobacco hornworm, *Manduca sexta* (Johannson). *Steroids* 14:469–76

74. Svoboda, J. A., Robbins, W. E. 1967. Conversion of beta sitosterol to cholesterol blocked in an insect by hypocholesterolemic agents. *Science* 156:1637–38

75. Svoboda, J. A., Robbins, W. E. 1968. Desmosterol as a common intermediate in the conversion of a number of C_{28} and C_{29} plant sterols to cholesterol by the tobacco hornworm. *Experientia* 24:1131–39

76. Svoboda, J. A., Robbins, W. E. 1970. The inhibitive effects of azasterols on sterol metabolism and growth and development in insects with special reference to the tobacco hornworm. *Am. Oil Chemist' Soc., 61st meeting, New Orleans, 1970*

77. Svoboda, J. A., Robbins, W. E., Thompson, M. J. Unpublished observations

78. Svoboda, J. A., Thompson, M. J., Robbins, W. E. 1967. Desmosterol, an intermediate in dealkylation of β-sitosterol in the tobacco hornworm. *Life Sci.* 6:395–404

79. Svoboda, J. A., Thompson, M. J., Robbins, W. E. 1968. 3β-Hydroxy-24-norchol-5-en-23-oic acid—A new inhibitor of the Δ^{24}-sterol reductase enzyme system(s) in the tobacco hornworm, *Manduca sexta* (Johannson). *Steroids* 12:559–70

80. Svoboda, J. A., Thompson, M. J., Robbins, W. E. Manuscript in preparation: Identification of fucosterol as a metabolite and a probable intermediate in the conversion of β-sitosterol to cholesterol in an insect

81. Svoboda, J. A., Womack, M., Thompson, M. J., Robbins, W. E. 1969. Comparative studies on the activity of 3β-hydroxy-Δ^5-norcholenic acid on the Δ^{24}-sterol reductase enzyme(s) in an insect and the rat. *Comp. Biochem. Physiol.* 30:541–49

82. Takemoto, T., Arihara, S., Hikino, H.

1968. Structure of ponasteroside A, a novel glycoside of insect-moulting substance from *Pteridium aquilinum* var. *Latiusculum*. *Tetrahedron Lett.* (39):4199–4202

83. Takemoto, T., Hikino, Y., Hikino, H., Ogawa, S., Nishimoto, N. 1968. Structure of rubrosterone, a novel C_{19} metabolite of insect-moulting substances from *Achyranthes rubrofusca*. *Tetrahedron Lett.* (26):3053–56

84. Takemoto, T., Ogawa, S., Nishimoto, N. 1967. Studies on the constituents of *Achyranthis radix*. III. Structure of inokosterone. *Yakugaku Zasshi* 87:1474–77

85. Thompson, M. J., Dupont, J., Robbins, W. E. 1963. The sterols of liver and carcass of 20,25-diazacholesterol-fed rats. *Steroids* 2:99–104

86. Thompson, M. J., Kaplanis, J. N., Robbins, W. E., Yamamoto, R. T. 1967. 20,26-Dihydroxyecdysone, a new steroid with moulting hormone activity from the tobacco hornworm, *Manduca sexta* (Johannson). *Chem. Commun.* (13):650–53

87. Thompson, M. J., Robbins, W. E., Kaplanis, J. N., Cohen, C. F., Lancaster, S. M. 1970. Synthesis of analogs of α-ecdysone. A simplified synthesis of $2\beta,3\beta,14\alpha$-trihydroxy-7-en-6-one-5β-steroids. *Steroids* 16. In press

88. Thomson, J. A., Siddall, J. B., Galbraith, M. N., Horn, D. H. S., Middleton, E. J. 1969. The biosynthesis of ecdysones in the blowfly *Calliphora stygia*. *Chem. Commun.* (12):699–70

89. Williams, C. M., Robbins, W. E. 1968. Conference on insect-plant interactions. *BioScience* 18:791–92, 797–99

90. Winter, J. S. D., Bongiovanni, A. M. 1968. Identification of cholesterol sulfate in urine and plasma of normal and hypercholesterolemic subjects. *J. Clin. Endocrinol. Metab.* 28:927–30

ARTHROPOD SILKS: THE PROBLEM OF FIBROUS 6003 PROTEINS IN ANIMAL TISSUES

K. M. Rudall and W. Kenchington

The University, Leeds, England

Animals produce quantities of fibrous protein as in muscle, connective tissue and keratin. In muscle and keratinous tissue this protein remains within the cytoplasm while in connective tissue it is secreted and extracellular in position. Muscle and connective tissue fibrous proteins are found in all metazoan groups while true keratin proteins seem restricted to vertebrates. The important animal function of producing fibrous protein is expressed in a very great variety of ways in the process of silk production in arthropods. These silk proteins are secreted and have an "extramural function" as cocoons, webs, and various attachment threads. Once produced they do not have contact with the humors of the animal body, and it is probably for this reason that such great variety of constitution and molecular form occur among the silks. Because they are completely detached from the body they can have molecular variations that would not be tolerated within the body. This could imply that at some stage in evolution cumbersome fibrous proteins arose within cells and that such cells could not survive unless these products were "excreted."

The study of silk proteins greatly extends our knowledge of the biological possibilities of fibrous protein production. The fibrous proteins of silks are in plan very closely related to synthetic homopolymers of amino acids (4). They generally show a large content of the regular protein conformations known as the parallel-β, the cross-β, the α-helical, and the collagen types. Known synthetic polypeptides such as polyglycine, polyalanine, etc., are valuable models and, in fact, segments of both these named polymers actually occur in silk proteins (20).

Fibrous proteins.—The major examples, keratin, light meromyosin, and collagen, have considerable lengths of regular chain conformation in them under physiological conditions. This confers mechanical strength to a fibrous framework, a fluid holding capacity, and capacity to bind ions, but no apparent catalytic function. They probably have some kind of repeated sequence of amino acids.

Corpuscular proteins.—The chain molecules are bent or folded into ovoid

73

or spherical corpuscles; the runs of regular conformation, α or β in type, are of very short length. In the native state, the molecules can be crystallized, each molecule having a specific conformation. With a change of physiological conditions (heating, pH change), the molecules transform to a random coil state. This state would not seem to be a single definable conformation. On further denaturation, the molecules interact to produce large insoluble cross-linked aggregates of partly β-pleated sheet conformation. In this end result boiled egg albumin is not greatly different from boiled myosin but the two molecules are completely different in shape and form under physiological conditions.

Whereas corpuscular proteins are universally present in living organisms as enzymes and other active molecules, the fibrous proteins serve structural and mechanical functions and are apparently restricted to animals, first becoming an integral part of the body in primitive metazoa. The fibrous proteins of muscle are conceived as being homologous in all muscles; the collagens appear to be related from jellyfish to man; within the vertebrate series, the keratins or prekeratins would seem related in having a common origin from the tonofibrils of the cell. The silk proteins of arthropods are far more varied but as they are the expression of the special capacity of animal cells to produce fibrous protein they are regarded as being related to one another and to the fibrous proteins found in muscle, keratin, and connective tissue.

Definition of silk.—In common usage, the term refers to the fine glossy threads which are extruded by certain moth larvae to form their relatively massive cocoons. Likewise, the fine threads of spider webs are silk. The process of extrusion is referred to as "spinning," or the cocoon as being "spun," but there is no obvious twisting of the filaments in the process. A silk filament is often round or oval in cross section and very uniform in diameter. But the spinneret aperture seems capable of much change in shape. A common phenomenon is for the first fibers to be coarse or very coarse, and within this initial framework the next production is of fine fibers which then changes to closely compacted broad ribbons. Silk proteins may thus be laid down as fibers, as ribbons, or as sheets. Parchment-like structures are just as truly made of silk as the recognizable silken form of separate fibers. Some apparent silk fibers are quite different in chemical constitution from the classical fibrous proteins, the predominant polymeric substance being polysaccharide or even hydrocarbon.

SOURCES OF INFORMATION

Ideally we would like to trace all aspects such as the general and macromolecular anatomy of cells producing silks, the nature of the "soluble" materials stored in "reservoirs," and the transformation into relatively anhydrous cocoon or web. To do this for just one silk would be a large task. Our aim at present is to give a broad picture of problems relating to silk proteins. Infrared absorption, X-ray diffraction analysis, and amino acid analysis give the major clues. We tend to use mainly X-ray diffraction at this stage as it defines a major structural component and informs us about its conformation and dimensional features. Unless pursued in extensive detail with deuteration, infrared absorption studies mainly serve to confirm the conformation type deduced from an X-ray study. But they are of major significance in detecting other components such as polysaccharide and lipids. General amino acid analysis has given substantial agreement with X-ray results. We have now reached the stage where the individual proteins should be purified from the soluble secretion in the glands in order to determine the amino acid sequence and measure the molecular weight.

Highly satisfactory X-ray diffraction information is obtained when we have fibers of uniform length as in silkworm cocoons and spider drag lines. In the majority of cases, high quality fibers are not readily obtained from the cocoon; it is then desirable to draw one's own fibers from the spinnerets or from freshly dissected glands. Fresh larvae of many interesting insect groups are available only at limited seasons and are found by chance. However, preserved larvae are obtainable from a number of institutes undertaking entomological surveys and the study of these has opened up many new profitable endeavors in this field.

THE SILK-FORMING GLANDS

Adult insects rarely produce silk, a few noteworthy examples being the egg stalks of *Chrysopa,* the cocoon of *Hydrophilus piceus* and other Neuroptera and Coleoptera closely related to these types. The fibrous protein of the parchment-like ootheca of mantids is regarded as being related to silks. All these examples arise from colleterial glands accessory to the genital system of the female. Male insects are said to produce "silk" in the case of *Lepisma* (Thysanura) in which it is used to form threads which restrict the movement of the female during mating (7). See also pages 407–46 in this volume.

Silk production is very characteristic of adult spiders. So-called ampullate, aggregate, and aciniform glands serve to maintain the web and catch the prey. Cylindrical glands produce the cocoon (40). Both sexes must form a small amount of silk to catch prey during their development to the adult stage. According to Sekiguchi (see 40), the aggregate glands (viscid thread producer) become vestigial in the adult male and males do not spin a web after the last moult. The cylindrical glands which produce cocoon threads are absent in the adult male.

By far the greatest variety of silk production occurs in insect larvae. Here, the silk is predominantly a salivary gland product and the secretion is extruded from spinnerets in the head region. Some species produce a typical cocoon from the Malpighian tubules, e.g., *Chrysopa,* or from a modified peritrophic membrane secretion, as in *Rhynchaenus fagi, Ptinus tectus,* and *Prionomerus calceatus.*

As we shall see, the peritrophic membrane silk and the Malpighian tubule silk are specialized. But adult spiders, adult insects, and insect larvae can produce protein silks which are similar in having the parallel-β type of fibroin structure. That is, ampullate, aggregate, and cylindrical glands of spiders can produce materials similar and related to those produced by adult colleterial and larval salivary glands of insects. The type of silk shared by spider, adult insect, and insect larva is a Group 3 fibroin, perhaps the commonest basic structure found in silks (20).

THE CLASSICAL FIBROINS
X-RAY STUDIES

X-ray diffraction by silk proteins reveals certain dimensions and structural features which are very important to an understanding of silks, and about which we are unlikely to learn by other methods of study. The studies by Marsh, Corey & Pauling (22, 23) on the silks of *Bombyx mori* and *Antheraea pernyi* were classical in establishing the pleated sheet structure for aggregates of β- or extended protein chains. The pleated sheets consisted of antiparallel protein chains, and using agreed interatomic distances and bond angles, the sheets, as seen in models, were corrugated or pleated in the plane of the sheet. In the plane of a sheet the chains are joined together by -CO··HN- hydrogen bonds. The distance between the chain in this plane is the "interchain distance" and has a value of from 4.72–4.77 Å according to the protein. The separation between superposed pleated sheets, the "intersheet distance," is directly related to the size of the side chains of the constituent amino acid residues. This is excellently brought out in Warwicker's study (38) which establishes that the fiber axis period and the interchain distance is the same in a wide variety of silks from Lepidoptera, braconids, and spiders, but the average intersheet distance varies from about 4.5 Å to nearly 8 Å. An X-ray study of these fibroins under favorable conditions would therefore inform us beyond doubt about the average size of the amino acid residues in the organized "crystallites" of the silk.

We can only interpret the diffraction data correctly if we know the space group or the way the chains are packed in the unit cell (the repeating unit of structure). We give in Figure 1 a cross section of a crystallite (viewed along the length of the extended polypeptide chains). The X means the chain is directed down, the · that the chain runs up. PS is the plane of successive pleated sheets and ac is the cross section of the unit cell. Because of halving, the first "reflection" in the direction a will have a spacing $a/4$ and the first "reflection" in direction c will have a spacing of $c/2$. Using this cell, Marsh et al (22) obtained reasonable agreement between the measured intensities and the calculated intensities for Tussah silk. Arnott, Dover & Elliott (1) made a detailed study of β-poly-L-alanine and found a unit cell of only half the a value of the structure proposed by Marsh et al. This arose because the sheets in Figure 1 were randomly displaced $\pm a/2$ so that there

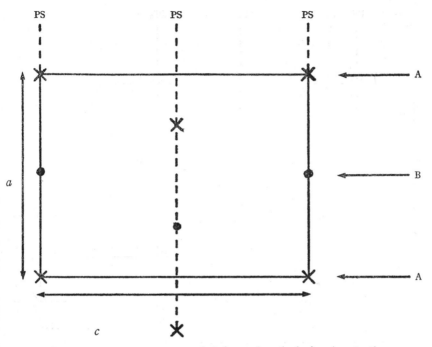

FIGURE 1. The arrangement of chains as in polyalanine (see text).

was no difference between the planes A & B. They proposed that this randomness also occurred in the natural Tussah silk (1).

A more general picture of the possibilities is shown in Figure 2 where we take into account the sequence of amino acids, particularly where alternate amino acids are of different size, e.g., glycine and alanine. This is the model for *B. mori* silk. The intersheet distance is now alternating between large PS_1-PS_2 and small PS_2-PS_3 values. The main effect is that the *001* reflection is present, not being halved, while *002* is not as strong as in the structure of Figure 1. Modifications in Figure 2 are possible in which the chains in PS_1 and PS_3 remain as drawn but all the chains in PS_2 and PS_4 are rotated through 180° to bring a and b side chains together at intersheet faces. Another possibility is random rotation of 180° about the chain axes in Figure 2 so that all interfaces contain a random mixture of a and b side chains.

Alternate interfaces having a - a and b - b side chains is the accepted structure for Group 1 silk (23) and for synthetic (Ala-Gly)$_n$ (11). Mixed population of a - b side chains in all the interfaces seems to be the explanation of the Group 2 silk structure (19). Insertion or deletion of amino acids in the Group 1 chains could bring about this change.

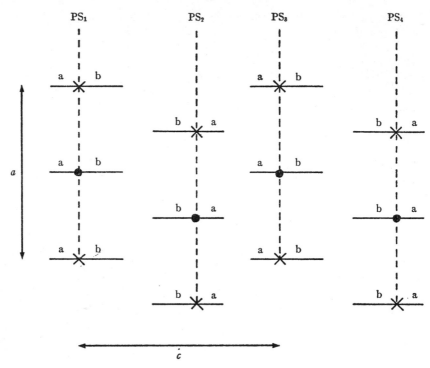

FIGURE 2. The arrangement of chains as in *B. mori* silk. a = small side chains, e.g., glycine; b = larger side chains, e.g., alanine.

Before leaving these classical fibroins whose structure is based on the cells depicted in Figures 1 and 2, we should draw attention to the special situation in the Argidae. Six species have been examined and the diffraction diagram is of the type Group 6 as illustrated and dtfined (20). The inter-sheet distance corresponds to a structure with equal quantities of alanine and glutamine which are the values obtained from amino acid analyses. The primary structure seems to be $(Ala-Gln)_n$. In proceeding from Group 1 to Group 6 there is a steady reduction in the quantity of glycine from 450/1000 to only 21/1000. Group 6 is suitably numbered in that it is at the lower end of a series of decreasing glycine contents. But, in order of intersheet sepa-ration it comes between Groups 3 and 4. Perhaps the "glycine value" is of more interest as a feature of the order in group number.

NEW TYPES OF PARALLEL-β FIBROINS

The classical fibroins described above are all interpreted as consisting of one type of chain, but this chain in its extended form may have quite differ-ent side chains on opposite sides, i.e., there is alternation in the sequence of large and small residues. Or there may be no evidence for different sized

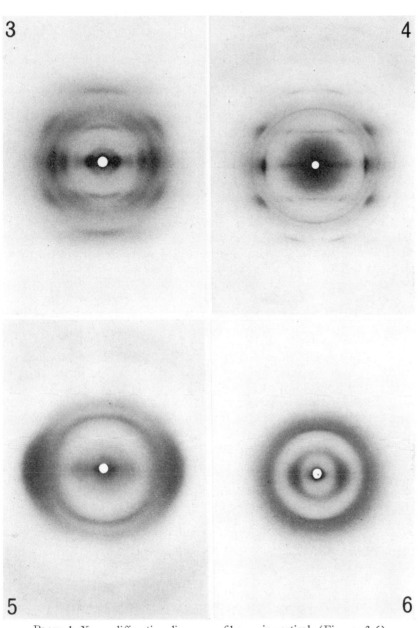

PLATE 1. X-ray diffraction diagrams; fiber axis vertical. (Figures 3–6)

FIGURE 3: The larval case of the caddisfly *Olinga feredayi*.

FIGURE 4: Spun fibers from *Cladius* larvae; the picture also shows a weak KCl pattern and presence of lipid.

FIGURE 5: Polglycine I in *Phymatocera* silk.

FIGURE 6: Slightly stretched cocoon of the flea *Xenopsylla cheopis*.

residues on opposite sides but different silks vary considerably in the average residue weight of the constituent chain.

We have found several examples in which it is possible that two or more dissimilar chains pack together in the unit of structure. The clearest example of this new phenomenon is found in an insect order, Trichoptera, not previously examined by X-rays to determine the nature of its silk.

The X-ray diagram of the larval case of the caddisfly *Olinga feredayi* (family Sericostomatidae?), a species obtained from New Zealand, is shown in Figure 3 (Plate I). Layer lines 1 and 2 are readily seen and the fiber axis period is 6.9 Å. On the equator four clear diffraction maxima appear at 4.54 (s), 5.7 (m) 8.4 (w) and 17 Å (vs). On the first layer line there are maxima on the row lines through these equatorial spots. The dimension 4.54 Å is the correct value for *120* in Warwicker's series with $a = 17$ Å and $b = 9.44$ Å. The three inner equatorials are clearly 1st, 2nd, and 3rd orders of 17 Å. The pattern suggests a periodicity every three superposed pleated sheets. This could arise with three different pleated sheets placed in regular order, or from two "similar" and one "different" in regular order. The considerable intensity of the 17 Å diffraction is greatly reduced by treating the material in N HCl and even more so with the chelating agent EDTA at pH 7. This suggests that in every third pleated sheet there are acid groups binding an ion such as calcium.

We have found two other examples of this same enlarged repeat of structure in the intersheet direction. One is a braconid and the other a sawfly; in both, the main structure of the cocoon wall is of the sheet or membranous type. [The first example, the braconid *Macrocentrus thoracicus* has been illustrated previously (20, Fig. 15). Improved diffraction patterns have been obtained from the cocoon wall of *Macrocentrus resinelli*.] In these sheet-like structures the fiber axis and interchain periods lie in the plane of the membrane while the intersheet period is perpendicular to the plane. Hence, with the X-ray beam directed parallel to the plane of the cocoon wall, intersheet diffractions appeared greatly enhanced compared with "fiber" photographs. There is a major diffraction intensity at 20 Å, indicating repetition at this distance in layers perpendicular to the surface. (A lesser diffraction at ca. 13 Å could be the second maximum of a Fraunhofer diffraction.) The strong area of diffraction in the 6–7 Å region is likely to be the third order of 20 Å enhanced by scattering matter at the position of pleated sheets separated by approximately 20/3 Å.

The third example of this type of structure is found in the tribe Cladiini, subfamily Nematinae of the Tenthredinidae (Symphyta). This tribe has attracted our attention because all the other members of the Nematinae examined have shown the silk to have a collagen-type conformation.

The *Cladius* cocoon resembles that of *Macrocentrus* in the general nature of its diffraction pattern when the beam is directed parallel to the surface of the cocoon. There is a well-defined but broadened reflection at 6.3 Å and an intense area of diffraction at small angles. This latter appears to consist of a discrete, somewhat broadened spot at 19–20 Å and a rapidly rising intensity to the center. The general character of this diffraction is that which would be expected from some close packing of 19–20 Å layers, with a major fraction of the material consisting of regular slabs of about 20 Å thickness, but not closely packed.

If *Cladius* silk is drawn as fibers from the spinnerets, excellent fiber patterns are obtained. It is characteristic of both species of Cladiini examined (i.e., *Cladius* and *Trichiocampus*) that a clearly defined pattern of crystalline potassium chloride is seen in the case of unwashed stretched gland material. However, Figure 4 (Plate I) is chosen to show the relatively crystalline nature of the fibroin pattern. The sharp arcs observed are all *hkO* reflections with $100 = 9.1$ Å and $010 = 6.9$ Å. By comparison *hkl* reflections are very weak, broad and defy accurate measurement at this stage. The general nature of the cell cross-section, which we are contemplating, is shown in Figure 7. The main requirements are to have no halving of the planes (200) separated by 4.55 Å, good reason for sharp reflections of the *hkO* type, but relative absence of *hkl* reflections. This latter feature is achieved by having the intersheet distances PS_1-PS_2, PS_2-PS_3, PS_3-PS_1 all different.

The cocoon wall "silks" of the above three examples (caddisfly, braconid, and sawfly) all support the view that pleated sheets can sometimes be arranged in groups of three as if different chains formed the separate sheets. The sheets may not be all different, i.e., A, B, and C in type, but may be A, A, and B in type. Nevertheless, it is implicit in our description that A B C or A A B form a unit of structure.

The amino acids are in no way unusual. We have single analyses for the small amount of material available in the case of *Olinga* and *Macrocentrus*. *Olinga*[1] showed over 70 percent of the residues as serine, glycine, and alanine in that order of abundance. The acidic residues amounted to about 80/1000. Other residues occurred in the amounts often found in other silks. In the case of *Macrocentrus*,[2] serine, alanine, and glycine (Gly a little less than Ser and Ala) account for over 80 percent of the residues. Acidic residues are very low. At present, no analyses are available for the *Cladius* fibroin.

COLLAGEN AND POLYGLYCINE SILKS

The finding of collagen-type structure and polyglycine conformations in silks were very striking observations and altered our concepts of the significance of silk proteins (20, 32). Collagen structure was first observed in the cocoon silk of *Nematus ribesii* and high quality diffraction diagrams were obtained by drawing out uniform fibers from the silk glands of the prepupal stage. These diagrams always showed an additional feature corresponding to a cross-β conformation superposed on the collagen pattern. But when fibers were drawn from the last larval instar (crop-full) the cross-β struc-

---

[1] Estimations by G. S. Bell.
[2] Estimations by Dr. S. Hunt.

ture was insignificantly present. This suggests that another protein may be secreted during the prepupal stage along with continued production of "collagen." Nevertheless, it could be that some of the collagen changes to a cross-β conformation during the prepupal stage.

We have succeeded in extracting and purifying the collagen from prepupal glands. It dissolves readily in 0.2 M borate buffer, pH 8, and precipitates in 25 percent ethanol. Solution and reprecipitation can be effected many times. The amino acids of these more purified products have been studied and we give in Table 1 the key values obtained by Dr. F. Lucas on analyses of four separate preparations. No hydroxyproline has been found (19) but there is a very considerable hydroxylation of the lysine. Our preparations are being examined by Dr. R. Spiro to see if he can find hydroxylysine-linked carbohydrate units as found in other Lys-OH-containing collagens (35).

TABLE 1. Purified Silk Collagens from *Nematus ribesii*

Amino acids Residues per 1000	Preparations			
	1	2	3	4
Glycine	348	346	335	313
Alanine	123	124	120	120
Serine	25	24	33	50
Proline	98	102	102	97
Lysine-OH	37	38	33	40
Lysine	12	12	20	10

The approximate one-third content of glycine and the near 50 percent content of glycine, alanine, and serine are characteristic of known natural collagens. The proline content is considerably low at 100/1000, yet the structure does not seem to undergo a helix-coil transition at normal room temperatures. However, as mentioned above, the presence of cross-β diffractions could mean a change of the protein from its normal triple helical conformation.

The collagen, precipitated from solution with acetic acid vapor, gives a banded structure of period ca. 550 Å and we show some fine electron micrographs taken by Dr. N. E. Flower (Fig. 9A and B).

Polyglycine II was first observed in the silk of *Phymatocera aterrima* (20, 32). The constitution of the whole gland protein gave 660/1000 residues of glycine (19). The highest content of glycine previously found was 480/1000 in a Group 1 silk (20, Table II). Thus, if the particular group numbers have meaning in terms of decreasing glycine content the *Phymatocera* silk could be called Group 0; the lower the group number, the lower is the content of nonglycine residues.

TABLE 2. DISTRIBUTION OF FIBROUS PROTEIN TYPES IN SYMPHYTA SILKS

	Fibre Type[a]				Fibre Type[a]		
	F	G	C		F	G	C
MEGALODONTOIDEA				Tribe Nematini			
Macroxyela	+			*Hemichroa*			+
Neurotoma	+			*Pristiphora*			+
TENTHREDINOIDEA				*Pachynematus*			+
Fam. Pergidae	+			*Pikonema*			+
Fam. Argidae	+[b]			*Nematus*			+
Fam. Cimbicidae	+[c]			5. Allantinae			
Fam. Diprionidae	+[c]			*Empria*		+	
Fam. Tenthredinidae				*Monostegia*	+		
Sub-families				*Allantus*			
1. Selandriinae				*Empyhytus*			
Strongylogaster	+			6. Blennocampinae			
2. Dolerinae				*Tomostethus*		+	+
Dolerus	+			*Tethida*		+	+
3. Heterarthrinae				*Phymatocera*		+	
Endelomyia		+		*Ardis*		+	
Caliroa		+		*Blennocampa*		+	
Fenusa		+		*Monophadnoides*		+	
4. Nematinae				*Erythraspides*		+	
Tribe Cladiini				7. Tenthredininae			
Cladius	+			*Macrophya*	+		
Trichiocampus	+			*Aglaostigma*	+	+	

[a] F = Fibroin or β pleated sheet; G = Polyglycine; C = Collagen.
[b] Group 6.
[c] Group 3.

The finding of collagen in Nematinae and polyglycine in the related sub-family Blennocampinae led us to explore the variation of silk types in many genera of the Tenthredinidae. It was feasible to do this using the silk glands of preserved larvae obtained from various institutions. It is clear that the preservation does not prevent us from defining the presence of collagen in many cases. Collagen, polyglycine, and β-fibroins were defined by the X-ray diffraction patterns obtained.

Polyglycine may be detected either as polyglycine II or as polyglycine I; these alternative structures were recognized by Meyer & Go (24) and given the numerals I and II at a later date (9). In the best studied material, *Phymatocera* silk, polyglycine II is the form shown by the trailing threads which a prepupa leaves on the surface of a petri dish, by the gland contents when the prepupa is prevented from spinning, and by the naturally formed cocoon in earth (19). However, if fresh "healthy" glands from a prepupa (*Phymatocera*) are placed in absolute ethanol the structure of the silk changes from the amorphous type characteristic of the gland silk (20, Fig. 7A) and shows the development of a well-defined diffraction ring of spacing 4.4 Å. On pressing this material in steam and taking photographs with the beam parallel to the films, a comparatively well-defined pattern of the polyglycine I type is obtained (Fig. 5, Plate I). A structure for polyglycine I was proposed by Astbury (2) and was found generally acceptable (4). The pattern in Figure 5 corresponds to an orientation of thin laminar crystallites in the plane of the pressed film, the 4.4. Å period being oriented on the meridian while the 3.4 Å (intersheet) repeat is on the equator. The considerable broadening of this latter diffraction indicates that very few pleated sheets are regularly superposed.

Because we know that *Phymatocera* silk normally contains polyglycine II we can confidently state that the diffraction pattern in Figure 5 means that polyglycine I is present. Often in preserved glands of museum specimens we find this same polyglycine I structure and would suppose that the fresh glands, if treated differently, would show the polyglycine II structure. Nevertheless, we have found the following situation in larvae given to us in one preservation fluid. The crop-full stage showed the polyglycine II structure while the crop-empty stage showed polyglycine I.

Our studies on Symphyta silks are brought together in Table 2. It is not specified whether the polyglycine is I or II as both types can be obtained from the same material under different conditions. The scheme of classification follows the key for North-American genera (28). It is clear that collagen is characteristic of the Tribe Nematini, while polyglycine is found in Heterarthrinae but is especially characteristic of Blennocampinae. However, the two leading genera in Blennocampinae have collagen without polyglycine in the anterior half of the gland and polyglycine in the posterior half; we cannot say polyglycine only in the posterior half, since it is usual to see traces of collagen as well. In the genus *Ardis* polyglycine was strongly present in both anterior and posterior portions of the gland. These studies establish that there is a close relationship between the presence of collagen and polyglycine, and successfully locate the species that will be of most interest in the further study of this association (See 20, 32).

Fibroin or β pleated sheet structure, i.e., F, is often present together with polyglycine, G, or collagen, C; to give emphasis to the presence (+) of G and C in Table 2, F is not shown. An exception is the silk of *Aglaostigma* which shows weak collagen diffraction and strong β pleated sheet diffraction. Preserved larvae were usually at the crop-empty stage, but not always. We suspect that in some cases a marked change in the gland contents occurs at the end of the last larval instar or during the prepupal stage.

The suggestion that the fibroin of *Cladius* may be composed of two or three types of chain, p. 81, may prove of interest in relation to the finding of collagen in the neighboring tribe Nematini. It is characteristic of vertebrate and invertebrate connective tissue collagens that they consist of α and β type chains which are slightly different in amino acid composition.

THE α-HELICAL SILKS

The first known case of an α-helical "silk" was that of the ootheca protein of praying mantids (30). As an α-protein, its glycine and glutamic acid contents are of particular interest; we give some values in Table 3. The mantis ribbons were viewed as consisting of uniform protofibrils which were two-strand ropes (30).

The first case of α-type silk of salivary gland origin was found in the cocoons of aculeate Hymenoptera (31). Like the mantis ribbons, these silks are low in glycine and high in acidic residues compared with β-fibroins (20). The protofibrils are regarded as consisting of four-strand ropes (3).

A key amino acid in silk proteins is obviously glycine and its widely varying content seems especially significant. The Group 6 fibroins, consisting largely of $(Ala-Gln)_n$, have a very low glycine content. We found well-defined α-structure in the case of *Arge ustelata*, although the principal structure is that of a parallel-β fibroin (20). We regard the α-helical form in silk proteins as being a consequence of reduced glycine and increased content of acidic residues. We will call these α-helical forms Group 7 silks.

TABLE 3. AMINO ACIDS OF SEPARATED MANTIS RIBBONS

Residues per 1000[a]							
Glycine	55	Serine	54	Lysine	101	Tyrosine	38
Alanine	147	Threonine	31	Arginine	50	Phenylalanine	19
Valine	32	Aspartic	95	Histidine	27	Proline	19
Leucine	65	Glutamic	214				
Isoleucine	24						

[a] Estimations by Dr. M. M. Attwood.

A new case of an α-structure in silk.—This came to light because Professor Glenn Richards was insistent that we study the flea. We fortunately obtained the help of the Hon. Miriam Rothschild who provided clean cocoons and fresh larvae. The cocoon fibers were partly oriented by moderate stretching and gave the typical α-pattern shown in Figure 6. There is a slight amount of β structure, presumably resulting from the stretching. The dimensions are scarcely distinguishable from those given by keratin and myosin—the main equatorial at ca. 9.5 Å in Figure 6 seems slightly less. Studies on fibers drawn from the larval glands will doubtless show whether this flea silk is more like keratin and myosin than it is like the aculeate silks (3, 31).

THE CROSS-β SILKS

A cross-β → parallel-β transformation was discovered in the egg-stalk silk of *Chrysopa* species. This was the first satisfactory evidence that protein chains could fold upon themselves and have "bend" regions. A detailed study proposed an exact model (13) and this involved sequences of amino acids along the chain in groups of 16. Cross-β structures seem comparatively common in silk proteins and perhaps this is due to regularly repeating sequences with special amino acids at bend regions. We should note that certain sequences in *B. mori* silk run as tetra- and octapeptides (12, 20).

CHITINOUS SILKS

The first case of chitin in apparent silk fibers was found in the hatching threads of praying mantids. These threads look like a "silk"; however, they are not formed from a secretion but are composed of cells (17).

The first case of chitin in a secreted "silk" was found in the cocoon threads of *Ptinus* species. Later, chitin was observed in the cocoon of the weevil *Prionomerus calceatus,* all these determinations being made by X-ray diffraction. In another weevil *Rhynchaenus fagi,* Streng finds chitin in the cocoon fibers and traces the origin of these from the peritrophic membrane (36).

Our X-ray examination of *Ptinus tectus* fibers from the cocoon established the aminopolysaccharide as being γ-chitin (34, Fig. 7). Chitin in the γ-form has not previously been found in insects or other arthropods. How-

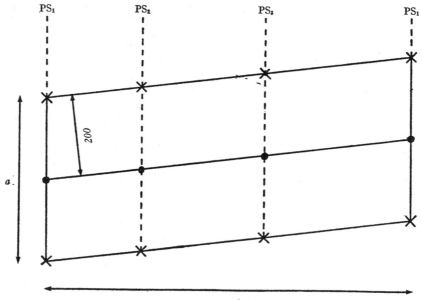

FIGURE 7. Possible arrangement of chains in the silk of *Cladius* (see text).

ever, there is reason to believe that units resembling those of γ-chitin, and measuring 28 Å × ca. 40 Å, are present in the larval cuticle of blow flies (33). Streng's electron micrograph studies (36) describe microfibrils of chitin from the peritrophic membrane material of 30–40 Å in width; our X-ray studies define this material from *R. fagi* as γ-chitin.

The γ-chitin structure is characterized by -CO··HN-linked sheets being superposed in groups of three, probably two "up" and one "down." It bears a formal resemblance to the new β-fibroin structure we have described (Fig. 4) in which the pleated sheets are apparently arranged in groups of three. Previous studies on the peritrophic membrane of adult insects (locusts, mantids) were inconclusive in that they contain the normal α-chitin which is found in arthropod cuticles. Evidence is available that larval and adult peritrophic membranes can be very different (27).

CUTICULIN SILK

We had said that the larval silk of *Chrysopa* "does not appear to be a protein" (20, p. 479); the small sample used did not "dissolve" during the hydrolysis procedure and the loading of any amino acids present in the hydrolysate was too low for meaningful recording on the analyzer. Nevertheless, infrared absorption showed a significant "protein" content; however, there was also a very large absorption at C-H stretching frequencies. Formally the material could be a kind of nylon with condensation of dicarboxylic and diamino hydrocarbons, corresponding to some hypothetical structure like a nylon 15 (5).

The cocoon wall consists of an outer felt of well-defined fibers and then a highly compact system of layers which form the bulk of the structure. (The inner surface of the cocoon bears a thin layer of specialized fibers.) However, both the external feltwork and the compact cocoon wall behave similarly to Wiggleworth's cuticulin (39), being insoluble in warm nitric acid but readily soluble in warm nitric acid saturated with sodium chlorate. (Indeed, pieces of nylon fabric behave similarly.)

Many years ago, Dr. K. D. Parker examined isolated "outer epicuticle" of *Calliphora* larvae (8, Fig. 2) and he has obtained a very similar infrared absorption spectrum using *Chrysopa* larval cocoon. The N-H absorption is much stronger in the latter, indicating four times as much protein compared with the epicuticle material. One notes, however, that to obtain the epicuticle from *Calliphora* larvae, treatment in concentrated HCl for several days was necessary, with the possibility of removing protein during the process.

The solid cocoon wall is formed of layers which show a "high" orientation of fibrous "protein" parallel to the circumference around the minor axis of the slightly ellipsoid cocoon. The infrared absorption frequencies and dichroism suggest that this oriented fibrous "protein" is α-helical; the nature of the 'polymerization' of the hydrocarbon is at present unresolved.

The existence of chitinous silk and "cuticulin" silk indicates that the known ability of insect epithelial cells to secrete chitin in procuticles, and cuticulin in epicuticles, can be modified to produce "silk" fibers composed of polysaccharide and hydrocarbon, respectively.

INSECT SILK GLAND MORPHOLOGY

In insects we recognize three main body regions which may produce silk: salivary glands, colleterial glands, and Malpighian tubules. Salivary gland silk is the most common and is characteristic of larval stages, while colleterial silk is restricted to adult insects, usually the females. Malpighian tubule silk may be produced by either stage.

SALIVARY GLANDS

Basically these are paired tubes lying freely in the haemocoel, and composed of a single-layered epithelium enclosing a storage lumen. A common feature of salivary silk glands is their division into two or more morphologically and functionally distinct regions. In some Lepidoptera, the posterior parts of the gland secrete the silk, the middle region is a reservoir, and the anterior ducts deliver the product towards the spinneret. Lepidopteran silk glands are often lined by a "cuticular"[3] intima which is about 20 μ thick in the posterior secretory regions of *Bombyx mori* glands (37), and attains a thickness of about 100 μ in the same region of *Cossus cossus* glands (6). In *B. mori,* microvilli are situated on small regularly spaced cytoplasmic processes projecting into the lumen (25, Fig. 6). More prominent microvilli, forming a substantial "brush" border, are found in the salivary silk glands of larvae in other insect orders, for example, in *Sciara* (26) and *Simulium* (21) in the Diptera, and in *Apis, Bombus,* and *Vespa* in the aculeate Hymenoptera (10).

Our own studies on larval silk glands have been concentrated on Aculeata and Symphyta (Hymenoptera) (10, 16). In Aculeata there are three gland regions: anteriorly there is a common duct, followed by a columnar cell region, and finally a cubical cell region. The anterior part of the common duct does not appear to secrete silk, and is lined by a thin intima. The rest of the common duct and a short length of the paired tubes is a region of tall columnar cells; in *Apis* and *Vespa* the cell apices are dome-shaped, but in *Bombus* they are drawn out into slender processes each of which terminates as an expanded bulb (10, Fig. 2). In all these cases the apical cytoplasm contains strongly acidophil granules, less than 1 μ in diameter. An intima is absent from the columnar and cubical cell regions, and here the luminal surface of the cells is organized into a substantial microvillous border. Silk is secreted in the cubical cell region and stored in the lumen. The columnar cells may secrete additives, possibly tanning precursors and other substances, which help to form strong insoluble fibers from the semisoluble stored protein.

A separation of secretory and storage regions is very distinct in the salivary glands of larval Symphyta. In these, the silk-secreting cells do not

[3] Constitution unknown.

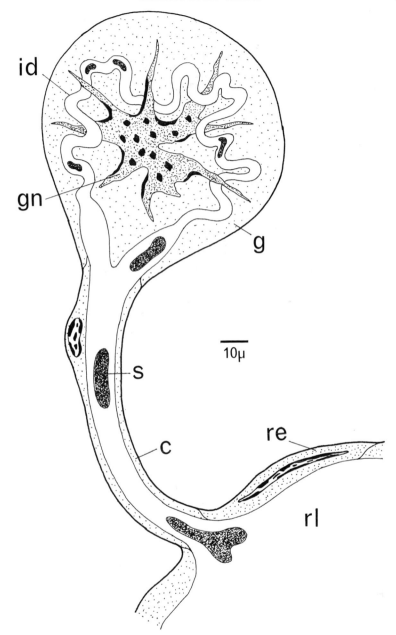

FIGURE 8. Generalized silk-secreting cell of larval Symphyta salivary glands. A slender secretion canal (c) joins the secretory gland cell (g) to the reservoir lumen (rl). gn: gland cell nucleus; id: intracellular duct; re: reservoir epithelium cell; s: globule of secreted silk. From Kenchington (16).

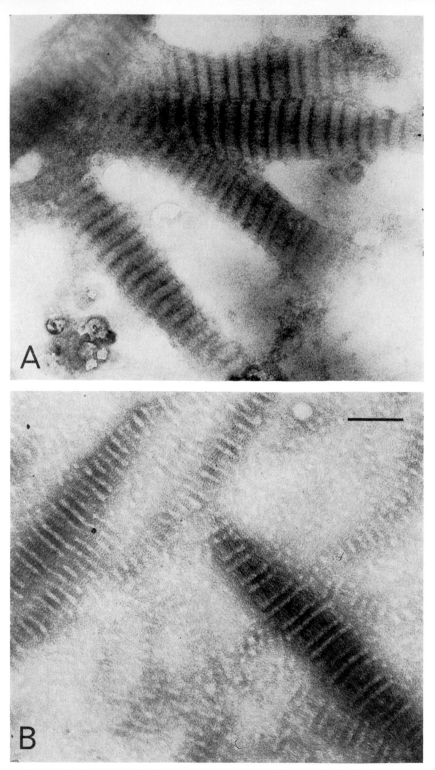

PLATE II. FIGURE 9 A,B. Electron micrographs of *N. ribesii* silk. Scale line represents 0.1 μ. A: Precipitated silk, positive staining in phosphotungstic acid, pH 5.6. B: Preparation similar to A, but negative staining in uranyl acetate.

form a continuous epithelium, but are usually disposed singly, in pairs or triplets, connected by slender canals to the thin-walled reservoirs. The basic cytology of a gland cell is shown in Figure 8. Silk synthesized in the cytoplasm passes into a meandering intracellular duct, the two ends of which join to form a secretion canal opening into the reservoir. The nuclei of these cells are large, irregular, and rather similar to the lobed nuclei in the silk glands of Lepidoptera. The cells of the reservoir epithelium are large, thin, roughly hexagonal in surface view, with sparsely staining cytoplasm. There is no microvillous border, nor is an intima developed. Individual silk-secreting cells are similar in all the Symphyta studied; but there are variations in the method of their arrangement and distribution on the reservoirs, and some variation in reservoir morphology (16).

In general, salivary gland silk is secreted as small globules which coalesce as ill-defined masses within the lumen. In the Aculeata, however, the secretions become arranged into a number of regular structures. *Apis* silk is stored as microscopic tactoids oriented approximately parallel to the length of the gland. These tactoids show regular cross-striations (20, Fig. 17A), and in the electron microscope the bands are resolved as regions of fibrillar and granular material. The fibrils appear to have their ends all at one level across the tactoid, forming a band approximately 0.3 μ wide (10). When treated in 0.3 M KCl solution, tactoids disperse as microfibrils, almost 0.3 μ long and 40–50 Å wide (10, Fig. 9). In the case of *Bombus,* nonstriated tactoids are produced in small numbers in the anterior parts of the cubical cell region, but most of the silk is stored as a bulky system of rod-like bundles of fibrils, or "fibrous bars" (20, Fig. 17B). The width of these bars is approximately 1 μ, and they are regularly spaced at a periodicity of about 1.5 μ. In polarized light the bars are positively birefringent parallel to their width, suggesting a substructure of microfibrils stacked with their ends in register across the width of the bars (a situation reminiscent of the A bands in striated muscle). This picture is confirmed by electron optical studies. In 0.3 M KCl, the bars disaggregate into microfibrils, about 1 μ long and 40 Å wide (10, Fig. 10a).

We have studied the collagen-type silk of *Nematus ribesii* (Symphyta) in some detail, using the electron microscope to test for the presence of a cross-banded structure which is generally characteristic of vertebrate and invertebrate collagens. Cross-banded tactoid-like aggregations (Fig. 9) were obtained when the purified silk protein was slowly precipitated from borate buffer by exposure to acetic acid vapor. The periodicity is about 550 Å.

COLLETERIAL GLANDS

Most colleterial gland secretions are not silk-like in texture but form sticky, foamy, or hardened protective coverings for egg masses. The α-helical protein of the mantis ootheca is produced in a large, much-branched tubular gland opening into the genital atrium. A number of different secretory cell types occur along the gland length (18), but basically the epithelium is made up of gland cells interspersed with small cells which possibly secrete the intima. The luminal surface of each secretory cell is greatly expanded by multiple infoldings of the plasma membrane to form an "intracellular" microvillous border, the end-apparatus. Structural protein passes through the end-apparatus and forms long fibril-like structures which aggregate as large globules in the gland lumen. Microscopic studies suggest that in the globules parallel arrays of these "fibrils" form a succession of concentric "shells" arranged in a helicoidal manner similar to cholesteric phase systems (18).

An end-apparatus is also present in the gland cells of two other colleterial silk producers, *Chrysopa* (Neuroptera) and *Hydrophilus* (Coleoptera). In *Chrysopa,* the sac-like gland is composed of columnar secretory cells containing a small apical end-apparatus. A thin corrugated intima is present, possibly secreted by small cells between the gland cells. Unlike the mantis colleterial secretion, *Chrysopa* silk shows little detail in sectioned material. The silk readily disperses in water, and negatively stained preparations show thin, sheet-like micelles, several hundred Å wide, but only 25 Å thick (20, Fig. 12). This picture supports the results obtained from X-ray diffraction studies (13). In *Hydrophilus,* the epithelium of the colleterial gland tube is more than one layer thick and contains two easily definable types of cell—secretory cells and interstitial cells. Each secretory cell contains a thin-walled sac, or rudimentary end-apparatus, confluent with a long narrow duct which penetrates the intima and opens into the gland lumen. Thus, these cells recall the so-called dermal glands which secrete the cement layer of the exoskeleton. The smaller interstitial cells are scattered among the secretory cells, and also lie against the intima.

MALPIGHIAN TUBULES

Wigglesworth (39) has reviewed the accessory functions of Malpighian tubules, including silk secretion. In some insects, for example, the larvae of *Hypera* (Coleoptera, Curculionidae) and *Chrysopa,* undoubted "silk" is produced from the Malpighian tubules. The Ptinidae (Coleoptera) are generally considered to secrete Malpighian tubule silk (see 39), but our histological examinations of *Ptinus tectus* larvae have shown no evidence of this. X-ray diffraction studies have established that the silk of *Ptinus* is made largely of chitin. The cocoons are constructed from an unbroken thread of dried peritrophic membrane which gradually issues from the anus. Streng (36) has found that *Rhynchaenus fagi* (Curculionidae) spins cocoons of peritrophic membrane "silk" which contains chitin; our studies on the "silk" of another weevil *Prionomerus calceatus* had shown it to be chitinous in structure and peritrophic membrane in origin.

The onset and duration of silk secretion varies according to the requirements of a particular species. In mantids the colleterial glands are active throughout the adult stage, during which time several oothecae are constructed. Silk gland activity is maintained during the juvenile stages of some Trichoptera which use silk in their shelters as well as in the pupal cocoons. The New Zealand glowworm larva *Arachnocampa luminosa* (Diptera) uses salivary gland silk to form food-trapping threads during larval life; later, it pupates on a stout "stalk" of this silk (31). In *B. mori* larvae, which produce large cocoons, build-up of silk secretion extends over the fourth and fifth larval, and the prepupal instars. Larvae of most bumble bee species produce small amounts of silk in the early instars to fabricate loose open "cells"; the cocoon silk is secreted during the final larval instar. By contrast, honey bee and wasp larvae produce pupation cocoons only, and silk secretion is limited to the final larval instar. Sawfly larvae usually begin to secrete a small amount of silk during the last larval instar, but most of the protein is produced rapidly by the early prepupa. However, in web-spinning and leaf-rolling sawflies (e.g., Pamphilidae) silk is produced throughout larval life.

SUMMARY

The fibrous proteins of insect silks are arranged in a series from Group O to Group 7. The series is characterized mainly by a steady reduction in glycine residues from two-thirds in Group O to less than 10 percent in Groups 6 and 7. This series is regarded as arising from one locus for the production of fibrous protein. The series does not seem to be unique to arthropods since the molluscan periostracum (15, p. 273) shows the same range of glycine residues from two-thirds to less than 10 percent. The hypothetical locus for fibrous protein production is conceivably universally present in animal cells. Extracellular fibers in the protozoan *Haliphysema* (Foraminifera) are quite possibly collagen in type (14); in another protozoan, *Stannophylum* (Rhizopoda), extracellular fibers, given to us by Dr. R. H. Hedley, are so very high in glycine content that they seem likely to be akin to the Group O silk proteins.

The origin of collagen-type structure is of special interest. Because of its one-third glycine content, it was first thought that "collagen" might have arisen from Group 3 fibroins (32). But an alternative was given associating its origin with polyglycine-rich proteins (20, 32). The data of Table 2 very strongly support a close association of "polyglycine" silk and "collagen" silk. Indeed, we have cases in which these two proteins occur together within the same gland, e.g., *Tomostethus* and *Tethida*.

A locus for fibrous protein production does not appear to be present in plant cells. If such a locus developed through stages where polyglycine and β-fibroins were the first products, then the wall of plant cells would seem to prevent these "awkward" proteins from escaping and such cells or the locus would tend to be suppressed. This implies that a fiber-forming locus became established because the materials initially produced by it could be secreted. It seems possible that among the very great variety of fibrous proteins which could be formed and safely removed from the cell as secretions, some few, related to collagen, myosin, and prekeratin, proved useful as an integral part of the animal body, and came to form a large part of connective, muscle, and keratin tissue.

Among insects there is a frequently occurring instinct to manipulate fibers, or be "textile operatives" (e.g., 29). We can see this in the similar construction of cocoons independent of the nature of the "yarn." Like Man who spins wool, cotton, or nylon, insect larvae will use protein, polysaccharide, or hydrocarbon fibers. Generally, these are synthesized by the insect but various extraneous materials are also used as illustrated by bagworms and caddisfly larvae.

LITERATURE CITED

1. Arnott, S., Dover, S. D., Elliott, A. 1967. Structure of β-poly-L-alanine: refined atomic co-ordinates for an anti-parallel beta-pleated sheet. *J. Mol. Biol.* 30:201–8
2. Astbury, W. T. 1949. Structure of polyglycine. *Nature* 163:722
3. Atkins, E. D. T. 1967. A four-strand coiled-coil model for some insect fibrous proteins. *J. Mol. Biol.* 24: 139–41
4. Bamford, C. H., Elliott, A., Hanby, W. E. 1956. *Synthetic Polypeptides.* New York: Academic. 445 pp.
5. Bradbury, E. M., Elliott, A. 1963. Infra-red spectra and chain arrangement in some polyamides, polypeptides and fibrous proteins. *Polymer* 4:47–59
6. Bradfield, J. R. G. 1951. Phosphatases and nucleic acids in silk glands: cytochemical aspects of fibrillar protein secretion. *Quart. J. Microscop. Sci.* 92:87–112
7. Chapman, R. F. 1969. *The Insects.* London: The English Universities Press, Ltd. 819 pp.
8. Dennell, R., Malek, S. R. A. 1955. The cuticle of the cockroach *Periplaneta americana.* II. The epicuticle. *Proc. Roy. Soc. London, Ser. B* 143:239–57
9. Elliott, A., Malcolm, B. R. 1956. Structure and properties of synthetic polypeptides and silk proteins. Infra-red studies of polypeptides related to silk. *Trans. Faraday Soc.* 52:528–36
10. Flower, N. E., Kenchington, W. 1967. Studies on insect fibrous proteins: the larval silk of *Apis, Bombus* and *Vespa. J. Roy. Microscop. Soc.* 86: 297–310
11. Fraser, R. D. B., MacRae, T. P., Stewart, F. H. C., Suzuki, E. 1965. Poly-L-alanylglycine. *J. Mol. Biol.* 11:706–12
12. Geddes, A. J., et al. 1969. Mass-spectrometric determination of the amino acid sequences in peptides isolated from the protein silk fibroin of *Bombyx mori. Biochem. J.* 114:695–702
13. Geddes, A. J., Parker, K. D., Atkins, E. D. T., Beighton, E. 1968. "Cross-β" conformation in proteins. *J. Mol. Biol.* 32:343–58
14. Hedley, R. H., Wakefield, J. St. J. 1967. A collagen-like sheath in the arenaceous foraminifer *Haliphysema. J. Roy. Microscop. Soc.* 87: 475–81
15. Hunt, S. 1970. *Polysaccharide-Protein Complexes in Invertebrates.* London and New York: Academic. 329 pp.
16. Kenchington, W. 1969. Silk secretion in sawflies. *J. Morphol.* 127:355–62
17. Kenchington, W. 1969. The hatching thread of praying mantids: an unusual chitinous structure. *J. Morphol.* 129:307–16
18. Kenchington, W., Flower, N. E. 1969. Studies on insect fibrous proteins: the structural protein of the ootheca in the praying mantis, *Sphodromantis centralis* Rehn. *J. Microscop.* 89:263–81
19. Lucas, F., Rudall, K. M. 1967. Variety in composition and structure of silk fibroins: some new types of silk from the Hymenoptera. *Symposium on fibrous proteins, Canberra, 1967,* 45–55. Australia: Butterworths
20. Lucas, F., Rudall, K. M. 1968. Extracellular fibrous proteins: the silks. In *Comprehensive Biochemistry,* eds. M. Florkin, E. H. Stotz, Chap. 7, 26B:475–558. Amsterdam, London, New York: Elsevier. 298 pp.
21. Macgregor, H. C., Mackie, J. B. 1967. The fine structure of the cytoplasm in salivary glands of *Simulium. J. Cell Sci.* 2:137–44
22. Marsh, R. E., Corey, R. B., Pauling, L. 1955. The structure of Tussah silk fibroin (with a note on the structure of β-poly-L-alanine). *Acta Cryst.* 8:710–15
23. Marsh, R. E., Corey, R. B., Pauling, L. 1955. An investigation of the structure of silk fibroin. *Biochim. Biophys. Acta* 16:1–34
24. Meyer, K. H., Go, Y. 1934. Observations roentgénographiques sur des polypeptides inférieurs et supérieurs. *Helv. Chim. Acta* 17: 1488–92
25. Morimoto, T., Matsuura, S., Nagata, S., Tashiro, Y. 1968. Studies on the posterior silk gland of the silkworm *Bombyx mori.* III. Ultrastructural changes of posterior silk gland cells in the fourth larval instar. *J. Cell Biol.* 38:604–14
26. Phillips, D. M., Swift, H. 1965. Cyto-

plasmic fine structure of *Sciara* salivary glands. I. Secretion. *J. Cell Biol.* 27 :395–409

27. Richards, A. G., Richards, P. A. 1969. Development of microfibers in the peritrophic membrane of a mosquito larva. *Ann. Proc. E.M.S.A., 27th, 1969.* 256–57

28. Ross, H. H. 1937. A generic classification of the nearctic sawflies. *Ill. Biol. Monogr.* 15 :1–173

29. Ross, H. H. 1964. Evolution of caddisworm cases and nets. *Am. Zoologist* 4 :209–20

30. Rudall, K. M. 1956. Protein ribbons and sheets. In *Lectures on the Scientific Basis of Medicine,* Chap. 12, 5 :217–30. London : Athlone Press. 473 pp.

31. Rudall, K. M. 1962. Silk and other cocoon proteins. In *Comparative Biochemistry,* eds. M. Florkin, H. S. Mason, Chap. 9, 4 :397–433. New York : Academic. 841 pp.

32. Rudall, K. M. 1968. Comparative biology and biochemistry of collagen. In *Treatise on Collagen,* ed. B. S. Gould, Chap. 2, 2A :83–137. London and New York : Academic. 434 pp.

33. Rudall, K. M. 1967. Conformation in chitin-protein complexes. In *Conformation of Biopolymers,* ed. G. N. Ramachandran, 2 :751–65. London and New York : Academic. 785 pp.

34. Rudall, K. M. 1963. The chitin/protein complexes of insect cuticles. In *Advances in Insect Physiology,* eds. J. W. L. Beament, J. E. Treherne, V. B. Wigglesworth, 1 :257–313. London : Academic. 512 pp.

35. Spiro, R. G. 1969. Characterisation and quantitative determination of the hydroxylysine-linked carbohydrate units of several collagens. *J. Biol. Chem.* 244 :602–12

36. Streng, R. 1969. Chitinhaltiger Spinnfaden bei der Larve des Buchenspringrüsslers (*Rhynchaenus fagi* L.). *Naturwissenschaften* 56 :333–34

37. Tashiro, Y., Morimoto, T., Matsuura, S., Nagata, S. 1968. Studies on the posterior silk gland of the silkworm *Bombyx mori.* I. Growth of the posterior silk gland cells and biosynthesis of fibroin during the fifth larval instar. *J. Cell Biol.* 38 :574–88

38. Warwicker, J. O. 1960. Comparative studies of fibroins. II. The crystal structures of various fibroins. *J. Mol. Biol.* 2 :350–62.

39. Wigglesworth, V. B. 1965. *The Principles of Insect Physiology.* London : Methuen ; New York : E. P. Dutton. 741 pp.

40. Witt, P. N., Reed, C. F., Peakall, D. B. 1968. *A Spider's Web.* Berlin, Heidelberg, New York : Springer-Verlag. 107 pp.

BIOLUMINESCENT COMMUNICATION IN INSECTS 6004

JAMES E. LLOYD

Department of Entomology, University of Florida, Gainesville, Florida

The phenomenon of living light has fascinated man for centuries. It has figured in his mythology, folk-lore, superstition, high-fashion, and poetry (90), it once affected the outcome of a military conquest (79), and he has even used it for illumination (9). That living organisms can emit light is somewhat incomprehensible—an out-of-the-ordinary event that in the past has demanded supernatural explanation, and today prompts most adults (but not children) to ask, "how do they do it" before "why do they do it? The "how" of bioluminescence has dominated scientific literature—few studies have centered upon function while hundreds of papers have dealt with (*a*) the kinds of organisms that emit light, (*b*) the structure of light organs, (*c*) the mechanisms of light production, and (*d*) the mechanisms of control. One worker stated "The *reason* [my emphasis] for the rhythmic flashing of the firefly is a subject of considerable interest and is dealt with in the present paper." In his summary he concludes "Cogent evidence is presented to show that flashing is the result of a rise and fall in the osmotic pressure . . ." (121). A taxonomist, after being bewildered by the relative development of female luminosity and male eyes in various firefly species, and stating that it ". . . does not seem to me explicable either on grounds of teleology or natural selection . . . ," went on to say "The luminous powers of these insects suggest three distinct investigations . . . spectroscopic . . . arrangement of cells . . . chemical . . ." He later invited younger entomologists to study habits of the different species ". . . before they can ascend with profit to the higher retirement of the museum and library . . ." (104).

Function as an object of scientific investigation has been neglected, yet it has received its share of speculative attention. Among functions of luminescence that have been suggested in addition to sexual communication, are prey attraction, warning, intimidation of predators, illumination, camouflage, and population regulation (21–23, 26, 55, 56, 95, 99, 100, 108, 110, 141, 147, 149, 164, 171, 174, 177 and others; see Harvey 89).

Technical developments of the past few years should provide impetus in the study of bioluminescent communication. Portable recorders and light sensors, such as photocells (108), photomultiplier systems (35, 112, 166), panning still cameras and cinematic cameras, especially when used in conjunction with image intensifiers (16, 35, 36, 85), and spectrometers (167) now permit precision analysis of luminescent signals.

The following comprehensive syntheses, specialized syntheses, and symposia provide an introduction to a study of bioluminescence and a historical perspective (31, 39, 40, 42–54, 59, 64, 87–90, 98, 118, 120, 138, 139, 151, 164, 168). Many of these have brief discussions of function, and some make at least passing comment on bioluminescent communication in insects. There are also a number of nontechnical or popular treatments of luminescent insects (1, 21, 25, 26, 38, 57, 60, 66, 70, 71, 82, 100, 113, 131, 133, 145, 149, 154, 157, 161, 172).

OCCURRENCE OF BIOLUMINESCENCE IN INSECTS

A great diversity of plants and animals possesses the ability to emit light. Among animals, in a classification of 25 phyla, 12 or 13 phyla have luminous species (89). McElroy and Seliger (140, 168) have suggested that present-day bioluminescence is the result of secondary developments of a vestigial system that evolved originally in the early history of life as a mechanism for removing toxic oxygen, ". . . an offshoot of a chemical reaction fundamental to all organisms." With the evolution of aerobic metabolism, the oxygen-removing light reaction was not easily lost since it was intimately associated with the primitive electron transport process. Among the arthropods, luminescent forms are found in the Pycnogonida, Eucrustacea, Diplopoda, Chilopoda, and Insecta (89). Insects positively known to be self-luminescent occur in the families Poduridae (Collembola), Mycetophilidae (Diptera), Lampyridae, Elateridae, Phengodidae, Drilidae, and Teleguesidae (Coleoptera), and Fulgoridae (Homoptera). There are also reports of doubtful validity for the orders Lepidoptera, Isoptera, Orthoptera, Ephemeroptera, and Hymenoptera, and the coleopteran families Buprestideae, Paussidae, and Carabidae. Growth or infections of luminescent bacteria may be confused with self-luminescence (150, 162). Harvey (89) compiled and discussed an extensive list of reports and references to the occurrence of luminescence in insects and other organisms.

COMMUNICATION

The literature on animal communication is voluminous (see 165), yet there is no consensus as to how it should be defined: definitions vary among researchers depending in part upon their orientation and the behavioral repertoires of the animals they study. Brief and useful discussions of definition have been written by Frings & Frings (63), Alexander (3), and Marler (122). Important to any definition, I believe, are the requirements (a) that the recipient of the signal show some observable response and (b) that natural selection has acted upon (to produce or enhance, and maintain) the signals and the mechanisms of production and reception in the communicative context. The first requirement is restrictive for operational purposes and the second for philosophical or theoretical validity. Midges in New Zealand caves for hundreds of generations have been attracted to and snared by luminescent glowworm larvae (Mycetophilidae). I would not consider this communication. The glowworm is obviously exploiting midge responses that were developed and are maintained in some other context; perhaps, for example, in spite of the loss to predation, successful midges are still those that respond positively to the light of dawn or a cave entrance. It is also desirable to exclude from communication bat echolocation (78) and its luminescent analogue in fireflies, illumination flashes (110).

Whether or not one considers aggressive mimicry in fireflies communication will depend upon what elements one stresses in his definition. *Photuris* females, mimicking the mating flashes of females of other species, attract and eat the males of these species (106, 112). Natural selection, in the context of conspecific communication, has brought about the responses of the males (prey) and much of the answering behavior of the predaceous females. But, while natural selection results in continued adjustment and refinement in the mimicry of the predator, its action on prey species is to eliminate this interaction. This relationship is analogous to that of midges and glowworms, the fundamental difference being that in aggressive mimicry the exploited responses of the prey are elements of an intraspecific, communicative system. Klopfer & Hatch (101) suggest, ". . . communication, *sensu stricto,* necessitates the existence of a code shared between two or more individuals whose use is mutually beneficial to its possessors, i.e., increases fitness." Regardless of definition, study of aggressive mimicry in fireflies is pertinent to bioluminescent communication and its evolution.

MAJOR CONSIDERATIONS IN BIOLUMINESCENT COMMUNICATION

Since all known bioluminescent communication in insects is associated with mating pair formation, I will direct my discussion to elements of communication relevant to this situation. At pair formation, luminescent signals, or exchanges of luminescent signals, function in mate identification and location. Of special interest are (*a*) the parameters of the signals that can encode species information, (*b*) the nature of the behavioral interaction of communicating individuals, and (*c*) the aspects of the signals, light organs, eyes, and behavior, that increase efficiency in locating and identifying mating partners.

CONSTRUCTION OF UNIQUE SIGNALS: ENCODING SPECIES INFORMATION

An astronomical number of different luminescent emissions is conceivable. The restricted abilities of biological systems limit this potential in nature. If we knew the limitations of the signal production and reception system for a group of bioluminescent insects, it would be possible to give a theoretical estimate of the number of species that could be sympatric and synchronic (the number would probably be so large that the ambient light level would make luminescent signals impractical). Unfortunately, few of these limitations have been determined.

Bioluminescent emissions can differ with respect to frequency or combination of frequencies (spectra), and intensity, and the distribution of these parameters in time and space. While it is not practical or worthwhile to make an exhaustive list that considers all parameters varying in all possible ways, it is useful to describe some categories, and exemplify when possible. Most examples will be drawn from fireflies since these are the only luminescent insects that have been studied in detail. In simplest and most appropriate terms the elements to be considered are spectral composition, brightness (intensity per unit area), shape and size of luminescent surface, timing, and movement; however, it will become obvious that in operation these categories are not independent. I will contrast signals that are identical except for the coding mode being illustrated.

Spectral composition.—Example: two stationary, continuous light sources, identical in size, shape and brightness, with differing spectral compositions. Males of *Lampyris noctiluca,* a European firefly, can perceive blue, red, green and yellow light but can be attracted only with yellow light (163).

Brightness.—Example: two stationary, continuous light sources, identical in size, shape and spectral composition, with differing brightness. Since apparent brightness is a function of distance there are several limitations on coding with this parameter. Brightness could be significant if (*a*) a light exceeded some absolute value or (*b*) at a measured distance (determined by angular measurements or by use of some other sensory modality) the brightness was not of some particular value. There have been no careful studies of this parameter. Males of *L. noctiluca* approached decoys that were brighter than normal but at "a certain distance" they suddenly turned around and went the other way, or if they had reached the decoy they remained motionless with their heads withdrawn under their pronota (163).

Shape.—Example: two stationary, continuous light sources, identical in size, brightness, and spectral composition, with differing spatial configurations. The shapes of light organs vary considerably among and within groups of luminescent insects (Figs. 1–23, Plate I). Males of the firebeetle *Pyrophorus atlanticus* (Elateridae) will approach and land on decoys with two lights simulating the twin prothoracic lights found in males and females (Fig. 19), but after inspecting decoys with single lights, fly off (107).

Size.—Example: two stationary, continuous light sources, identical in shape, brightness, and spectral composition, one with greater area. *L. noctiluca* males discriminate against very large or very small decoys which otherwise resemble females (163).

Timing.—Example: two stationary light sources, identical in size, shape,

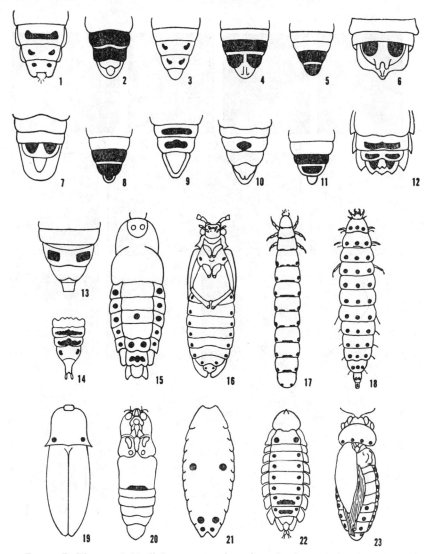

PLATE I. Figures 1–23, light organs of various luminescent beetles, shown in black. 1. *Lychnuris rufa*, female. 2. *Photinus scintillans*, male. 3. *Pyractomena* sp., female. 4. *Luciola* sp., male. 5. *Luciola chinensis*, male. 6. *Callopisma* sp., female. 7. *Robopus montanus*, male. 8. *Luciola cruciata*, male. 9. *Photuris* sp., female. 10. *Photinus scintillans*, female. 11. *Luciola lateralis*, male. 12. *Pleotomus* sp., female. 13. *Luciola lusitanica*, female. 14. *Lampyris noctiluca*, female. 15. *Lamprohiza splendidula*, female. 16. *Harmatelia bilinea*, male. 17. *Phengodes* sp., female. 18. *Diplocladon hasselti*, female. 19. *Pyrophorus* sp., male and female, dorsal. 20. *Pyrophorus* sp., male and female, ventral. 21. *Phausis reticulata*, female. 22. *Lamprohiza mulsanti*, female. 23. *Dioptoma adamsi*, male. Figures 1–5, 8–11, 13–15, 19, 20, and 22 redrawn from Buck (31) from various sources; Figures 7 and 17 redrawn from Buck (31); Figures 12 and 21 redrawn from Lloyd (107); Figures 16 and 23 redrawn from Green (77); Figure 18 redrawn from Harvey (89); Figure 6 original. Figures by P. Laessle.

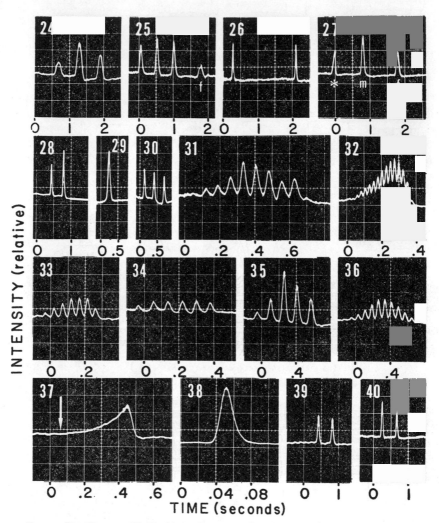

PLATE II. Figures 24–40, bioluminescent signals of various Lampyridae. 24. Flash pattern of *Photinus ardens*. Female answers about 6 sec after last pulse of this signal. 25. Flash pattern and female answer (f) of *Photinus obscurellus*. Note differences between this species and its close relative *P. ardens* with respect to pulse duration, pulse rate, and female delay. 26. Two-pulse flash pattern of *Photinus macdermotti*. Female delay is like that of *P. greeni* (below). 27. Experiment with *Photinus greeni* males. After seeing artificial two-pulse flash pattern *and* female answer, a male *P. greeni* then flashes (m) at the appropriate time and provides the second pulse after a single artificial flash (*). The female answers (f) at the species-specific delay following the "mixed" stimulus. 28. Two-pulse flash pattern of *Photinus consanguineus*. Female delay is like that of *P. greeni*. 29. Flash pattern of *Photinus ignitus*. 30. Flash pattern of *Photinus lineellus*. 31. Flash pattern of *Photuris* "B," a flicker of a different frequency from that of

brightness, and spectral composition, differing in the timing of their flashes. This category includes such features as flash number, flash duration, flash rate, flicker frequency, flash form, and duration of the intervals between the signals of interacting males and females (Figs. 24–40, Plate II). Males of *Photinus pyralis* can be attracted to artificial flashes only if they occur at approximately two seconds after the male's flash (29). Females of *Photinus macdermotti* will flash in response to two pulses only if they are approximately two seconds apart (108). Females of *Photinus scintillans* answer flashes that are 0.13–0.16 seconds in duration but not those that are 0.20–0.34 seconds in duration (108). The values given above are for specific temperatures; temperature has significant and predictable influence on timing.

Movement.—Example: two continuous light sources, identical in size, shape, brightness, and spectral composition, with differing movements or gestures. There are no known examples of movement coding information in luminescent insects although in nonluminescent animals, both vertebrates and invertebrates, this is an extremely important source of distinct visual signals in a variety of behavioral interactions (122). There are many examples of fireflies performing species-typical movements while glowing or flashing (15, 108, 133, 147, 163), but none of those tested has been found to be necessary to elicit the species-specific response to the signal.

Other possibilities; combinations.—Some additional categories are worth mentioning since they may occur. In the discussions of timing and shape, the parameter that was altered through time and space was light intensity. Color can also be varied in these dimensions. The railroad worm, *Phrixothrix* (Coleoptera; Phengodidae), has both greenish-yellow and red lights that can be activated independently (89, 172). One can imagine insect signals involving lights of various colors or rhythms, or both, such as those humans use along railways and at airports. The relative brightness of portions of a luminescent surface could also be incorporated in a coding sys-

---

its sympatric and synchronic congener *Photuris* "D," Figure 32, 33 and 34. Flash patterns of two cryptic species, *Photinus evanescens* complex. 35 and 36. Flash patterns of two closely related New Guinea *Luciola*. 37. Crescendo flash of *Photuris* "WM." Arrow indicates beginning of flash. 38. Flash pattern of *Photuris salinus*. Note the asymmetry and the transient at the beginning. 39. Two-pulse flash pattern of *Photuris* "DM" (Photurinae). Female response, as well as male flash pattern, is similar to that of *Photinus consanguineus*. 40. Two-pulse flash pattern of New Guinea *Luciola* sp. (Luciolinae). Female response, as well as male flash pattern, is similar to that of *Photinus consanguineus*.

tem, with one portion of the surface serving as a standard for comparing the brightness of another portion.

SIGNAL SYSTEMS: THE INTERACTION OF COMMUNICATING INDIVIDUALS

Identification signals may appear in a variety of protocols or "systems." McDermott (129) and Blair (21–23) first discussed the two major systems although Mast (124) was certainly aware of them, and much earlier (Gorham, 74) had realized that different systems were possible. Until naturalists became aware of this they questioned the sexual function of luminescence in those species in which both sexes are luminescent (26, 100, 104 162). The basic systems in fireflies appear to be: I. one sex (usually the female) is stationary or sedentary, broadcasts a species-specific signal, and the other is attracted to it (e.g., *Lampyris noctiluca,* Lampyridae; *Phengodes laticollis,* Phengodidae; *Arachnocampa luminosa,* Mycetophilidae) ; and II. one sex (usually the flying male) broadcasts a species-specific signal, the other responds with a species-specific signal, and the advertiser (male) is then attracted to the response (e.g. *Photuris, Photinus, Aspisoma, Pyractomena*). These systems are essentially the same as the two Alexander (5) considered the basic ones in the acoustical and visual signalling of arthropods; "those in which the signal pattern of one individual carries all the species distinctiveness and those in which the timing of responses between individuals is species-specific." Alexander's system II, however, places emphasis on the timing between the emissions of the communicants. While, in most fireflies, with system II this timing is species-specific and in some cases has been found to encode species information (29, 108, 147), the timing or some other parameter within the respondent's signal may alone encode the important information. Such is the case in the firefly *Luciola discicollis* (99). The critical timing in the signals of *Photuris brunnipennis floridana* is perhaps within as well as between the emissions of males and females; in this species sexual recognition may depend upon the synchronization of the communicant's flashes (112).

Some lampyrids have systems that appear to be intermediate to systems I and II. In the North American *Phausis reticulata,* glowing males are attracted to larviform, glowing females (175). Although the females may turn on their lights at low ambient light levels, nonglowing females will turn on their lights in response to the glows of males. The behavior of this species fits an evolutionary scheme that derives system II from system I (107, 108). In *Dioptoma adamsi* of Ceylon, glowing females attract nonglowing males, but these males display ". . . under sexual excitement series of lights of an emerald green colour" (77).

Some fireflies may have compound signal systems in which elements of both basic systems occur. In the synchronously flashing fireflies (*Pteroptyx*) of southeast Asia, the males and females are attracted to aggregations of synchronizing males (system I) (34, 117, 152). Then, some species apparently use system II. In one New Guinea *Pteroptyx* the roles of males and females during the system II phase apparently reverse during the evening and perched males flash in response to the light of flying females. This behavior may be homologous with the flaring of *L. discicollis* males described by Kaufmann (99), and similar patterns in other Luciolini (117). In *Pteroptyx tener* and *Pteroptyx malaccae* "chance collisions between a male's head and a conspecific female's abdomen" rather than luminescent signals bring pairs together after they reach the firefly trees (152). In the system of *Luciola obsoleta,* a New Guinea firefly, the male response to the flying, glowing female has become an aerial chase. During the first two hours after sunset males and females flash from perches with individuals of both sexes occasionally taking flight; each sex has a distinctive flight flash. After this early phase, which may result in aggregation and identification, when a female takes flight she is pursued by one or more flashing males and eventually one lands with her and mounts her (117). The Japanese fireflies *Luciola cruciata* and *Luciola lateralis* may have similar systems (143).

Pair formation in some animals involves complex sequences of signal exchanges, or reaction chains (17). A signal of one individual stimulates a response signal from its potential mate and this in turn stimulates the first animal to produce yet a different signal, etc. Such sequences would seem to be possible in luminescent insects using system II. The flash pattern of males of a New Guinea firefly *Pteroptyx* sp. is a single flash that is emitted each 3–4 seconds. After receiving a response flash, males immediately emit 2–4 rapid flashes (117) but the communicative significance of these pulses is not known.

INCREASING EFFICIENCY IN THE SIGNAL CHANNEL

The use of light signals imposes certain restrictions upon organisms. For one thing, background "noise" in the form of ambient light precludes the use of such signals during most of the daylight hours except in special habitats, for example caves, ravines, leaf litter, logs, burrows, and deeper portions of lakes and seas. Nocturnal insects frequently encounter activity-limiting temperatures. In temperate latitudes, even during summer, nocturnal temperatures may be too low for mating activity several days in succession. In such an exigency nocturnal insects, such as crickets, that use a different signal channel, can switch their mating to daylight hours (7) but this is impossible with bioluminescent signals. Perhaps this partially explains why so many North American fireflies are active during evening twilight (108, 133), but none restrict their mating activity to the hours after midnight as do some tropical fireflies (19, 61, 62, 114). *Photinus obscurellus,* a boreal, early-summer species, commonly encounters temperatures too low for flight (ca. 54°F), but males and females exchange flashes from perches and mating pairs are common (see *P. ardens* in 108). "Spurious" signals, and noise from moon- and starlight reflections may also interfere with luminescent communication, and mating activity is much reduced in some fireflies during a bright or full moon (119, 137, 170).

The directionality of light is an important consideration (34, 108). Once visual contact has been established precise orientation is possible but during the broadcast-search phase of mating behavior, line-of-sight signal propagation has serious limitations. Fireflies appear to have a number of morphological and behavioral characteristics that are adaptive in this regard. Males of some species (system II) emit several pulses in quick succession even though the morpheme (communicative unit) is a single pulse. This redundancy would appear to enhance the male's chances of being seen by a female that is perched in dense vegetation such as tall marsh-grass. After visual contact has been made, males of one species omit the supernumerary pulses (112).

Males and females or some species respond to flashes that are deficient or incorrect (108). Since critical signal parameters such as pulse length, or entire pulses, can be obscured by vegetation, these responses permit additional signal exchanges during which conspecificity may be established.

Flight attitudes of flying, flashing males are such that their light is directed toward the ground in front of them (124, 132, 146, 147, 174). Males of some species follow regular search patterns as they fly and flash, and males of some twilight species may actually direct their flashes at those places in a site that are apt to harbor females (108, 109). Flying males of many species execute aerial maneuvers during their flashes such as turns, twists, sudden stops, and ascensions that would seem to enhance their chances of seeing or being seen by females (108).

Females of most fireflies climb up on perches during the hours of mating activity (18, 21, 29, 99, 108, 124, 147, 155, 159, 163). This exposes them and their luminescent signals to searching males. A number of features found in the females of various species prevent their own bodies or their perches from occluding their light. Many have translucent tissues through which light can pass, and some curl or rotate their abdomens and their light is directed to the side, forward or upward (14, 21, 76, 77, 107, 145, 148, 155, 159, 163). Females of some species (e.g. *Photinus*, *Aspisoma*) aim their flashes at males that they are answering (61, 92, 108, 124), *Lampyris* and *Pleotomus* sometimes wiggle their abdomens when glowing (21, 107, 163), the light organs of *Pyractomena* are ventro-lateral and not obstructed by the stems on which females climb (108), and *Lampyris* females "scintillate" (pulse?) their light at intervals (145). The pulsing of *Arachnocampa* (Diptera) females (156) perhaps makes them more conspicuous in a cave of glowing larvae.

Phengodes (Phengodidae) females are luminescent and attract males to their light (10, 13). Their males have plumose antennae, and some observations (103) suggest that pheromones are important in their mating behavior, as they are in other phengodidis, *Zarhipis* and *Phrixothrix* (171, 172). By using a pheromone and a light phengodids would have the benefit of both the long-range potential of chemical communication and the directionality of light for the final approach and landing (27).

The eyes of luminescent fireflies usually are conspicuously sexually dimorphic, with those of males being larger and having more facets than those of females. In his studies on *Lampyris* and *Lamprohiza,* Schwalb (163) found a number of adaptations in the eyes of males for detecting and localizing the lights of females. These adaptations increase the visual field (large subtended angles, variations in focal radii, positioning of ommatidia), increase acuity in certain directions (small ommatidial angles on certain axes), permit distance discrimination (binocular fields), and increase sensitivity ("superposition," tracheal tapeta). The eyes of *Photuris* fireflies are of the "superposition" type, a receptor with increased sensitivity found in nocturnal insects. The classical hypothesis for the operation of such eyes has been questioned (142); and Horridge (94), after studying the eye of *Photuris* (*versicolor?*), concluded that the classification itself is invalid. He suggested that the increased sensitivity of the firefly eye is due to the presence of crystalline threads and to the peculiar construction of the corneal cone, the optical behavior of which (when detached), originally gave rise to the superposition hypothesis.

The pronota of fireflies extend forward and frequently cover the eyes. These may simply shade the eyes from skylight. In some species, e.g. male *Phausis reticulata,* areas over the eyes are transparent and these permit males to see females that are perched above them (107).

Males of some species of fireflies recognize the flash exchanges of a communicating pair of their species (Fig. 27) and approach the female flash (28, 115, 116, 124, 125). Since this places the communicating male at a disadvantage (i.e. it invites competition), there has been a selection for a "tight beam" or directed signal, and males of some species dim their flashes as they approach answering females. (116). In *Luciola lusitanica* (system II), during a male approach a female will sometimes stop answering; the male will sometimes then land and emit female-like responses to the flashes of passing males (147). This may give a rejected male a chance to see and approach the female's flashed answers to another male.

When males of *L. lusitanica* approach females they frequently hover and spiral above the female as they continue to exchange flashes. Papi (147) suggests that this "inspection-dialogue" provides the male with information on the landing area. Males and females of several species sometimes emit glows, flashes, or flickers when landing (110, 112, 114).

Since the timing of female responses with respect to male flashes is critical in some species, it is important that both male and female flashes have a distinctive marker to start the female timing mechanism and to stop the male's. The mating flashes of most fireflies (system II) have distinctive transients. While these are usually found at the beginning or both the beginning and the end of the flashes of most species, in some species they are found only at the end (Figs. 37 and 38, Plate II). One way of quickly recognizing the nominating flashes of flying *Photruis* females is by the absence of transients. In experiments with *L. lusitanica,* Papi (147) determined that the precision of the female response time is dependent upon the slope of the ON transient of the stimulus flash, and the female response flash could not only be altered or displaced by decreasing the slope, it could be inhibited.

Finally, ecological adaptations can enhance bioluminescent communication in insects, and again most examples pertain to fireflies. Orientation to a mating site, such as the tops of trees as in some *Pteroptyx* and *Photuris,* or habitat specificity, as found in most fireflies, restrict the area that males patrol. Differences among species reduce background noise and the chances of interspecific interaction. Restricted periods of activity, seasonaly or diurnally, have similar advantages (108) in this regard. With respect to habitat specificity, a coral-shore *Luciola* (?) of New Guinea is particularly interesting. Its salt-spray habitat is not one that is readily entered and has perhaps buffered this insect against signal competition from other lampyrids. This may have been important in its retention of a primitive signal system (117).

CURRENT STATUS IN SELF-LUMINESCENT INSECTS

COLLEMBOLA

There are nearly two dozen references to luminescent Collembola, and none directly concern communication. Nothing is known of the chemistry or the anatomy of the light-producing structures; although luminous fungi and bacteria have been suggested as the source, self-luminescence seems probable in at least some cases. Some species glow continuously while others emit flashes when stimulated. For details and bibliography see Harvey (89).

HOMOPTERA: FULGORIDAE

The literature on self-luminescence in fulgorids is extensive, considering that there still is a slight question of its occurrence. This subject has been discussed in nearly one hundred publications but few facts are known. Some "eye witness" reports probably pertain to the elaterid *Pyrophorus.* The species involved are *Fulgora laternaria,* the South American lanternfly; *Pryops candelaria,* the Asian candlefly; and *Xosophara* (=*Rinortha*) *guttata* of South Africa. Harvey (89) summarized the history and current status of scientific knowledge and concluded that *Fulgora* is self-luminescent, but reserved judgement on *Pryops* and *Xosophara.*

The reports that at least temporarily settled the *Fulgora* question indicate that bioluminescent communication occurs in a mating context. Presumably the head emits a bright, white light that is turned on only when males and females are together late at night (and flying?).

DIPTERA: MYCETOPHILIDAE

Self-luminescence has been found in the genera *Ceroplatus, Orfelia* (= *Platyura,* not Meigen), and *Arachnocampa* (=*Bolitophila*). The larvae are luminescent, and spin webs upon which they live and with which they snare prey that is presumably attracted to their light (67–69, 80, 89, 156). The evolution of glowing-snaring behavior of fungus gnats has been discussed by Goldschmidt, first in support of evolution by "macromutation" (72) and then (73) as a "complicated adaptation" that may be understood through a comparative approach.

Bioluminescent communication apparently occurs in the New Zealand glowworm fly *Arachnocampa luminosa*. The pupae are luminescent and those of females particularly so during the latter part of their development. When a male fly lands on a female pupa, or the pupa is touched, it turns on its light. When a female is about to emerge males commonly cling to it and females often glow brilliantly when males attempt to mate with them (156). If no males are present when they emerge, females apparently attract males to them with their lights; "She used her light to attract a male fly, flashing it on and off till one arrived. Then she extinguished it and flew a short distance away. Having settled, she exhibited her light till she succeeded in attracting him again." This was repeated several times until finally they mated (156). Hudson (96) suspected the female light was used for mating but was unable to demonstrate this. There have been no experiments with artificial lights or decoys to examine the details of the signalling.

The light organ of adult females is composed of the distal ends of the four Malpighian tubules and has a tracheal reflector (67). Adult males are also luminescent but their light does not appear to be used in mating (156).

COLEOPTERA: PHENGODIDAE
(INCLUDING WHAT WAS FORMERLY PART OF DRILIDAE

While bioluminescent communication probably occurs in this family, in most cases it may play a role secondary to chemical communication. Males have large plumose antennae and are brightly, faintly, or not at all luminescent, depending upon age and genus. Faint luminescence has been observed in male *Phengodes* (102, 135) and *Phrixothrix* (89), *Zarhipis* males luminesce briefly after eclosion (171), males of two species of *Phrixothrix* glow brightly when captured or disturbed (171), and males of an improperly identified Brazilian phengodid (*Ptenophengus;* no such genus) are also luminous (27). Females are larviform and brightly luminescent, and the patterning of luminescent areas varies (Figs. 17 and 18). Females of one species of *Phrixothrix* have eight fewer pairs of lights than do those of another species although their larval stages have identical light arrangements (89). This suggests a mating function and spatial coding.

In some genera males may first be attracted by a pheromone and then at close range determine the female's exact location by her luminescence (27). Atkinson (10) and Barber (13) observed the attraction of male *Phengodes laticollis* to captive glowing females; these were apparently in unstoppered containers so that a pheromone may have been involved. On the basis of a limited number of experiments, Atkinson concluded that attraction took place only at night.

Tiemann (171) and Rivers (158) concluded that the bright luminescence of *Zarhipis* females is not associated with mating and that they attract males with pheromones. Rivers found that *Zarhipis* males were attracted to females during daylight, that they fly only from 9 A.M. to 4 P.M. in temperate heat but in hot weather do not appear "until the sun is declining." Tiemann (171) notes that "Both Linsdale (1961) and Williams (unpublished notes) observed that males will approach females during the day in the humid coastal area. Although males were not observed to approach females during the day in the low relative humidities of the desert environment, they did come to the females at dusk . . . males were attracted to the females within 10 minutes . . . the glow from the females was barely visible in the twilight." From the experiments reported by Tiemann it is quite obvious that pheromones are involved in *Zarhipis* mating behavior and it would seem that luminescence is not necessary for male attraction. However, if males orient visually at close range to the pale-banded females, luminescence would enhance the conspicuousness of the females (and the pale bands) and facilitate male orientation. This would occur especially during periods of low humidity and high temperature when diurnal attraction was prevented and the insects became active at twilight. These considerations lead to a new view of the evolution of bioluminescent communication in Cantharoid beetles. It is not necessary to postulate a presexual function for luminescence in adults (108) for if it first occurred in the context just described it would be adaptive and selected for in a role intimately associated with mating from the start. The behavior of *Phengodes* suggests a subsequent stage with obligatory nocturnal activity and luminescent signalling. These developments may parallel (or be the same as?) those that occurred in the Lampyridae: a relationship between antennal, eye and light-organ development has been noted (21, 74).

Another possible instance of bioluminescent communication occurs in *Phrixothrix* during copulation. Although Tiemann (172) believes pheromones to be involved in attraction, he notes that while mating they "stage a pyrotechnic display"; *Phrixothrix* males, unlike *Phengodes* and *Zarhipis* males, are brightly luminescent. Male luminescence in *Phrixothrix* may identify them to females and have some courtship function, although Tiemann suggests that "they glow while mating—and also while attacking prey—to warn off predators while preoccupied."

COLEOPTERA: ELATERIDAE

The function of luminescence in firebeetles is not known but sexual communication has been suggested (24, 65, 107, 129) and seems probable. There is little direct evidence. Males of various species of *Pyrophorus* are attracted to artificial lights, lighted cigarets, captive specimens and decoys (32, 58, 93, 107, 118, 119). I was unable to attract *Photophorus bakewelli* of the New Hebrides to artificial lights, captive specimens or decoys (117). With *Pyrophorus atlanticus* in Florida, males can be attracted to decoys with twin lights (simulating prothoracic light organs) but not those with single lights (107). The most extensive behavioral observations made on *Pyrophorus* were by Withecombe, reported by Blair (24), in Trinidad on *P. pellucens* and *P. extinctus*. He observed the attraction of males of both species to their respective females. Withecombe also noted species and sexual differences in glowing and flight behavior. While species with brightly luminescent females may use bioluminescent signal system I, others may use pheromones (24).

Pyrophorus and *Photophorus* have two kinds of light organs; a pair is located on the pronotum and a single organ is situated ventrally on the abdomen (Figs. 19, 20, Plate I). There apparently are differences among species in the function of these organs: in some species of *Pyrophorus* only the abdominal organ glows during flight while in others, the prothoracic lights glow (also?). When roughly handled *Pyrophorus* and *Photophorus* turn on their prothoracic lights. The organs of *Pyrophorus* are controlled independently and have different mechanisms of control (83). They do not flash but emit long-sustained glows that vary in intensity (91, 166). The flashing reported by some observers may have been due to changes in flight angle of the beetles they observed. The spectra of the two organs differ in some species with the abdominal light being more red (75, 86, 93, 160, 167). Spectral differences have also been noted within species (20). Harvey (89) summarized the literature on firebeetles and discussed their histology, physiology, and biochemistry.

COLEOPTERA: LAMPYRIDAE
(INCLUDING WHAT WAS FORMERLY RHAGOPHTHALMIDAE)

The sexual significance of the bioluminescence of adult fireflies has been recognized for well over a century (89), but with a few notable exceptions the details of their communications have only recently been examined. Mast (124) was the first to attempt a functional analysis of firefly signals, and Buck (29), in 1937, demonstrated the communicative significance of the female delay in *Photinus pyralis,* thus being the first to "crack the code" of any species. The early cataloguing of signals by McDermott (126–130) also deserves mention since his chart of firefly signals was the only one available until 1951, and is still being used for illustrative purposes (41, 123). In the mid-thirties, Buck (28, 30) collected and discussed the published accounts of the synchronous flashing of fireflies, thus beginning an investigation that is still active and that, in 1969, placed several investigators in New Guinea on the Alpha Helix Expedition. The research reports from this expedition will be published independently but some of the findings have been mentioned in this review.

Recent studies have added to the catalogue of firefly signals, have determined and discussed coding elements in firefly signals, and described behavior associated with signalling (33, 34, 37, 61, 62, 105, 107–109, 111, 112, 114–117, 137, 147, 163); discussed firefly communication in terms of natural selection and evolution (34, 108), explored the nature of the synchrony mechanism (16, 35, 84, 152), and made use of signals in taxonomic studies (15, 108, 111, 112, 114–116). Other studies have been less directly related to communication and have dealt with neurophysiology, histology, and bio-chemistry. For bibliographies and general coverage of these latter subjects see Harvey (89), McElroy (138), Carlson (39), and Buck (31).

When compared with what is known of other luminescent insects the extent of knowledge on lampyrids appears impressive; in fact, it was possi-ble to illustrate the first section of this paper almost exclusively with firefly examples. However, these few examples virtually exhaust the information on the coding and systems of firefly communication. Nothing is known of the behavior of at least 95 percent of the described species of fireflies (I am sure most firefly species have yet to be discovered and described) and repre-sentatives of only a few luminescent genera have been carefully studied; indeed, the ability to luminesce is usually inferred from the presence of yel-low or pale cuticle in appropriate places. This is unfortunate for it means that a valuable research potential has been overlooked. I can best illustrate this by posing some questions which stem from specific observations.

The signal of male *Photinus consanguineus,* a species of eastern North America, is composed of two pulses with an interval of about one-half sec-ond (Fig. 28, Plate II). Females answer only this signal; their response flash is emitted about one second after the second male pulse. Two close relatives of this species have similar signals but the timing of the male pul-ses is different (Figs. 26 and 27, Plate II). Quite probably these species share a recent common ancestor that had a two-pulse flash pattern. What was the raw material upon which natural selection acted, during or subse-quent to speciation, which resulted in these timing differences, and can such variation be found in extant species? Did these timing differences occur gradually, millisecond by millisecond, or has the central nervous system an oscillator that doubled or halved its period? What is the nature of the neural coding of this timing, and of the differences? Can hybrids be pro-duced—what of their timing—how many allele substitutions are required to produce an X millisecond timing difference? What is the relationship be-tween the two pulses of the flash patterns of these males and the single pulse in the flash pattern (Fig. 29, Plate II) of a sibling species, *P. ignitus,* or the 1–4 pulses of another relative, *P. lineellus* (Fig. 30, Plate II)? And lastly, what is the ecological and evolutionary significance, if any, of the fact that a signal like that of *P. consanguineus* (Lampyrinae) occurs in *Photuris* (Photurinae) of Florida and *Luciola* (Luciolinae) of New Guinea (Figs. 28, 39, 40, Plate II). Surely these signals can't be homologous, but are the flashing-type light organs or signal system (II) that they possess?

In addition, fireflies provide an excellent opportunity for the experimen-tal study of species and species interaction since the signals that bring about mating can easily be recorded, analyzed, and simulated (29, 108, 147). These signals, just as those of acoustical insects, can permit the familiarity with species that is necessary to conduct field investigations in zoogeography, ecology, taxonomy, evolution, and behavior (2, 4, 6, 8, 173). Because male flashes are conspicuous and usually permit rapid field identification, they can be used to locate demes of tree-top or "rare" species even from a moving vehicle.

Firefly flashes as epideictic displays.—Wynne-Edwards (177) has suggested that the bioluminescent signals of fireflies function as part of population regulation mechanisms; according to his hypothesis the flashing of males serves as the indicator portion of a homeostatic system through which "population-density is constantly adjusted to match the optimum level of exploitation of available food resources." Recruitment, i.e. reproduction, is adjusted according to information derived by males from their "communal displays." In addition to the flashing of fireflies, Wynne-Edwards implicated various behavior patterns of a diversity of animals. This hypothesis runs counter to knowledge of the operation of Darwinian selection and has been rigorously and severely criticized (176). I will comment here on the current validity of some of the statements made by Wynne-Edwards in his discussion of the Lampyridae since (*a*) epideictic displays would involve communication, and (*b*) he supports his argument for epideictic displays by generally dismissing the mating function of luminescence.

Wynne-Edwards states "There have been two common theories as to the function of the flashing: first, that it is a defence against aggressors—an old and now discredited view . . . [There has been no experimental work done relative to this hypothesis and it is no more or less valid than it ever was] and second that it attracts the sexes together for mating. The latter is 'rather universally accepted at the present time' (Harvey, 1952, p. 403) although, to judge from the reservations expressed by some writers (e.g. Skaife, 1953, p. 230), largely for the want of anything better." The sexual function of bioluminescence has been found in species of the genera *Luciola, Photinus, Lamprohiza, Phausis, Aspisoma, Pyractomena, Micronaspis, Photuris, Pteroptyx, Dioptoma,* and *Lamprophorus.* These genera represent four of the seven subfamilies of Lampyridae; the three remaining include only 61 of the 1891 described species (136). In no species that has been carefully studied has the luminescence of adults been found not to function in mating attraction. Sexual communication via bioluminescence is widespread in the family and probably occurred in the lampyrid progenitor. Skaife (169), whose reservations were noted, placed fireflies in the family Cantharidae in his popular book *African Insect Life* and questioned the sexual function for the same reason early writers had (26, 100, 162), i.e. ignorance of signal system II. He gave virtually no data on communication in African fireflies since he obviously had only casually observed them.

Following the above reference to Skaife, a few lines are quoted from Hudson (97) concerning the behavior of *Cratomorpheus* sp.: diurnal activity includes "wooing . . . and feeding," and evening activity involves 2–3

California Baptist College
Riverside, California

hours of (flashing) and "flitting aimlessly . . . apparently for amusement." The quote ends with the sentence "Thus, the more closely we look at the facts, the more unsatisfactory does the explanation seem." Hudson was not referring to the mating function of luminescence; his subsequent sentence reads: "That the firefly should have become possessed of so elaborate a machinery, producing incidentally such splendid results, merely as a protection against one set of enemies for a portion only of the period during which they are active, is altogether incredible."

Another feature mentioned to support the epideictic display hypothesis was the combination of large eyes and a bright luminescent emission in males. This combination is also consistent with the mating function and the broadcast, search, and orientation behavior of males (see above) using system II.

Wynne-Edwards also discussed the synchronizing fireflies of southeast Asia. Since there was no information on mating behavior in these fireflies and, to the contrary, the mating function had been dismissed by those who had seen them (144, 170), he considered the synchronizing swarms to be another example of epideictic display. These fireflies (*Pteroptyx*) have recently received considerable study. A taxonomic revision is in press and several papers have been published or are in preparation on their behavior and physiology (12, 16, 34–36, 81, 117, 152, 153). The early criticisms of the mating "hypothesis," e.g., females are wingless and not found in the display trees with the males, are groundless. No known *Pteroptyx* (or *Luciola* from this area, 11, 134) has wingless females and although they do not participate in the synchronous flashing, the females are present in the display trees and can be seen flying in and out of the trees (34, 81, 117). Copulating pairs are found in and near display trees (34, 81, 117), attraction and mounting has been observed, and males of at least two synchronizing species of *Pteroptyx* have been experimentally attracted to a penlight while they were in synchrony with other fireflies (117).

Wynne-Edwards' conclusion that "The 'mating' theory can in fact provide an adequate explanation only in a minority of fireflies," is not in the least supported by the evidence. With respect to the statements ". . . that in the great majority the primary function of flashing is epideictic . . . ," and "That it should often serve also the secondary function of drawing the sexes together. . . ." I can't imagine anything more primary than mating, and there is no evidence for the function he suggests.

GLOSSARY

CANDLEFLY *Pryops candelaria*. An Asian homopteran of the family Fulgoridae that is presumed to be luminescent.

CUCUJO See firebeetle.

DELAY See response delay time.

FIREBEETLE A luminescent beetle of the family Elateridae. This usage is occasionally found in the older literature and is recommended here.

Among other terms that have been used for these beetles include glow-fly, cucujo, pyrophore, firefly, night-lighting elater and peeny wally.

FIREFLY   A beetle of the family Lampyridae, not necessarily luminescent. In older literature this term was used for luminescent click-beetles (Elateridae) to distinguish them from Lampyridae, then termed lightning bugs. See firebeetles.

FLASH   A single emission of light of short duration.

FLASH PATTERN   The species-typical unit of light emission of the male that is repeated at somewhat regular time intervals by advertising males and that is or contains the unit (morpheme) that stimulates the female response flash.

GLOW   A steady emission of light; the term by itself does not indicate intensity level.

GLOWWORM   1. A luminescent larva or larviform female, usually of the family Lampyridae but also other beetle families. In older literature this term commonly referred to species of Lampyridae that had larviform females, especially the well-known European *Lampyris noctiluca*. 2. Luminescent larva of Mycetophilidae (Diptera).

GLOWWORM FLY   Adult fly of family Mycetophilidae whose larva is luminescent.

INSPECTION-DIALOGUE   The variable phase of the male approach flight that occurs when the male is responding to the female signal and hovers and signals near her.

INTERVAL   The time between the beginning of an event (flash pattern, pulse) and the beginning of the next such event. PERIOD is undoubtedly a better term.

LANTERNFLY   *Fulgora laternaria*. A tropical New World homopteran of the family Fulgoridae that apparently is luminescent.

LATENCY   See response delay.

MORPHEME   The unit that conveys species information that cannot be divided into smaller, independently functional, communicative units.

PAUSE   The time between the end of an event (flash pattern, pulse) and the beginning of the next such event.

PEENY WALLY   See firebeetle.

PHRASE   See flash pattern.

PULSE   See flash.

RESPONSE DELAY, RESPONSE DELAY TIME   The time between the beginning of the last pulse of a male flash pattern and the beginning of the female response flash.

RESPONSE FLASH   In signal system II, the flashed signal emitted by a female following stimulation by the flash pattern of a conspecific male.

RESPONSE INTERVAL   See response delay.

SYSTEM   Method, arrangement, order of signalling or signal exchanging between communicating individuals; protocol. System I, one individual (sex) broadcasts a signal, the other receives and approaches. System

II, one individual broadcasts a signal, the other receives and responds with a signal, the first approaches.

TRANSIENT An abrupt change in intensity, usually at the beginning or beginning and end of an emission.

ACKNOWLEDGMENTS

I gratefully acknowledge the help and assistance I have received from several friends and colleagues: John B. Buck, William H. Biggley, Albert D. Carlson, Peter C. Drummond, Yata Haneda, Frank E. Hanson, Karl V. Krombein, Floriano Papi, Ivan Polunin, Darwin L. Tiemann, and Robert E. Woodruff, provided me with reprints, taxonomic assistance, and unpublished manuscripts and information. Thomas J. Walker read the manuscript at several stages and provided much helpful discussion. Without the help of my assistant Susan Griffis I would be at the library still. This review was written during the tenure of NSF Grant GB-7407.

LITERATURE CITED

1. Acloque, A. 1907. La lumiere des insects. *Cosmos* 56:624–27
2. Alexander, R. D. 1962. The role of behavioral study in cricket classification. *Syst. Zool.* 11:53–72
3. Alexander, R. D. 1967. Acoustical communication in arthropods. *Annu. Rev. Entomol.* 12:495–526
4. Alexander, R. D. 1967. *Singing Insects*. Chicago: Rand McNally. 86 pp.
5. Alexander, R. D. 1968. Arthropods. In *Animal Communication, ed.* T. A. Sebeok, Chap. 10, 167–216. Bloomington: Indiana Univ. Press. 686 pp.
6. Alexander, R. D., Bigelow, R. S. 1960. Allochronic speciation in field crickets, and a new species, *Acheta veletis. Evolution* 14:334–46
7. Alexander, R. D., Meral, G. H. 1967. Seasonal and daily chirping cycles in the northern spring and fall field crickets, *Gryllus veletis* and *G. pennsylvanicus. Ohio J. Sci.* 67:200–9
8. Alexander, R. D., Moore, T. E. 1962. The evolutionary relationships of 17-year and 13-year cicadas, and three new species. *Misc. Publ. Mus. Zool., Univ. Mich.,* No. 121:1–59
9. Allen, I. M., Wootton, A. 1963. Man's use of fireflies for light. *Entomol. Monthly Mag.* 99:27–30
10. Atkinson, G. F. 1887. Observations on the female form of *Phengodes laticollis* Horn. *Am. Natur.* 21:853–56 Also in *J. Elisha Mitchell Sci. Soc.* 4:92–95
11. Ballantyne, L. A. 1968. Revisional studies of Australian and Indomalayan Luciolini. *Univ. Queensland Papers* 2:105–39
12. Ballantyne, L. A., McLean, M. R. Revisional studies on the firefly genus *Pteroptyx* Olivier. In press
13. Barber, H. S. 1906. Note on *Phengodes* in the vicinity of Washington, D.C. *Proc. Entomol. Soc. Wash.* 7:196–97
14. Barber, H. S. 1923. A remarkable wingless glow-worm from Ecuador. *Insecutor Inscitiae Menstruus* 11:191–94
15. Barber, H. S. 1951. North American fireflies of the genus *Photuris. Smithson. Misc. Coll.* 117:1–58

16. Bassot, J. M., Polunin, I. V. 1967. Synchronously flashing fireflies in the Malay Peninsula. Symposium on bioluminescence in the Pacific area of the 11th Pacific Sci. Congr., 1967, *Sci. Rep. Yokosuka City Mus.* No. 13:5–28
17. Bastock, M. 1967. *Courtship: an Ethological Study.* Chicago: Aldine Publ. Co., 220 pp.
18. Beebe, W. 1949. *High Jungle.* New York: Duell, Sloan, and Pearce. 379 pp.
19. Bengry, R. P. 1944. Living light. *Natur. Hist. Notes Natur. Hist. Soc. Jamaica* 15:39–40, 42
20. Biggley, W. H., Lloyd, J. E., Seliger, H. H. 1967. The spectral distribution of firefly light. II. *J. Gen. Physiol.* 50:1681–92
21. Blair, K. G. 1915. Luminous insects. *Natur* 96:411–15. Also in *Proc. Soc. London Entomol. Natur. Hist. Soc.* 1914–1915:31–45
22. Blair, K. G. 1924. Some notes on luminosity in insects. *Entomol. Monthly Mag.* 60:173–78
23. Blair, K. G. 1924. [Remarks on the luminosity of the Lampyridae.] *Trans. Roy. Entomol. Soc. London* 7
24. Blair, K. G. 1926. On the luminosity of *Pyrophorus. Entomol. Monthly Mag.* 62:11–15
25. Blakeslee, A. L. 1948. Stars of death. *Natur. Hist.* 57:75–77
26. Bowles, G. H. 1882. On luminous insects. *Rep. Entomol. Soc. Ontario* 34–37
27. Bristowe, W. S. 1924. Notes on the habits of insects and spiders in Brazil. *Trans. Entomol. Soc. London* 475–504
28. Buck, J. B. 1935. Synchronous flashing of fireflies experimentally induced. *Science* 81:339–40
29. Buck, J. B. 1937. Studies on the firefly II. The signal system and color vision in *Photinus pyralis. Physiol. Zool.* 10:412–19
30. Buck, J. B. 1938. Synchronous rhythmic flashing of fireflies. *Quart. Rev. Biol.* 13:301–14
31. Buck, J. B. 1948. The anatomy and physiology of the light organ in fireflies. In *Bioluminescence,* eds. E. N. Harvey et al. *Ann. N.Y. Acad. Sci.* 49:397–482
32. Buck, J. B. 1964. The march of

science Jamaican style. *Johns Hopkins Mag.* April :4–8

33. Buck, J. B., Buck, E. M. 1965. Photic signalling in the firefly *Photinus consanguineus. Ann. Zool.* 5 :682

34. Buck, J. B., Buck, E. M. 1966. Biology of synchronous flashing of fireflies. *Nature* 211 :562–64

35. Buck, J. B., Buck, E. M. 1968. Mechanism of rhythmic synchronous flashing of fireflies. *Science* 159 :1319–27

36. Buck, J. B., Hanson, F. E., Hopkins, T. Types of synchrony in New Guinea fireflies. In preparation

37. Buschman, L. L. 1968. *A study of flash communication in the firefly* Photuris divisa. M.S. thesis. Kansas State Teachers College of Emporia, 1966

38. Butler, E. A. 1895. The glowworm. *Sci. Am. Suppl., No. 994,* 39 : 15883–84

39. Carlson, A. D. 1969. Neural control of firefly luminescence. *Advan. Insect Physiol.* 6 :51–96

40. Chase, A. M. 1964. Bioluminescence —production of light by organisms. In *Photophysiology,* ed. A. C. Giese, Chap. 21, 2 :389–421. New York and London : Academic Press. 441 pp.

41. Curtis, H. 1968. *Biology.* New York : Worth Publs. Inc. 854 pp.

42. Dahlgren, U. 1915. The production of light by animals. Introduction. *J. Franklin Inst.* 180 :513–37

43. Dahlgren, U. 1916. The production of light by animals. Luminous plants. *J. Franklin Inst.* 181 :109– 25

44. Dahlgren, U. 1916. The production of light by animals. Porifera and Coelenterata. *J. Franklin Inst.* 181 : 243–61

45. Dahlgren, U. 1916. The production of light by animals. Luminosity in echinoderms. *J. Franklin Inst.* 181 : 377–400

46. Dahlgren, U. 1916. The production of light by animals. Light production in cephalopods. *J. Franklin Inst.* 181 :525–56

47. Dahlgren, U. 1916. The production of light by animals. Light production as seen in worms. *J. Franklin Inst.* 181 :659–96

48. Dahlgren, U. 1916. The production of light by animals. The luminous

crustaceans. *J. Franklin Inst.* 181 : 805–44

49. Dahlgren, U. 1917. The production of light by animals. The luminous insects. *J. Franklin Inst.* 183 :79– 94

50. Dahlgren, U. 1917. The production of light by animals. Light production by elaterid beetles. *J. Franklin Inst.* 183 :211–20

51. Dahlgren, U. 1917. The production of light by animals. The fire-flies or Lampyridae. *J. Franklin Inst.* 183 :323–48

52. Dahlgren, U. 1917. The production of light by animals. Histogenesis and physiology of the light tissues in lampyrids. *J. Franklin Inst.* 183 :593–624

53. Dahlgren, U. 1917. The production of light by animals. Tunicates. *J. Franklin Inst.* 183 :735–54

54. Dahlgren, U. 1917. Investigations of light organs of arthropods. *Proc. Am. Soc. Zool.* 11 :481–83

55. Dahlgren, U. 1922. Phosphorescent animals and plants. *Natur. Hist.* 22 :5–26

56. Dillon, L. S. 1967. *Animal Variety.* Dubuque : Wm. C. Brown Co. Publ. 180 pp.

57. Domestica, A. 1851. Stars of the earth. In *Episodes of Insect Life,* Ser. 3, 160–77. New York : J. S. Redfield, Clinton-Hall. 432 pp.

58. Drummond, P. C., Woodruff, R. E. Personal communication

59. Dubois, R. 1914. *La vie et la lumiere.* Paris : Alcan. 338 pp.

60. Evans, H. E. 1968. In defense of magic : the story of fireflies. In *Life on a Little-Known Planet,* Chap. 6, 102–15. New York : E. P. Dutton. 318 pp.

61. Farnworth, E. G. 1969. Flash communication in *Aspisoma* sp. *Entomol. News* 80 :249–51

62. Farnworth, E. G. Bioluminescent communication in Jamaican fireflies. Ph.D. dissertation in preparation

63. Frings, H., Frings, M. 1964. *Animal Communication.* New York : Blaisdell Publ. Co. 204 pp.

64. Gadeau de Kerville, H. 1890. *Les vegetaux et les animaux lumineaux.* Paris. 327 pp.

65. Gahan, J. C. 1924. Living fireflies from Argentina. *Trans. Roy. Entomol. Soc. London* 5

66. Gannon, R. 1967. Fireflies: nature's light fantastic. *Frontiers* June: 138–41
67. Gatenby, J. B. 1959. Notes on the New Zealand glowworm, *Bolitophila* (*Arachnocampa*) *luminosa*. *Trans. Roy. Soc. New Zealand* 87 :291–314
68. Gatenby, J. B. 1960. The Australasian mycetophilid glowworms. *Trans. Roy. Soc. New Zealand* 88 :577–93
69. Gatenby, J. B., Cotton, S. 1960. Snare building and pupation in *Bolitophila luminosa*. *Trans. Roy. Soc. New Zealand* 88 :149–56
70. Gauroy, P., Grive, J. 1964. Les organes lumineux des etres vivants. *Nature* (Paris) 92 :98–104
71. Girard, M. 1873. Les taupins lumineux. *Nature* (Paris) 1 :337–39
72. Goldschmidt, R. B. 1948. Glow-worms and evolution. *Rev. Sci.* 86 :607–12
73. Goldschmidt, R. B. 1951. Eine weitere Bemerkung über "Glühwürmer und Evolution" *Naturwissenschaften* 38 :437–38
74. Gorham, H. S. 1880. On the structure of the Lampyridae, with references to their phosphorescence. *Trans. Roy. Entomol. Soc. London* 63–67
75. Gosse, P. H. 1848. On the insects of Jamaica. *Ann. Mag. Natur. Hist.* (Ser. 2) 1 :109–15, 197–202, 268–70, 349–52
76. Gravely, F. H. 1915. Notes on the habits of Indian insects, myriapods, and arachnids. *Rec. Indian Mus.* 11 :483–539
77. Green, E. E. 1912. On some luminous Coleoptera from Ceylon. *Trans. Roy. Entomol. Soc.* 4 :717–19
78. Griffin, D. R. 1968. Echolocation and its relevence to communication behavior. In *Animal Communication*, ed. T. A. Sebeok, Chap. 9, 154–64. Bloomington: Indiana Univ. Press. 686 pp.
79. Guenther, K. 1931. *A Naturalist in Brazil.* Boston and New York: Houghton Mifflin. 400 pp.
80. Haneda, Y. 1957. Luminous insects of Hachijo Island, Japan. *Sci. Rep. Yokosuka City Mus.* 2 :24–26
81. Haneda, Y. 1966. Synchronous flashing of fireflies in New Guinea. *Sci. Rep. Yokosuka City Mus.* No. 12 :4–8
82. Haneda, Y. 1968. Luminescence of firefly. *Nature and Insects* 3 :7–12
83. Hanson, F. E., Buck, J. B. 1965. Photic control in *Pyrophorus*. *Am. Soc. Zool.* 5 :682
84. Hanson, F. E., Case, J. F., Buck, J. B. Pacemaking and synchrony in New Guinea fireflies. In preparation
85. Hanson, F. E., Miller, J., Reynolds, G. T. 1969. Subunit coordination in the firefly light organ. *Biol. Bull.* 137 :447–64
86. Harrington, W. H. 1880. On the Elateridae or click-beetles. *Annu. Rep. Entomol. Soc. Ontario* 77–84
87. Harvey, E. N. 1920. *The Nature of Animal Light.* Philadelphia: J. B. Lippincott. 182 pp.
88. Harvey, E. N. 1940. *Living Light.* New York: Hafner Publ. Co. 328 pp.
89. Harvey, E. N. 1952. *Bioluminescence.* New York: Academic. 649 pp.
90. Harvey, E. N. 1957. *A History of Luminescence.* Philadelphia: The American Philosophical Soc. 692 pp.
91. Hastings, J. W., Buck, J. B. 1956. The firefly pseudoflash in relation to photogenic control. *Biol. Bull.* 3 :101–13
92. Hess, W. N. 1920. Notes on the biology of some common Lampyridae. *Biol. Bull.* 38 :39–76
93. Heward, R. 1840. Memorandum on the fireflies of Jamaica. *Entomologist* 1 :42–43
94. Horridge, G. A. 1969. The eye of the firefly *Photuris*. *Proc. Roy. Soc. Ser. B* 171 :445–63
95. Hudson, G. V. 1887. On New Zealand glow-worms. *Trans. New Zealand Inst.* 19 :62–64
96. Hudson, G. V. 1926. "The New Zealand glow-worm." *Bolitophila* (*Arachnocampa*) *luminosa:* summary of observations. *Ann. Mag. Natur. Hist.* 17 :228–35
97. Hudson, W. H. 1922. *The Naturalist in La Plata.* New York: E. P. Dutton. 394 pp.
98. Johnson, F. H., Haneda, Y., Eds. 1966. *Bioluminescence in progress; proceedings.* Princeton: Princeton Univ. Press. 650 pp.
99. Kaufmann, T. 1965. Ecological and biological studies on the West African firefly *Luciola discicollis*. *Ann. Entomol. Soc. Am.* 58 :414–26

100. Kirby, W., Spence, W. 1858. On luminous insects. In *Introduction to Entomology,* Letter XXV, 503–15. London

101. Klopfer, P. H., Hatch, J. J. 1968. Experimental considerations. In *Animal Communication,* ed. T. A. Sebeok, Chap. 3, 31–43. Bloomington: Indiana Univ. Press. 686 pp.

102. Knab, F. 1905. Observations on Lampyridae. *Can. Entomol.* 37: 238–39

103. Knaus, W. 1907. Notes and news. *Entomol. News* 18:318–19

104. LeConte, J. L. 1880. On lightening bugs. *Proc. Am. Assoc. Advan. Sci.* 29:650–59

105. Lloyd, J. E. 1964. Notes on flash communication in the firefly *Pyractomena dispersa. Ann. Entomol. Soc. Am.* 57:260–61

106. Lloyd, J. E. 1965. Aggressive mimicry in *Photuris:* firefly femmes fatales. *Science* 149:653–54

107. Lloyd, J. E. 1965. Observations on the biology of three luminescent beetles. *Ann. Entomol. Soc. Am.* 58:588–91

108. Lloyd, J. E. 1966. Studies on the flash communication system in *Photinus* fireflies. *Misc. Publ. Mus. Zool.,* Univ. Mich. No. 130: 1–95

109. Lloyd, J. E. 1966. Signals and mating behavior in several fireflies. *Coleopt. Bull.* 20:84–90

110. Lloyd, J. E. 1968. Illumination, another function of firefly flashes? *Entomol. News* 79:265–68

111. Lloyd, J. E. 1968. A new *Photinus* firefly, with notes on mating behavior and a possible case of character displacement. *Coleopt. Bull.* 22:1–10

112. Lloyd, J. E. 1969. Flashes of *Photuris* fireflies: their value and use in recognizing species. *Fla. Entomol.* 52:29–35

113. Lloyd, J. E. 1969. Flashes and behavior of some American fireflies. *N.Y.S. Conservationist* 23:8–12

114. Lloyd, J. E. 1969. Signals and systematics of Jamaican fireflies: notes on behavior and on undescribed species. *Entomol. News* 80:169–76

115. Lloyd, J. E. 1969. Flashes, behavior additional species of nearctic *Photinus* fireflies. *Coleopt. Bull.* 2: 29–40

116. Lloyd, J. E. Systematics and behavior of nearctic *Photuris* fireflies. Work in progress

117. Lloyd, J. E. Signals and behavior of some New Guinea and Pacific fireflies. In preparation

118. Lund, E. J. 1911. On the structure, physiology and use of photogenic organs, with special reference to the Lampyridae. *J. Exp. Zool.* 11: 415–67

119. Lund, E. J. 1911. On light reactions in certain luminous organisms. *Johns Hopkins Univ. Circ.* NS 2: 10–13

120. Macartney, J. 1810. Observations upon luminous animals. *Phil. Trans. Roy. Soc.* 100:258–93

121. Maloeuf, N. S. R. 1938. The basis of the rhythmic flashing of the firefly. *Ann. Entomol. Soc. Am.* 31:374–80

122. Marler, P. 1968. Visual systems. In *Animal Communication,* ed. T. A. Sebeok, Chap. 7, 103–26. Bloomington, Indiana: Univ. Press. 686 pp.

123. Marler, P. R., Hamilton, W. J., III. 1966. *Mechanisms of Animal Behavior.* New York: John Wiley. 771 pp.

124. Mast, S. O. 1912. Behavior of fireflies (*Photinus pyralis?*) with special reference to the problem of orientation. *J. Anim. Behav.* 2:256–72

125. Maurer, U. M. 1968. Some parameters of photic signalling important to sexual and species recognition in the firefly *Photinus pyralis.* M.S. thesis, State University of New York at Stony Brook, 1968

126. McDermott, F. A. 1910. A note on the light-emission of some American Lampyridae. *Can. Entomol.* 42:357–63

127. McDermott, F. A. 1911. Some further observations on the light-emission of American Lampyridae: the photogenic function as a mating adaptation in the Photinini. *Can. Entomol.* 43:399–406

128. McDermott, F. A. 1912. Observations on the light-emission of American Lampyridae. *Can. Entomol.* 44: 309–11

129. McDermott, F. A. 1914. The ecologic relations of the photogenic function among insects. *Z. Wiss. Insekt.* 10:303–7

130. McDermott, F. A. 1917. Observations

on the light-emission of American Lampyridae: the photogenic function as a mating adaptation. *Can. Entomol.* 49 :53–61

131. McDermott, F. A. 1948. *The Common Fireflies of Delaware.* Wilmington: Soc. Nat. Hist. Delaware. 19 pp.

132. McDermott, F. A. 1954. Firefly in flight. *Coleopt. Bull.* 8 :1

133. McDermott, F. A. 1958. *The Fireflies of Delaware.* Wilmington: Soc. Nat. Hist. Delaware. 36 pp.

134. McDermott, F. A. 1964. The taxonomy of the Lampyridae. *Trans. Am. Entomol. Soc.* 90 :1–72

135. McDermott, F. A. 1965. The Pterotinae. *Entomol. News* 76 :98–104

136. McDermott, F. A. 1966. *Coleopterorum Catalogus Supplementa.* Pars 9, W. Junk. 149 pp.

137. McDermott, F. A., Buck, J. B. 1959. The lampyrid fireflies of Jamaica. *Trans. Am. Entomol. Soc.* 85 :1–112

138. McElroy, W. D. 1964. Insect bioluminescence. In *Physiology of the Insecta,* ed. M. Rockstein, Chap. 11, 463–508. New York and London: Academic. 640 pp.

139. McElroy, W. D., Glass, B. 1961. *A Symposium on Light and Life.* Baltimore: The Johns Hopkins Press. 934 pp.

140. McElroy, W. D., Seliger, H. H. 1962. Biological luminescence. *Sci. Am.* 207 :2–14

141. Meyrick, E. 1886. A luminous insect larva in New Zealand. *Entomol. Monthly Mag.* 22 :266–67

142. Miller, W. H., Bernard, G. D., Allen, J. L. 1968. The optics of insect compound eyes. *Science* 162 :760–67

143. Minamo, K. 1961. Hotaru no kenkyu (A study of fireflies). Published by the author. 321 pp. In Japanese

144. Morrison, T. F. 1929. Observations on the synchronous flashing of fireflies in Siam. *Science* 69 :400–1

145. Newport, G. 1857. On the natural history of the glow-worm (*Lampyris noctiluca*). *J. Proc. Linn. Soc. London Zool.* 1 :40–71

146. Olsen, C. E. 1958. The fireflies' lovelight. *Illus. Lib. Natur. Sci.* 2 : 980–85

147. Papi, F. 1969. Light emission, sex attraction and male flash dialogues in a firefly, *Luciola lusitanica* (Charp.). *Monitore Zool. Ital.* (N.S.) 3 :135–84

148. Parfitt, E. 1880. On the phosphorescence of the glow-worm. *Entomol. Monthly Mag.* 17 :94

149. Perkins, G. A. 1869. The cucuyo; or, West Indian fire beetle. *Am. Natur.* 2 :422–33

150. Pfeiffer, H., Stammer, H. J. 1930. Pathogenes Leuchten bei Insekten. *Z. Morphol. Oekol. Tiere* 20 :136–71

151. Phipson, T. L. 1862. *Phosphorescence; or, the emission of light by minerals, plants and animals.* London and New York: Lovell Reeve & Co. 210 pp.

152. Polunin, I. Social behavior of *Pteroptyx* fireflies with special reference to aggregation and synchronous flashing. In preparation

153. Polunin, I., Hon, Y. Investigations on synchronous and other light patterns of *Pteroptyx* fireflies using an artificial light source. In preparation

154. Poole, L., Poole, G. 1965. *Fireflies in Nature and the Laboratory.* New York: Thomas Y. Crowell. 149 pp.

155. Priske, R. A. R. 1910. Notes on the glow-worm (*Lampyris noctiluca*). *Proc. S. London Entomol. Natur. Hist. Soc.* 74–76

156. Richards, A. M. 1960. Observations on the New Zealand glow-worm *Arachnocampa luminosa* (Skuse) 1890. *Trans. Roy. Soc. New Zealand* 88 :559–74

157. Richards, A. M. 1964. The New Zealand glow-worm. *Stud. Speleology* 1 :38–41

158. Rivers, J. J. 1886. Description of the form of the female Lampyrid (*Zarhipis riversi* Horn). *Am. Natur.* 20 :648–50

159. Roth, V. D. 1969. Feature photograph. Bioluminescent insects. *Ann. Entomol. Soc. Am.* 62 :680

160. Sanderson, I. T. 1939. *Caribbean Treasure.* New York: Viking. 292 pp.

161. Schaller, F. 1963. Das Licht der Tiere. *Umschau Wiss. Tech.* 21 : 663–65

162. Schmidt, P. 1894. Uber das Leuchten der Zuckmücken. *Zool. Jb. Abt. Syst.,* 8 :58–66. Transl. in *Ann. Mag. Natur. Hist.* 15 :133–41

163. Schwalb, H. H. 1960. Beitrage zur Biologie der einheimischen Lampyriden *Lampyris noctiluca* Geoffr. und *Phausis splendidula* Lec. und

experimentelle Analyse ihres Beutefang- und Sexualverhaltens. *Zool. Jb. Syst. Bd.* 88:399–550

164. Seaman, W. H. 1891. On the luminous organs of insects. *Proc. Am. Soc. Microsc.* 13:133–62

165. Sebeok, T. A., Ed. 1968. *Animal Communication.* Bloomington, Indiana Univ. Press. 686 pp.

166. Seliger, H. H., Buck, J. B., Fastie, W. G., McElroy, W. D. 1964. Flash patterns in Jamaican fireflies. *Biol. Bull.* 127:159–72

167. Seliger, H. H., Buck, J. B., Fastie, W. G., McElroy, W. D. 1964. The spectral distribution of firefly light. *J. Gen. Physiol.* 48:95–104

168. Seliger, H. H., McElroy, W. D. 1965. Bioluminescence—enzyme-catalyzed chemiluminescence. In *Light: Physical and Biological Action,* Chap. 4, 168–205. New York and London: Academic. 417 pp.

169. Skaife, S. H. 1953. *African Insect Life.* London: Longmans Green. 387 pp.

170. Smith, H. M. 1935. Synchronous flashing of fireflies. *Science* 82:151–52

171. Tiemann, D. L. 1967. Observations on the natural history of the western banded glowworm *Zarhipis integripennis* (LeConte), *Proc. Calif. Acad. Sci.* 35:235–64

172. Tiemann, D. L. 1970. Nature's toy train, the railroad worm. *Nat. Geogr.* 138(1):56–67

173. Walker, T. J. 1963. The taxonomy and calling songs of United States tree crickets. II. The nigricornis group of the genus *Oecanthus. Ann. Entomol. Soc. Am.* 56:772–89

174. Waller, R. 1685. Observations of the *Cicindela volans,* or flying glowworm. *Phil. Trans. Roy. Soc.* 15:841–45

175. Wenzel, H. W. 1896. Notes on Lampyridae, with the description of a female and larva. *Entomol. News* 294–96

176. Williams, G. C. 1966. *Adaptation and Natural Selection.* Princeton: Princeton Univ. Press. 307 pp.

177. Wynne-Edwards, V. C. 1962. *Animal Dispersion in Relation to Social Behavior.* New York: Hafner Publ. Co. 653 pp.

SORPTIVE DUSTS FOR PEST CONTROL   6005

WALTER EBELING

Department of Agricultural Sciences (Entomology),
University of California, Los Angeles, California

The insecticidal effect of inert dusts such as road dust and powdered clay has long been known by civilized peoples as well as by isolated primitive tribes. Mammals and birds taking "dust baths" get the dust well distributed in their fur or feathers for an apparently instinctive protection against ectoparasites. Road dust drifting into orchards and fields can upset the balance of insect pests and their natural enemies. The hymenopterous parasites in particular are more susceptible than most of their insect hosts to the insecticidal action of inert dusts. In insectaries, tubes used for collecting parasites are well cleaned and stored in plastic bags to prevent contamination by dust. Even those insects not killed by dust clinging to their bodies and wings "are weakened and engage in continuous cleaning activity to the exclusion of normal mating and oviposition" (48). As might be expected, many insects are repelled by dusts. Cockroaches are a conspicuous example (36, 47). Drywood termites will not bore into wood covered with a barely perceptible film of sorptive dust and they soon pick up a lethal quantity (40).

The small size of the insect body and the long, slender appendages result in a great surface area and consequently a large evaporative surface, per unit of volume. Insects and other arthropods are protected from a lethal rate of desiccation by a superficial lipid water barrier averaging only about 0.25 μ in thickness (8). In view of the ease with which this protective film can be absorbed or otherwise disrupted, it is remarkable that relatively little effort has been directed toward exploring the economic potential of this obvious method of attack against insect pests.

The purpose of this paper is to discuss the epicuticular lipid as a water barrier, to explain the physicochemistry of its removal by means of sorptive dusts, and to discuss the current uses and the outlook for these dusts in pest control.

THE EPICUTICLE AS A WATER BARRIER

In the insect cuticle a lipid film covers a thin lipoprotein cuticulin which in turn covers a chitin-protein procuticle. The surface of the cuticulin, initially hydrophobic, becomes hydrophilic when tanned by quinones originating at least in part from the oxidation of "polyphenols" issuing from pore canals arising from epidermal cells (117). In some insect species a tanning-hormone secretion is under nervous control and tanning can be stopped by

123

severing the nerve cord (49, 85). The cuticulin, "polyphenol layer," lipid, and usually a shellac-like material called "cement" (9, 117) or "tectocuticle" (96) make up what is known as the epicuticle.

Composition and organization of the lipid layer.—Epicuticular lipid is generally a solid wax, the mobile "grease" of the cockroach being a rarity among insects. The lipid consists of hydrocarbons, wax acids, esters, alcohols, diols, and sometimes aldehydes, phospholipids, and other minor constituents, but lipids vary greatly among different species in the relative percentages of these compounds (5, 15, 16, 21, 23, 51, 111). Bursell & Clements (21) point out that the long-chain alcohols appear to predominate in the hard cuticle waxes and the proportion of hydrocarbon is high in the soft waxes and greases.

Alexander et al (1) applied beeswax, which consists mainly of molecules with hydrophilic and lipophilic ends (111), on a very thin membrane of celluloid (hydrophilic) resting on water. Although they applied successive layers of wax to a total thickness of 50 μ, maximum impermeability to water was obtained with the first 0.02 μ, calculated to be the equivalent of 5 monolayers of C_{30} hydrocarbon chains if they were vertically oriented. The rest of the wax layer made no further detectable contribution to impermeability. In contrast, as up to 50 μ of paraffin wax was gradually added to the celluloid substrate, impermeability continued to increase at an exponential rate. They concluded that the polar groups in the celluloid substrate probably caused orientation of a very thin layer of beeswax molecules above the hydrophilic substrate, but that the paraffin-wax molecules, having no polar groups, crystallized at random and possessed no special impermeability. Beament (8), by means of a biophysical model employing a butterfly wing as the water-permeable membrane, obtained results similar to those of Alexander et al. Further evidence of a special water barrier of organized lipid was obtained in various ways by several investigators (8, 10–14, 72–74, 76, 90, 115, 118).

Beament (14) now believes the organized lipid consists of "tilt-packed" long-chain molecules with alkyl ends outward, and with the molecules inclined at about 24.5°. He and other investigators formerly considered them to be vertically oriented. Judging from the zig-zag configuration of the carbon atoms, and with the polar groups larger in cross section than the chains, the molecules would provide the closest contact of chains and greatest impermeability to water when inclined at 24.5°. Beament (14) now believes that the sudden and marked increase in water loss of insects at a definite "critical temperature" for each species (30°C for the American cockroach) is caused, not by a reorientation of the organized lipid water barrier as formerly believed, but by a change in the position of the molecules so that they occupy a *mean* vertical position, resulting in a much greater space for the escape of water from the water-bearing cuticulin substrate. On the other hand, Locke (74) suggests that the lipid monolayer on the cuticulin might be composed of "liquid crystals," and that phase changes in the liquid-crystalline systems could explain the sudden loss of water at "critical temperatures."

Beament (12) washed off 95% of the grease of cockroaches with water, leaving what he calculated to be a monolayer, yet this resulted in only a threefold increase in permeability to water instead of the expected twentyfold increase. Locke (72) observed that only the very thin oriented lipid layer of the insect species he investigated remained in place in sections prepared for the electron microscope, while the relatively thick unorganized lipid and cement layers were lost. Olson & O'Brien (90) were able to distinguish between the lipid removed from the American cockroach, *Periplaneta americana,* by 1 ml of water rinse (the "A" layer) and the lipid that was subsequently removed by 1 ml acetone rinse (the "B" layer) and concluded that the "B" layer, with the alkyl ends of the molecules directed outward, was particularly hydrophobic.

The electron microscope may have given us new insight into the organization of the water barrier at the cuticulin/lipid interface. Using this instrument, Locke (73, 74) and Gluud (52) found filamentous "wax canals" (Fig. 1), 60–130 Å in diameter, in the cuticle and penetrating the cuticulin. Wax canals have been noted in the pore canals of *Rhodnius, Galleria,* and *Tenebrio* and in various species of Hemiptera (52). Pore canals end under or in the "dense layer" of the cuticulin. From that point the wax canals branch out from the pore canals to the surface of the cuticulin and their contents are continuous with the oriented wax layer on that structure (52, 73, 74). Wax canals provide a route to the surface for liquid waxes similar in appearance to the liquid crystalline phases of lipid-water systems. Locke believes that upon reaching the surface of the cuticulin, lipid-water liquid crystals spread out to cover it with a monolayer which is not static, but is being continuously "moved, lost, and replaced."

An insect's resistance to water loss is not entirely passive. Desiccation via the cuticle can be increased by topical application of insecticides (35, 40, 59, 63, 83, 99). Arthropods killed by some toxicants remain turgid and retain a lifelike appearance while the same species killed by certain other toxicants become completely shrivelled (35). Ingram (59) obtained water loss by abdominal injection or topical application of pyrethrins and concluded that the loss was caused by a neurotoxic action of the pyrethrins on the secretory activity by epidermal cells. This conclusion is corroborated by the observation that the great difference in water content of blood and cuticle of insects could not be maintained passively and indicates that a "water pump" is located in the epidermis (119, 120).

Topically applied petroleum oil causes water loss probably by disrupting the epicuticular lipid (114). Rate of water loss can be greatly increased by adding surface-active solutes to the oil (40). Rate of penetration of aqueous solutions through insect wax increases with increasing pH, as the result of saponification of acids and esters following the hydrolysis of ester linkages

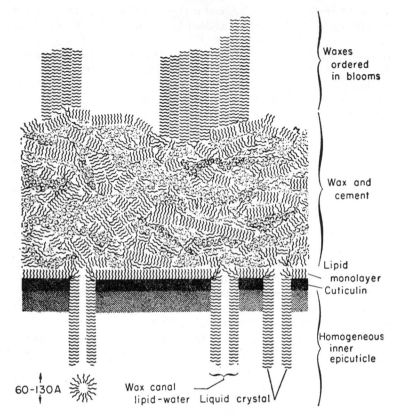

FIGURE 1. Structure of the cuticle of those insects possessing all cuticular structures and constituents. From Locke (74).

(34). This was especially well illustrated with films of beeswax in which, according to Warth (111), acids and esters constitute about 85% of the total wax.

Molecular mobility at the cuticulin/lipid interface.—While there is general agreement that a relatively thin layer of lipid, resting on the cuticulin, provides the principal water barrier of the epicuticle, opinions vary as to its organization. Biophysical models provide useful approximations to biological systems but sometimes fail to explain what is actually observed in the living, or even in the dead organism. For example, Beament found that an oriented and possibly fully compressed monolayer of cockroach grease on butterfly-wing membrane could not be removed by alumina dust, based on its inability to increase water permeability significantly. He concluded that adsorption by inert dusts "cannot overcome the orientational bonds of a wax layer deposited on a membrane." Alumina is a relatively ineffective lipid adsorbent, but Wigglesworth (115) found that when sprinkled on cockroaches that had been killed, to eliminate the possibility of abrasion through the insect's movements, the powder caused a rapid water loss (46.8% in 48 hr compared with 32.0% when the insects were rubbed with the alumina). One must conclude that the constitutents of the water barrier in the cuticle of the cockroach are much more amenable to adsorption than an oriented and compressed monolayer of cockroach grease on butterfly-wing membrane.

The writer's interest in what transpires at the cuticulin/lipid interface derives principally from his interest in the mechanism of water loss through adsorption. A water barrier as conceived by Locke, comprising mobile "liquid crystals," would be readily removed by adsorption. In fact, he suggests that the influence of a sorptive dust on the surface of the epicuticular wax and porous cement would be expected to extend down into the wax canals (Fig. 1), withdrawing the lipid-water liquid crystals. These would be replaced by water, facilitating a lethal rate of water loss (74).

On the other hand, let us presume that the classic view of the water barrier is correct, namely, that it consists of a monolayer or more of packed, oriented, and tilted lipid molecules. The descending order of absorption of the common wax constituents in protein is acid > alcohol > ester > hydrocarbon, so the water barrier is likely to consist of wax acids. Alexander et al (1) found acid obtained from beeswax to be an effective water barrier (about as effective as stearic acid) when applied to a very thin celluloid membrane situated on water, but alcohol obtained from beeswax was even more effective. Neither was as effective as whole beeswax. Without discounting the probable complexity of the cuticulin surface, the likely variability between insect species, and the paucity of detailed information on cuticle chemistry, it is nevertheless reasonable to assume a linkage of lipid molecules and those of the proteinaceous substrate via hydrogen bonds. Hydrogen bonds provide a relatively weak affinity between lipids and their tanned protein substrate. The facility with which lipid molecules might be "moved, lost, and replaced" as postulated by Locke (73) is readily visualized. A relatively small increase in intermolecular spacing can result in a disproportionately great weakening of the lipid barrier, for intermolecular forces of attraction (van der Walls forces) vary inversely as the sixth power of the distance between molecules.

Current theories on the nature of the lipid/cuticulin interface should not be accepted without reservations, for they are all speculative and, in fact, the theory of a monolayer of lipid as the principal water barrier is not accepted by all authorities. For example, Gilby & Cox (51) argued that there are physicochemical considerations that militate against the "monolayer barrier hypothesis" and currently reiterate this view (correspondence). They suggest the possibility of other types of barriers, but point to the lack of experimental evidence. However, a discrete and extremely thin barrier, whatever its composition, has been identified in electron micrographs (52, 73, 74).

The above theories concerning the physicochemistry of lipid/tanned protein interface are of great academic interest, but may not be relevant to the problem of how sorptive dusts cause a lethal rate of water loss. Now that the presence of wax canals has been established (52, 73, 74), it is evident that a considerable portion of the lipid barrier rests over the mouths of the myriads of wax canals, 60 to 130 Å in diameter, that penetrate the cuticulin. These portions of lipid are probably much more readily disrupted than that which is bound to tanned protein. An accentuated withdrawal of "middle-phase filaments" (74) of the lipid-water liquid crystals in the wax canals, through forces of attraction initiated by a sorptive powder, is readily visualized. The lipid-water liquid crystals would be replaced by water and, in accordance with the "pinhole effect" (20), allow for a very rapid movement of water out of the cuticle.

Rate of lipid replacement after it is removed.—Since the principal water barrier on the cuticulin is conceived to be continuous with the wax canals, one might expect that if the wax molecules were removed from this barrier they would normally be rapidly replaced, thus preventing excessive desiccation. At least with some insect species, cuticular transport of wax is rapid. This was indicated by an experiment in which the writer (unpublished data) fed Purina Dog Chow saturated with C^{14}-labeled sucrose to adult German cockroaches (*Blattella germanica* L.). Within 24 hr the radioactive carbon atoms were absorbed with epicuticular wax, by means of silica aerogel. The silica was washed off and dried and the absorbed radioactive atoms were detected in great abundance with a Geiger counter. The epicuticular lipid of cockroaches consists principally of hydrocarbons (51). Nelson (88) found that within $3\frac{1}{2}$ hr C^{14}-labeled acetate and palmitate injected into the abdomens of adult American cockroaches was synthesized to lipid, principally hydrocarbons, by the epidermal cells of the integument. These cells may be under the control of one or more hormones, for intermolt wax secretion in larvae of *Calpodes* was found to be under continuous control of a hormone from the head (75). Despite the remarkable speed with which it is synthesized, lipid in the principal water barrier is not replaced rapidly enough to prevent a lethal rate of water loss if the epicuticle is covered with a highly sorptive dust.

Wigglesworth (115) found that after being rubbed with alumina, *Rhodnius* nymphs lost 31.1% of their weight when left in dry air for 24 hr at 30°C, but lost only 5.2% when they were rubbed with alumina, kept in moist air at 25°C for 24 hr and then placed in dry air. Evidently much of the damage to the epicuticular water barrier was repaired in 24 hr. Ebeling & Wagner (40) observed that when full-grown drywood termite nymphs, *Incistermes minor*, were dusted with clay and placed in covered stender dishes on moist blotter paper, they soon lost the dust particles and survived in the water-saturated atmosphere, otherwise they were soon desiccated because of the absorption of a portion of the lipid water barrier by the dust. Unlike *Rhodnius* nymphs, the termites had to remain in the moist environment for three days before there was sufficient repair of their epicuticular water barriers to allow them all to survive after the covers of the stender dishes were removed.

In larvae of *Sarcophaga*, the lipid appears to be distributed throughout the cuticle (31, 97) and to bring about a lethal rate of water loss by the usual experimental method of cuticular abrasion, the entire epicuticle must be abraded in depth.

Sorption Versus Abrasion for Removal of the Lipid Water Barrier

Early theories that finely divided powders kill insects by (*a*) blocking the spiracles, (*b*) becoming lodged between segments of the cuticle and thereby increasing water loss, (*c*) directly absorbing water from the cuticle like a blotter, or (*d*) being ingested, were proved to be erroneous (2, 19, 24, 57, 66). Yet, for three decades an equally erroneous belief persisted in the minds of most entomologists, viz, that with most insects abrasion was the only mechanism by which finely divided powders removed epicuticular wax to the extent of causing desiccation. Investigators were not able to bring about water loss of motionless or dead[1] insects with inert dusts, with the exception of the cockroach (115), at least not with the powders they tested and during the periods allotted for a test (2, 8, 115).

Some investigators found that the insecticidal efficacy of dusts against certain adult grain-infesting beetles crawling among kernels of dusted wheat increased with increasing hardness and abrasiveness of the particles (19, 30, 50, 60, 64). Alexander et al (2) observed that, when various dusts were mixed with wheat, hard abrasive dusts such as diamond, carborundum, and wet-ground silica were the most effective against *Sitophilus granarius*, in either "dish tests" or when the beetles were placed in the dusted wheat. But against *Sitophilus oryzae* the adsorbant dusts such as activated gas mask charcoal and alumina were approximately as effective as the abrasive dusts and against *Tribolium castaneum* the adsorbent dusts killed much more rapidly. Against the larvae of *T. castaneum*, *Tenebrio molitor*, and *Anagasta kühniella*, carborundum was not effective, but the adsorbents, alumina and a colloidal silica, were highly effective.

[1] Ebeling & Wagner (40) found that dead drywood termite nymphs dusted with a montmorillonite clay and decolorizing carbon lost only 0.5% of their body weight in 3 hr, compared with no detectable loss for undusted termites. During the following 14 hr the dusted dead termites lost 15.2% and 12.9%, respectively, of their body weight from desiccation and the undusted termites lost 1.1%. Nair (87) in an experiment made in dry air and at 27.2°C found that dead adult beetles (*Tribolium*, *Rhizopertha*, and *Bruchus*) that had been dusted with magnesite dust lost weight at an average rate that during a 20-hr period was 24.2% as rapid as for the dusted live insects.

The team of British investigators (P. Alexander, H. V. A. Briscoe, and J. A. Kitchener) who published the above findings possessed an insight into the physicochemical mechanisms indicated by their research, but unfortunately they proposed as a theory something that could have been easily demonstrated as fact. Kitchener et al (64) state: "It is possible that the epicuticle fat film may be preferentially attracted to the crystalline forces at the surface of a solid particle, and adheres and orientates itself on the crystal rather than on the relatively structureless surface of the cuticle. The continuity of the film might then be interrupted over submicroscopic areas. The fat film might even spread over the particle by surface migration (since there is much evidence of the mobility of adsorbed layers) and this would certainly account for the high effectiveness of finely powdered activated charcoal, which has a large available surface." The theoretical basis for productive scientific research on sorptive dusts was indicated in the above statement, but for many years it led to no sustained interest. Yet, the hard wax occurring on the epicuticle of most insect species (not a "fat" as indicated in the above quotation) can be absorbed by suitable sorptive dusts. The absorbed wax can be easily seen with the aid of a microscope. Absorption of insect wax at ordinary temperatures depends entirely on the physical properties of the sorptive dust.

Probably the early investigators would have become aware of the superiority of sorption as a mechanism for the removal of epicuticular wax if they had worked with a wider variety of dusts, including silica aerogels, activated charcoal, commercial bleaching earths, and acid-activated clays. The usual "adsorbent" dusts used by these investigators were Almicide (aluminum oxide or alumina) and Neosyl (amorphous precipitated silica). The writer dusted the surface of a layer of dyed beeswax with finely divided alumina, as used for chromatography. In two days no appreciable absorption of beeswax could be detected. However, Neosyl was visibly sorptive and there was complete saturation of films of certain silica aerogels and of acid-activated kaolin prepared by J. S. Venugopal.

Among nonsorptive dusts, insecticidal efficacy depends largely on abrasiveness, but to be effective the dust must be very finely divided. Parkin (91) treated grain with 1 percent by weight quartz dust ranging in mean particle diameter from 0.5 to 15 μ and found that 1.8 μ dust was the most effective against *Sitophilus granarius*. Among sorptive (generally nonabrasive) dusts, insecticidal efficacy is closely correlated with specific surface, provided the pores are sufficiently large to admit wax molecules (33, 84). Almicide (specific surface 2.95 m^2/g) and Neosyl (5.3 m^2/g) (30) compare unfavorably with good wax adsorbents like the silica aerogels SG-67 and SG-68 (300 m^2/g). Relatively nonabrasive clays (2.5 on Moh's scale for hardness), initially less insecticidal than finely ground flint, carborundum, quartz, etc. (7.0 on Moh's scale) against grain-infesting beetles, can be made far superior to these hard, abrasive dusts by acid and heat activation, by means of which specific surface and sorptivity are greatly increased (80, 82) although the abrasive properties of the clays are not changed.

The habitat of grain-infesting beetles makes them particularly susceptible to the action of abrasive powders with which the kernels of grain are coated. By crawling among dusted kernels, grain-infesting beetles cannot avoid some degree of abrasion (116), yet under such conditions the most effective dust desiccants have been nonabrasive, highly sorptive powders, particularly the silica aerogels (61) and acid- and heat-activated clays (66, 79–82). Diatomaceous earth, a moderately effective dust desiccant against most grain-infesting beetles, is both abrasive and sorptive. It has been found to be inferior to nonabrasive amorphous silicas for protection of beans (29, 92), shelled corn (93, 95), and wheat (70a) against various species of grain-infesting beetles, even when present in as much as sevenfold greater quantity.

The abrasion concept had become so firmly established that most investigators assumed abrasion to be the only mode of action of inert dusts in the treatment of grain- or pulse-infesting beetles even though in some cases (28, 58) a consideration of the physical properties of the dusts that were used would have revealed that the most effective dusts were the least abrasive and most sorptive. Although some striking examples of the efficiency of sorption as a mechanism for the removal of insect wax by finely divided powders were recognized by some investigators (2, 33, 39, 53, 56, 80, 81, 87), these did not appear to change the generally prevailing belief that abrasion was necessary.

Prolonged residual efficacy of undisturbed insecticide-dust deposits.— Blocks of wood were treated with five chlorinated hydrocarbon insecticides (lindane, chlordane, dieldrin, DDT, and toxaphene) applied (*a*) as solutions, emulsions, mists, or vapors which deposited undiluted toxicants, or (*b*) as dilute dusts, which deposited 95 to 98% of inert diluent along with the toxicant (40, 109). After being suspended in an attic for 17 months, the treated blocks of wood were taken to the laboratory and were exposed to full-grown nymphs of the drywood termite, *Incistermes minor.* Only the dusts were insecticidal, indicating that their insecticidal action resulted from the diluents rather than the toxicants. The toxicants had lost their insecticidal activity. Further tests showed that the efficacy of the diluents was directly proportional to their sorptiveness. Termites crawling over a barely visible film of highly sorptive dust became desiccated in a few hours and never attempted to bore into the wood as they did in untreated checks. Certain montmorillonite and attapulgite clays and silica aerogels were particularly effective. Of nine chlorinated hydrocarbon or organophorphorus insecticides (as 12% to 75% wettable powders) that were compared with two fluorinated silica aerogels (SG-67 and SG-77) against the drywood termite, German cockroach, and *Drosophila,* some caused a more rapid knockdown, but none resulted in as rapid death. Likewise Greening (54) reports that M. J. Watt obtained a more rapid kill of the American cockroach *Periplaneta americana* with SG-67 than with diazinon, malathion, and BHC dusts.

Regardless of the period required to kill an insect species, death occurred when 28 to 35% of the body weight (about 60% of the water content) was lost (40). A lethal rate of water loss can be caused by sorptive powders even at the highest humidities. When drywood termite nymphs were treated with sorptive powders, the loss of body weight at 100% RH was 56% of what it was at 20% RH (40). With cockroaches, desiccation caused by sorptive powders is sometimes almost as rapid at 100% as at 25% RH (38). The percent mortality of adult *Tribolium castaneum* beetles after 30 hr in a mixture of 0.05% acid-activated kaolin in wheat at 27.8°C and at different relative humidities was as follows: 25% RH, 92.3; 50% RH, 73.0; 80% RH, 67.6; approx. 0% RH, 41.2; and 100% RH, 2.8. Only at 100% RH was mortality less than in practically dry air (38).

PHYSICOCHEMICAL MECHANISMS FOR THE SORPTION OF EPICUTICULAR LIPID

Maximum progress in the search for dust desiccants as well as in the treatment of the dusts to increase their insecticidal activity, as in the acid-activation of clays or the synthesis of new products with optimum physical properties, depends on an understanding of the physicochemical mechanisms involved in the sorption of epicuticular lipid, particularly the hard wax of most insect species.

A photomicrographic study of wax sorption.—A photomicrographic record of beeswax sorption (adsorption and absorption) by a silica aerogel (SG-67) from layers of wax on microscope slides was obtained by Ebeling (33). The dust was allowed to settle on the slides from a cloud formed in an inverted 1000-ml beaker. The sorption of beeswax was most easily observed when the wax was dyed with Oil-Soluble Red, but sorption of undyed wax was also readily observed with the aid of a microscope. The translucent silica aerogel aggregates, when saturated with wax, changed to shiny white and became crystalline in appearance. Many of the smaller aggregates became wax-impregnated within 15 min at 22°C; within 2 hr the wax had migrated upward 10 to 20 μ into the silica aerogel layer and eventually to a distance of 30 to 40 μ. The rate of migration of beeswax into the silica aerogel layer decreased with increasing humidity, but proceeded slowly even at 100% RH, when the silica was visibly moist. The sorption of wax from the bodies of fleas, termites, and beetles that had been dyed with Oil-Soluble Red powder (with the excess powder removed) was observed with the aid of a microscope and was photomicrographically recorded. Wax was not adsorbed to any detectable extent by powders that were known to be relatively ineffective in killing insects by desiccation, such as pyrophyllite, talc, sulfur, or walnut-shell flour.

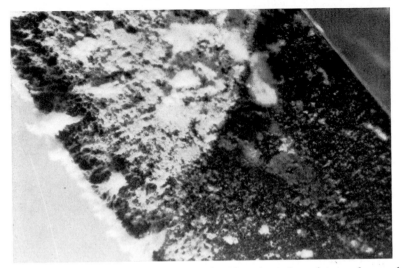

FIGURE 2. Absorption of lipid from the dorsal surface of an elytron of a carabid beetle by silica aerogel. The aggregates of silica are translucent and invisible against the black background until they become saturated with lipid. (Original.)

Ebeling & Reierson (38) observed that the elytra of carabid beetles, *Pristonychus complanatus*, lend themselves particularly well for the photography of wax sorption by certain dusts. A silica aerogel (Dri-die® 67) was allowed to settle on either the dorsal or ventral surfaces of the elytra of these beetles in the manner described in the previous paragraph. The silica film was somewhat translucent and could hardly be distinguished from the black substrate. Upon absorbing wax it turned white. Figure 2 shows the strong contrast in appearance on the dorsal surface of a beetle elytron, between areas in which absorption has taken place and those in which it has not. Absorption begins in certain areas and spreads out from these foci as can be seen on the left half of Figure 2. Or it may begin at the anterior end of the elytron and spread posteriorly, sometimes forming a sharp line of demarcation as it advances over the elytron. That tendency is shown to some extent in Figure 2 where the advancing front of wax-saturated silica is moving from left to right.

The aggregates of wax-saturated silica, clearly defined on the margin of the elytron in Figure 2, were as much as 0.16 mm (160 μ) in depth. They did not become as heavily saturated with wax with a concomitant change in form, as was observed with silica films on beeswax (33). The delicate, lacy structure of the silica aggregates was preserved. It appears that the large quantity of lipid absorbed by the silica film could not have been supplied solely from the insect epicuticle, for such large quantities would not be available. Apparently large quantities of lipid are drawn from deep within the cuticle.

The ease with which lipid can be withdrawn from *P. complanatus* would indicate that this insect might be easily killed by means of dust desiccants and this proved to be the case in an experiment. Although it weighs about 31 times more than *Tribolium confusum,* it was killed in an average of 6 hr, compared with 17 hr for *T. confusum,* when both were placed in petri dishes containing 2 cc of the silica aerogel Dri-die 67® (38).

Using a mixture of paraffin wax and C^{14}-labeled stearic acid, Rideal & Tadayon (98) found that stearic acid molecules can be "overturned" and anchored to new surfaces (various metals) and that the molecules can also migrate along the metal or mineral surfaces for considerable distances. For example, diffusion of C^{14}-labeled stearic acid over a surface of mica proceeded at the rate of 5.3×10^{-8} cm²/sec at 35°C. After the first monolayer was deposited, diffusion of stearic acid molecules over this monolayer was more rapid than over the original surface of mica. This may explain the saturation of the silica aerogel film by beeswax as observed by Ebeling (33) and the obviously large quantities of lipid observable in Figure 2.

The great distance the wax molecules migrated up into the silica film is a striking demonstration of the remarkable mobility of adsorbed films. Such films are too often regarded by biologists as static systems.

Wax sorption with successive exposures to abrasive and desiccant dusts. —Ebeling (33) immersed plastic vials with an outside surface area of about 380 cm², and coated with layers of beeswax about 65 μ thick, in jars containing fluorinated and unfluorinated silica aerogels, montmorillonite clay, pyrophyllite clay, and carborundum. These jars were agitated mechanically for periods of 1 hr. After each of these periods the vials and untreated controls were lightly swabbed with cotton under a stream of cold water and were air dried and weighed. Curves depicting the rate of weight loss showed that the removal of wax by powders that act primarily as absorbents was an exponential function, whereas removal of wax by carborundum (primarily abrasive) was a straight-line function. Weight losses caused by sorptive powders exponentially diminished, with successive exposures, until no further sorption was possible. On the other hand, carborundum could remove wax with successive exposures until the wax had completely disappeared from the vials. In the first 10 to 15 exposures, however, weight loss caused by the most efficient sorptive powders was much greater than with carborundum.

Experiments were also made with plastic vials covered with beeswax films only 1 μ thick. These coated vials were placed in jars containing the silica aerogel SG-67, for successive periods of 10 min, 1 hr, or 24 hr and were not agitated. After each period of immersion the powder was washed off and the weight losses were compared with the controls. Again, the exponential rate of net weight loss was noted. The maximum total loss ranged from 20 to 46%, depending on the periods of exposure to the sorptive dusts. Maximum loss was obtained with long, continuous exposures, such as 24-hr exposures. With repeated exposures of 1 hr, 30% of the wax was absorbed in 24 hr, but about 16% of it was absorbed in 2 hr.

When absorption of wax from a given layer is no longer possible, the remaining wax may be dissolved in CCl_4 and reapplied to the plastic vials. Presumably a redistribution of wax molecules takes place and something approaching the original ratio of amorphous and crystalline material is again formed and wax can be absorbed from the reorganized layer.

X-ray crystallography, according to the powder method, revealed that beeswax contains both crystalline and amorphous material (33). Therefore, it might be expected that sorption of beeswax with sorptive powders could be selective, removing amorphous material and leaving behind those molecules of any compound that are mutually bound in the form of crystallites. When surface molecules are adsorbed, those below take their place and they in turn are replaced by molecules at a still lower level. In this manner minute channels are formed in the otherwise predominantly crystalline matrix. In the case of insects, such channels might allow for the escape of water from the water-bearing cuticulin substrate. Judging from the rapidity with which some insects such as ants and *Drosophila* are killed by sorptive powders (40), only a small portion of the epicuticular wax needs to be removed to create a sufficient number of channels to allow for a lethal rate of water loss.

Abrasion changes the appearance of a wax film from shiny to dull. Sorption does not change its appearance, but decreases its opacity as determined by means of a light meter (33). This is further evidence that the structure of a layer of insect wax is not changed by sorptive powders except for the formation of minute channels that allow for the passage of light (and water).

The physical properties of 16 powders used in the above investigation ranged from abrasive and feebly sorptive carborundums to nonabrasive and highly sorptive silica aerogels. The quantity of beeswax sorbed from a wax-coated vial (and its insecticidal efficacy against four species of insects) was inversely related to the abrasive index[2] of the powder, whether the vial was in motion or motionless when in contact with the powder; on the other hand, with the exception of powders known as "Molecular Sieves," wax sorption and insecticidal efficacy were directly related to the specific surface (surface area per unit mass) of the powder. The two types of Molecular Sieves used in the experiments had a very high specific surface, but had small pores with uniform diameters of 5 or 13 Å. The pores were not suffi-

---

[2] The "abrasive index" for each of the 16 powders was determined by rubbing each powder for 1 min on each side of a sheet of vinyl plastic, then passing light through the abraded area and recording its intensity by means of an exposure meter.

ciently large to admit the large molecules composing the solid wax of most insects, although the Molecular Sieves with the greater pore diameter were fairly effective against cockroaches, which have a protective film of mobile "grease" instead of hard wax. To be efficient in absorbing solid wax, particles should have pore diameters of at least 20 Å. Another drawback with Molecular Sieves is that they absorb much water from the atmosphere, which interferes with lipid sorption.

Melichar & Willomitzer (84) referred to the above investigation as having provided an explanation for the physicochemical mechanisms of wax removal by finely divided powders, greatly facilitating their choice of useful dust desiccants. They found that the efficacy of 17 silica dioxides and 4 silicates against the chicken mite, *Dermanyssus gallinae,* was highly significantly correlated ($r_s = 0.957$) with their specific surfaces, which ranged from 2 to 200 m²/g (Fig. 3).

A possible direction of future research on sorptive powders is indicated by the performance of a silica aerogel (AL-1), which caused a complete knockdown of adult male German cockroaches in 2 min compared with 39 min for Dri-die (33). Like Molecular Sieves, AL-1 has an enormous specific surface (700 m²/g) but the pores have a greater average diameter (22 Å). The powder was spectacularly effective against cockroaches, which possess a protective barrier of mobile grease, but was no better than Dri-die against insects protected against desiccation by a film of hard wax. An additional disadvantage is that AL-1 absorbs too much water from the air and gradually declines in insecticidal efficacy after it is removed from its original air-tight container. The development of a silica aerogel with somewhat larger pores (possibly ca 30–40 Å) and with less water sorptivity, might result in a substantial "breakthrough" in the search for superior dust desiccants.

The cement layer as a protection against sorptive dusts.—If a cement layer covers epicuticular lipid it may retard the action of sorptive dusts, but it evidently cannot completely protect the dusted insect from desiccation, for the efficient wax-adsorbent inert dusts have invariably resulted in the death of the many insect species that have been investigated by the writer. Wax penetrates through the cement layer of some species of insects to form a "bloom" on its surface. Therefore, it is not likely that this layer would offer more than moderate impedance to wax sorption. Collins (26) found a drywood termite, *Incistermes minor,* to have a "heavy" cement layer, yet this species is highly susceptible to the desiccating effect of sorptive dusts (40).

Cement is said to be thickest on the cuticles of beetles (73). Nair (87) investigated four species of rice- or grain-infesting beetles: *Sitophilus oryzae, Tribolium castaneum, Rhizopertha dominica,* and *Bruchus chinensis.*

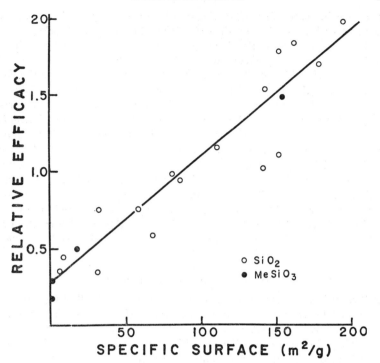

FIGURE 3. Relation of the efficacy of silicas and silicates to specific surface. Redrawn from Melichar & Willomitzer (84).

He found that *Sitophilus* had its wax layer protected by a cement layer; *Tribolium* had no cement layer, but a very hard wax; *Rhizopertha* had a softer wax than *Tribolium;* and *Bruchus* had a very soft wax. The rate of desiccation caused by three sorptive dusts was in the descending order *Bruchus* > *Rhizopertha* > *Tribolium* > *Sitophilus,* indicating that the cement layer on *Sitophilus* gave some protection against desiccation.

The most effective dust used by Nair was magnesite. Yet, the weight loss of live *Sitophilus* beetles in 20 hr when placed in magnesite dust in dry air over $CaCl_2$ at 26.7°C, was only 5.4%. Under identical conditions but using the silica aerogel SG-67 as the dust desiccant, the writer found that in the 15 hr required for 100% mortality, *Sitophilus oryzae* adults lost 24.4% of their body weight compared with 8.2% in the controls, and in 24 hr they lost 37.9% compared with 12.5% in the controls. Nair concluded that beetles without a cement layer could be desiccated with sorptive dusts but that *Sitophilus* required "highly abrasive and adherent dusts." However, the writer's experience indicates that *Sitophilus* can lose water relatively rapidly when in contact with a nonabrasive dust if it is an efficient wax absorbent.

The susceptibility of the four species to the dusts used by Nair (87), based on time-mortality curves, was in the descending order *Bruchus* > *Rhizopertha* > *Sitophilus* > *Tribolium*. *Sitophilus*, the species with the cement layer, was not the most resistant to the insecticidal action of the dusts but this may have been in part, as suggested by Nair, because water comprises a 24% greater porportion of the body weight of *Tribolium* than of *Sitophilus*. However, it does not appear to the writer that this alone could account for the fact that it required only three days for a 100% kill of *Sitophilus*, compared with five days for *Tribolium*. A possible explanation is suggested by an experiment made by Carlson & Ball (22). They could not account for the mortality of *Sitophilus oryzae* from diatomaceous earth on the basis of desiccation, for weight loss was insignificant. They suggested "the possibility of an additional but unkown lethal factor in the dust." For seven other beetle species used in their experiments, they attributed mortality to "excessive loss of water following epicuticular damage."

FLUORINATED SILICA AEROGELS

Silica gels are adsorbents, dehydrating agents, and flatting agents that are formed by reacting sodium silicate and sulfuric acid. They are amorphous, nonabrasive, chemically inert, and are not usually injurious to humans. They do not cause silicosis. Silica gels of smallest particle size and lowest bulk density and consequently of greatest porosity, are called aerogels. SG-68 is an extremely light, fluffy silica aerogel with a specific surface of 300 m^2/g and can absorb 300% of its weight of linseed oil. The silica aergoel that has been designated as SG-67 for experimental purposes but which is known to the insecticide trade as Dri-die 67® is similar to SG-68 except that it contains 4.7% ammonium fluosilicate present in somewhat less than a continuous monolayer (121). SG-67 is produced by adding ammonium fluosilicate to the silica "hydrogel" during the process of manufacture in order to decrease the caking characteristic of the gel to make it easier to grind to a powder, and improve it as a flatting agent. SG-68 and SG-67 owe their outstanding insecticidal efficacy not only to their large specific surface and adequate pore diameter (115 Å), but also their low sorptivity for water —none at 40% RH, 5% of their weight at 80% RH, and 80% of their weight at 100% RH.

The safe use of many thousands of pounds of Dri-die by the pest control industry for more than a decade, as well as no adverse effect among those who manufacture and package the powder for pesticidal and other industrial uses, testifies to its harmlessness to humans as ordinarily used. According to a report of Hazelton Laboratories, Falls Church, Virginia, dated August 5, 1958, the acute oral LD_{50} of Dri-die 67 for male albino rats is greater than 3160 mg/kg of body weight. There was no evidence of systemic toxicity from percutaneous absorption. Mice, rats, and guinea pigs were exposed for 6 hr to Dri-die dust at a concentration of 0.0473 mg/l (19.3 ppm) without evidence of toxicity. Male albino rats were fed 2000, 10,000, and 50,000 ppm (changed to 25,000 beginning on 5th day) of Dri-die with their food for a period of 28 days with no effects at the lower doses. The highest dose caused alteration of body weight and food consumption, apparently as a result of food refusal. Autopsies revealed no gross pathology attributable to Dri-die.

Krishnakumari (65) administered Dri-die 67 suspended in distilled water to the stomachs of male albino rats by means of a catheter attached to a syringe. She found the LD_{50} of Dri-die to be 2000 mg/kg of body weight when the animals were observed for 48 hr. She noted that symptoms were similar to those of fluorine poisoning. An anomalous feature of Krishnakumari's results is that she found no mortality in 48 hr from Drione® at 8000 mg/kg. Drione, beside containing 1% pyrethrins and 10% piperonyl butoxide, would supply 60% more ammonium fluosilicate at 8000 mg/kg then Dri-die at 2000 mg/kg.

Electrostatic charge.—Fluorides, when present as monolayers, greatly improve silica aerogels as dust desiccants by causing particles of the aerogel to have a positive electrostatic charge, thus enhancing their deposition and adherence on most dusted surfaces. When insects crawl over equivalent deposits of SG-68 and SG-67, the latter adheres to their cuticles in much greater quantity. This results in a more rapid mortality of the treated insects. SG-67 is the most insecticidally effective of hundreds of dust desiccants the writer has investigated.

SG-67 retains its positive charge indefinitely when kept in a closed container or when kept in an open container if there is a considerable amount of dust present. A layer of SG-67 in a petri dish (2 cc per dish) retains its charge and original insecticidal efficacy if the dish is covered. If it is not covered, the powder loses its charge within two to three months or longer, depending on humidity and the rate of air movement. After the dust has lost its charge, its adherence to insect cuticle, wood, or other surfaces, as well as its insecticidal efficacy, is no greater than that of the unfluorinated SG-68 (38).

The special efficacy of fluoride monolayers on silica particles at high humidities.—Ebeling & Wagner (40) demonstrated that ammonium and magnesium fluosilicates, existing as partial monolayers on silica aerogel particles, contributed to a degree of toxicity that increased with increasing humidity and thereby partially balanced the adverse effect of reduced mortality rate that would otherwise be expected from decreased water loss. They

TABLE 1. Effectiveness of Dust Desiccants at 4 Relative Humidities Against Alate Drywood Termites. From Ebeling & Wagner (40)

Material	Period (hr) for 100% mortality at relative humidities of			
	20%	60%	90%	100%
SG-68 (silica aerogel)	1.1	1.7	2.7	7.5
SG-67	0.75	0.85	0.96	1.0
Ammonium silicofluoride	2.0	3.6	3.8	4.0
Pyrophyllite+5% chlordane	1.9	2.2	3.0	9.2
Pyrophyllite	2.2	3.5	10.2	88.5

determined the weight loss of drywood termite nymphs in constant contact with various desiccating dusts at 20, 60, 90, and 100% RH. Weight loss decreased with increasing humidity and in 6 hr the average loss caused by two silica aerogels and a montmorillonite clay averaged only 46% as great at 100% RH as at 20% RH. It was noteworthy, however, that desiccation took place even in a water-saturated atmosphere. The period required for the unfluorinated SG-68 to result in a 100% mortality ranged from 1.1 hr at 20% RH to 7.5 hr at 100% RH. The corresponding periods for the fluorinated SG-67 ranged from 0.75 hr to 1.0 hr (Table 1). SG-67 caused only slightly more rapid water loss than SG-68 and this difference was probably not significantly relevant to the difference in rate of mortality. Since the only difference in the two powders is the 4.7% ammonium fluosilicate in SG-67, the inference to be drawn from this experiment is that with increasing humidity the fluoride must have become more insecticidally active. Therefore, there was little change in rate of mortality with SG-67 despite the decrease in rate of desiccation as humidity increased.

The same relationship between SG-68 and SG-67 was shown by Ebeling & Reierson (38) when using the German cockroach as the test insect (Table 2). In this experiment the rate of water loss at 100% RH was 74.3% of what it was at 50% RH and the rate of mortality from SG-68 was decreased at the higher humidity. Whenever a fluoride was present, as a fine powder mixed with SG-68 or as a monolayer in SG-67, death was more rapid in the water-saturated atmosphere despite the decreased rate of water loss. The fluoride was more effective when present as a monolayer.

SG-67 resulted in more rapid desiccation of German cockroaches than SG-68. This was apparently because of its better adherence to the cuticle, which resulted in a more complete coverage of dust. The more rapid water loss when SG-67 was used did not appear to be the result of its more rapid insecticidal action, for within the period of the experiment depicted in Table 2, no difference could be shown in the rate of water loss of live adult male German cockroaches and insects from the same lot that were killed

TABLE 2. EFFECT OF HUMIDITY AND THE ADDITION OF AMMONIUM FLUOSILICATE ON THE RATE OF DESICCATION AND MORTALITY OF GERMAN COCKROACHES IN SILICA AEROGEL POWDER. FROM EBELING & REIERSON (38)

Material	% wt loss in 90 min		50% knockdown in hr	
	50% RH	100% RH	50% RH	100% RH
SG-68 (silica aerogel)	12.8	8.7	5.2	7.0
SG-68+4.7% ammonium fluosilicate (400 mesh)	14.0	10.3	4.6	3.3
SG-67	16.1	12.9	2.4	1.8

with HCN gas. Also, when cockroaches that had been confined to SG-67 or SG-68 for one-half hour were killed with HCN and kept in contact with the dust, they lost water no more rapidly than cockroaches in the same dust that had not been killed.

Ebeling & Wagner (40) suggested that aqueous pathways facilitate the movement of fluoride into the insect's body after a portion of the wax is absorbed by the silica. This might explain the fact that an increase in relative humidity from 20 to 100% increased the period for 100% mortality of drywood termites by only one-third when SG-67 was used (Table 1). The greater availability of dissolved fluoride at higher humidity is most advantageous when a water continuum to the interior is provided by the absorption of epicuticular wax. O'Brien (89) believed this to be an unlikely explanation in view of the fact that the fluorinated silica aerogel particles adsorb epicuticular wax and the fluoride is not apt to be available to the insect because it is bound in wax-encapsulated particles. However, even on a motionless insect, only a small portion of the dust film may be wax-impregnated at least in the early stages of adsorption, and in the case of fluorinated silica aerogels the remainder will contain available fluoride for which access through the epicuticular wax will have been provided by the removal of a portion of the wax by sorption. Moreover, ordinarily an insect will crawl over the dust film and a continuous succession of new particles will contact certain areas of its body surface. The wax available for sorption is soon removed and a water continuum through the cuticle for the transport of dissolved fluoride from wax-free particles is provided.

The ability of fluoride to partly or even completely compensate for the otherwise adverse effect of increased humidity on the insecticidal efficacy of silica aerogels against drywood termites and cockroaches (Tables 1 and 2) does not extend to all arthropod species. For example, Ridgway (100) found that the period required for 100% mortality of scorpions, *Centruroides vittatus*, exposed to Dri-die, ranged from 17.5 hr at 5% RH to 95.5 hr at 100% RH.

ACID-ACTIVATED AND CALCINED CLAYS

Comprehensive investigation of acid-activated and calcined clays as dust desiccants had its origin in the research program initiated by Dr. S. K. Majumder at the Central Food Technological Research Institute in Mysore, India, ultimately involving cooperative work among entomologists, chemists, geologists, and others. Majumder et al (81) found that activated clays, silica aerogels, and coconut-shell carbons were highly insecticidal against *Tribolium castaneum*. The insecticidal efficacy of activated clays was proportional to their oil-bleaching properties (79, 82). Upon storage, freshly prepared activated carbon gradually became less insecticidal, but activated clays retained their initial efficacy.

Investigations at the Institute showed that activated clay could be used effectively for the protection of stored wheat, rice, sorghum, corn, and pulses against insect infestation. Dr. Majumder (personal communication) prefers kaolinic to montmorillonite clay. The latter swells and holds more moisture, yet it adsorbs more lipid than does kaolin at a given relative humidity. However, the spaces between kernels of grain provide microclimates for insects. Because of its higher water content, activated montmorillonite clay results in a higher humidity in these microclimates than does activated kaolin and therefore desiccation of insects proceeds more slowly.

Sulfuric acid is preferred to hydrochloric acid for the activation of clays. Krishnamurthy et al (67) obtained best results with kaolinic clay by first disintegrating it to pass through a 100 mesh sieve and activating it with 10 N sulfuric acid (3 times the weight of clay) at 15 lb pressure for 1 hr. The clay was then washed free of acid and dried at 110°C. The dried clay was again disintegrated to pass through 100 mesh and heated to 410–450°C for 3 hr. For each type of clay the treatment needs to be modified for optimum results.

Through acid activation the spaces in the clay are enlarged because various salts are dissolved and washed away. By controlling acid concentration, time, and pressure, the magnitude of the spaces can be controlled so as to meet the requirements for maximum insecticidal efficacy. Venugopal & Majumder (108) found that changes in physicochemical and optical characteristics of kaolin during acid and heat activation indicated a transformation from kaolinite to H-halloysite and then to meta-H-halloysite. This was confirmed by X-ray diffraction patterns, observations made with the aid of the electron microscope, differential thermal analysis, and solid state reactions. Lipid sorptivity was then optimum for insecticidal activity.

H-halloysite is formed during the first stage of the activation process (temp. 110°C). During the second stage, when the clay is heated to 410–450°C, meta-H-halloysite is formed. If heat is increased to 900°C, there occurs a further transformation to mullite. The pore structure is destroyed and the mineral becomes hard. When ground, it is an abrasive powder, but it is no longer sorptive and is worthless as an insecticide (108).

The difference in the insecticidal efficacy of kaolinite and the products derived from this mineral by acid and heat activation, using adult *Tribolium confusum* as the test insect, is shown below (38).

Mineral	LT_{50} (hours)
Kaolinite	67.0
H-halloysite	8.4
Meta-H-halloysite	6.3
Mullite	89.3

When the powders were blown into inverted 1000-ml beakers and the dust cloud was allowed to settle on glass slides coated with beeswax to form a thin film of powder, H-halloysite and meta-H-halloysite were found to absorb large quantities of wax within 24 hr. Some of the smallest aggregates of kaolinite adsorbed visible quantities of wax but even the smallest aggregates or particles of mullite showed no evidence of having adsorbed wax.

CURRENT USES AND FUTURE PROSPECTS FOR SORPTIVE DUSTS

The best criterion for the practical usefulness of an idea is its acceptance and usage by industry. Sufficient time has elapsed since the demonstration of lipid sorption as a means of combating insects with inert or nonhazardous powders to indicate its potentialities and limitations in pest control under the conditions of commercial usage.

Prevention of drywood termite infestation.—As discussed previously, the discovery of the efficacy of sorptive insecticide diluents in preventing infestation of dusted wood by drywood termites was made during an investigation of the relative efficiency of toxicants in a variety of insecticide formulations (40, 109). Further tests of inert sorptive dusts in preventing feeding on small wood blocks by drywood termites was then determined by suspending these blocks from rafters in an attic and dusting the attic by means of an electric blower. In one such test, the dusts used were fluorinated silica aerogel Dri-die 67® at 1 and 2 lb, a montmorillonite clay (Olancha Tox®, also known as Olancha Clay) at 6.4 lb, and 5% chlordane dust at 2 and 4 lb/ 1000 ft² of attic space (39). The following day the blocks of wood were removed from the attic, placed in 100-ml beakers (3 for each dust), and 10 full-grown drywood termite nymphs were placed in each beaker. (Nymphs succumb to sorptive powders more slowly than alates). The periods for 100% knockdown and 100% kill were determined. Results are shown in Table 3. The termites were knocked down and killed more rapidly with Dri-die than with either Olancha Tox or 5% chlordane dust, even though the latter two were applied at higher dosages. Table 3 shows that a month after treat-

TABLE 3. COMPARISON OF DRI-DIE 67 WITH OLANCHA TOX AND 5% CHLORDANE AS ATTIC DUSTS FOR THE PREVENTION OF DRYWOOD TERMITES. FROM EBELING & WAGNER (39)

Material	Pounds per 1000 ft	Date treated	Date exposed to termites	100% Knockdown (hr)	100% Lethal time (hr)
Dri-die 67	1	5–22–58	5–23–58	1.5	4.8
Olancha Tox	6.4	5–22–58	5–23–58	6.0	12.5
Dri-die 67	1	9–9–58	9–10–58	3.5	14.3
Chlordane	2	9–9–58	9–10–58	35.0	63.0
Dri-die 67	2	9–9–58	9–10–58	1.7	4.7
Chlordane	4	9–9–58	9–10–58	8.5	24.0
Dri-die 67	2	9–9–58	10–14–58	1.9	3.9
Chlordane	4	9–9–58	10–14–58	24.3	33.3

ment the Dri-die deposits were as effective as they were 24 hr after treatment, but 5% chlordane dust had lost much of its original efficacy. In the high temperatures of an attic, the effect of the toxicant had probably been lost completely, leaving only the insecticidal action of the diluent, pyrophyllite, which is a relatively ineffective desiccant.

Dri-die has an extremely low bulk density (4.5 lb/ft^3 as packed) and can be uniformly distributed throughout the attic, even though it is blown in from only one point. Despite the very small size and light weight of the Dri-die particles, they adhere well to wood surfaces and insects that crawl over these surfaces, because each particle has a positive electrostatic charge. The charge is eventually lost in exposed films, as explained previously, and Dri-die is then about as insecticidal as the corresponding unfluorinated silica aerogel, SG-68, but it is sufficiently insecticidal to prevent infestation by drywood termites. Termites were not able to infest dusted blocks of wood that had remained suspended from rafters in an attic for six years, although when the barely visible film of dust was removed, they readily burrowed into the wood. It is likely that the dusted wood members of an attic remain immune to termite attack indefinitely.

In order to justify the use of less sorptive dusts than Dri-die, it has been argued that the speed of kill is not important as long as the termites eventually die. However, particularly with dusts of high bulk density, not all parts of an attic receive as much dust as the test blocks described above, and in areas of minimal deposit the heavier and less insecticidal dusts have been found to fail while Dri-die has been effective in the same locations when applied at the same volume (but less weight) per unit of attic area.

Many termite operators in California follow a termite treatment (fumigation or "drill-and-treat") with Dri-die dust. Some refuse to fumigate without a subsequent dusting when they are providing a control contract, for they have found from their years of experience that attic dusting can eliminate the costly "call backs" characteristic of the years preceding the use of Dri-die (44).

Dri-die has also been blown into attics, wall voids, soffit voids, voids under cabinets and built-in kitchen appliances, and other out-of-the-way places, during construction of a building, in what is known as "insect proofing" or "built-in pest control" (42–44) in order to eliminate permanently harborage and breeding places for cryptobiotic insects.

Control of household pests.—Tarshis (103) reported good results against German, brown-banded, and oriental cockroaches when dusting over 60 houses, apartments, restaurants and other types of buildings with Dri-die (SG-67) and another fluorinated silica aerogel (SG-77). The buildings were dusted so thoroughly that no place of refuge was left untreated, and the dust was effective despite its repellency. However, pest control operators who have attempted to control cockroaches with Dri-die report that it is so light that if floats throughout the building, leaving a film on furniture, curtains, draperies, etc. Even with the most thorough treatment, the kill is considered to be too slow.

Dri-die has never become popular for use in controlling cockroaches or any other household pest when treatment is necessary within the "living space" of a dwelling. On the other hand, it is a useful dust to control or repel insects and mites that inhabit attics, wall voids, soffit voids, voids under cabinets and built-in appliances, or any enclosed space (37). Ebeling et al (45, 46) had success when blowing Dri-die into wall, soffit, and subcabinet voids, followed by the treatment of the "living space" with Drione® (1% pyrethrins, 10% technical piperonyl butoxide, 38.12% amorphous silica aerogel, 1.88% ammonium fluosilicate, 49% petroleum base oil). Drione kills cockroaches more rapidly than any insecticide with which it has been compared and, because it contains a large quantity of petroleum in the silica aerogel, it does not float about excessively.

The effect of the pyrethrins in Drione lasts much longer than it does in most other formulations. A deposit from a Drione dust blown into a wall void was found to be highly potent against German cockroaches 146 days after application, giving a 100% kill in 7 min compared with 3 min for the same type of deposit of fresh powder. When deposits of Drione (2 cc per open petri dish) were kept for 3 months in bright light (during daylight hours) in a glasshouse, they resulted in a 50% knockdown of German cockroaches in an average of 42 min compared with 102 min for Dri-die. (The two powders cause about the same rate of desiccation of German cockroaches.) In Drione a large proportion of the oil-pyrethrin-piperonyl butoxide solution is protected from light and air in the pores of the silica particles, prolonging the insecticidal activity of the toxicant.

Ebeling et al (36, 47) showed by means of laboratory devices called "choice boxes" that cockroaches are strongly repelled by most finely divided powders, including the silica aerogel Dri-die 67. The least repellent powder tested was boric acid. Cockroaches crawled about freely in deposits of boric acid in the dark halves of the choice boxes, but avoided contact with Dri-die. Although boric acid is a weak insecticide, German cockroaches succumbed in treated boxes in four to six days, depending on the distribution of the dust deposit. On the other hand, when highly toxic insecticides were applied as dusts or liquids, some insects survived for more than a month, having learned to avoid the deposits before picking up a lethal dose of toxicant. German cockroaches were so successful in avoiding Dri-die deposits that little or no mortality occurred despite the fact that Dri-die kills cockroaches much more rapidly than boric acid when the insects are forced to remain in continuous contact with the deposits.

In cockroach control the greatest limitation of Dri-die, Drione, Baygon, diazinon, and in fact most insecticides, is their repellency. The efficacy of a highly toxic insecticide, when repellency is eliminated, was indicated by the treatment of sewerage systems with Drione. One-third lb of Drione, blown into a manhole with a high-capacity mechanical blower, resulted in complete elimination of American cockroaches throughout the sewerage system for a block in each direction. The cockroaches were in a closed system and could not escape. Drione is currently used for control of cockroaches in sewerage systems by the Los Angeles Sanitation Division (110).

Dust desiccants have been recognized by the pest control industry as being useful in structural pest control, particularly when applied in wall voids, etc., in pet shops, and research laboratories where experimental animals are reared, and in areas where contamination of the treated surfaces by highly toxic and chemically stable insecticides, or the presence of toxic vapors in the air, cannot be tolerated(3).

When sorptive dusts were applied in an aqueous suspension, or if they became wet after they were applied, they adhered too firmly to the substrate and were not picked up as readily by insects crawling over the treated surface. When aqueous suspensions were applied directly to the body of the insect, after the water had evaporated the silica aerogel SG-68 was found to be as insecticidal as an equivalent amount applied as a dust, but sorptive clays lost some of their insecticidal efficacy (38).

Control of ectoparasites.—Tarshis (104, 105) found in laboratory and field tests that Dri-die 67 killed more than 30 species of arthropods in various stages of development. Most of these were ectoparasites of man or animals. Lice and fleas were particularly susceptable. However, competition with effective and reasonably safe organic insecticides is so great that in the United States the snake mite, *Ophionyssus natricis,* appears to be one of the few ectoparasites investigated by Tarshis for which a sorptive dust is universally used for control. Tarshis (105) recommends that, if the cages are heavily infested, the reptiles should be removed and the cages treated by dusting all surfaces with Dri-die and then working the dust into the sand, vermiculite, gravel, etc. covering the floor. For light infestations, reptiles and floor-covering material can be dusted while the animals are in the cage. Dri-die provides better and longer lasting control of snake mite than previously used acaricides and without the adverse effects against reptiles.

In buildings infested with rats, the tropical rat mite, *Ornithonyssus bacoti,* may attack the human occupants. When the rats are killed, the mites leave the bodies of these rodents and may travel great distances, particularly along heating pipes in the walls, and maybe found anywhere in the building. Their bites may produce irritation and sometimes painful dermatitis that may continue for two or three days, leaving red spots similar to those caused by fleas. Ebeling (32) found that Dri-die applied in attics of apartment houses and residences at the rate of 1 lb/1000 ft^2, as well as in appropriate places in the living spaces, resulted in apparently permanent control of tropical rat mites, for no recurrence of infestation has been reported in any of nine treated properties. The dusting of the attic apparently contributed most to the long residual efficacy of this treatment. Pest control operators generally apply Dri-die in the attic for control of the tropical rat mite and the house mouse mite, *Allodermanyssus sanguineus,* although they may use other acaricides in the living space of the infested dwelling. The northern fowl mite, *Ornithonyssus sylvarium,* infests buildings and affects the occupants in much the same way as the tropical rat mite and a similar treatment with Dri-die was effective in control (106).

The use of sorptive dusts for control of ectoparasites of poultry has been investigated in various parts of the world. Melichar & Willomitzer (84) in Czechoslovakia investigated sorptive dusts ("physikalische Insektizide") against the chicken mite, *Dermanyssus gallinae.* Using 17 silicas and 4 silicates, they showed a high correlation between the specific surface of these powders and their efficacy against mites (Fig. 3), confirming results of an earlier investigation by Ebeling (33) regarding the relation of specific surface and insecticidal efficacy of finely divided powder used against four insect species.

Melichar & Willomitzer (84) believe one of the important advantages of "physical insecticides" is their high chemotherapeutic index resulting from their selective action against ectoparasites in contrast to their safety to warmblooded hosts. In view of the mode of action involved, they believed the development of resistance to such insecticides was not likely.

In Russia good results against lice and mites were reported by Nabokov et al (86) with Dri-die and a Russian-made silica gel. Acid-activated clay is used for control of poultry lice and the chicken mite, as well as for the control of the human head louse in India. For the control of lice and mites on poultry, the dust is applied directly to the body and below the wings by means of small rotary or other types of hand blowers (Majumder, personal communication).

In California, sorptive dusts appear to be practical for control of poultry ectoparasites only in dust-bath boxes, and these are not used in modern large-scale operations. In parts of the world where chickens are in farm yards or in pens on the ground, dust baths could be practical. The dust should not be too light, otherwise the action of the chickens dusting themselves carries most of the dust into the air rather than into the feathers.

Tarshis & Blinstrub (107) reported good results from treatment of humans with Dri-die for control of crab lice, *Phthirus pubis*, and head lice, *Pediculus humanus capitus*. All active stages and eggs ("nits") were desiccated and killed. Tarshis (104) also found Dri-die to be effective against the short-nosed cattle louse, *Haematopinus eurysternus* (Fahrenholz), at 30 to 60 g per animal. Shull (101) obtained control of the cattle-biting louse, *Bovicola bovis*, with one or two applications of diatomaceous earth at 90 g per animal.

Protection of stored food products.—The beginnings of stored-food protection with inert dusts probably antedate recorded history. The California Indians protected acorns from insect attack by mixing them with "red earth" (27). Entomologists at Louvanium University in the Republic of Congo (personal communication) inform us that primitive peoples in the Congo region not only mix clay dust with stored beans, to protect them against insects, but also recognize the superiority of clays obtained from certain areas. Apparently the first attempts at practical use of chemically inert dusts in modern times, i.e., after the disclosure by Kühnelt (69) of the nature of the epicuticular lipid layer protecting insects from desiccation, were in connection with the protection of stored food products, particularly grain, pulses, and corn.

In the United States, diatomaceous earths and silica aerogels have been tested in large bins under simulated commercial storage conditions and one diatomaceous earth has had limited commercial use (22, 70, 71, 93, 95, 102). Diatomaceous earths are both abrasive and slightly sorptive, and both types of action can result in desiccation when insects crawl among dust-coated kernels, thus rubbing all parts of their bodies against films of dust. Diatomaceous earth apparently acts principally as an abrasive, for its specific surface is relatively small (ca 3 m^2/g compared with 300 m^2/g for silica aerogel SG-68). Its sorptivity for insect wax can be shown to be very low merely by dusting it onto the surface of dyed beeswax and observing the slow rate and limited extent to which a film of the dust absorbs the wax, compared with the rapid rate of sorption and large quantities absorbed with SG-68. Both dusts are principally silica and the significant difference, with regard to wax absorption, appears to be in their specific surface.

Redlinger & Womack (95) found SG-68 at 1 lb/ton (0.05 wt/%) to be more effective in keeping insects (*Sitophilus oryzae, Tribolium confusum* and *T. castaneum*) from infesting shelled corn than diatomaceous earth at 7 lb/ton, but slightly less effective than malathion 57E at 1 pint/1000 bu. Quinlan & Berndt (93) noted that a part of the insecticidal action of the inert dusts resulted from their repellency, the diatomaceous earths being more repellent than the silica aerogels. There is also a certain degree of insecticidal action of some inert dusts, at least against certain species, that cannot be entirely attributed to desiccation (22, 25, 40). Headlee (55) attributed the effect of colloidal clay in protecting beans against *Bruchus obtectus* to the fact that it caused the larva to slip and prevented it from obtaining sufficient purchase on the surface to enable it to drill its way into the bean.

When diatomaceous earth was mixed with wheat at 4 lbs/ton, grain-infesting beetles ranked in order of decreasing susceptibility as follows: *Cryptolestes pusillus, Sitophilus oryzae, S. granarius, Rhyzopertha dominica, Oryzaephilus surinamensis, Trogoderma parabile, Tribolium castaneum,* and *T. confusum* (22). In another investigation, when used against six species of grain pests, diatomaceous earth at 4 lb/ton of wheat prevented infestation for 6 months, 6 lb/ton for 9 months and 8 lb/ton for 12 months (102). La Hue (70) found a diatomaceous earth at 4 lb/ton to be less effective after 1 year than malathion at 1½ or even at 1 pint of 57E/ 1000 bushels (30 tons) of wheat, but somewhat superior to synergized pyrethrins (Pyrenone® 60–6 O.T.) at 1 qt/1000 bu. In a later investigation, two brands of diatomaceous earth (Kenite 2-I® and Perma-Guard®) at dosages of 120, 210, and 300 lb/1000 bu protected the quality of wheat better than 1 pint of malathion 57E and much better than two brands of unfluorinated silica gel (SG-68 and Cab-O-Sil®) at 45 lb/1000 bu (71). Wheat treated with malathion was graded No. 3, but when treated with diatomaceous earth it was down-graded to No. 4, because of decreased test weight, and when treated with the silica aerogels it was reduced to Sample Grade (very low quality) because of the presence of foreign substances.

In the most recent paper on the relative efficacy of toxicants and dust desiccants, a highly sorptive and nonabrasive silica aerogel (Cab-O-Sil) at 60 lb/1000 bu (2 lb/ton or 0.1%) afforded nearly complete protection from damage to wheat by beetles (*Rhyzopertha dominica, Cryptolestes pusillus,* and *Sitophilus oryzae*) for 12 months. It was superior to diatomaceous earth (Kenite 2-1) at 210 lb/1000 bu. Both dusts were superior to malathion at 1 pint (0.63 lb active ingredient) and far superior to diazinon at 0.5 pint (0.25 lb active)/1000 bu (70a).

Reduction in "test weight" of grain and the fact that the addition of foreign material[3] to grain reduces the commercial grade, has resulted in

---

[3] The reason diatomaceous earth is favored among inert dusts is that it can easily be identified when examined with a microscope. The presence of an unknown substance reduces grain to "Sample Grade" unless the substance can be identified as diatomaceous earth.

comparatively little commercial use of chemically inert dusts for grain protection in the United States. However, in many countries grain grading standards are less stringent and commercial interest is greater. Interest in silica aerogels has been manifested in Egypt (61), Mexico (94), and Romania (16, 17). In Egypt, Kamel et al (61) found that the fluorinated silica aerogel Dri-die 67® at 0.1 wt/% in wheat gave 100% kill of grain-infesting beetles in from three to five days at 20.8°C and 51.5% RH. At 29.4°C and 59.8% RH, this period was reduced to two to four days. The beetles ranked in order of decreasing susceptibility as follows: *Tribolium castaneum* = *Rhizopertha dominica* > *Sitophilus oryzae* > *S. granarius*. Dri-die at 0.1 wt/% resulted in 100% kill more rapidly than Drione® (Dri-die plus petroleum, pyrethrins, and synergist), which is to be expected because with mixtures of Dri-die and toxicants, mortality depends on rate of desiccation and not on the action of the toxicant (41). When Dri-die and Drione are used at the same wt/%, the quantity of sorptive dust is less in the latter.

For many years the Egyptians have used a relatively inefficient mixture of finely ground rock phosphate and sulfur for grain protection. In Madagascar an amorphous aluminum pentasilicate dust is officially prescribed for treatment of beans which are intended for export to France.

Apparently the greatest commercial acceptance of sorptive dusts for grain protection has occurred in India, where acid-activated kaolin clay is used in seed- and food-grain stores. About 70% of the annual crop of grain (wheat, rice, sorghum, maize, etc.) and 40% of the pulses, crops that total approximately 100 million tons annually, is retained by the growers for their own use or to sell locally. The remainder finds its way through the usual marketing channels to the nonfarming population. The growers may store grain and pulses for long periods up to a year awaiting higher prices. Some authorities believe that as much as half of the grain crop may be lost each year after harvest because of rats, insects, spoilage, spillage, etc.

Unlike the situation in the United States, grains and pulses are grown on millions of small farms. The government of India does not believe that it would be safe to entrust the measurement of malathion dosages to the millions of farmers, most of them illiterate, and does not allow its use. However, the urgent need for a grain protectant is indicated by the use of castor oil for this purpose in Gujarat State. Workers rub the grain between their hands, moistened with castor oil, and this has resulted in good protection against insects. As might be expected, farmers who have used this method of grain protection greatly appreciate the decreased cost and greater practicability of sorptive dusts and have eagerly purchased all that could be prepared by the limited experimental facilities at the Central Food Technological Research Institute in Mysore, where the idea of using acid-activated clay originated. Currently, a pilot plant is being built to produce the initial commerical quantities of activated clay and to serve as a model for business enterprises interested in commercial production of this grain protectant.

Insecticidal efficacy can be increased by substituting tricalcium phosphate (TCP) for a portion of the acid-activated kaolin. TCP is insecticidal to the larvae of many pests of stored grain, causing a variety of aberrations in metabolism and metamorphosis (6, 7, 77, 78). Distortion and degeneration of the alimentary canal, dwarfed larvae, and misshaped pupae with nodular outgrowths are characteristic visible evidence. The insecticidal effect of TCP to insects is markedly increased by the addition of most sugars such as glucose, sucrose, fructose, lactose, xylose, mannose, and rhamnose, at the rate of 1 part of sugar to 19 parts of TCP, but trehalose, the major sugar in insect plasma, completely inhibits the action of TCP at the same concentration (78). Apparently TCP causes an imbalance in the energy metabolism and this is accelerated by glucose and other sugars but is corrected by trehalose.

By combining acid-activated kaolin and TCP, greater insecticidal efficacy is obtained against both adults and larvae of grain-infesting insects, for TCP is particularly effective against lepidopterous larvae. In a visit to the CFTRI laboratory in July 1969, the writer observed that the addition of 25% of TCP to the grain protectant resulted in somewhat improved adherence of the powder to *T. castaneum* beetles. The current recommendation is to mix with the grain 1% of a powder consisting of 25% acid-activated kaolin, 25% TCP, and 50% unactivated kaolin. The latter has little insecticidal value, but is added as a diluent. This mixture has protected grain for a year, not only in laboratory tests at 80% RH, but under different conditions of commercial storage.

Protection of pulses is less difficult than protection of grain, for the only insect pest of importance is *Callosobruchus chinensis,* and it is more easily controlled than the more resistant of the grain-infesting beetles such as *Tribolium* spp. A 50/50 mixture of activated and unactivated kaolin at 1 wt/% is recommended.

Activated clay will not protect grain indefinitely, indicating that it gradually declines in insecticidal efficiency. An experiment was made in which 1% activated kaolin was mixed with wheat in quart jars that were either tightly sealed with a screw-top lid or covered with a screen. The treated wheat was kept at 26.7°C and 65% RH. In three months the powder was separated from the wheat and tested against *Tribolium confusum,* using powder from a sealed jar as a control. The powder from the sealed jar of wheat showed a 26.4% decrease and the powder from the jar with a screen top a 47.2% decrease in insecticidal efficacy based on the period required for 50% kill. The powders were then placed in a desiccating jar over $CaCl_2$ for one week, insuring the same amount of moisture in each sample. This had no effect on the relative efficacy of the powders, indicating that some change takes place in the physical structure of the powder when it is exposed to the atmosphere in a thin film for extended periods (38).

In India, millions of farmers, each with only a few acres of grain, have their own small storage bins. A grain protectant that can be applied by un-skilled labor and by simple mechanical means is highly desirable. Ordinary drums with lids, as used for seed dressing, or commercial seed dressing ma-chines or poultry feed mixers or power-operated concrete mixers, have been found to be satisfactory for mixing grain and clay. In large-scale opera-tions, the grain protectant has been added to the grain at the terminal end of the cleaning operation where it was automatically mixed with the grain after it was cleaned. Although the powder is added to the grain at 1 wt/%, after normal handling and processing it does not exceed 0.01% and normal dirt and foreign matter in market samples of grain often exceed that limit.

In Australia a machine is being marketed with which protectant dusts can be applied to grain as it enters the bin of a harvester. It consists of a hopper of 20-lb capacity with a V-shaped trough at its base and is attached to the side of the header bin. The header driver operates the applicator by pulling a lever once every 2 to 8 min depending on the speed of the header and the yield of grain per acre (4).

Kaolin is generally regarded as nontoxic and is prescribed for control-ling diarrhea and for adsorbing intestinal toxins (113). Krishnamurthy et al (68) summarized their intensive investigation of the possible effects of clay ingestion along with food by mammals: "The growth of weanling rats is not adversely affected by the presence of insecticidal clay (activated kao-linic clay) at 0.1 or 1% level, or of raw clay at 1% level in their diets. The presence of insecticidal clay at 1% level in the diet has no adverse effect on growth and the digestibility of protein, fat, calcium and phosphorus in young rats. In adult rats the digestibility of protein and fat is not signifi-cantly affected by the inclusion of insecticidal clay even at 5% level in the diet."

Tricalcium phosphate is considered to be a useful food amendment and, when added to flour, it has dietary value.

Silica aerogels showed promise for the control of dermestid beetles, *Der-mestes frischii, D. ater,* and *Necrobia rufipes,* infesting dried fish, but their efficacy was adversely affected when the fish was oily or inadequately dried (62).

Watters (112) investigated the possibility of protecting packaged food in warehouses by impregnating the storage sacks with the fluorinated silica aerogel SG-67 (Dri-die 67®). He concluded that in clean warehouses, at 70% RH or lower, securely sealed packages treated with this silica aerogel "may offer safe, effective means of protecting foods against stored-product insects for long periods."

Plant protection.—To date, the most insecticidal sorptive dusts have had a very low bulk density. For agricultural usage these have not been practical because of their tendency to float far beyond the treated area. Yet the desire to avoid toxicants on certain crops has been very great, along with a corre-sponding desire to develop an effective and usable sorptive dust. Diatomaceous earth has a desirable bulk density but its insecticidal efficacy is relatively low. W. R. Belford has used diatomaceous earth as a base on which to build a thin layer of a highly porous and sorptive substance of much greater insecticidal efficacy than diatomaceous earth and the coated particles have good dusting properties and adherence. Using *Tribolium confusum* and *T. castaneum* as test insects, the writer has found the coated diatomaceous earth to be approximately equivalent to acid- and heat-activated kaolin in insecticidal efficacy. Field tests on important agricultural crops will be made as soon as sufficient dust is available.

LITERATURE CITED

1. Alexander, P., Kitchener, J. A., Briscoe, H. V. A. 1944a. The effect of waxes and inorganic powders on the transpiration of water through celluloid membranes. *Trans. Faraday Soc.* 40:10–19

2. Alexander, P., Kitchener, J. A., Briscoe, H. V. A. 1944, Inert dust insecticides. *Ann. Appl. Biol.* 31:143–59

3. Anonymous. 1960. Osmun featured at Ill. PCA spring meet. *Pest Control* 28(6):57

4. Anonymous. 1967. Cooper's Grain Guard Applicator. *Pybuthrin Dig.* 4(1):15

5. Baker, G., Pepper, J. H., Johnson, L. H., Hastings, E. 1960. Estimation of the composition of the cuticular wax of the Mormon cricket, *Anabrus simplex* Hald. *J. Insect Physiol.* 5:47–60

6. Bano, A., Majumder, S. K. 1965. Pathological changes induced by tricalcium phosphate in insects. *J. Invert. Pathol.* 7:384–87

7. Bano, A., Majumder, S. K. 1968. Tricalcium phosphate as an insecticide. In *Pesticides,* ed. S. K. Majumder, 177–85. *Acad. Pest Cont. Sci.,* Mysore, India

8. Beament, J. W. L. 1945. The cuticular lipids of insects. *J. Exp. Biol.* 21:115–38

9. Beament, J. W. L. 1955. Wax secretion in the cockroach. *J. Exp. Biol.* 32:514–38

10. Beament, J. W. L. 1958. The effect of temperature on the waterproofing mechanism of an insect. *J. Exp. Biol.* 35:494–515

11. Beament, J. W. L. 1959. The waterproofing mechanism of arthropods. I. The effect of temperature on cuticle permeability in terrestrial insects and ticks. *J. Exp. Biol.* 36:391–422

12. Beament, J. W. L. 1960. Wetting properties of insect cuticle. *Nature* 186:408–9

13. Beament, J. W. L. 1961. The water relations of insect cuticle. *Biol. Rev.* 36:281–320

14. Beament, J. W. L. 1964. The active transport and passive movement of water in insects. *Advan. Insect Physiol.* 2:67–129

15. Beatty, I., Gilby, A. R. 1969. The major hydrocarbon of a cockroach cuticular wax. *Naturwissenschaften* 56:373

16. Beratlief, C. 1966. Protection of stored grain with dusts. *Probl. Agr. (Bucharest)* 18:65–70

17. Beratlief, C., Popescu, C., Cojocaru, M. 1965. Observations on the efficiency of silica gel in the insect pest control of stored wheat. *Bucharest Inst. Cent. Ceccet. Agr. Sect. Prot. Plant An.* 3:363–67

18. Bowers, W. S., Thompson, M. J. 1965. Identification of the major constituents of the crystalline powder covering the larval cuticle of *Samia cynthia ricini* (Jones). *J. Insect Physiol.* 11:1003–11

19. Briscoe, H. V. A. 1943. Some new properties of inorganic dusts. *J. Roy. Soc. Arts* 91:593–607

20. Brown, H. T., Escombe, F. 1900. Static diffusion of gases and liquids in relation to the assimilation of carbon and translocation in plants. *Trans. Roy. Soc. London* 193:223–91

21. Bursell, E., Clements, A. N. 1967. The cuticular lipids of the larvae of *Tenebrio molitor* L. (Coleoptera). *J. Insect Physiol.* 13:1671–78

22. Carlson, S. D., Ball, H. J. 1962. Mode of action and insecticidal value of diatomaceous earth as a grain protectant. *J. Econ Entomol.* 55:964–70

23. Chibnal, A. C., Piper, S. H., Pollard, A., Williams, E. F., Sahai, P. N. 1934. The constitution of the primary alcohols, fatty acids and paraffins present in plant and insect waxes. *Biochem. J.* 28:2189–2208

24. Chiu, S. F. 1939a. Toxicity studies of so-called "inert" materials with the bean weevil, *Acanthosceles obtectus* (Say). *J. Econ. Entomol.* 32:240–48

25. Chiu, S. F. 1939b. Toxicity studies of so-called "inert" materials with the rice weevil and the granary weevil. *J. Econ. Entomol.* 32:810–21

26. Collins, M. S. 1969. Water relations in termites. In *Biology of Termites, eds.* K. Krishna, F. M. Weesner, 433–58. New York, London: Academic

27. Cool, L. F. 1966. From acorns to

supermarkets. *Calif. Dept. Agr. Bull.*, 55 :1–10

28. Cotton, R. H., Frankenfeld, J. C. 1949. Silica aerogel for protecting stored seed or milled cereal products from insects. *J. Econ. Entomol.* 42 :553

29. Coulon, J., Barres, P. 1966. Efficacité de la silice hydratic amorphe vis-a-vis de la bruch du haricot *Acanthoscelides obtectus* (Say). *Phytiat.-Physopharm.* 15 :99–104

30. David, W. A. L., Gardiner, B. O. C. 1950. Factors influencing the action of dust insecticides. *Bull. Entomol. Res.* 41 :1–61

31. Dennell, R. 1946. The larval cuticle of *Sarcophaga falculata* Pand. *Proc. Roy. Soc. London, Ser. B* 133 :348–73

32. Ebeling, W. 1960. Control of the tropical rat mite. *J. Econ. Entomol.* 53 :475–76

33. Ebeling, W. 1961. Physicochemical mechanisms for the removal of insect wax by means of finely divided powders. *Hilgardia* 30 : 531–64

34. Ebeling, W. 1961. The permeability of insect cuticle. In *The Physiology of Insects,* ed. M. Rockstein, 507–56. New York, London : Academic

35. Ebeling, W., Pence, R. J. 1957. The use of pesticides in the preparation of insects and mites for photography. *Ann. Entomol. Soc. Am.* 50 : 637–38

36. Ebeling, W., Reierson, D. A. 1969a. Cockroaches learn to avoid insecticides. *Calif. Agr.* 23 :(2), 12–15

37. Ebeling, W., Reierson, D. A. 1969b. Insect proofing during building construction. *Calif. Agr.* 23(5) : 4–7

38. Ebeling, W., Reierson, D. A. 1969. Effect of humidity on the insecticidal efficacy of sorptive dusts. Unpublished manuscript on file in the Dept. of Agr. Sci. (Entomology), Univ. of California, Los Angeles

39. Ebeling, W., Wagner, R. E. 1959a. Control of drywood termites. *Calif. Agr.* 13(1) :7–9

40. Ebeling, W., Wagner, R. E. 1959b. Rapid desiccation of drywood termites with inert sorptive dusts and other substances. *J. Econ. Entomol.* 52 :190–207

41. Ebeling, W., Wagner, R. E. 1961. Relation of lipid adsorptivity of powders to their suitability as insecticide diluents. *Hilgardia* 30 : 565–86

42. Ebeling, W., Wagner, R. E. 1964a. Built-in pest control. *Pest Contr.* 32(2) :20–22, 24, 26, 28, 31–32

43. Ebeling, W., Wagner, R. E. 1964b. The treatment of voids under cabinets. *PCO News* 24(4) :8–11

44. Ebeling, W., Wagner, R. E. 1964c. Built-in pest control for wall and cabinet voids in houses and buildings under construction. *Calif. Agr.* 18 (11) :8–12

45. Ebeling, W., Wagner, R. E., Reierson, D. A. 1965a. Cockroach control with Dri-die and Drione. *PCO News* 25(10) :16–22; 25 (11) :16–19, 25

46. Ebeling, W., Wagner, R. E., Reierson, D. A. 1965b. Cockroach control in public housing. *PCO News* 25 (12) :21–24

47. Ebeling, W., Wagner, R. E., Reierson, D. A. 1966. Influence of repellency on the efficacy of blatticides. I. Learned modification of behavior of the German cockroach. *J. Econ. Entomol.* 59 :1374–88

48. Finney, G. L., Fisher, T. W. 1964. Culture of entomophagous insects and their hosts. In *Biological Control of Insect Pests and Weeds,* ed. P. DeBach, 328–53. New York : Reinhold

49. Fraenkel, G., Hsiao, C. 1969. Hormonal and nervous control of tanning in the fly. *Science* 138 :27–29

50. Germar, B. Versuche zur Bekämpfung des Kornkäfers mit Staubmitteln. *Z. Angew. Entomol.* 22 : 603–30

51. Gilby, A. R., Cox, M. E. 1963. The cuticular lipids of the cockroach *Periplaneta americana* (L.). *J. Insect Physiol.* 9 :671–81

52. Gluud, A. 1968. Zur Feinstruktur der Insektencuticula. Ein Beitrag zur Frage des Eigengiftschutzes der Wanzencuticula. *Zool. Jahrb. Abt. Anat. Ontog. Tiere* 85 :191–227

53. Glynne Jones, D. 1955. The cuticular waterproofing mechanism of the worker honeybee. *J. Exp. Biol.* 32 :95–109

54. Greening, H. G. 1967. Sorptive dusts, a new approach to household pest

control. *Pybuthrin Dig.* 4(1) :9–11, 14

55. Headlee, T. J. 1924. Certain dusts as agents for the protection of stored seeds from insect infestation. *J. Econ. Entomol.* 17 :298–307

56. Helvey, T. C. 1952. Insecticidal effect of inert solid diluents. *Science* 116 : 631–32

57. Hockenyos, C. L. 1933. Effects of dusts on the oriental roach. *J. Econ. Entomol.* 26 :292–94

58. Hunt, C. R. 1947. Toxicity of insecticide dust diluents and carriers to larvae of the Mexican bean beetle. *J. Econ. Entomol.* 40 :215–19

59. Ingram, R. L. 1955. Water loss from insects treated with pyrethrum. *Ann. Entomol. Soc. Am.* 48 :481–85

60. Kalmus, H. 1944. Action of inert dusts on insects. *Nature* 153 :714–15

61. Kamel, A. H., Fam, E. Z., Ezzat, T. M. 1964. Studies on Drione dust as a grain protectant. *Agr. Res. Rev.* (UAR, Ministry Agr.) 42 :52–69

62. Kane, J. 1967. Silica based dusts for the control of insects infesting dried fish. *J. Stored Prod. Res.* 2 :251–55

63. Kitaoka, S., Yajima, A. 1958. Effects of insecticides on ticks. II. Effect on loss of body weight and oviposition of ticks. *Jap. Appl. Entomol. Zool.* 2 :11–16

64. Kitchener, J. A., Alexander, P., Briscoe, H. V. A. 1943. A simple method of protecting cereals and other stored foodstuffs against insect pests. *Chem. Ind.* 62 :32–33

65. Krishnakumari, M. K. 1968. Comparative acute oral toxicity of some mineral pesticides to albino rats. In *Pesticides,* ed. S. K. Majumder, 335–38 ; Acad. Pest. Cont. Sci., Mysore, India

66. Krishnakumari, M. K., Majumder, S. K. 1962. Modes of action of active carbon and clay on *Tribolium castaneum* (Hbst.) *Nature* 193 :1310–11

67. Krishnamurthy, K., Godavari Bai, S., Majumder, S. K. 1962. Insecticidal substances from kaolinic clays. *Res. Ind.* 7 :267–69

68. Krishnamurthy, K., Subramanyaraj Urs, T. S., Majumder, S. K.

1965. Studies on the effect of insecticidal residues in food : Part II Effect of insecticidal clay in the diet of albino rat on the growth and digestibility of food constituents. *Ind. J. Exp. Biol.* 3 : 171–73 (1965)

69. Kühnelt, W. 1928. Über den bau des insektenskelletes. *Zool. Jahrb. Abt. Anat.* 50 :219–78

70. La Hue, D. W. 1965. Evaluation of malathion (synergized pyrethrum, and diatomaceous earth as insect protectants in small bins. *U.S. Dept. Agr. ARS Market. Res. Rep. No. 726,* 13 pp.

70a. La Hue, D. W. 1970. Evaluation of malathion, diazinon, a silica aerogel, and a diatomaceous earth as protectants on wheat against lesser grain borer attack in small bins. *U.S. Dep. Agr. ARS Market. Res. Rep. No. 860.* 11 pp.

71. La Hue, D. W., Fifield, C. C. 1967. Evaluation of four inert dusts on wheat as protectants against insects . . . in small bins. *U.S. Dept. Agr. ARS, Market. Res. Rep. No. 780.* 29 pp.

72. Locke, M. 1961. Pore canals and related structures in insect cuticle. *J. Biophys. Biochem. Cytol.* 10 : 589–618

73. Locke, M. 1964. The structure and formation of the integument in insects. In *The Physiology of Insects,* ed. M. Rockstein, 379–470. New York, London : Academic

74. Locke, M. 1965. Permeability of insect cuticle to water and lipids. *Science* 147 :295–98

75. Locke, M. 1965. The hormonal control of wax secretion in an insect, *Calpodes ethlius* Stoll *(Lepidoptera, Hesperiidae). J. Insect Physiol.* 11 :641–58

76. Loveridge, J. P. 1968. The control of water loss in *Locusta migratoria migratorioides* R & F. I. Cuticular water loss. *J. Exp. Biol.* 49 :1–13

77. Majumder, S. K., Bano, A. 1964. Toxicity of calcium phosphate to some pests of stored grain. *Nature* 202 :1359–60

78. Majumder, S. K., Bano, A. 1968. Reversion of the toxicity of tricalcium phosphate to insects by trehalose. *Nature* 210 :1052–53

79. Majumder, S. K., Krishnamurthy, K. 1960. Bleaching earths as insecticides. *J. Sci. Ind. Res.* 19c :29–30

80. Majumder, S. K., Krishnamurthy, K., Krishnakumari, M. K. 1962. Processed clays as insecticidal substances for pest control. *Proc. I Intern. Congr. Food Sci. Tech. (London)* 2:285–92

81. Majumder, S. K., Narasimhan, K. S., Subrahmanyan, V. 1959. Insecticidal effects of activated charcoal and clays. *Nature* 184:1165–66

82. Majumder, S. K., Venugopal, J. S. 1968. Pesticidal minerals. In *Pesticides*, ed. S. K. Majumder, 190–99. Acad. Pest Cont. Sci., Mysore, India

83. Mansingh, A. 1965. Water loss in malathion-intoxicated German cockroaches. *J. Econ. Entomol.* 58:162–63

84. Melichar, B., Willomitzer, J. 1967. Bewerting der physikalischen insektizide. *Sci. Parmaceut.* 2:589–97. Proc. 25th Congr. Pharmaceut. Sci., Prague, 1965

85. Mills, R. R. 1967. Control of cuticular tanning in the cockroach: bursicon release by nervous stimulation. *J. Insect Physiol.* 13:815–20

86. Nabokov, V. A., Turnich, M. L., Mitrofanov, A. M., Uspenskii, I. V. 1964. Use of sorptive powders-desiccants in the control of arthropods. (Rus) *Med. Parazitol. Parazit. Bolez.* 33:515–18

87. Nair, M. R. G. K. 1957. Structure of waterproofing epicuticular layers in insects in relation to inert dust action. *Indian J. Entomol.* 10:37–49

88. Nelson, D. R. 1969. Hydrocarbon synthesis in the American cockroach. *Nature* 221:854–55

89. O'Brien, R. D. 1967. *Insecticides, action and Metabolism,* 200. New York: Academic

90. Olson, W. P., O'Brien, R. D. 1963. The relation between physical properties and penetration of solutes into the cockroach cuticle. *J. Insect Physiol.* 9:777–86

91. Parkin, E. A. 1944. Control of granary weevil with finely ground mineral dusts. *Ann. Appl. Biol.* 31:84–88

92. Parkin, E. A., Bills, G. T. 1955. Insecticidal dusts for the protection of stored peas and beans against bruchid infestation. *Bull. Entomol. Res.* 46:625–41

93. Quinlan, J. K., Berndt, W. L. 1966. Evaluation in Illinois of four inert dusts on stored shelled corn for protection against insects—a progress report. *U.S. Dept. Agr., Agr. Res. Serv.* 51–56, 20 pp.

94. Ramirez Ginel, M. 1961–62. Two new grain protectors for temperate and tropical zones (Sp) *Agr. Tec. en Mex.* 12:25–27

95. Redlinger, L. M., Womack, H. 1966. Evaluation of four inert dusts for the protection of shelled corn in Georgia from insect attack. *U.S. Dept. Agr., Agr. Res. Serv.* 51–57, 25 pp.

96. Richards, A. G. 1951. *The Integument of Arthropods.* Minneapolis, Minn: Univ. Minnesota Press

97. Richards, A. G., Clausen, M. B., Smith, M. N. 1953. Studies on arthropod cuticle. X. The asymmetrical penetration of water. *J. Cell. Comp. Physiol.* 42:395–414

98. Rideal, E. K., Tadayon, J. 1954. On overturning and anchoring of monolayers. II. Surface diffusion. *Proc. Roy. Soc. London Ser. A* 225:357–61

99. Roan, C . C., Hopkins, T. L. 1961. Mode of action of insecticides. *Ann. Rev. Entomol.* 6:333–35

100. Ridgway, R. L. 1962. Laboratory studies of the effect of Dri-Die 67 on the scorpion, *Centruroides vittatus. J. Econ. Entomol.* 55:1014

101. Shull, W. E. 1932. Control of the cattle louse, *Bovicola bovis* Linn. (Mallophaga, Trichodectidae). *J. Econ. Entomol.* 25:1208–11

102. Strong, R. G., Sbur, D. E. 1963. Protection of wheat seed with diatomaceous earth. *J. Econ. Entomol.* 56:372–74

103. Tarshis, I. B. 1959. Sorptive dusts on cockroaches. *Calif. Agr.* 13(2):3–5

104. Tarshis, I. B. 1961. Laboratory and field studies with sorptive dusts for the control of arthropods affecting man and animal. *Exp. Parasitol.* 11:10–33

105. Tarshis, I. B. 1962. The use of silica aerogel compounds for the control of ectoparasites. *Proc. Anim. Care Panel* 12:217–58

106. Tarshis, I. B. 1964. A sorptive dust for the control of the northern fowl mite, *Ornithonyssus sylvarium,* infesting dwellings. *J. Econ. Entomol.* 57:110–11

107. Tarshis, I. B., Blinstrub, R. 1963. Preliminary studies on the use of

sorptive dusts for the control of
the human lice *Phthirus pubis* (L.)
and *Pediculus humanus capitis*
De Geer. *Am. J. Trop. Med. Hyg.*
12 :91–95

108. Venugopal, J. S., Majumder, S. K.
1968. Active mineral in insecticidal clays. In *Pesticides,* ed. S. K.
Majumder, 200–9. Acad. Pest
Cont. Sci., Mysore, India

109. Wagner, R. E., Ebeling, W. 1959.
Lethality of inert dust materials
to *Kalotermes minor* Hagen, and
their role as preventives in structural pest control. *J. Econ. Entomol.* 52 :208–12

110. Wagner, R. E., Ebeling, W., Reierson, D. A. 1966. Control of cockroaches in sewers. *Pub. Works
Mag.* 97 :82–84

111. Warth, A. H. 1956. *The Chemistry
and Technology of Waxes.* New
York: Reinhold

112. Watters, F. L. 1966. Protection of
packaged food from insect infestation by the use of silica gel. *J.
Econ. Entomol.* 59 :146

113. White, W. H. 1949. *Materia Med.,*
381. London : J & A Churchill

114. Wigglesworth, V. B. 1942. Some

notes on the integument of insects
in relation to the entry of contact
insecticides. *Bull. Entomol. Res.*
33 :205–19

115. Wigglesworth, V. B. 1945. Transpiration through the cuticle of insects. *J. Exp. Biol.* 21 :97–114

116. Wigglesworth, V. B. 1947. The site
of action of inert dusts on certain
beetles infesting stored products.
*Proc. Roy. Entomol. Soc. London,
Ser. A* 22 :65–69

117. Wigglesworth, V. B. 1947. The
epicuticle of an insect *Rhodnius
prolixus. Proc. Roy. Entomol. Soc.
London, Ser. B* 134 :163–81

118. Wigglesworth, V. B. 1948. The insect
cuticle. *Biol. Rev. Cambridge Phil.
Soc.* 23 :408–51

119. Winston, P. W. 1967. Cuticular water
pump in insects. *Nature* 214 :283–
84

120. Winston, P. W., Beament, J. W. L.
1969. An active reduction in water
level in insect cuticle. *J. Exp. Biol.*
50 :541–61

121. Young, L. O. 1953. Silica flatting
agent and a method of manufac-
turing it. *U.S. Patent No. 2,625,492*

HOST SPECIFICITY DETERMINATION OF INSECTS 6006 FOR BIOLOGICAL CONTROL OF WEEDS

H. Zwölfer

Commonwealth Institute of Biological Control, Delémont, Switzerland

AND

P. Harris

*Research Institute, Canada Department of Agriculture,
Belleville, Ontario*

Huffaker (39) and Wilson (74) have reviewed the biological control (biocontrol) of weeds in earlier numbers of this volume. The present review is restricted to a single but crucial aspect of this topic—that of determining whether an insect can be introduced to control a weed without also damaging desirable plants. There is no absolute test for this—tests used show only that particular insects are "unsafe," and the best criterion of safety still represents the accumulation of many pieces of circumstantial evidence. Unless the mechanism of host selection is understood, individual pieces of evidence are often contradictory and misleading. Realization of this led to the recent liberation against *Xanthium* in Australia of two insects that had previously been excluded from consideration because they had damaged plants of minor economic importance in some tests (67, 73). This change in thinking also reflects the difficulty of distinguishing between "safe" and "unsafe" insects. The point of division made on the gradation between these two extremes reflects differing needs and situations. For example, Hawaii can take less strict precautions than Australia or North America because it grows a smaller range of crops and is isolated. Also, the more urgent the weed problem, the greater are the risks that are justified in any course of action against it. Thus, the decision to liberate a particular insect is the result of many considerations, and may vary for different species even if the interpretation and the results of host specificity tests are identical.

There are many technical difficulties attending host specificity tests. An important one is the discrepancy between insect behavior in captivity and in nature: tests done in cages, petri dishes, and other artificial environments do not allow the insect to use a full repertoire of host-finding maneuvers (47). This leads to the acceptance of a broader range of plants than in nature. Other factors to be considered are the physical and genetic variability of both test insect and plants, the choice and interpretation of tests for different types of insects, the design of tests that will be convincing to both entomologist and informed layman, and the possibility of mutations leading to host transfer. Solutions to these problems are still being developed. Indeed, the procedures used in almost every study are an evolutionary advance over those used previously and the final form that the tests will take is still not clear. It is clear, however, that all the evidence must be viewed as a whole, as host range is determined by many other factors beside the ability of an insect to survive on a plant. In recognition of this, Harris & Zwölfer (34) proposed that prerelease studies should be broadened to include (*a*) study of the insect's biology with particular attention to adaptations likely to restrict its host range; (*b*) review of the plants attacked by related insects; (*c*) determination of the laboratory host range of the insect; (*d*) investigation of the chemical and physical basis of host recognition; and (*e*) starvation tests on economic plants to confirm the limits of the laboratory host range. This review summarizes current practices in these five study areas and concludes with a discussion of host transference by phytophagous insects.

INVESTIGATION OF THE BIOLOGY AND HOST-DETERMINED ADAPTATIONS OF PHYTOPHAGOUS INSECTS

Positive and negative host records.—Field records of host plants are always of consequence for deciding whether to introduce an insect. Evidence that an insect is a pest of a crop grown in the country of introduction, of course, precludes its use for biocontrol. Such records are obtained from field and literature surveys. Ideally, field surveys include plant species and genera related to the target weed: a survey for thistle and knapweed insects in Europe covered all available members of the tribe Cynareae (77); for insects of *Rhamnus cathartica,* additional species and genera of Rhamnaceae were searched (51); for insects of *Cytisus scoparius,* a broad range of Leguminosae were included (56); and to obtain information on insects of *Cuscuta,* 10 species of the genus were surveyed (5). Broad surveys such as these reveal the preferences and host range of the insects in the study area and the probable host range in the country of introduction. The absence of the insect from plants related to the weed after a thorough field survey is the best single index of host specificity.

A literature search for host records is always made before a weed insect is used for biocontrol. If the insect is an important pest, this soon becomes apparent although individual literature records are not necessarily reliable. Incorrect records result from transient insects and from misidentification of either insect or host. For example, a conspicuous chrysomelid *Chelymorpha cassidea* has been recorded from maize, cabbage, raspberry, pea, and mustard although it does not feed on plants outside the Convolvulaceae (55). Alternatively, the absence of records of attack on economic plants may be meaningless in countries where the insect fauna is poorly known, or where specific identification of the insect is difficult. In these circumstances, the generic host range is often used. Thus, the absence of any record of the arctiid genus *Ammalo* as a serious pest was used as one indication of the suitability of *A. insulata* for release against *Eupatorium odoratum* (9). Identification of species in the trypetid genus *Urophora* is difficult, and therefore host records are often incorrect. However, as the genus is easily recognized, the lack of its mention as a pest in the *Review of Applied Entomology* was used as an index of safety for *Urophora sirunaseva* and *U. affinis* (80, 83). On the other hand, records of *Chaetorellia carthami* as a pest of safflower excluded the use of an unidentified *Chaetorellia* sp. for biocontrol of *Centaurea solstitialis* in California (84). Indeed, morphometric, behavioral, and cross-breeding studies indicated that the insects represent different biotypes or subspecies of a polytypic *Chaetorellia*.

The most desirable literature record is that of the safe use of the insect for biocontrol. On this basis, insects used against *Opuntia, Lantana, Hypericum, Senecio,* and other weeds have been subsequently released in other parts of the world with little or no additional testing (10, 11, 24, 25, 64).

Host range and insect taxonomy.—The release of unidentified insects for biocontrol of weeds is a hazardous practice. It is particularly undesirable if there are closely related species that are pests, as in the genus *Chaetorellia* (84). If literature or museum material is inadequate for identification, taxonomic studies are necessary. These are usually undertaken but in our opinion there is often insufficient investigation of biotypes, host races, and sibling species. Attempts should be made to determine their existence, their taxonomic relationships and exact host ranges. For example, the sesiid *Chamaesphecia empiformis* includes two biotypes or perhaps sibling species differing in host preference, oviposition site, and size of egg (62). To release the wrong one would compromise the possibility of controlling the weed, and to release an unstudied biotype negates the reason for making prerelease studies. The trypetid *Urophora affinis* contains several biotypes with different host ranges, but clinal variation makes their separation difficult (83). Hence, the source of *U. affinis* used for biocontrol should be restricted to the geographical area from which they were studied. The noctuid *Eulocastra argentisparsa* which attacks witchweed (*Striga* sp.) has two larval color forms, one feeding on the seeds and the other on the leaves (58). This polymorphism of color and feeding behavior should be elucidated before the insect is released as it probably affects its value as a biocontrol agent.

Physiological adaptations.—Physiological adaptations of phytophagous insects often express dependence on a host. The restriction of oogenesis to adult feeding on particular plants has been used to demonstrate an obligate host dependence and hence safety of the insect for biocontrol. Adults of

Microlarinus lareynii and *M. lypriformis* (two weevils successfully used as control agents for *Tribulus terrestris*) fed in tests on a broad range of plant families. The damage to nonhost plants was not regarded as serious. Furthermore, oocyte development occurred only in females fed on *Tribulus* or the closely related genus *Zygophyllum,* and oocytes retrogressed when the weevils were transferred to other plants (1). A similar dependence has been shown between the weevil *Rhinocyllus conicus* and thistle (78), between the chrysomelid *Longitarsus jacobaeae* and the plant tribe Senecioneae with the species *Senecio jacobaea* being the most favorable (23), and between the agromyzid *Melangromyza cuscutae* and *Cuscuta* (4). On the other hand, the chrysomelid *Gastrophysa cyanea* matured oocytes after feeding on plants unrelated to *Rumex,* its host (20).

The host range of gall-forming insects is a striking example of a physiological host-dependent adaptation. A close interrelationship between oviposition and larval feeding on the one hand and specific host responses on the other, are required for formation of a histioid gall, without which the insect cannot subsist. The intricacy of this special form of insect-plant relationship attests to a high degree of host specificity. Indeed, it is relatively rare that an insect can cause galls on more than one species of plant (52) although sometimes species groups may be suitable. Therefore, the production of galls in the receptacle of *Centaurea* spp. by the trypetids *Urophora sirunaseva* and *U. affinis* was an important index of their "integrity" for biocontrol (80, 83).

Behavioral adaptations.—Behavioral adaptations usually provide a less decisive measure of host dependence than do physiological adaptations: insects often lay mature eggs or do a small amount of feeding on many substrates if the host is not present. However, oviposition tests on the cerambycid *Plagiohammus spinipennis* provided convincing evidence of its host specificity (29). Most phytophagous insects have an obligate manner of feeding (leaf mining, stem boring, seed mining) and this may restrict their potential host range. Behavior dependent on structures characteristic of particular plants is of special value for demonstrating host specificity—for example, the use of spines as oviposition sites by many cactus-feeding insects (16, 27). Behavioral adaptations that restrict an insect to plants with stimuli for host finding, oviposition, or feeding are discussed under "host plant acceptance."

Morphological adaptations.—Morphological adaptations in the females of many insects relate to a specific oviposition site, and physically prevent utilization of other sites. Numerous species in the genus *Urophora* breed in the flower heads of the tribe Cynareae (thistles and knapweeds). The flower heads of different species range from 3 to 50 mm in diameter, and there is a similar diversity in the ovipositor length of *Urophora* spp. *U. sirunaseva* and *U. affinis* are adapted to the comparatively small heads of

Centaurea solstitialis or *C. maculosa* and cannot lay in the large heads of safflower or artichoke because their ovipositors are too short (80, 83). Physically this prevents them from becoming a pest on these economic plants even if the larvae could survive in them. Similarly, the flat ovipositor sheath of *Chaetorellia* is adapted to oviposition on flower heads with a certain type of bract (84). Also, in contrast to the majority of grasshoppers which lay in the soil near their food plants, the ovipositor of *Gesonula punctifrons* is adapted to spongy plant tissue which limits its potential host range (59).

The weevil genus *Larinus* breeds in flower heads of the tribe Cynareae (Compositae). The females bore oviposition holes with their rostra and then push their eggs into the interior of the flower head. The marked difference in size and shape of the female rostrum between *Larinus* species corresponds to the size and structure of the host flower head. For example, the closely related species *L. sturnus, L. jaceae,* and *L. carlinae,* all of which breed on species of *Cirsium* and *Carduus,* are segregated by requirements for large, medium, or small flower heads, respectively (75).

Broad ecological adaptations.—The dependence of an insect on temperature, moisture, or other ecological thresholds often restricts distribution and hence host range. Usually these are not convincing safety criteria for biocontrol agents. There are exceptions such as the adaptation of an insect to an aquatic environment which obviously prevents breeding on a terrestrial crop. Thus, Bennett (8) tested the aquatic weevil *Cyrtobagous singularis,* the grasshopper *Paulinia acuminata,* and the pyralid *Samea multiplicalis* on only five plants before releasing them against *Salvinia* because few aquatic plants were of economic importance. Similarly, the breeding range of the grasshopper *Gesonula punctifrons,* which attacks *Eichornia crassipes* in India, is restricted by its obligatory association with an aquatic environment (59).

Phenological adaptations.—Specialized phytophagous insects are normally closely synchronized with the phenology of their host. Asynchrony with leaf burst, flowering, or other event may confer partial or total immunity to otherwise suitable plants. This is not a reliable criterion of safety as it may be easily modified on transfer to a new environment, but it is worth noting. For example the ragwort seed-fly *Hylemya seneciella* is so closely synchronized with flowering of its host *Senecio jacobaea* that similar but later flowering *Senecio* spp. are not attacked. Moreover, *H. seneciella* is adapted to the slow maturation of *S. jacobaea* heads, and *Senecio* species with more rapid maturation did not allow larval development (24).

Often phenological adaptation is a liability rather than an asset in biocontrol. For example, the chrysomelid *Chrysolina hyperici* failed to control *Hypericum* in California because it was poorly synchronized there with its host (41), and possibly the same applies to the trypetid *Euaresta aequalis* on

Xanthium in Australia (73). Synchronization of life cycle and diapause behavior with host phenology has led to speciation in some stenophagous moths attacking species of Rhamnaceae (51) and in many trypetid genera (86). Such insects are only useful for biocontrol if the synchrony is retained in the new environment.

PLANTS ATTACKED BY RELATED INSECTS

Harris & Zwölfer (34) suggested that the host range of species related to the biocontrol candidate should be investigated. Often it can be shown that the weed insect belongs to a taxon (species group, section, subgenus, or higher taxon) which is restricted to plants in one genus, subtribe, tribe, or family. This indicates that the insect taxon has speciated on the plant taxon; hence, over a long time and usually a wide geographical area, the insect taxon has not exploited any other group of plants. A high degree of reliability can consequently be placed on their restriction to the plants concerned. Examples are the weevil genera *Larinus* and *Rhinocyllus* which are restricted to the tribe Cynareae (75), and *Microlarinus* which is confined to Zygophyllaceae (1); the trypetid genera *Urophora* (80) and *Chaetorellia* (84) are both restricted to the Cynareae, and the noctuid genus *Calophasia* to the plant genera *Linaria* and *Antirrhinum* (50). The information on host plants can also be used in other ways. Frick (24) helped to justify the use of the seed-fly *Hylemya seneciella* for biocontrol by pointing out that well-known pest species of *Hylemya* are in a different subgenus. A favorable prognosis exists for the use of the chrysomelid genus *Longitarsus* for biocontrol, even though it attacks taxonomically diverse plants, as 85.5 percent of European species for which hosts are known are monophagous or oligophagous within a single family (23).

The hosts of related insects should be determined even though the plants are taxonomically unrelated, so that they can be included in prerelease feeding tests (34). The propensity for feeding by stenophagous insects is usually greatest on plants having the greatest taxonomic affinity with the host. However, host recognition may depend largely on a single chemical or physical stimulus which has arisen independently in several plant taxa. Insect speciation tends to follow the character to give an apparently disjoined distribution of hosts (17, 18). In these circumstances, if the weed insect accepts plants unrelated to its host they are likely to be plants attacked by related insects. The feeding by partly grown larvae of *Celerio euphorbiae*, a sphingid closely associated with *Euphorbia*, on *Vitis*, *Oenothera*, and other plants associated with related sphingids, is an example (54).

LABORATORY TESTS TO DETERMINE POTENTIAL HOST RANGE OF AN INSECT

The procedure in early biocontrol attempts was to use feeding tests to demonstrate that economic plants were immune to attack by the weed insect to be introduced. For reasons discussed later, this approach was unsatisfactory. As an alternative we advocated that feeding tests should determine the range of plants acceptable to the insect (34). The difference is in the plants used; the tests are done in the same way. Plants tested should include those related to the known hosts, host plants of related insects, plants from which occasional records have been obtained, and plants which have characteristics in common with the host that may make them acceptable. A greater range of plant taxa should be tested if there is little or no information on the hosts of related insects, and especially if they have no special host adaptations. The number and range of plants tested in relation to previously available information on the insects are summarized in Table 1.

Methods.—The responses of an insect in a feeding test depend on the type of insect and on the test used. The results can be scored subjectively but it is preferable that they be quantified. This has been done by counting feeding holes or "feeding units" (23, 75, 76, 81, 87); measuring the leaf area eaten from a photographic enlargement (31) or with photometric techniques (48); counting or weighing fecal pellets (45); and measuring the length of larval mines (62). In another approach, suitability of the plant as a host is related to insect longevity on it (78, 81) or to the proportion of individuals surviving in a given period (31). The rate of larval development at a constant temperature also has been used (30). However, relative larval weight gain has apparently not been used, although for many insects this is probably a more sensitive index than some of those mentioned. The rate of oogenesis (1), egg production (82), numbers of eggs deposited (24, 36, 80, 82, 84), or viability of eggs (23) offer opportunities to rate test plants. The frequency of behavioral responses such as "attempts at probing," "probing," or "examination" by trypetids (80, 83, 84) reflected the acceptability of the plants offered. Normal versus atypical deposition of eggs (24, 82) or different types of feeding behavior as in *Rhinocyllus conicus* where feeding on nonhosts was restricted to leaf veins, stems, or bracts (82) may be used.

Tests may be made on single plants or as multiple choice trials. In multiple choice trials in which the host is offered with other plants, the insect is not under a hunger stress. As a result, the insect is more selective than if the host were not available.

Factors influencing results of host tests.—Attempts are usually made to standardize the insects used in plant acceptability tests. However, it is not always realized that insects often become conditioned to the plant species they have (been) fed on and that this affects their subsequent preferences. This has been observed both for larvae (36, 44, 61, 66) and for newly emerged adults (36). Conditioning is usually minimized by using newly hatched larvae: newly hatched larvae of the chrysomelid *Altica carduorum* fed on several thistle genera which were refused after feeding for several hours on *Cirsium arvense* (46). Oviposition tests are often misleading because insects tend to lay mature eggs regardless of the substrate (23, 82) or fail to orient to host oviposition stimuli in captivity. Similar difficulties are encountered with feeding tests.

Geographical races or biotypes of phytophagous insects often respond differently (23, 62), and even within a population individuals differ (31).

TABLE 1. EXAMPLES OF THE RANGE OF PLANTS TESTED AND ACCEPTED IN FEEDING TESTS ON INSECTS

Authority	Insect species studied	Normal host plant	Information affecting range of plants tested	Range of test plants	Test results
60	*Altica caerulea*	*Jussiaea* spp.	A	47f, 110g, 123sp.	L on 2 unrelated sp.(2f); af+Lf also on other g+f
31 46	*Altica carduorum*	*Cirsium arvense*	A, F	9f, 25g, 31sp.	af, ov, Lf, L restricted to 3 related g
23	*Longitarsus jacobaeae*	*Senecio jacobaea*	F	5f, 37g, 65sp.	L+oo only on a few related sp.(2g); af+ov less specific
20	*Gastrophysa cyanea*	*Rumex* spp.	A, F	5f, 9g, 19sp.	L restricted to few sp.(1g); Lf, af, ov, oo less specific
19	*Lema trilineata*	*Solanum* spp.	F	6f, 13g, 19sp.	af, oo, ov+L restricted to 2–3 related g (1f); Lf less specific
1	*Microlarinus* spp.	*Tribulus terrestris*	A, D, F	21f, 34g, 39sp.	ov+oo restricted to 2–3g (same f); af unspecific (19f)
78	*Rhinocyllus conicus*	*Carduus* spp	A, D, F	8f, 47g, 55sp.	oo, ov+L only on few related sp.+g; af less specific
88	*Ceutorhynchus litura*	*Cirsium arvense*	A. F	9f, 33g, 48sp.	af, ov+L only on few related sp.+g
59	*Gesonula punctifrons*	*Eichhornia*	C, D	27f, 39g, 42sp.	ov on few unrelated sp (different f); Lf+af on numerous f
12	*Tyria jacobaeae*	*Senecio jacobaea*	E, F	1f, 13g, 17sp.	L only on 2 related g; Lf slightly less specific
3	*Oeobia verbascalis*	*Xanthium* spp.		39f, 78g, 90sp.	ov+L on 5 unrelated g (2f)
7	*Diastema tigris*	*Lantana camara*		49f, 101g, 111sp.	L restricted to host sp.; slight Lf on 7 unrelated sp.(6f)
5	*Herpystis cuscutae*	*Cuscuta* spp.		28f, 54g, 59sp.	L on 5 unrelated sp.(3f); Lf on additional g+f
11	*Ammalo insulata*	*Eupatorium odoratum*		16f, 39g, 51sp.	L only a few sp. of host g; Lf less specific (3g of 2f)
24	*Hylemya seneciella*	*Senecio jacobaea*	A, D, E, F	1f, 20g, 48sp.	ov-preference for host sp.; slight ov on other sp.+g
4	*Melanagromyza cuscutae*	*Cuscuta* spp.		42f, 75g, 81sp.	oo+L restricted to host g; occasional ov on 2 other sp.(2f)
80	*Urophora sirunaseva*	*Centaurea solstitialis*	C, A, B, D, F	1f, 20g, 37sp.	ov only on few related sp.
83	*Urophora affinis*	*Centaurea maculosa*	A, B, C, D, F	1f, 17g, 36sp.	ov only on few related sp.

A=related insect spp. on same or related hosts; B=gall producers; C=morphological adaptations; D=special host structures attacked; E=previous use in biocontrol; F=available literature information. f=family; g=genus; sp.=species.
L=complete larval development; Lf=larval feeding; af=adult feeding; oo=oogenesis; ov=oviposition.

Karny (45) and Harris (30) obtained different results when they forced the noctuid *Calophasia lunula* to feed on *Antirrhinum* instead of the host genus *Linaria* because there was a differential mortality of larvae infected with a nonvirulent virus. Diseased insects are likely to feed on a narrower range of plants than healthy ones and should not be used for host-range determination (12).

Evaluation of results.—The results of host tests vary widely between insect species with identical host range, and even within the same species different types of test may be contradictory. The host range is usually determined by the most restricted stage of the insect which means that each species must be evaluated individually. For example, there may be a high longevity and rate of feeding on plants not accepted for oviposition (20, 36). A greater range of plants was accepted for oviposition than was suitable for development in *Altica caerulea* (60), *Octotoma scabripennis* (28), *Gastrophysa cyanea* (20), *Melanagromyza cuscutae* (4), *Eulocastra argentisparsa* (58), and *Leucoptera spartifoliella* (56) but gravid females refused plants suitable for larval development in *Gesonula punctifrons* 59). The ability to complete larval development is usually restricted to a smaller range of plants than are acceptable for larval feeding (Table 1). Examples among Coleoptera are *Altica caerulea* (60), *Uroplata girardi* (10), *Octotoma scabripennis* (28), *Gastrophysa cyanea* (20), and *Agasicles* sp. (63).

Causal Analysis of Host Plant Specificity

Host plant selection has been reviewed by Thorsteinson (68), Beck (6), Kennedy (47), and House (37). Three aspects of particular interest for host specificity studies on weed insects are: host plant finding, host plant acceptance for feeding and oviposition, host plant suitability for development and reproduction. The first two aspects and, to some extent, the third, involve the insect's response to specific visual, chemical, or tactile stimuli. The host range of an insect can be precisely defined and explained if these stimuli are identified; but this is often difficult, especially if a mixture of chemical substances is involved. To disallow liberation of weed-feeding insects without knowledge of their host-recognition stimuli would largely terminate biocontrol of weeds. However, it is desirable (34) to establish that the insect is using stimuli characteristic of the host and whether they are physical or chemical. This confirms that the insect is selecting its host on a positive basis, rather than attacking all plants that lack deterrents.

Host plant finding.—The stimuli used by stenophagous insects in host plant finding may result in a directed movement to the host. Alternatively, they may increase or arrest random movement and in either way improve the insect's chance of encountering its host (47). Such host-finding maneuvers are difficult to quantify and have been neglected in investigation of weed insects. However, success in reaching a host can be easily scored and some assumptions can be made about the stimuli involved. For example, *Tyria jacobaeae* larvae ceased wandering on encountering *Senecio jacobaea* foliage, even if it was dead and dry, presumably because it contained a larval arrestant (35).

In some stenophagous chrysomelids, curculionids, trypetids, and other insects, the host plant serves as a rendezvous for males and females. This is pronounced in many trypetids: interspecific mating may be induced in the laboratory but in nature the rendezvous is an effective isolating device (79). Mating on the host plant is a behavioral adaptation tending to restrict the insect to this plant. It is also possible that where male trypetids establish a territory on the host plant (13), the females may be orienting to an odor produced by the male as a long-range method of host plant location. The panicle shape in *Centaurea maculosa* appeared to be a short-range visual stimulus eliciting landing by the trypetid *Urophora affinis*. Once on the plant, the stereotyped behavior of the fly and the spatial arrangement of the flower stalks lead the female to the buds which are its specific oviposition site (83). Experiments with *U. sirunaseva* and *Chaetorellia* sp. on dummy and mutilated hosts indicated that short-range host recognition depended on the long spines on the upper bracts of *Centaurea solstitialis* (80, 84).

Host plant acceptance.—Feeding and oviposition in stenophagous insects is elicited by stimuli characteristic of the host plant, but not necessarily the same ones as those used in host plant finding. Indeed, feeding by stenophagous insects is often in response to chemical stimuli that are not detected at a distance. A compound in the cuticular waxes of *Hypericum* induced larval feeding by the geometrid *Anaitis plagiata* (32). A feeding stimulant for the noctuid *Calophasia lunula* was extracted from *Linaria* foliage (30), and for the chrysomelid *Agasicles* sp. from the juice of *Alternanthera* (63). Larvae and adults of *Gastrophysa cyanea* responded to host plant extracts incorporated into an artificial diet (20) and *Longitarsus jacobaeae* responded to a water extract of *Senecio* foliage infiltrated into lettuce by vacuum (23).

The presence of chemical feeding stimulants has been demonstrated by bioassay of host plant extracts on pith, charred filter paper, agar diet, or vacuum infiltration into lettuce (33). Feeding deterrents, on the other hand, are more simply demonstrated with the "sandwich test," in which the test leaf is offered between the leaves of the host. The presence of deterrents in the test plant inhibits feeding on the sandwich (43).

Visual and tactile recognition stimuli elicit probing and oviposition in several Diptera that attack Compositae. Currie (14) showed that in *Euaresta aequalis,* a trypetid associated with *Xanthium,* oviposition responses were elicited by the curled ends of burr spines. Chemical properties of *Senecio* buds induced examination by female *Hylemya seneciella,* but physical characters of *S. jacobaea* buds such as size, color, arrangement and stiffness of the bracts are required for oviposition (24). The trypetid *Urophora affinis* probed buds on *Centaurea maculosa* covered in wax provided they were a certain size, shape, and had a particular arrangement and color of bracts (83).

Host plant suitability.—The requirements for oogenesis and larval growth of weed insects has received less attention than host acceptance. Fraenkel (21) suggested that since basic nutritional requirements of most insects are similar, food selection and specificity must be determined almost entirely by nonnutritional factors. This hypothesis is supported by the successful rearing of many stenophagous phytophagous insects on standard artificial diets. Also, Waldbauer (72) showed that tobacco hornworm, *Protoparce sexta,* induced to feed on *Taraxacum* by maxillectomizing them, grew as well and the adults produced as many eggs as they did on their normal host. Tests on several weed insects (1, 82, 88, 89) showed that oogenesis and successful larval development are correlated with the level of adult and larval feeding, respectively. As feeding in these insects is related to the presence of feeding stimulants, host suitability appears to be largely a function of host acceptance. However, there are insects in which intensive adult feeding on nonhost plants does not lead to oocyte development (36) or in which substantial larval feeding does not allow complete development. It is likely that these plants lack special nutrients required by the insect, or that nutrients are not supplied in suitable ratios (37).

Biophysical features may also determine host suitability (6). The survival of insects which mine in roots, stems, leaves, or fruits often depends on the physical properties of their feeding site, but little precise information is available. Experiments with *Urophora affinis* suggested that the thickness and toughness of the bracts of some *Centaurea* spp. prevented oviposition (83). Similarly, a pilose leaf surface deterred feeding by some phytophagous beetles (31). The size of achene and the presence of a latex-like substance determined the suitability of Cynareae for larval development of *Chaetorellia* spp. The larvae of *C. carthami* developed in plants with large achenes (*Cnicus, Carthamus, Arctium*) but survived poorly in *Centaurea solstitialis* which has comparatively small ones. On the other hand, *Chaetorellia* sp. bred in *C. solstitialis* but not *Carthamus* in which the first-instar larvae were overcome by latex present in the receptacle and walls of the achenes. Hybrids of *C. carthami* and *Chaetorellia* sp. developed readily in both hosts (84).

Single components in a behavior sequence may be important for determining host suitability. This is shown by the weevil *Rhinocyllus conicus* which, after laying its eggs on a thistle, covers them with masticated particles of thistle leaves. The cover is an essential abuttment used by the newly hatched larva to bore into plant tissue. In the laboratory, *R. conicus* laid on a wide range of plants, but unless the female had fed on thistle immediately prior to oviposition, its eggs were not covered and the larvae perished on hatching (82). Similarly, on nonhosts the eggs of the ragwort seed-fly, *Hylemya seneciella,* are usually not positioned so that the larvae can enter the plant (24).

STARVATION AND NEGATIVE-OVIPOSITION TESTS ON ECONOMIC PLANTS

The "starvation test" consists of confining the insect to a test plant until it feeds on it or dies. The method was used in early biocontrol attempts to show that economic plants were immune to attack. However, there was no assurance that untested plant species would not be accepted and it was obviously impossible to test all plants. Starvation and negative-oviposition tests are easily done and would be more than justified for public reassurance (they are often superficially convincing) if a host-specific insect could be relied on not to accept nonhost plants in the test. Some economic plants, especially lettuce and pea, are eaten by many stenophagous insects when they are confined to them. Other acceptable nonhost plants are scattered through various families and are not necessarily the same for different species of insect (43). Acceptance apparently depends on the absence of feeding inhibitors to which the insect is sensitive. Usually the insect does not feed when first confined to such plants; but gradually over several days it eats increasing amounts. This was observed for *Rhinocyllus conicus* confined on lettuce and artichoke bracts (82) and for *Microlarinus* spp. on alfalfa (1). The amount of feeding was less and the life span of the adult shorter than on the host (82). Larvae were either unable to complete development (9) or suffered high mortality and prolonged development (5).

Tests on confined insects often give unnatural results because by preventing dispersal they eliminate a large part of the host selection process (19, 24, 47, 73). Phytophagous insects normally become restless when put on nonhost plants (*Tyria jacobaeae,* 35, *Calophasis lunula,* 30, *Microlarinus* spp., 1) because, in the absence of primary host-recognition stimuli, hunger is almost invariably redirected into a stimulus for dispersal (88). Consequently, though an insect may nibble on various plants in the absence of its host, it does not remain to feed extensively. The value of this behavior to the insect is to increase its chances of finding the host plant and hence its own chances of survival.

At best the starvation test is inconsistent with establishing the safety of a weed insect for release; but more important, it may result in the rejection of a valuable species for biocontrol (73). The alternative (31, 34) is to show that the insect's host falls into a discrete and predictable category that excludes economic plants. Starvation tests on economic plants, from as many families as possible, are useful to confirm this host range; but they cannot be the prime means of showing that the insect is safe to introduce. However, tests should be made on economic plants that normally grow in close association with the target weed. This is not necessarily a large task, as most weeds are associated with particular crops. Sudden destruction of the weed could result in damage to the crop from the starving insect unless it contains a feeding deterrent. This is discussed under "Risk of host transference." Laboratory feeding on economic plants should not be the sole cause for rejection of the insect for biocontrol unless the plants eaten contain the host-recognition stimuli, support complete development of the insect, or normally occur in close proximity to the target weed.

Negative-oviposition tests can be even more misleading than feeding tests. For example, many Lepidoptera in captivity lay on cage walls even when the host is present. This is so common that unless the results of laboratory oviposition tests show clear differences between host and nonhost plants they are normally disregarded.

Field and field-cage studies.—The difficulties arising from laboratory feeding on economic plants can be avoided by testing for host specificity under field conditions. There are, however, serious limitations to the use of this method. It can be applied only where the insect is already present. Also, the responses of the insect are more difficult to follow than in the laboratory and it is often impossible to distinguish damage caused by the insect from that of other species in the habitat. Ordinarily, surveys of crops where the insect occurs will accomplish (and at less cost) as much as a field test.

Field tests have been made in Hawaii on insects for the control of *Lantana* in Australia: 70 plant species representing 64 genera and 32 families were interplanted or occurred naturally in the vicinity of *Lantana* heavily infested with the cerambycid *Plagiohammus spinipennis*. None of them was eaten or laid on, although several had been accepted for oviposition in laboratory tests (29). A similar test was also made with the hispine leaf miners *Octotoma scabripennis* and *Uroplata girardi* on 68 plant species (62 genera, 29 families). The former species of miner was completely host-specific under these conditions and the latter was fairly specific (28). Field tests made with the weevils *Microlarinus lareynii* and *M. lypriformis* to simulate conditions in which the host *Tribulus terrestris* is suddenly removed or killed, showed that apart from some nibbling, no feeding occurred on the economic plants (1).

The use of a field cage for host specifity tests is sometimes a satisfactory compromise between open field and laboratory studies. Conditions are more natural than in the laboratory and there is room for some mobility by the insect. Also, the number of insects in the test can be controlled and other phytophagous species excluded. However, some insects are not amenable to cage studies: few matings of the fly *Hylemya seneciella* occurred in captivity (22). The five insect species mentioned in connection with open field tests were generally less selective in field cages but accepted fewer plants than they had in the laboratory (1, 28, 29). On the other hand, field and field-cage results with the pyralid *Oeobia verbascalis* on carrot interplanted with its host *Xanthium* were identical. In both there was limited oviposition but no larval development on carrot although in laboratory tests the insect had laid and developed readily on it (3).

Risk of host transference.—The bad dream of people concerned with bio-control is that an introduced weed insect may transfer to a new host, and so become a pest itself. The risks of this have been discussed and discounted in all reviews on the principles of biocontrol of weeds (34, 38, 39, 40, 74) but it is still a live issue with many people. Host transference involves acquisition by the insect of the ability to find and survive on a new host (85) and it must be stressed that both these attributes are necessary. The difficulty is that demonstration that an insect species is presently host-specific does not guar-antee that it will continue to be so in the future, because both the insect and the environment are subject to change. Feeding on unusual or nonhosts may occur for various reasons beside host transference but propensity for this can be demonstrated in prerelease tests. For example (*a*) overpopulation and starvation: *Archips cerasivorana* fed on pine trees after the supply of its host (*Prunus*) was exhausted (2); (*b*) host range is a function of population density: the number of tree species attacked by *Euproctis chrysorrhoea* in-creased in a regular and predictable sequence during the course of an out-break (71); (*c*) host range is a function of insect developmental stage: during the course of larval development the range of hosts acceptable to *Dasychira selenitica* expanded from *Medicago* to include deciduous and coniferous trees (26); (*d*) incomplete knowledge of host range: apparent extension in host range is often attributable to a lack of knowledge of its actual range and does not necessarily reflect anything unusual or any change in the insect's biology.

An insect's host range can be assumed to be the result of selection to maximize the production of offspring. If this premise is sound it can be shown mathematically that monophagy or polyphagy are favored depending on the proportion of an extended diet that would be suitable if chosen ver-sus the difficulty in finding the most suitable food (49). In other words, extension of an insect's host range (host transference) is increasingly fa-vored by the insect's decreasing ability to find or survive on its present host. For this reason, host transference is usually correlated with one of the fol-lowing events: (*a*) disappearance of the original host plant; (*b*) introduc-tion of the insect into a new environment where the original host plant does not occur; (*c*) confrontation with large populations of plants hitherto lack-ing in the insect's environment; and (*d*) presence of ecologically homologous competitors.

The ability of insects to develop on nonhost plants in laboratory feeding tests is often regarded as a preadaptation for host transference. However, behavioral responses of insects to host plant finding and acceptance stimuli are more stable than physiological adaptation to a food (53). Hence, host transference by stenophagous insects occurs on to plants that provide similar recognition stimuli, even though they are at present nutritionally poor hosts, rather than vice versa. The fewer the stimuli used by an insect for host recognition, the greater are its chances of encountering the correct combination in unrelated plants. Thus, the distinctive selection of hosts by *Urophora dzieduszyckii* may have arisen in this manner (80). Another example may be the transference of North American *Rhagoletis* spp. to new hosts (13). It is this rigidity of behavior and relative plasticity of physiological adaptation that make it desirable for host specificity determination of weed insects to be based primarily on an understanding of the host-selection mechanism rather than on the results of starvation tests. Other methods of host transference are possible. Hybridization in aphid species has produced populations with host ranges different from either parent (42, 57). Also, new host races of *Papilio aegeus* appear to have resulted from a combination of preimaginal conditioning and genetic variability (66).

Host transference is a rare event although there are several recent examples in which it has resulted in an insect becoming a pest (13, 65, 69, 70). It has not occurred in any insect used for biocontrol of a weed and it is not likely to do so if the insects used have host ranges that are integrated with many features of their biology. The greater the integration, the more improbable the already remote risk of host transference. The following criteria (1, 34, 78, 80, 83, 88) indicate a stable and highly integrated insect-host plant relationship and should be looked for in prerelease studies: (*a*) restriction of the insect taxon to a plant taxon; (*b*) occurrence of the same host range in an insect species over widely separated or extended geographical areas; (*c*) presence of host-determined adaptations.

Although host transference has not occurred among insects used for biocontrol of weeds, there have been reports of local and temporary damage to nonhost plants. These are the result of overpopulation and starvation following sudden destruction of the weed. Under these circumstances the insect may feed on adjacent nonhost plants if they do not contain deterrents. Such behavior is often anticipated, as with the weevils *Microlarinus* spp., the adults of which do a small amount of feeding distributed among many plant species in the absence of their host (1). Provided prerelease studies indicate the probable occurrence of such damage, they have fulfilled their function. It is only the unexpected damage to a crop that is alarming. The best-documented example of this is attack by the tingid *Teleonemia scrupulosa* on *Sesamum* in Uganda, a plant on which it had not been tested (15). The attack coincided with the sudden collapse of *Lantana,* as a result of its successful control by the insect. The immediate consequence was a large population of starving *T. scrupulosa* that fed and laid on *Sesamum*. The eggs hatched but few of the nymphs matured into adults and such individuals did not produce eggs. The result expected and the one that has occurred with other weed insects (28) is a rapid decline of the insect population to a level commensurate with its host density and then little or no further damage. The best way to avoid similar occurrences with other weed insects is to subject them to extensive "starvation tests" on crops associated with the weed. Indeed, this constitutes the main value of the test. It will not, however, eliminate complaints that involve the mere presence of large numbers of the insect on surrounding vegetation after its host has disappeared. There seems to be no way to counteract this except for fostering good public relations.

LITERATURE CITED

1. Andres, L. A., Angalet, G. W. 1963. Notes on the ecology and host specificity of *Microlarinus lareynii* and *M. lypriformis* and the biological control of puncture vine, *Tribulus terrestris. J. Econ. Entomol.* 56 :333–40
2. Atwood, C. E. 1966. A change of feeding habit in *Archips cerasivorana* (Fitch). *Proc. Entomol. Soc. Ontario* 97 :115–17
3. Baloch, G. M., Din, I. M., Ghani, M. A. 1966. Biology and host-plant range of *Oeobia verbascalis* Schiff.; an enemy of *Xanthium strumarium* L. *Tech. Bull. Commonw. Inst. Biol. Control* 7 :81–90
4. Baloch, G. M., Mohyuddin, A. I., Ghani, M. A. 1967. Biological control of *Cuscuta* spp. II. Biology and host plant range of *Melanagromyza cuscutae* Her. *Entomophaga* 12 :481–89
5. Baloch, G. M., Mohyuddin, A. I., Ghani, M. A. 1969. Biological control of *Cuscuta* spp. III. Phenology, biology and host-specificity of *Herpystis cuscutae* Bradl. *Entomophaga* 14 :119–28
6. Beck, S. D. 1965. Resistance of plants to insects. *Ann. Rev. Entomol.* 10 :207–32
7. Bennett, F. D. 1963. Feeding tests with the lantana noctuid *Diastema tigris* Guen. *Tech. Bull. Commonw. Inst. Biol. Control* 3 :83–97
8. Bennett, F. D. 1966. Investigations on the insects attacking the aquatic ferns, *Salvinia* spp. in Trinidad and northern South America. *Proc. S. Weed Control Conf.* 19 :497–504
9. Bennett, F. D., Cruttwell, R. E. Undated. Host specificity tests with *Ammalo insulata* (Walk.), a potential agent for the control of *Eupatorium odoratum* L. *Rep. Commonw. Inst. Biol. Control*
10. Bennett, F. D., Maraj, S. 1967. Host specificity tests with *Uroplata girardi* Pic., a leaf-mining hispid from *Lantana camara* L. *Tech. Bull. Commonw. Control* 9 :53–60
11. Bucher, G. E., Harris, P. 1961. Foodplant spectrum and elimination of disease of cinnabar moth larvae, *Hypocrita jacobaeae* (L.). *Can. Entomol.* 93 :931–36
12. Bucher, G. E., Harris, P. 1968. Virus diseases and their interaction with food stress in *Calophasia lunula*. *J. Invert. Pathol.* 10 :235–44
13. Bush, G. L. 1969. Sympatric host race formation and speciation in frugivorous flies of the genus *Rhagoletis*. *Evolution* 23 :237–51
14. Currie, G. A. 1932. Oviposition stimuli of the burr-seed fly *Euaresta aequalis* Loew. *Bull. Entomol. Res.* 23 :191–93
15. Davies, J. C., Greathead, D. J. 1967. Occurrence of *Teleonemia scrupulosa* on *Sesamum indicum* L. in Uganda. *Nature* 213 :102–3
16. Dodd, A. P. 1940. *The biological campaign against prickly pear*. Brisbane, Australia : Commonwealth Prickly-Pear Board, 177 pp.
17. Ehrlich, P. R., Raven, P. H. 1964. Butterflies and plants : a study in coevolution. *Evolution* 18 :586–608
18. Forbes, W. T. M. 1958. Caterpillars as botanists. *Proc. Int. Congr. Entomol., 10th, Montreal, 1956,* 1 :313–17
19. Force, D. C. 1966. Reactions of the three-lined potato beetle, *Lema trilineata,* to its host and certain nonhost plants. *Ann. Entomol. Soc. Am.* 59 :1112–19
20. Force, D. C. 1966. Reactions of the green dock beetle, *Gastrophysa cyanea,* to its host and certain nonhost plants. *Ann. Entomol. Soc. Am.* 59 :1119–25
21. Fraenkel, G. S. 1959. The raison d'être of secondary plant substances. *Science* 129 :1466–70
22. Frick, K. E. 1970. Behaviour of adult *Hylemya seneciella,* an anthomyiid used for the biological control of tansy ragwort. *Ann. Entomol. Soc. Am.* 63 :184–87
23. Frick, K. E. 1970. *Longitarsus jacobaeae,* a flea beetle for the biological control of tansy ragwort. 1. Host plant specificity studies. *Ann. Entomol. Soc. Am.* 63 :284–96
24. Frick, K. E., Andres, L. A. 1967. Host specificity of the ragwort seed fly. *J. Econ. Entomol.* 60 :457–63
25. Fullaway, D. T. 1964. Biological control of Cactus in Hawaii. *J. Econ. Entomol.* 47 :696–700
26. Ghazi-Bayat, A. 1967. Zur Morphologie, Biologie und Oekologie des Mondfleckbürstenspinners, *Dasy-*

chira selenitica Esp. *Z. Angew. Entomol.* 60 :467–92

27. Goeden, R. D., Fleschner, C. A., Ricker, D. W. 1967. Biological control of prickly pear Cacti on Santa Cruz Island, California. *Hilgardia* 38 :579–606

28. Harley, K. L. S. 1969. The suitability of *Octotoma scabripennis* Guer. and *Uroplata girardi* Pic. for the control of *Lantana* in Australia. *Bull. Entomol. Res.* 58 :835–43

29. Harley, K. L. S., Kunimoto, R. K. 1969. Assessment of the suitability of *Plagiohammus spinipennis* (Thoms.) as an agent for control of weeds of the genus *Lantana* II. Host specificity. *Bull. Entomol. Res.* 58 :787–92

30. Harris, P. 1963. Host specificity of *Calophasia lunula* (Hufn.). *Can. Entomol.* 95 :101–5

31. Harris, P. 1964. Host specificity of *Altica carduorum* Guer. *Can. J. Zool.* 42 :857–62

32. Harris, P. 1967. Suitability of *Anaitis plagiata* for biocontrol of *Hypericum perforatum* in dry grassland of British Columbia. *Can. Entomol.* 99 :1304–10

33. Harris, P., Mohyuddin, A. I. 1965. The bioassay of insect feeding tokens. *Can. Entomol.* 97 :830–33

34. Harris, P., Zwölfer, H. 1968. Screening of phytophagous insects for biological control of weeds. *Can. Entomol.* 100 :295–303

35. Hawkes, R. B. 1968. The cinnabar moth, *Tyria jacobaeae*, for control of tansy ragwort. *J. Econ. Entomol.* 61 :499–501

36. Hilterhaus, V. 1965. Biologisch-ökologische Untersuchungen an Blattkäfern der Gattungen Lema und Gastroidea. *Z. Angew. Zool.* 52 : 257–95

37. House, H. L. 1966. The role of nutritional principles in biological control. *Can. Entomol.* 98 :1121–34

38. Huffaker, C. B. 1957. Principles of biological control of weeds. *Hilgardia* 27 :101–57

39. Huffaker, C. B. 1959. Biological control of weeds with insects. *Ann. Rev. Entomol.* 4 :251–76

40. Huffaker, C. B. 1962. Some concepts on the ecological basis of biological control of weeds. *Can. Entomol.* 94 :507–14

41. Huffaker, C. B., Kennett, C. E. 1952. Ecological tests on *Chrysolina*

gemellata (Rossi) and *C. hyperici* Forst. in biological control of Klamath weed. *J. Econ. Entomol.* 45 :1061–64

42. Iglisch, I. 1968. Über die Entstehung der Rassen der "Schwarzen Blattläuse" (*Aphis fabae* Scop. und verwandte Arten), über ihre phytopathologische Bedeutung und über die Aussichten für erfolgversprechende Bekämpfungsmassnahmen. *Mitt. Biol. Bundesanst. Land-Forstwirtsch. Berlin-Dahlem.* 131, 34 pp.

43. Jermy, T. 1966. Feeding inhibitors and food preferences in chewing phytophagous insects. *Entomol. Exp. Appl.* 9 :1–12

44. Jermy, T., Hanson, F. E., Dethier, V. G. 1968. Induction of specific food preferences in lepidopterous larvae. *Entomol. Exp. Appl.* 11 : 211–30

45. Karny, M. 1963. The possibilities of *Calophasia lunula* Hufn., in the biological control of toadflax, *Linaria vulgaris* Mill. *Tech. Bull. Commonw. Inst. Biol. Control* 3 :1–26

46. Karny, M. 1963. *Haltica carduorum* Guer., attacking *Cirsium arvense. Tech. Bull. Commonw. Inst. Biol. Control* 3 :99–110

47. Kennedy, J. S. 1965. Mechanisms of host plant selection. *Ann. Appl. Biol.* 56 :317–22

48. Kogan, M., Goeden, R. D. 1969. A photometric technique for quantitative evaluation of feeding preferences of phytophagous insects. *Ann. Entomol. Soc. Am.* 62 :319–22

49. Levins, R., MacArthur, R. 1969. An hypothesis to explain the incidence of monophagy. *Ecology* 50 :910–11

50. Malicky, H. 1967. Climatological and ecological aspects of the geographical distribution of the genus *Calophasia* Steph. *Tech. Bull. Commonw. Inst. Biol. Control* 8 : 103–16

51. Malicky, H., Sobhian, R., Zwölfer, H. 1970. Investigations on the possibilities of a biological control of *Rhamnus cathartica* L. in Canada ; Host ranges, feeding sites, and phenology of insects associated with European Rhamnaceae. *Z. Angew. Entomol.* 65 :77–97

52. Mani, M. A. 1964. *Ecology of plant galls.* (Monographiae Biologicae).

The Hague: Junk. 434 pp.

53. Mayr, E. 1958. Behaviour and systematics. In *Behaviour and Evolution*, 341–62. New Haven: Yale Univ. Press. 557 pp.

54. Merz, E. 1959. Pflanzen und Raupen. Über einige Principien der Futterwahl bei Grosschmetterlingsraupen. *Biol. Zbl.* 78 :152–88

55. Mohyuddin, A. I. 1969. Insects from *Calystegia* spp. and *Convolvulus* spp. *Tech. Bull. Commonw. Inst. Biol. Control* 11 :93–104

56. Parker, H. L. 1964. Life history of *Leucoptera spartifoliella* with results of host transfer tests conducted in France. *J. Econ. Entomol.* 57 :566–69

57. Prior, R. N. B., Stroyan, H. L. G. 1960. A new subspecies of *Acyrthosiphon malvae* (Mosl.) from *Poterium sanguisorba* L. *Proc. Roy. Entomol. Soc. London, Ser. B* 33 :47–49

58. Sankaran, T., Rajendran, M. K., Ranganath Bath, Y. 1969. The biology of *Eulocastra argentisparsa* Hamps. with special reference to the occurrence of dimorphism. *Tech. Bull. Commonw. Inst. Biol. Control* 11 :31–43

59. Sankaran, T., Srinath, D., Krishna, K. 1966. Studies on *Gesonula punctifrons* Stål attacking water-hyacinth in India. *Entomophaga* 11 :433–40

60. Sankaran, T., Srinath, D., Krishna, K. 1967. *Haltica caerulea* 01. as a possible agent of biological control of *Jussiaea repens* L. *Tech. Bull. Commonw. Inst. Biol. Control* 8 : 117–37

61. Schoonhoven, L. M. 1967. Loss of hostplant specificity by *Manduca sexta* after rearing on an artificial diet. *Entomol. Exp. Appl.* 10 :270–72

62. Schröder, D. 1969. Studies on phytophagous insects of *Euphorbia* spp. *Chamaesphecia empiformis* (Esp.). *Progr. Rep. Commonw. Inst. Biol. Control* 23, 13 pp.

63. Simons, J. N., Silverstein, R. M., Bellas, T. 1968. Tests of a feeding stimulant in alligatorweed for *Agasicles* n.sp. *J. Econ. Entomol.* 61 :1448–49

64. Smith, J. M. 1958. Biological control of Klamath weed, *Hypericum perforatum* L. in British Columbia. *Proc. Int. Congr. Entomol., 10th, Montreal, 1956,* 4 :561–65

65. Southwood, T. R. E. 1961. The numbers of species of insects associated with various trees. *J. Anim. Ecol.* 30 :1–8

66. Stride, G. O., Straatman, R. 1962. The host plant relationship of an Australian swallowtail, *Papilio aegeus,* and its significance in the evolution of host plant selection. *Proc. Linnean Soc., N.S. Wales* 87 :69–78

67. Stride, G. O., Straatman, R. 1963. On the biology of *Mecas saturnina* and *Nupserha antennata,* cerambycid beetles associated with *Xanthium* species. *Australian J. Zool.* 11 :446–69

68. Thorsteinson, A. J. 1960. Host selection in phytophagous insects. *Ann. Rev. Entomol.* 5 :193–218

69. Tischler, W. 1955. Ist der Begriff "Kultursteppe" in Mitteleuropa berechtigt? *Forsch. Fortschr.* 29 : 353–56

70. Uvarov, B. 1967. Problems of insect ecology in developing countries. *Pest. Abstr.* (A) 13 :202–13

71. Voute, A. A., v.d.Linde, R. J. 1962. Die Reihenfolge der Wirtspflanzen beim Massenauftreten von *Euproctis chrysorrhoea.* *Z. Angew. Entomol.* 51 :215–17

72. Waldbauer, G. P. 1962. The growth and reproduction of maxillectomized tobacco hornworms feeding on normally rejected non-solanaceous plants. *Entomol. Exp. Appl.* 5 :147–58

73. Waterhouse, D. F. 1967. The entomological control of weeds in Australia. *Pacific Sci. Congr., 11th Symposium No. 28: Natural enemies in the Pacific area. Mushi* 39 :109–18 (Suppl.)

74. Wilson, F. 1964. The biological control of weeds. *Ann. Rev. Entomol.* 9 :225–44

75. Zwölfer, H. 1964. *Larinus-Rhinocyllus. Progr. Rep. Commonw. Inst. Biol. Control* 10, 28 pp.

76. Zwölfer, H. 1965. Observations on the distribution and ecology of *Altica carduorum* Guer. *Tech. Bull. Commonw. Inst. Biol. Control* 5 : 129–41.

77. Zwölfer, H. 1965. Preliminary list of phytophagous insects attacking wild cynareae species in Europe. *Tech. Bull. Commonw. Inst. Biol Control* 6 :81–154

78. Zwölfer, H. 1967. The host-range,

distribution, and life-history of
Rhinocyllus conicus Froel. *Progr.
Rep. Commonw. Inst. Biol. Control* 18, 21 pp.

79. Zwölfer, H. 1968. Untersuchungen zur
biologischen Bekämpfung von *Centaurea solstitialis* L. Strukturmerkmale der Wirtspflanze als Auslöser des Eiablageverhaltens bei
Urophora siruna-seva Hg. *Z.
Angew. Entomol.* 61 :119–30

80. Zwölfer, H. 1969. *Urophora siruna-seva* (Hg.), a potential insect for
the biological control of *Centaurea
solstitialis* L. in California. *Tech.
Bull. Commonw. Inst. Biol. Control* 11 :105–55

81. Zwölfer, H. 1969. Experimental feeding range of species of Chrysomelidae associated with Cynareae in
Europe. *Tech. Bull. Commonw.
Biol. Control* 12 :115–30

82. Zwölfer, H. 1969. Additional feeding
and oviposition tests with *Rhinocyllus conicus* Froel. *Progr. Rep.
Commonw. Inst. Biol. Control* 24,
9 pp.

83. Zwölfer, H. 1970. Investigations on
the host specificity of *Urophora
affinis* Frfld. *Progr. Rep. Commonw. Inst. Biol. Control* 25, 28
pp.

84. Zwölfer, H. 1970. Investigations on
Chaetorellia sp. attacking *Centaurea solstitialis* L. *Rep. Commonw. Inst. Biol. Control*

85. Zwölfer, H. 1970. Der "Regionale
Futterpflanzenwechsel" bei phytophagen Insekten als evolutionäres
Problem. *Z. Angew. Entomol.* 65 :
233–39

86. Zwölfer, H. 1970. Unpublished observations

87. Zwölfer, H., Eichhorn, O. 1966. The
host ranges of *Cassida* spp. attacking Cynareae in Europe. *Z. Angew.
Entomol.* 58 :384–97

88. Zwölfer, H., Harris, P. 1966. *Ceutorhynchus litura* (F.) a potential
insect for the biological control of
thistle, *Cirsium arvense* (L.) Scop.
in Canada. *Can. J. Zool.* 44 :23–30

89. Zwölfer, H., Patullo, W. 1970. Zur
Lebensweise und Wirtsbindung des
Distel-Blattkäfers *Lema cyanella*
L. (*puncticollis* Curt.) *Anz. Schädlingskunde.* 43 :53–59

THE BIONOMICS OF LEAFHOPPERS 6007

DWIGHT M. DeLONG

Academic Faculty of Entomology,
The Ohio State University, Columbus, Ohio

The recent publication of the Metcalf catalogue of world Cicadelloidea (128) listed 4378 species, placed in 591 genera and 54 subgenera. These were assigned to the following 17 different families: Tettigellidae, Hylicidae, Gyponidae, Ledridae, Ulopidae, Evacanthidae, Nirvanidae, Aphrodidae, Hecalidae, Euscelidae, Coelidiidae, Eurymelidae, Macropsidae, Agalliidae, Iassidae, Idioceridae, and Cicadellidae. The catalogue includes all of the literature on the leafhoppers to 1942. Since that time at least 1200 additional species and 100 additional genera have been described. The world literature at the present time therefore covers some 5500 to 6000 species and probably 700 or more genera.

Morphological characters have been used predominantly for the classification of the leafhoppers, particularly the type, form, and venation of wings, structures of the head and crown, position of the ocelli, general shape of the body, and the genital structures of both sexes. The character of the structures of the male genital chamber, the styles, aedeagus, connective, plates, pygofer, and the apodemes on the anterior portion of the abdomen have been used especially for the separation of species. In most groups these characters are excellent for specific identification. In a few groups, *Oncopsis* Burmeister, for instance, they have proved to be of little or no value.

ISOLATED POPULATIONS, MUTANTS AND SPECIES CONCEPTS

Certain species of leafhoppers such as *Macrosteles fascifrons* have a wide geographical range through different life zones extending from Mexico and Puerto Rico to Alaska, from tropical to arctic habitats. As now recognized by taxonomists, this species is a complex of genetic mutants ranging in size from 2.0 to 5.1 mm and occurring on a great variety of host plants. Economic and ecologic studies have revealed isolated populations of this species which were biologically, ecologically, or physiologically specific but could not be separated morphologically.

Beirne (6), after a detailed study of all of the material in the genus *Macrosteles,* has affirmed that physiologic species and isolated ecologic populations undoubtedly exist and that the progeny of certain populations are sedentary, which would favor isolation. Severin (180) failed in his attempt to mate a long-winged form from the Montara Mountains of California with a short-winged form from the eastern United States. With the exception of the wing condition these were not different morphologically. The occurrence of the same insect *M. fascifrons* on *Puccinellia nutkaensis* in the tidal flats of Muir Inlet, Alaska, which are submerged twice daily with water at 1°C and containing numerous icebergs (44, 47), is an excellent example of ecologic isolation.

When mating does not take place between two similar populations they must be considered as specific biologic and genetic entities. The most important characters for separating species are those which the animal uses for itself and these populations have separated themselves even though man may not be able to separate them with visible characteristics. The geneticist, biologists, and ecologists recognize living breeding, isolated field populations as genetic species but the taxonomist is inclined to recognize species only when he can find morphologic characters on preserved specimens.

BIOLOGY

Reproduction.—This is usually bisexual and eggs are normally produced as a result of mating. Black & Oman (10) showed that *Agallia quadrinotata* reproduces parthenogenetically, and the progeny, insofar as is known, are females. Bisexual reproduction has not been observed in this species. In other species, if a male attempts to mate with a female and she is not receptive to his approach, she may kick viciously with her hind legs, and if he persists, she may walk or fly away (141, 173).

Male adults normally appear in the field a short time before females of the same brood and disappear before the females at the end of the brood. In the case of overlapping broods, the males of one brood will mate with females of another brood. Species that hibernate as adults may mate either in the autumn or the spring, but apparently these mate predominantly before hibernation. However, in certain hibernating species which have been observed, the males are present in abundance in the spring, but usually are greatly outnumbered, 4 to 1 (127), or 20 to 1 (130). If a species migrates, mating is common during migration, and the number of gravid females usually increases from near 10, to 85 or 90 percent (112). In other cases males rarely, if ever, survive the hibernation period.

Normally, one mating is sufficient to fertilize the eggs of the female during her lifetime (39). Where comparisons have been made (137), once-mated females live longer and produce more progeny than multi-mated females of the same brood.

The duration of the mating function is usually from a few minutes to an hour. In the case of *Colladonus geminatus,* 2 of 10 pairs were in coition for a period of 5 and 6 days, respectively (185).

Preoviposition.—The preoviposition period will vary with the species, their seasonal habits and climatic conditions. In several species which have been observed, the oviposition period is from 1.5 to 10 days (39, 89, 137, 175, 178, 185, 211), but in hibernating species this may be several months in duration. The preoviposition period of *Empoasca fabae,* based on many females, studied during 3 years, averaged 6.4 days (39). The preoviposition period for the same species was previously recorded as 23.3 days (63), 12.8 days for *Endria inimica* (25, 26), and 7 to 13 days for *C. geminatus* (185).

Oviposition.—The period of oviposition of individual females will vary from a few days to a few months even in the same species, and will average from 1 to 2 months. The number of eggs laid will usually not exceed 200 to 300.

Biological records show a great diversity in both egg laying conditions and the number of eggs produced by a single female, in different species and under different conditions. Insofar as is known, all eggs are inserted beneath the epidermis of leaves or stems of plants in slits made by the ovipositor. Although eggs are usually laid singly and a few at a time, often only one or two to a leaf (39, 84, 144, 211), they may be laid in clusters of about 20 (84) under the leaf epidermis (159, 211) or in rows (84, 183) in the petiole. The under surface of the leaf or petiole is usually selected (39, 144); the upper surface is used by certain species for a small portion of the eggs (106). The selection of an oviposition site on a suitable plant by *E. fabae* depends upon leaf characteristics and physiological age of the leaves. Youthful tissue is avoided; terminal leaflets received more eggs than subterminal leaflets and most eggs were placed between the apical and basal leaves of an axillary branch (131).

Eggs of *Colladonus clitellarius* are deposited in leaves of trees, and the newly hatched nymphs drop to herbaceous vegetation beneath, feeding preferably upon dandelion (77). *Gyponana mali, G. contractura,* and probably other species of *Gyponana* have a similar habit (41).

Certain species are known to cover their eggs when laid with a secreted liquid which becomes white or chalky when dry, forming a covering of thread-like papillae over the egg puncture (183), or spread on by combs of long tibial spines (157).

The number of eggs produced will average between 100 and 200 per female, but will vary greatly with the species. The lowest recorded fecundity is an average of 27 eggs per female (83), 16 eggs per female (204), and 44 eggs per female (25).

Probably the highest recorded fecundity for any leafhopper is 1170 viable eggs in 65 egg clusters produced in 87 days by one female of *Homalodisca insolita,* and several other females of the same species produced from 600 to 800 eggs each (159, 211). The maximum record for a female of *Circulifer tenellus* is 675 eggs, with the average overwintering female depositing 300 to 400 eggs each (89).

Records of *E. fabae* for 3 years gave a fecundity of about 200 eggs per female (39). These were sometimes laid in a month at the rate of 6 or more eggs per day or over a period of 92 days at the rate of 2.1 eggs per day. All females for all generations averaged 2.7 eggs per day.

In the case of *C. tenellus,* although egg laying continues throughout life it declines after 2 or 3 weeks (171). Records of *Deltocephalus sonorus* show that the peak of oviposition is reached in 72 to 96 hr after mating and the females tend to oviposit on alternate days thereafter (83). In species such as *E. fabae* the oviposition seems to remain more constant throughout the life of the female (39). In *C. tenellus,* oviposition occurs at a minimum temperature of 50°F, is accelerated by increasing temperatures to approximately 100°F, and declines at temperatures above 100°F (89).

As many as 35 eggs may be deposited in a single growing leaf and may cause it to become distorted and curled. Apple fruits may be injured by egg punctures of *Draeculacephala mollipes.* As many as 125 egg punctures per apple have been counted, each egg puncture containing more than 20 eggs (19).

Incubation.—The period of incubation is dependent upon the accompanying temperature. During the summer months, the number of days necessary for incubation will vary from 2.5 to 6 (29). Several other species will average about 10 days (9, 23, 39, 62, 63, 137, 178). Certain species require from 13 to 21 days (25, 26, 183, 185, 211). In the case of hibernating insects which lay their eggs either late in the autumn or early spring, or of overwintering eggs, the incubation period is often of one to several months duration (164, 178, 199, 211).

Hatching.—The process of hatching usually requires about 30 minutes, and often the nymphs are unable to free the abdomen and die during the hatching process (39, 178). Eggs laid by second-brood females did not hatch in late summer or early autumn, suggesting a diapause condition (144). A cold shock is necessary for egg hatching when laid in woody stems or twigs (91).

Nymphal development.—Leafhopper nymphs, after hatching normally pass through five instars (39, 84, 89, 132, 158, 183). Certain conditions cause the insect to pass through only four instars or as many as six (182, 185).

The number of days necessary for the completion of nymphal development varies considerably. From available records, *E. fabae* seems to require the shortest amount of time to complete summer development, 7 or 8 days, an average of 15 and a maximum of 26 (1, 39, 62). Several species require about 30 days (25, 77, 91, 130, 174, 185) ; from 30 to 66 days is required for several species (83, 182, 183, 211). The longest period recorded for nymphal development is 10 months in the case of *Dorycephalus platyrhynchus* (4).

The food plant upon which the nymphs are reared seems to influence the rate of development of the two sexes. The female of *Hordnia circellata* developed more rapidly on grape and the male developed more rapidly on alfalfa (182). When nymphs of *Texananus incurvatus* were reared on healthy celery and diseased (curly top) celery, there was a higher mortality of nymphs on the healthy celery (183).

Number of generations.—The number of generations per year of a given species will depend to a large extent upon the geographic area in which it occurs, the type of host plant upon which it lives, the type of plant tissue in which the eggs are placed, and similar pertinent factors. The length of the life cycle, period of egg laying, and the number of generations may be determined by the physiology of the host plant and the prevailing temperature, and may vary from one to five or even six in Florida (9).

If the eggs are laid in a tree or shrub, green tissue alone can be used and one generation a year is produced (170) as in *Macropsis trimaculata* on plum and peach (91), *Empoasca maligna* on apple (2), and *Typhlocyba quercus* on prune (132). If the host plant is an annual, the number of successive seasonal plantings, or cuttings in the case of alfalfa, furnishes a long season of available green tender foliage upon which several seasonal generations may be produced.

The number of seasonal generations in *C. tenellus* is variously reported by several authors. This is probably due to its range over widely different geographic areas and associated climatic conditions. In California and Arizona, nymphs occur in the field in every month except December and January; in this area five generations may occur, one or two in the desert before migration, and two or more later in the season. In the northern portion of the range not more than three occur (23, 175, 178). Three or four generations are common for *E. fabae* in the eastern United States (39). The adult peak emergence of developing potato leahoppers has been placed in the accumulation of day-degrees which is 480 to 500 over a base of 52.5°F (110). Five generations a year have been reported for *M. fascifrons* (117).

Several species have two seasonal generations (2, 77, 84, 144, 204). Four of the Proconiini have two and a partial third generation (211). Other species have three seasonal generations (2, 211).

Longevity.—Longevity is dependent upon the biological habits of the species and the season of the year under consideration. In the case of hibernating adults, longevity may be considered as several months (199), as much as 9 or 10 months in some cases (175); but longevity in the summer months will vary from a few days to 3 or 4 months (39, 104, 211). Several of the common pest species observed, such as *E. fabae* (39), *C. tenellus* (89), *Graminella nigrifrons* (204), *M. fascifrons* (130), *Dalbulus maidis* (29), and *D. sonorus* (83) have an average longevity during the growing season of 30 to 50 days.

In general the females, especially if mated, live longer than males of the same age (89, 130, 175). On the other hand, males in some cases have been observed to live longer than females (137, 178).

Certain important factors that may influence longevity are available moisture, type of food plant, temperature, and seasonal conditions. The influence of the host plant upon longevity is seen in the case of *C. tenellus* where females lived longer on flixweed than on sugar beet (89).

Although sugar beet leafhopper females lived longer than males under all temperatures observed, they lived three times as long as males under temperatures of 40° to 60° F. This probably has some significance in respect to the successful hibernation of the species. *Texananus spatulatus* lived longer on healthy celery plants than on those infected with aster yellows and the time required for the completion of the nymphal development was less on sugar beet infected with curly top than on healthy sugar beets (181).

Overwintering.—Leafhoppers may overwinter in any life stage, but few species pass the winter in more than one life stage, at least in the same geographic area. The adult stage is a common form of overwintering. Certain of the empoascans such as *Emposaca recurvata* (45) all species of *Erythroneura* (2), and species of many other genera (45, 84, 94, 182, 211) pass the winter in this manner. Certain species such as *Empoasca fabae, Macrostetes fascifrons,* and *Circulifer tenellus* pass the winter in the adult stage but in an area remote from their summer breeding ground, sometimes a milder area, so that migration is necessary in the spring or early summer.

A large number of species of leafhoppers including all of the *Typhlocyba* (2, 132), several species of *Empoasca* (45), *Idiocerus, Macropsis,* and other genera pass the winter in the egg stage in plant tissues. The eggs of *E. inimica* laid in soil at the base of the plant often dessicate and will not hatch (26). *C. clitellarius* eggs are reported as overwintering in fallen leaves and a cold shock is necessary for hatching (106). The eggs of *Scaphytopius acutus* overwinter in apple and stone fruit leaves and there is evidence of egg diapause (144).

Although *M. fascifrons* normally does not overwinter in the northern United States and southern Canada except in Washington (85), adults of this species are able to survive in frozen sod if snow covered (130). This is a rather unusual situation in leafhoppers. Prolonged exposure of adults at −12°C and −15°C gave 100 percent mortality in a few minutes (170). *Dorycephalus platyrhynchus* has been reported wintering as a partly grown nymph (4).

Conditions may favor year-round activity of certain leafhoppers. This is apparently true of *E. fabae* where there is a permanent subtropical breeding ground. Also wintering adults of *C. tenellus* in California feed during the winter months when the sun warms the habitat (174).

LEAFHOPPER FEEDING HABITS

Several investigators have made detailed studies of the feeding habits of leafhoppers. Smith & Poos (195) studied the feeding habits of six species of *Empoasca* and found that those species which produce white spotting on the leaves, e.g., *Empoasca bifurcata* fed on the underside of the leaves and produced the white spotting on the upper surface. The palisade cells are torn, the cell walls of the spongy mesophyll are broken and the cell contents emptied. On apple, mesophyll-feeding leafhoppers remove the contents of the cells in the mesophyll layer (122). Other workers observing the same type of feeding attributed the white spots to empty air spaces (97), or to the destruction of chlorophyll in the mesophyll cells and the collapse of the overlying membrane (197). In Hawaii, *Empoasca solana* produced a white stipling effect on the leaves of *Amaranthus* spp. but hopperburn resulted from its feeding on castor bean and other hosts (15).

The feeding habits of *E. fabae* have been studied by several workers and all agree that it feeds on phloem tissue, the cells of which are punctured, torn, and distorted. When leafhoppers puncture plant tissue during feeding they normally secrete an apparent salivary product which forms a sheath about the stylets; this secretion is apparently protein or some pectinate substance (194). An abundance of this sheath material is often present in the phloem cells. The gelatinous sheath formed by the beet leafhopper encasing the stylets, seals off the penetrated cells external to the phloem (7). Certain of the species of the Proconiini also feed in the xylem (108).

The beet leafhopper *C. tenellus* (7) and other species (164, 205) are phloem feeders. From the limited work which has been done, the results seem to indicate that the vectors of plant viruses are all phloem feeders; the mesophyll feeders are apparently not disease vectors.

While all workers agree upon the method of feeding of *E. fabae,* there are two different explanations concerning the injury caused by this type of feeding. Certain workers (102, 195) assert that hopperburn and other symptoms are caused by interference with the translocation of food materials and water due to the physical plugging of the xylem and the destruction of the phloem cells. Other investigators (9, 15, 63, 81, 124) support the theory that a toxin is injected into the plant during or preceding insect feeding.

Although the entire compliment of salivary fluid is not known, recent investigations (8) showed that both amylase and invertase are injected into the plants with saliva by *E. fabae*. Hereford (92), working with *E. solana,* found that both diastase and invertase were injected into the plant by both nymphs and adults. The diastase, besides being secreted by the salivary gland, is apparently also formed by certain yeasts which are regurgitated by the insect gut.

Medler (124) describes plant injury by *Empoasca fabae* as a combination of the feeding in vascular tissue and the action of a specific compound injected during the feeding process. This salivary secretion causes hypertrophy of the affected phloem cells which, in turn, causes an interruption of the translocation of photosynthetic materials from the leaves to a degree that causes plasmolysis of parenchymal cells and the resulting hopperburn. Other secondary external symptoms which then appear are the red and yellow coloration of leaves, stunting and curling of bush bean leaves, and other characteristic injuries (39).

Microchemical tests of plant cells injured by *E. fabae* feeding showed a greater accumulation of starch grains and sugars, especially glucose and fructose (81). Histological studies showed disorganization and granulation of the plastids of affected tissue, clogging and isolation of vascular bundles, and complete disorganization of the phloem region of severely injured tissue (81).

Increases in respiration, accompanied by reduction in photosynthesis were usually noted after intensive feeding; the reduction of photosynthesis is related in degree to the extent of leafhopper feeding (113). Photosynthesis and transpiration in apple leaves fed upon by *E. fabae* may be reduced early in the growing season when the leafhopper population is usually low; thus the capacity of the injured leaves to function normally is permanently impaired (122).

Hibbs, Dahlman & Rice (93), working with *E. fabae* on potato foliage, showed that sugar content was normally higher in a tolerant than in a non-susceptible variety of potato. Normally in both varieties the sugar concentrations of the foliage were progressively less from terminal to mid-range to basal foliage. Following leafhopper feeding, the sugar concentrations became reversed and were progressively greater from terminal to basal foliage. The greatest elevation of sugar concentration occurred in physiologically aged basal leaves of the susceptible variety in which necrosis and destruction of primary foliage is severe.

While hopperburn is probably the result of the elevation of sugars to intolerable levels, the primary cause of sugar increase in plant leaves is unknown. Contributary causes are probably the injection of amylase and invertase which may act directly in sugar production and conversion, and the injury to the phloem cells which may initiate the release of plant hormones that elicit starch hydrolysis (9).

Although both the physical eruptions of plant cells and the feeding injections of body fluids may cause hopperburn in plants, the strongest evidence that toxins introduced by *E. fabae* are responsible, is the fact that hopperburn has been produced artificially by emulsions made from mascerated bodies of the nymphs and injected into the main ribs of the leaf; injections made in the same manner using various fluids, mascerated bodies of other leafhoppers, and punctures by various mechanical instruments while producing certain types of injury, did not produce hopperburn (58, 65–67). Further evidence of the introduction of leafhopper toxins in plant tissues is illustrated by the work of Fenton (60) who observed that nymphs of *Norvellina chenopodii* caused crimson "feeding" spots on pigweed *Cheopodium albun* L. but adult feeding with the same mouth structures did not.

Fife & Frampton (70) claimed to have shown that a pH gradient in the sugar beet leaf affects feeding behavior of *C. tenellus* and that this could be altered by the use of CO_2.

ECOLOGIC ASPECTS AND PLANT RELATIONSHIPS

Available water and humidity.—These factors are very important to the maintenance of life in leafhoppers which live entirely upon liquid foods, usually plant sap, and are not able to survive long without plant sap or an adequate substitute (43, 89). The process of feeding goes on almost continuously and excessive amounts of sap are taken from plant tissues and utilized in the process of nutrition.

Certain of the life stages are dependent upon high humidity, and two species are known to live normally under tidal submergence (44, 129) ; in one case, the water is approximately 1°C (44). Leafhopper eggs are normally deposited in tender plant stems or in the under sides of leaf ribs or veins. In this position the incubation period is passed in saturated humidity. After hatching, the nymphs and adults, if present, remain on and feed from the under surface of the leaf. Seventy-five to 100 percent of the stromata on the average leaf are found on the underside, and since all or the major part of the transpiration occurs through these stomata, the leafhoppers on the under surface of the leaf live continuously in a highly humidified atmosphere. High humidity is necessary for the survival of many species. Desert leafhoppers are adapted to more xeric conditions. In spite of this, the hibernating beet leafhopper females in California may die because of lack of moisture during certain winters (89). Many leafhoppers have specific plant host relationships; Oman (138), Poos & Wheeler (163), Ross (168), and DeLong (34, 36, 40, 42, 43) have cited food or host plants of several groups of leafhoppers. The relation of leafhoppers to plant associations and their successions in several plant formations in the United States has been discussed previously (34, 43).

Host selection and attractancy.—All degrees of difference occur between polyphagus and monophagus insects (48, 49). Dethier (49) designates three responses, orientation to food, the biting response, and continued feeding. Thorsteinson (209) states that the response stimulus may be an attractant odor, probably derived from a substance which is generally peculiar to the food plants of an insect.

Favorable food plants for the adults are not necessarily the plants on which successful development occurs. At least 33 different kinds of plants will serve as food for *Dalbulus maidis,* adults, but only 20 species of Graminacae were found suitable for oviposition; eggs hatch on 16 hosts, but nymphs developed successfully only on corn (155).

When *D. maidis* was given a choice of corn hybrids, it did not select the corn-stunt-susceptible hybrid over the hybrid exhibiting resistance. The differences in egg laying on these hybrids in the field were insignificant (21) ; similar results were obtained with *Graminella nigrifrons* (22). On the basis of leafhopper counts and egg deposition, disease-susceptible hybrids in general were not preferred over or were more attractant than hybrids exhibiting disease resistance when leafhoppers are given a choice under field conditions.

Cereal crops are preferred by the six-spotted leafhopper *Macrosteles fascifrons* (117). In the field, spring oats are chosen over winter wheat; also spring lettuce is preferred to flax, but in the autumn the latter preference is reversed. This is apparently a case of seasonal physiological changes due to plant growth and maturity. The maturity of the crop affects the size of the population also. Bird weed serves only as a food plant for large populations of *M. fascifrons;* celeriac, celery, and carrot were preferred food plants. Cultivated lettuce is preferred to prickly lettuce (214).

The early spring climatic factors which preceded the spring migration of *E. fabae* determined the initial preferred food plant; the total sugar content of the leaves at the time of migration was an indicator (39).

MIGRATION

Migration is the phenomenon which causes the transfer of insect populations from place to place by mass flights (101). Migration is considered as an outgrowth of the ecological problem and is usually attributed to the development of unfavorable conditions in the insect habitat (31, 179). The cause of leafhopper migration is not known and may differ with the species concerned. Several causal factors such as hunger, overcrowding, drying of host plant, and formation of grain heads, have been suggested (14, 24, 31, 112, 126, 179). Certain workers (112, 173, 179) have observed migrations when the plants were not dry, and ample food still available. Migration in delphacids when there was no shortage of food was attributed to internal conditions (165). Carter (13) observed that beet leafhoppers tend inherently to migrate when they become adults and this may occur at a time when there is a change in osmotic concentration of the host plant and the plant begins to dry. Potato leafhoppers migrate only in May and June although plants and crops upon which they feed become dry continuously throughout the summer months and are unsuitable as sources of food (39, 43).

Only three or four species of leafhoppers are known to migrate periodically and the migrations occur at the same time each year or season. In the case of two of the major species, only a portion of the population leaves a permanent breeding area and does not return. All of the available evidence points strongly to a genetic factor in the life cycle pattern as in aphids and in migrating vertebrates.

Temperature is apparently an important factor as a stimulant in the timing of migration (112, 115) and the migrants are usually deposited or "pile up" at a weather front, as occurred with *Circulifer tenellus* in California (112) or with *Empoasca fabae* in Illinois 1951–59 (98) and in Wisconsin 1951–60 (151).

Migrations may occur at different altitudes and may extend for various distances; they may be caused by, or be due largely to air currents. For this reason many collections have been made at different altitudes by both airplanes and balloon. Glick (78) found leafhoppers at almost every altitude from 200 to 14,000 feet. The greatest number of specimens was collected between 200 and 3000 feet elevation. In 1960 (79), the predominant number of *Empoasca* were collected at 200 feet.

Most insects seem capable of withstanding extreme environmental conditions to which they might be exposed for a reasonable duration (215); 97 percent of 1610 insects caught in nets suspended from a balloon at altitudes of between 1000 and 5000 feet were alive and undamaged (208). Migration flights may require several days or weeks; the sugar beet leafhopper is known to have migrated 200 miles in 2 days on prevailing winds (50).

One of the important factors in flight survival is apparently the fat content of the body. Many beet leafhoppers migrate in excess of 100 miles in Utah and other western states (50), follow river valleys in flight and become fewer in numbers as they travel farther from the place of origin. The females survive better than the males. The fat content of migrating females decreases about 75 percent in migrations of 300 to 400 miles (74).

An interesting observation made at a lighthouse in Delaware Bay showed a total of 3466 leafhoppers belonging to 50 species and 33 genera that migrated from vegetation to a light at least 3 miles from the nearest shore (200).

Two patterns of migration have been observed among leafhoppers of economic importance. One is illustrated by the sugar beet leafhopper in which two migrations occur each year, one in autumn and one in spring; the migrations being in opposite directions. The other is represented by the potato leafhopper, in which, insofar as is known, migration occurs but once a year, in the spring, is always in one direction and arises from a permanent breeding area.

Beet leafhopper.—Although the beet leafhopper *Circulifer tenellus* is known to occur in Europe and Africa along the Mediterranean (72, 73), on desert plants in Florida (35), on field crops in Illinois (46) and South Dakota (43, 186), it is natively a desert, dry climate insect inhabiting the western United States. It has been studied particularly in California (7, 70, 73, 111, 112, 121, 177, 184, 187), Washington (24, 134), Idaho (14, 51, 52, 53, 74, 86, 142, 148, 149), New Mexico (23, 166), Utah (50, 87, 109a), and Arizona (23, 112, 166).

These leafhoppers live on desert plants, particularly those belonging to the mustard family, during the winter, upon which an early spring brood is usually produced. When these desert plants become unfavorable for feeding or some inherent factor in the cycle is triggered, there is a spring migration during which cultivated crops become infested. Most of these migrants become established upon extensive stands of Russian thistle which constitutes the principle weed host in Idaho, Utah, and other areas, so that only a small percentage find their way to distant sugar beet fields (71). The leafhopper produces enormous populations in either situation until autumn when the beets are harvested and the Russian thistle becomes dry. In both situations, the leafhoppers migrate again. If sufficient late summer precipitation has caused the mustard seeds to germinate, it is an attractive and acceptable green host plant. If rainfall has been deficient, the leafhoppers may be forced to accept an undesirable "holdover" green plant such as sage brush to obtain water. If rainfall is further delayed, a high mortality results and the large populations are greatly reduced. In view of the insect's wide geographic range over the western deserts and the varied conditions to which it is exposed, several types of winter or summer desert host plants may be selected (51).

In California, where desert saltbush occurs, this is the principal autumn host for the beet leafhopper. With the late annual sprouting of winter annuals, plantain, peppergrass, and other hosts of less importance, the leafhoppers migrate to and spend the winter and spring on these plants. In the areas where the desert saltbush is still succulent, many leafhoppers remain on this host and breed on the winter annuals in the spring. In drier seasons the winter annuals become dry in the spring before the leafhoppers can produce a spring generation. In this case the largest spring populations are produced in the hills below 1500 feet where the annuals remain succulent for a longer time. During the summer the leafhoppers breed in abundance on Russian thistle on abandoned land, which simulates the Idaho summer cycle.

From the ecological point of view, many factors or combinations of factors may determine sequence of food plants and population levels. In burned, overgrazed, or deteriorated rangeland, the most important food plants are flixweed, perfoliate pepperweed, and tumble mustard. In areas where sagebrush predominates, green tansy mustard is most important (148, 149). A seasonal sequence of host plants for egg laying may occur. The largest nymphal populations occur on fanweed in early summer and on papata in midsummer; in later summer smotherweed is preferred.

The acreage of Russian thistle is a good index to the size of the populations of beet leafhoppers. When Russian thistle acreage is reduced and replaced by brome grass which is not a host, the leafhopper populations are proportionately smaller. Summer drought accompanied with high temperatures usually causes drying of Russian thistle and destruction of both eggs and nymphs, reducing populations extensively. In general, climatic factors determine population levels which are reached at anytime with either irrigated plants or the native host plants.

Winged leafhoppers are continuously produced at a breeding site and mass migrations are usually associated with large populations. Prevailing temperatures and the time of sunrise and sunset determine the timing of the flight peaks. When the temperature is above the threshold of flight, the prevailing winds in the morning or evening, spring, summer, or autumn determine the direction of flight (112).

The potato leafhopper.—*Empoasca fabae* occurs over most of the eastern half of the United States. Although a few species of *Empoasca* are known to hibernate as adults in the northern states, Ohio, Indiana, Illinois, etc. (45) it has been impossible for *E. fabae* to hibernate in this area (39, 45, 160). Laboratory tests have demonstrated that without doubt *E. fabae* could not survive the winter months in the North Central Area (33). Certain species of *Empoasca* pass the winter in the egg stage on wild host plants. None of these when reared has proven to be *E. fabae* (45, 160). In view of the late spring (mid-May to early June) and the abrupt appearance of the adults of the potato leafhopper in abundance, a migration for *E. fabae* was proposed (39, 45, 160). Subsequent investigations have verified the migration theory but without finding the factor causing migration from what is apparently a permanent breeding area in the lower Mississippi Valley (31, 32, 79, 125, 152). Migration occurs only in late spring, is northward, and is apparently controlled by the prevailing winds under certain climatic conditions. The first adults arriving are predominately fertile females (126, 151).

The warm winds from the south pass over the Great Plains in the spring as fields are plowed in the Mississippi Delta. Convection currents from plowed soil can lift leafhoppers upward at the rate of several miles per hour. They are then transported northward on these warm winds and deposited when the warm air meets a cold air mass (151). No evidence has been presented that there is a fall migration to the breeding area.

Six-spotted leafhopper.—Spring migrations of the six-spotted leafhopper *Macrosteles facifrons* have been studied by several entomologists (115, 123, 126, 213, 216). Both eggs and adults may survive the winter in the southern United States. Movements of certain populations are from Louisiana and western Texas, beginning in late March and arriving at North Dakota, Montana, and Manitoba in early June. Small grain crops are the main food plants during migration (213). There is no evidence of overwintering either in egg or adult stages in the northern United States (213) except in Washington (85). Overwintered eggs were hatched in Minnesota but large field populations are attributed to migrating adults (123). It overwinters as adult in Kansas (198).

This species apparently migrates in the spring and early summer except in Washington (85), and migrations may occur at any time that light and temperature are suitable; leafhoppers may accumulate at the source before migration (18).

Six-spotted leafhoppers have been collected at several altitudes up to 13,000 feet (78). Study of air currents for several years revealed that the migrants arriving at Winnipeg traveled from South Dakota, Nebraska, and Kansas (136). The adult migrants which arrived in Canada were either predominantly (130) or entirely (116) females.

D. maidis is unable to survive the winter in Mississippi (154). Field collections have indicated that it migrates north and easterly from the southwest United States or possibly Mexico each year (156).

INSECT PARASITES

The leafhoppers are parasitized by species of several insect families, particularly the Eulophidae, Platygastridae, Mymaridae, Trichogrammatidae, Chalcididae, Encyrtidae, and the Dryinidae of the Hymenoptera; the Pipunculidae of the Diptera; the Epipyropidae of the Lepidoptera; and species of Strepsiptera.

Many parasites have been reared from leafhopper eggs by several workers, a few from sugar beet leafhopper eggs. The eggs on salt bush are more heavily parasitized than those on sugar beet (90, 176, 199). Several were reared from other economic leafhoppers (1, 2, 118, 133, 201, 202). Ten species of egg parasites were reared from five species of large leafhoppers (211).

In general, field observations show only 1 to 3 percent of egg parasitism. Several workers have reported rather high percentages of attacks. Sixty percent is reported for *Erythroneura plena* (118); heavy parasitism of *Typhlocyba prunicola* (133); 78 percent of second brood of *T. pomaria* (216); 22–35 percent for late summer and winter broods of the beet leafhopper (176, 179); 65–70 percent of winter eggs of *Typhlocyba pomaria* (1); *Aphelopus comesi* reduced materially the second brood of *Erythroneura maculata* (2).

An attempt was made in 1954 (99) to introduce five egg parasites of the beet leafhopper from the Mediterranean area into California. After colonization and release of these in several areas of leafhopper abundance, only two species were recovered from the field.

Pipunculidae.—Pipunculid larvae feed internally in the bodies of leafhopper nymphs and cause their death by rupturing the body wall, usually at the junction of the thorax and abdomen when attempting to leave the body (88). Their presence is not easily detected although the larvae are large and usually completely fill the abdomen of the leafhopper. Leafhoppers belonging to five subfamilies, the Dorycelphalinae, Macrospinae, Iassinae, Cicadellinae, and Deltocephalinae, are known to be parasitized (88, 138). Two species deposit eggs in the adults and nymphs of beet leafhoppers (90, 199). One of these becomes important in certain seasons. Dorst observed 60 percent parasitism in migrating beet leafhoppers (50) and Esaki & Hashimota reported 65 percent parasitism of *Nephotettix bipunctatus cincticeps* (55). Three species of *Pipunculus* are parasites of the sugarcane leafhopper, *Perkinsiella sacchiricida,* in Hawaii.

Dryinidae.—Many species of Dryinidae have been reported as parasites of leafhoppers. Fenton (59, 61) cited 77 species of leafhoppers belonging to

22 genera, mostly grass-feeding species, attacked by 42 species of dryinids belonging to the Anteoninae. Certain species of dryinids may confine their attacks to a specific leafhopper host and the parasite may be on a specific body region or area. Others may attack leafhoppers belonging to different genera. Normally, only one parasite is found in a host; Ainslie reported a specimen of *Macrosteles fascifrons* carrying two dryinid sacs (3). Species of dryinids have been on and/or reared from many species of leafhoppers (1, 59, 61, 76, 90, 130, 176, 199). Only adults of *Cicadula* and *Deltocephalus* were attacked by *Gonatopus ombrodes* (3) and only adults of *Typhlocyba* were parasitized by *Aphelopus melalevens* (3).

Most species of the family normally place their eggs within the body of the host (20). At least one placed her eggs on the host externally (59). The first instar larva usually lives in the host body, the second instar pushes out between the abdominal or thoracic segments and forms a pupal sac. The position on the host is relatively constant within a species and may be dorsal, lateral, or beneath a wing pad in different species. George & Davidson (77) reported 17 percent parasitism by dryinids on *Colladonus clitellarius*.

Strepsiptera.—Several species of Strepsiptera are considered beneficial because of their attacks upon many pest species of Cicadellidae. Oman reported 80 percent parasitism in a population of *Acertagallia fuscoscripta* (138). Thirty percent of *Idiocerus atkinsoni* in India were parasitized by *Pyrilloxenos compactus* (207).

Biologically considered, the strepsipterans do not kill their hosts directly, but leave an opening through the body wall when emerging, into which pathogenic bacteria or fungi may enter and thus kill the host.

Species of strepsipteran parasites belonging to three families, Triozoceridae, Halictophagidae, and Elenchidae, are known to attack leafhoppers. Bohart (11, 12) and Pierce (153) discussed and described several North American species which attack leafhoppers; nineteen species of Strepsiptera are listed which are known to attack some thirty species of leafhoppers. Members of several subfamilies of Nearctic leafhoppers are known to be attacked; these include Ledrinae, Hecalinae, Agallinae, Aphrodinae, Iassinae, Idiocerinae, Tettigellinae, Cicadellinae, and Deltocephalinae (138). Most of the species attacked feed upon grasses or low-growing vegetation.

NEMATODE PARASITES

A parasitic nematode belonging to the Goridiaceae (Nemataphorpha) was occasionally found attacking the beet leafhopper (179). Mermithid infestations in leafhoppers have been reported in Japan, ranging from 40 to 70 percent in certain years, which have given a practical control of certain species (203).

PREDATORS

In several cases spiders have been referred to as preying upon leafhoppers but without specific identification (2, 90, 176). Many nymphs of *Colladonus clitellarius* were destroyed by spiders (77). The nymphs seem to be attacked more than the adults. Both spiders and the red ant *Dorymyrmex pyramicus* prey upon *Empoasca* spp. (9).

In California the green lacewing *Chrysopa californica* has been observed feeding upon the nymphs and adults of the beet leafhopper; chrysopid larvae also prey upon young nymphs of leafhoppers in apple orchards (2, 219). When aphids and leafhoppers were feeding on the same leaf, the leafhoppers were practically ignored by the lacewing larvae (114); larvae and adults of at least three species of lady bird beetles avoided leafhopper nymphs (225). At least four species of predaceous Hemiptera are predators of the beet leafhopper in California (90, 176).

The mirid, *Diaphnidia pellucida* preys upon the white apple leafhopper in Pennsylvania orchards (202). *Triphleps insidiosus* preys on the nymphs of *E. fabae* and *E. maligna* (2). Both pentatomids and assassin bugs are predators of leafhoppers; the most important of these is the soldier bug *Hyaliodes vitripennis* which attacks both nymphs and adults of *Typhlocyba pomaria*. Predaceous mites also attack *T. pomaria* (2).

Species of several genera of sand wasps (Nyssoninae) provision their ground nests with many species of leafhoppers (56, 57). Although a cell frequently contains only nymphs or adults of one species of leafhopper, a wide variety of leafhoppers, both nymphs and adults, are used as prey. The species selection is probably due to two factors, the geographic location of the species of wasp and the proximity to a population of certain grass- or shrub-feeding leafhoppers.

Female wasps hunt their prey on herbs and low trees near their burrows, usually within a few meters. An observed provisioning with 21 species of 17 meadow genera indicated a grassy hunting ground; when 12 species were used belonging to 7 genera of woody plant-inhabiting forms, a shrub hunting ground was indicated (56).

The number of leafhoppers placed in a nest will depend upon the type of wasp concerned and the size of the prey; if small sized species such as *Empoasca* or *Typhlocyba* are used in provisioning the cell, then several must be supplied for the developing wasp larva, but if a species of *Gyponana* is used as in the case of *Gorytes simillimus,* only a few specimens are necessary. As few as 3 large adults were found in a fully provisioned nest and as many as 23 adults of a smaller species were used. The prey of *Psammaecius punctulatus* is always 4 or 5 adults of the leafhopper *Selenocephalus obsoletus* (68, 69). Other species such as *Dienoplus laevis,* a palaearctic form, uses both Cicadellidae and Cercopidae to provision its burrow. The wasp *Hapalomellinus albitomentosus* uses as prey both nymphs and adults of five families of Homoptera belonging to both the Auchenorrhyncha and the Sternorrhyncha. Nineteen cells of the wasp *Didineis lunicornis* contained 191 adults and 3 last instar nymphs of *Parabolocratus brunneus*. Most of the species of these wasps are not host-specific, but the species of *Agrogorytes* are highly host-specific.

The leafhoppers are stung by the alert wasps, the beak of each leafhopper is then grasped by the wasp's mandibles and carried to the burrow where the individuals are piled venter up and the egg is laid usually on the ventral thorax of the top adult leafhopper.

When stung the leafhoppers usually die and soon dessicate if placed in a rearing container. Some respond to stimuli up to 48 hr after being stung. The burrows where they are stored are usually moist at the bottom and the leafhoppers remain soft for several days. The provisioning of the nest is usually rapid. The egg is usually laid immediately after the burrow is provisioned and ready to close, but certain species, such as *Bembecinus prismaticus,* for instance, in which the female deposits her eggs erect on a pedestal, supplies leafhoppers to the larva as needed after the egg hatches; the first leafhopper is brought to the nest 24 hr after oviposition.

In Japan, a wasp was observed which constructs as many as 50 nests per square meter; each cell contains from 10 to 20 leafhoppers. These cells were provisioned with some 700 individuals belonging to 18 species of homopterans of the families Cicadellidae, Cercopidae, and Issidae (226).

Certain of the sphecid wasps of the Crabroninae build their nests in rotting timbers and provision them with leafhoppers. Examination of 47 cells made by the wasp *Crossocerus ambiguus* showed a total of 9588 leafhoppers averaging 20.4 per cell and belonging to 3 leafhopper genera and 31 species: 11 of *Empoasca,* 6 of *Typhlocyba,* and 14 of *Erythroneura* (28).

FUNGAL PARASITES

Several types of fungi are known to parasitize specific leafhoppers; other fungi attack several species belonging to different genera. *Empusa sphaerosperma* attacks a wide variety of leafhopper hosts, including the apple leafhoppers and potato leafhopper (9, 171). Other species of *Empusa, E. grilli* and *E. muscae,* are important parasitic fungi of leafhoppers.

Empusa apiculata attacks *Deltocephalus repletus* and *D. distinguendus* in Europe (210); the same fungus is known to attack species of *Typhlocyba* in the United States and *Empoasca papayae* Oman in tropical America.

Fungal diseases reduce the number of overwintering beet leafhoppers and spring migrants in the California fog belt in favorable years (53). Many species of fungi from at least six genera are known to attack several species of leafhoppers (17). Species of green fungi attack leafhoppers. One species of this group, *Metarrhizium album,* parasitic on a species of leafhopper, is white in color, and *M. brunneum,* parasitic on a cicadellid in the Philippine Islands, is yellow to brown.

LEAFHOPPER COMMUNICATIONS

As early as 1907, Kirkaldy (109) quoted a note from Muir regarding sound production in leafhoppers. Very little attention was directed to this subject until the work of Ossiannilsson in 1949. During his extensive studies of sound production in Homoptera, leafhoppers were used repeatedly as experimental subjects. Sound production is quite common in all the groups of Auchenorrhyncha. The leafhoppers' sounds are scarcely detected by the human ear unless amplified. Many species produce audible sounds if placed in a glass tube about a centimeter in diameter, together with a small portion of the host plant, one end of the tube plugged with cotton and the other inserted into the human ear.

The tymbal is the most complex sound-producing organ known. This mechanism varies greatly in the several groups of the Auchenorrhyncha, but the structure designated as the tymbal in the leafhoppper is both homologous to and analogous with the tymbal organ of the cicadas (141). Sound is produced by a pair of membranous structures which are on the dorsolateral surface of the first abdominal segment.

Several types of songs have been interpreted by sound recordings (139). The common song of most species consists of an unmusical drumming, buzzing, rattling, or croaking sound (141). The quantity of sound produced by leafhoppers will also vary as well as the quality. Species such as *Cicadella notata* and *Empoasca viridula* produce very weak songs, but *Agallia brachyptera* emits a particularly loud sound.

The common song in *Aphrodes trifasciata* consists of three parts. First, a long or short clacking sound, secondly, a rapid drumming for a few seconds, and thirdly, a series of trilling notes higher in pitch. Several males of the same species may sing in chorus and the pitch is usually very low, but may be high and occasionally shrill. Normally, two males when together produce alternating calls. The alternating singing usually ends with actions of rivalry, each male trying to push the other away by "kicking" actions with the hind legs. The alternate singing may thus be interpreted as a call of rivalry even though females may not be present (141).

Calls interpreted as courtship songs have been observed by Ossiannilsson which are quite different from the general call of the same species. These normally lead to mating and if the female can produce sound she answers with an invitation call. In *Doratura stylata* both sexes have well-developed sound-producing organs which can produce the same quality of sound (140). When together the male emits an enticing call which the female answers with her call of invitation. The male then searches for the female and mates with her. Calls of distress or dissatisfaction are interpreted as dying calls and are emitted when the leafhoppers are trapped, held, or are in the process of dying, and are often of very short duration.

Although sound is produced predominantly by males, some females, particularly in *Iassus, Oncopsis, Platymetopius,* and *Ulopa,* are known to have sound-producing mechanisms similar to the males or weaker. As mentioned, in *Doratura,* the female can produce sound equally as well as the male white in *Paropia* the striated, well-developed tymbal of the female is different from the poorly developed organ of the male and apparently represents a simpler and more primitive stage of development.

Often the mating or enticing call of the male is combined with a dance which is caused apparently by the vigorous movements of the abdomen and the necessity of moving the legs in order to maintain proper balance. Unlike the cicadas, the leafhoppers will sing in darkness, at almost any hour, even after they have remained in darkness for extended periods. In *Empoasca fabae* both species produce sound which has been recorded and analyzed. It seems to indicate intraspecies communication. A "laughing" call was common and has not been associated with mating (212). Also both sexes produce sound in the beet leafhoppper *Circulifer tenellus* (196). The male emits two precopulatory and one copulatory sound, the female emits only one precopulatory sound.

Another means of communication in the beet leafhopper is by means of the vibrations of a substrate (147). Sexually mature males perform a characteristic courtship dance in response to sexually mature virgin females when both sexes are on a common substrate. If the continuous substrate is severed, communication is no longer effected.

PLANT RESISTANCE

Insect resistance in general has been treated by Painter (143) and by Beck (5). Painter divided resistance into three major phases, thus, 1. preference for oviposition and food; 2. tolerance, ability to recover or withstand the infestation; and 3. antibiosis, the adverse affect of the plant on the insect. The exact combination of these three factors depends upon the plant variety and the species of insect concerned. Leafhopper resistance may be accomplished by any one of these factors or a combination of two or three. Resistance has been developed as a major factor in the control of certain crop leafhoppers; especially those attacking alfalfa, beans, cotton, potato, and sugar beets have been worked upon extensively. Resistant varieties of plants have been developed in each case.

Resistant varieties of the host plant may cause reduced fecundity, increased mortality, decreased size of individuals, and abnormal length of life. Certain resistant varieties may support a population as large as one which would kill a susceptible host and grow normally in spite of the insect attack. Cumulative resistance is possible by means of crosses and plant breeding. Often a wild plant, a single cultivated plant, or a small group of field plants will exhibit resistance and these are often selected and used as breeding stock for future field strains or varieties.

Potato.—Resistance to *Empoasca fabae* in *Solanum* results from all three factors, preference, antibiosis, and tolerance, the exact combination of the three factors depending on the plant variety. The resistance to leafhoppers present in the potato variety *Sequoia* has given nearly as much protection from the insect as has the use of insecticides. Resistance to *E. fabae* was recognized in potato varieties by Sleesman et al (188, 189, 191–193), some of which showed tolerance and others resistance (86). Sleesman also observed resistance in wild potato (190), one strain of which showed immunity; no young nymphs ever appeared. Crossing of wild potatoes with commercially important varieties has developed excellent resistance. Immunity can be developed only through species hybridization (191). Many seedlings selected for several years have been commercially resistant, but none of them is immune. Evidence has been presented that the solanaceous alkaloidal glycosides, tomatine, solanine, and leptine I, and the aglycons, solanidine and demissidine may be effective feeding deterrents in the production of resistance to *E. fabae* by certain varieties of potato (27).

If resistant varieties are not used and chemical controls fail, distinct losses are suffered in all crops attacked by *E. fabae*. In the case of potato, carbohydrates expended in increased respiration are not available for plant growth and tuber development (113).

Alfalfa.—Several investigators have cooperated in the development of resistant strains of alfalfa. Certain varieties, such as *Culver,* show tolerance to leafhopper populations, others show low damage ratings associated with low populations, suggesting unattractiveness; clones may differ in this respect (30, 100, 224). Certain data suggested that leafhopper damage was associated with stand reduction rather than yield production per se (30). Resistance to leafhopper attacks may be observed as tolerance to stunting, tolerance to color changes and associated reduction in forage quality. This, in turn, might reduce oviposition as the females prefer yellow color for egg laying (135). The extent of yellowing may be an index to the resistance (172). Resistance may also be observed by the antibiosis reaction. If nymphal development is retarded, nymphal survival may be reduced (135). On alfalfa, when leafhoppers were controlled before cutting, nine varieties averaged 45 percent more growth two weeks after cutting. Under the same conditions in the spring, ⅓ ton more hay per acre is produced (220). Timely cutting of alfalfa is a very important factor in population control (150).

Bean.—Investigations to produce resistant varieties of beans have shown that all crosses of resistant and susceptible varieties were susceptible to both nymphal infestations and hopperburn (222). The Bonavist variety was resistant to hopperburn but susceptible to nymphal infestations (222). Some of the common bean varieties show varying degrees of resistance to nymphal infestations, others exhibit no hopperburn (16, 119, 221). Certain flower, seed, and other plant characteristics can be associated with resistant varieties of beans.

Certain bush bean leaves have hooked hairs which trap and kill both nymphal and adult *E. fabae* leafhoppers (39, 75, 120 162). In general, simple pubescence is not correlated with resistance on crop plants, but in soy beans freedom from leafhopper injury is correlated with the occurrence of rough hairy pubescence (96, 103, 163). Other investigators believe that some other factor, at least in part, is responsible for resistance (162, 223). Experimental work showed 74.6 percent of *E. fabae* nymphs hatched from foreign, more glabrous, Italian strains of soy beans, while 25.4 percent hatched from native, rough hairy Ohio strains which were selected for vigor (161).

Leafhopper-infested lima beans produced 57 to 291 pounds (depending on the variety) less shelled beans per acre than did noninfested beans (54).

In south central Idaho where losses in bean yield have been very high in years when large beet leafhopper populations invaded this area, a bean variety, Great Northern U. I. 15, early gave excellent resistance and good yields. Selection and crossing have produced varieties resistant to leafhopper attacks in dry beans and in snap beans (86, 121, 134). Several varieties of squash and pumpkin have been developed which show considerable resistance to curly top.

Sugar beet.—Resistance of beets to the curly top virus apparently originated by accidental hibridization with wild beets (142). Plant selecting and further hybridizing have continued to produce resistant varieties combining curly top resistance with a degree of downy mildew resistance. Two varieties, U.S. 12 and U.S. 33, have been used widely; the former produced a satisfactory yield under a very high degree of leafhopper infestation (53, 142).

Cotton.—In the United States *E. fabae* causes only occasional damage to cotton but species of *Empoasca* are very important in several other countries. The three species that have caused the most injury to cotton and for which resistance has been reported, are *E. fascialis* in Africa, *E. devastans* in India, and *E. terra-reginae* in Australia. Resistant cotton varieties have been developed by plant breeding on all three continents.

The breeding of cotton varieties resistant to *E. fascialis* has been quite successful in South Africa where the first widespread and severe attacks occurred in 1922. The development of resistant varieties has been so successful that the present strain, U 4, grown commercially, is not entirely free from leafhopper attacks, but the yield is not reduced to any extent (145, 146). The feeding of small populations causes the edges of the leaves to turn yellow, then red, finally dying then slowly being shed. Also extreme stunting and reduction in yield follows leafhopper attacks. The cotton varieties vary considerably regarding leafhopper injury. The Sea Island varieties and Pima, an Egyptian variety (*Gossypium barbadense*) were the most severely injured. The American Upland cotton variety (*G. hirsutum*) was usually least injured. Without exception, the resistant varieties were more hairy than the susceptible varieties (143).

In a similar manner, resistant varieties for *E. devastans* in India were developed. The combination of factors producing resistance differed in these resistant varieties from those resistant to *E. fascialis,* one of which was the deficiency of hairiness. Both length and diversity of hairs is required for resistance. While the association of hairiness and resistance to *E. fascialis* is well established in cotton, there is little evidence of how hairiness affects the leafhopper. There appears to be a preference for glabrous leaves or lack of preference for hairy surfaces (143). The varieties resistant to *E. fascialis* were not resistant to *E. terra-reginae.* Several of the formerly mentioned leafhopper resistant varieties were crossed and a resistant strain has been produced which is acceptable. Similarly sized populations of nymphs of *E. fascialis* caused different amounts of injury to different varieties of cotton. At the same population level certain varieties appear more tolerant than others. Other species of *Empoasca* such as *E. bigutulla* in Taiwan are known to attack cotton.

The U 4 variety of cotton is resistant to *Empoasca* attacks in both South Africa and Queensland, but is not resistant to *E. devastans* in India. The variety which shows resistance to *E. devastans* is less hairy, fewer eggs are laid over a long period of time, and the plants are apparently antibiotic, causing slow fecundity to feeding adults. In Egyptian cotton, the correlations between toughness of leaf veins, hairiness, and leafhopper populations were all significant. In general, as the hairiness increases the leafhopper population decreases (107).

Cranberry.—No cranberry variety is entirely immune to feeding of *Scleroracus vaccinii.* The evidence indicates that it is entirely feasible to originate varieties from parental material now on hand which will have considerable more resistance to leafhopper attacks than any variety now grown (217). One named variety (Shaw's Success) and several hybrid seedlings are more resistant to vector attack than McFarlin, which possesses considerable resistance in plantings (217).

SAMPLING OF FIELD POPULATIONS

Attempts to estimate field population levels have been made in connection with all types of crop production studies. Several methods of sampling have been used in order to obtain at least comparable population densities. Criticisms and evaluations have been offered by a few persons regarding the accuracy of certain methods which have been used or proposed (37, 38, 64, 82, 95, 167, 169, 206).

LITERATURE CITED

1. Ackerman, A. J. 1919. Two leaf-hoppers injurious to apple nursery stock. *U.S. Dep. Agr. Bull.* 805: 1–35

2. Ackerman, A. J., Isley, D. 1931. The leafhoppers attacking apples in the Ozarks. *U.S. Dep. Agr. Tech. Bull.* 263:1–40

3. Ainslie, G. N. 1920. Notes on *Gonatopus ombrodes,* a parasite of Jassids. *Entomol. News* 31:169–73

4. Ball, E. D. 1920: The life cycle of Hemiptera (excluding aphids and coccids) *Ann. Entomol. Soc. Am.* 13:142–51

5. Beck, S. D. 1965. Resistance of plants to insects. *Ann. Rev. Entomol.* 10:207–32

6. Beirne, B. P. 1956. Leafhoppers of Canada and Alaska. *Can. Entomol.* 88:1–180

7. Bennett, C. W. 1934. Plant-tissue relations of the sugar-beet curly-top virus. *J. Agr. Res.* 48:665–701

8. Berlin, L. C. 1962. *The morphology and the salivary enzymes of the digestive system of the potato leafhopper,* Empoasca fabae (*Harris*). M.S. thesis, Iowa State Univ., Ames, Iowa

9. Beyer, A. H. 1922. Experiments on the biology and tipburn disease of the bean leafhopper, with methods of control, *Empoasca mali* LeBaron. *J. Econ. Entomol.* 15:298–302

10. Black, L. M., Oman, P. W. 1947. Parthenogenesis in a leafhopper, *Agallia quadripunctata* (Provancher). *Proc. Entomol. Soc. Wash.* 49:19–20

11. Bohart, R. M. 1941. A revision of the Strepsiptera with special reference to the species of North America. *Calif. Univ. Publ., Entomol.* 7(6):1–160

12. Bohart, R. M. 1943. New species of *Halictophagus* with a key to the genus in North America. *Ann. Entomol. Soc. Am.* 36(3):341–59

13. Carter, W. C. 1927. Populations of *Eutettix tenellus* (Baker) and the osmotic concentrations of the host plants. *Ecology* 8:350–62

14. Carter, W. C. 1930. Ecological studies of the beet leafhopper. *U.S. Dep. Agr. Tech. Bull.* 206:1–114

15. Carter, W. C. 1939. Injuries to plants caused by insect toxins. *Bot. Rev.* 5:273–326

16. Chalfant, R. D. 1965. Resistance of bunch bean varieties to the potato leafhopper, and relationship between resistance and chemical control. *J. Econ. Entomol.* 58:681–82

17. Charles, V. K. 1941. A preliminary check list of the Entomophageous Fungi of North America. *U.S. Dep. Agr. Insect Pest Surv. Bull.* 21:707–816

18. Chiykowski, L. N., Chapman, R. K. 1965. Migration of the six spotted leafhopper, *Macrosteles fascifrons* (Stal). Part 2. Migration of the six spotted leafhopper in Central North America. *Wis. Agr. Exp. Sta. Res. Bull.* 261:21–45

19. Claassen, P. W. 1933. *Draeculacephala mollipes* (Say) A cicadellid pest of apples. *J. Econ. Entomol.* 26:282

20. Clausen, C. P. 1940. *Entomophagus Insects,* 316–25. New York, London: McGraw-Hill

21. Collins, H. L., Pitre, H. N. 1969. Corn stunt vector leafhopper attractancy to and oviposition preference for hybrid dent corn I. *Dalbulus maidis. Ann. Entomol. Soc. Am.* 62:770–72

22. Collins, H. L., Pitre, H. N. 1969. Corn stunt vector leafhopper attractancy to and oviposition preference for hybrid dent corn. II *Graminella nigrifrons. Ann. Entomol. Soc. Am.* 62:773–75

23. Cook, W. C. 1941. The beet leafhopper. *U.S. Dep. Agr. Farmers' Bull.* 1884:1–21

24. Cook, W. C. 1967. Life history, host plants and migrations of the beet leafhopper in the Western United States. *U.S. Dep. Agr. Tech. Bull.* 1365:1–122

25. Coupe, T. R., Schulz, J. T. 1968. The influence of controlled environments and grass hosts on the life cycle of *Endria inimica. Ann. Entomol. Soc. Am.* 61:74–76

26. Coupe, T. R., Schulz, J. T. 1968. Biology of *Endria inimica* in North Dakota. *Ann. Entomol. Soc. Am.* 61:802–6

27. Dahlman, D. L., Hibbs, E. T. 1967. Responses of *Empoasca fabae* to tomatine, solanine, leptine I; to-

matidine, solanidine and demissidine. *Ann. Entomol. Soc. Am.* 60 : 732–40

28. Davidson, R. H., Landis, B. J. 1938. *Crabro davidsoni* Sandh. A wasp predaceous on adult leafhoppers. *Ann. Entomol. Soc. Am.* 30 :5–8

29. Davis, R. 1966. Biology of the leafhopper *Dalbulus maidis* at selected temperatures. *J. Econ. Entomol.* 59 :766

30. Davis, R. L., Wilson, M. C. 1953. Varietal tolerance of alfalfa to the potato leafhopper. *J. Econ. Entomol.* 46 :242–45

31. Decker, G. C. 1959. Migration mechanisms of leafhoppers. *Proc. N. Centr. Br. Entomol. Soc. Am.* 14 : 11–12

32. Decker, G. C., Cunningham, H. B. 1967. The mortality rate of the potato leafhopper and some related species when subjected to prolonged exposure at various temperatures. *J. Econ. Entomol.* 60 : 373–79

33. Decker, G. C., Maddox, J. V. 1967. Cold hardiness of *Empoasca fabae* and some related species. *J. Econ. Entomol.* 60 :1641–45

34. DeLong, D. M. 1923. The distribution of the leafhoppers of Presque Isle, Pennsylvania and their relation to plant formations. *Ann. Entomol. Soc. Am.* 16 :363–74

35. DeLong, D. M. 1925. The occurrence of the beet leafhopper, *Eutettix tenella* Baker in the Eastern United States. *J. Econ. Entomol.* 18 :637–38

36. DeLong, D. M. 1926. Food plant and habitat notes on some North American species of *Phlepsius*. *Ohio J. Sci.* 24 :69–72

37. DeLong, D. M. 1931. A successful method for sampling populations of quick-moving insects. *J. Econ. Entomol.* 24 :1108–9

38. DeLong, D. M. 1932. Some problems encountered in the estimation of insect populations by the sweeping method. *Ann. Entomol. Soc. Am.* 25 :13–17

39. DeLong, D. M. 1938. Biological studies of the leafhopper, *Empoasca fabae* as a bean pest. *U.S. Dep. Agr. Tech. Bull.* 618 :1–60

40. DeLong, D. M. 1941. The genus *Arundanus* in North America. *Am. Midl. Natur.* 25 :632–43

41. DeLong, D. M. 1942. A monographic study of the North American species of the subfamily Gyponinae exclusive of *Xerophloea*. *Ohio State Univ. Press Biol. Ser.* 5 : 1–187

42. DeLong, D. M. 1949. The leafhoppers or Cicadellidae of Illinois. *Ill. Nat. Hist. Surv. Bull.* 24 :93–376

43. DeLong, D. M. 1965. Ecological aspects of North American leafhoppers and their role in agriculture. *Bull. Entomol. Soc. Am.* 11(1) :9–26

44. DeLong, D. M. 1970. An Alaskan leafhopper which lives normally beneath icy tidal submergence. *Ohio J. Sci.* 70 :111–14

45. DeLong, D. M., Caldwell, J. S. 1935. Hibernation studies of the potato leafhopper, *Empoasca fabae* (Harris) and related species of *Empoasca* occurring in Ohio. *J. Econ. Entomol.* 28 :442–44

46. DeLong, D. M., Kadow, K. J. 1937. The sugar beet leafhopper appears in Illinois. *J. Econ. Entomol.* 30 : 210

47. DeLong, D. M. et al. 1966. Development and ecological succession in a deglatiated area of Muir Inlet, Southeast Alaska, *Ohio State Univ. Res. Found, Inst. Polar Studies, Rep.* 20 Part 5 ; 97–120

48. Dethier, V. G. 1947. *Chemical Insect Attractants and Repellants*, 1–289. Philadelphia : Blakiston

49. Dethier, V. G. 1953. Plant host perception in phytophagus insects *Int. Congr. Entomol., 9th, Trans. 1951,* 81–87

50. Dorst, H. E., Davis, E. W. 1937. Tracing long distance movements of beet leafhopper in the desert. *J. Econ. Entomol.* 30 :948–54

51. Douglass, J. R., Cook, W. C. 1954. The beet leafhopper. *U.S. Dep. Agr. Circ.* 943 :1–21

52. Douglass, J. R., Hallock, H. C. 1957. Relative importance of various host plants of the beet leafhopper in southern Idaho. *U.S. Dep. Agr. Tech. Bull.* 1155 :1–11

53. Douglass, J. R., Wakeland, C. W., Gillette, J. A. 1939. Field experiments for control of the beet leafhopper in Idaho 1936–39. *J. Econ. Entomol.* 32 :69–79

54. Eckenrode, C. J., Ditman, L. P. 1963. An evaluation of potato leafhopper damage to lima bean. *J. Econ. Entomol.* 56 :551–53

55. Esaki, T., Hashimota, S. 1936. Report of the leafhoppers injurious to the rice plant and their natural enemies. *Fukuoka Entomol. Labr. Dept. Agr. Kyushu Univ.* 7:1–31

56. Evans, H. E. 1966. *The Comparative Ethology and Evolution of the Sand Wasps*, 1–526. Cambridge: Harvard Univ. Press

57. Evans, H. E. 1968. Notes on some digger wasps that prey upon leafhoppers. *Ann. Entomol. Soc. Am.* 61:1343–44

58. Eyer, J. R. 1922. Notes on the etiology and specificity of the potato tip burn produced by *Empoasca mali* LeBaron. *Phytopathology* 12:181–84

59. Fenton, F. A. 1918. The parasites of leafhoppers. *Ohio J. Sci.* 18:177–296

60. Fenton, F. A. 1925. Notes on the biology of the leafhopper, *Eutettix strobi* Fitch. *Proc. Iowa Acad. Sci. 1924* 31:437–40

61. Fenton, F. A. 1927. New parasitic Hymenoptera of the subfamily Anteoninae from the Americas. *Proc. U.S. Nat. Mus.* 72:8

62. Fenton, F. A., Hartzell, A. 1920. The life history of the potato leafhopper (*Empoasca mali* LeBaron) *J. Econ. Entomol.* 13:400–8

63. Fenton, F. A., Hartzell, A. 1923. Bionomics and control of the potato leafhopper, *Empoasca mali* LeBaron. *Iowa Agr. Exp. Sta. Res. Bull.* 78:377–440

64. Fenton, F. A., Howell, D. E. 1957. A comparison of five methods of sampling alfalfa fields for arthropod populations. *Ann. Entomol. Soc. Am.* 50:606–11

65. Fenton, F. A., Ressler, I. L. 1921. Artificial production of tip-burn. *J. Econ. Entomol.* 14:510

66. Fenton, F. A., Ressler, I. L. 1922. Artificial production of hopperburn. *J. Econ. Entomol.* 15:288–95

67. Fenton, F. A., Ressler, I. L. 1922. Artificial production of tip-burn. *Science* 55:54

68. Ferton, C. 1901. Notes detachees sur l'instinct des hymenopteres melliferes et ravisseurs. *Ann. Soc. Entomol. France* 70:83–148

69. Ferton, C. 1905. Notes sur l'instinct des hymenopteres melliferes et ravisseurs. 3° ser. *Ann. Soc. Entomol. France* 74:56–101

70. Fife, J. M., Frampton, V. L. 1936. The pH gradient extending from the phloem into the parenchyma of the sugar beet and its relation to the feeding behavior of *Eutettix tenella. J. Agr. Res.* 53:581–93

71. Fox, D. E. 1938. Occurrence of the beet leafhopper and associated insects of secondary plant successions in Southern Idaho. *U.S. Dep. Agr. Tech. Bull.* 607:1–44

72. Frazier, N. W. 1953. A survey of the Mediterranean region for the beet leafhopper. *J. Econ. Entomol.* 46:551–54

73. Frietag, J. H., Frazier, N. W., Huffaker, C. B. 1955. Crossbreeding beet leafhopper from California and French Morocco. *J. Econ. Entomol.* 48:341–42

74. Fulton, R. A., Romney, V. E. 1940. The chloroform-soluable components of beet leafhoppers as an indication of the distance they move in the spring. *J. Agr. Res.* 61:737–43

75. Gandara, G. 1931. La Hierba de la langosta. *Acad. Noc. de Cien. "Antonio Abzate" Mem. y Rev.* 51:107–14

76. George, J. A. 1959. Note on *Epigonatopus plesius* (Fenton) a parasite of the six-spotted leafhopper, *Macrosteles fascifrons* (Stal) in Ontario. *Can. Entomol.* 91:256

77. George, J. A., Davidson, T. R. 1959. Notes on life history and rearing of *Colladonus clitellarius* (Say.). *Can. Entomol.* 91:376–79

78. Glick, P. A. 1939. The distribution of insects, spiders, and mites in the air. *U.S. Dep. Agr. Tech. Bull.* 673:1–151

79. Glick, P. A. 1960. Collecting insects by airplane with special reference to dispersal of the potato leafhopper. *U.S. Dep. Agr. Tech. Bull.* 1222:1–16

80. Granados, R. R. 1969. Pathogenicity of corn stunt to its vector, *Dalbulus maidis. Bull. Entomol. Soc. Am.* 15:195 (Abstract 274, Chicago meeting)

81. Granovsky, A. A. 1930. Differentiation of symptoms and effect of leafhopper feeding on histology of alfalfa leaves. *Phytopathology* 20:121

82. Gray, H. E., Treloar, A. E. 1933. On the enumeration of insect populations by the method of net collections. *Ecology* 14:356–67

83. Gustin, R. D., Stoner, W. N. 1968. Biology of *Deltocephalus sonorus*. *Ann. Entomol. Soc. Am.* 61 :77–82

84. Hackman, L. M. 1922. Studies of *Cicadella hieroglyphica* Say. *Univ. Kans. Sci. Bull.* 14 :189–209

85. Hagel, G. T., Landis, B. J. 1967. Biology of the aster leafhopper, *Macrosteles fascifrons*, in eastern Washington and some overwintering sources of aster yellows. *Ann. Entomol. Soc. Am.* 60 :591–95

86. Hallock, H. C. 1946. Bean leafhopper selection of bean varieties and its relation to curly top. *J. Econ. Entomol.* 39 :319–25

87. Hallock, H. C., Douglass, J. R. 1956. Studies of four summer hosts of beet leafhopper. *J. Econ. Entomol.* 49 :388–91

88. Hardy, D. E. 1943. A revision of the nearctic Dorilaidae. *Univ. Kans. Sci. Bull.* 29 :1–231

89. Harries, F. H., Douglass, J. R. 1948. Bionomic studies on the beet leafhopper. *Ecol. Monogr.* 18 :45–79

90. Hartung, W. J. 1919. Enemies of the leafhopper. Natural foes of *Eutettix tenella* in California. *Facts Sugar* VIII (24) :470–71, (26) : 492–93

91. Hartzell, A. 1937. Bionomics of the plum and peach leafhopper, *Macropsis trimaculata*. *Boyce Thompson Inst. Contrib.* 9 :121–36

92. Hereford, A. 1935. Studies of the secretion of diastase and invertase by *Empoasca solana* DeLong. *Ann. Appl. Biol.* 22 :301–9

93. Hibbs, E. T., Dahlman, D. L., Rice, R. L. 1964. Potato foliage sugar concentration in relation to infestation by the potato leafhopper, *Empoasca fabae*. *Ann. Entomol. Soc. Am.* 57 :517–21

94. Hildebrand, E. M. 1954. The generation time of the leafhopper *Baldulus maidis* in Texas. *U.S. Dep. Agr. ARS Plant Div. Rep.* 38 : 572–73

95. Hill, O. A. 1933. A new method for collecting samples of insect populations. *J. Econ. Entomol.* 26 :906

96. Hollowell, E. A., Johnson, H. W. 1934. Correlation between rough-hairy pubescence in soybeans and freedom from injury by *Empoasca fabae*. *Phytopathology* 24 :12

97. Horne, A. S., Lefroy, H. M. 1915. Effects produced by sucking insects and red spider upon potato foliage. *Ann. Appl. Biol.* 1 :370–86

98. Huff, F. H. 1963. Relation between leafhopper influxes and symoptic weather conditions. *J. Appl. Meteorol.* 2 :39–43

99. Huffaker, C. B., Hollaway, J. K., Doutt, R. L., Finney, G. L. 1954. Introduction of egg parasites of the beet leafhopper. *J. Econ. Entomol.* 47 :785–89

100. Jarvis, J. L., Kehr, W. R. 1966. Population counts versus nymphs per gram of plant material in determining degree of alfalfa resistance to the potato leafhopper. *J. Econ. Entomol.* 59 :427–30

101. Johnson, C. G. 1969. *Migration and Dispersal of Insects by Flight*, 1–763. Barnes & Noble Inc., Distrib. London : Methuen

102. Johnson, H. W. 1934. Nature of injury to forage legumes by the potato leafhopper. *J. Agr. Res.* 49 :379–406

103. Johnson, H. W., Hollowell, E. A. 1935. Pubescent and glabrous characters of soybeans as related to resistance to injury by the potato leafhopper. *J. Agr. Res.* 51 :371–81

104. Jensen, D. D. 1953. Longevity of the leafhoppers, *Collandonus geminatus* (Van Duzee) and *C. montanus* (Van Duzee) on peach. *J. Econ. Entomol.* 46 :1120–21

105. Jensen, D. D. 1959. A plant virus lethal to its insect vector. *Virology* 8 :164–75

106. Kaloostian, G. H. 1956. Overwintering habits of the geminate leafhopper in Utah. *J. Econ. Entomol.* 49 :272

107. Kamel, S. A., Farouk, Y. E. 1965. Relative resistance of cotton varieties in Egypt to spider mites, leafhoppers and aphids. *J. Econ. Entomol.* 58 :209–12

108. King, W. V., Cook, W. S. 1932. Feeding punctures of mirids and other plant sucking insects and their effect on cotton. *U.S. Dep. Agr. Tech. Bull.* 296 :1–11

109. Kirkaldy, G. W. 1907. Leaf-hoppers-supplement. *Rep. Exp. Sta. Hawaiian Sugar Planters Assoc. Div. Entomol. Bull.* 111

109a. Knowlton, F. F. 1929. The beet leafhopper in northern Utah. *Utah Agr. Exp. Sta. Bull.* 234 :5–64

110. Kouskolekas, C. A., Decker, G. C. 1966. The effect of temperature on the rate of development of the

potato leafhopper, *Empoasca fabae*. *Ann. Entomol. Soc. Am.* 59 :292–98

111. Lawson, F. R., Piemeisel, R. L. 1943. The ecology of the principal summer weed hosts of the beet leafhopper in the San Joaquin Valley, California. *U.S. Dep. Agr. Tech. Bull.* 848 : 1–37

112. Lawson, F. R., Chamberlain, J. C., York, G. C. 1951. Dissemination of the beet leafhopper in California. *U.S. Dep. Agr. Tech. Bull.* 1030 :1–59

113. Ladd, T. L., Jr., Rawlins, W. A. 1965. The effects of the feeding of the potato leafhopper on photosynthesis and respiration in the potato plant. *J. Econ. Entomol.* 58 :623–28

114. Lavallee, A. C., Shaw, F. R. 1969. Preferences of golden-eye lacewing larvae for pea aphids, leafhoppers and plant bug nymphs, and alfalfa weevil larvae. *J. Econ. Entomol.* 62 :1228–29

115. Leach, J. G., Mullins, J. R. 1942. The daily flight of aster leafhoppers as determined by a light trap. *W. Virginia Acad. Sci. Paper* 15 :93–95

116. Lee, P. E., Robinson, A. G. 1958. Studies on the six-spotted leafhopper *Macrosteles fascifrons* (Stal), and aster yellows in Manitoba, *Can. J. Plant Sci.* 38 :320–27

117. McClanahan, R. J. 1962. Food preferences of the six-spotted leafhopper, *Macrosteles fascifrons* (Stal). *Proc Entomol. Soc. Ontario* 93 :90–92

118. McConnell, H. S. 1931. A leafhopper injuring peach. *J. Econ. Entomol.* 24 :560

119. McFarlane, J. S., Rieman, G. H. 1943. Leafhopper resistance among bean varieties. *J. Econ. Entomol.* 36 :639

120. McKinney, K. B. 1938. Physical characteristics on the foliage of beans and tomatoes that tend to control some small insect pests. *J. Econ. Entomol.* 31 :630–31

121. Mackie, W. W., Esau, K. 1931. A preliminary report on the resistance to curly top of sugar beets in bean hybrids and varieties. *Phytopathology* 21 :997

122. Marshall, G. E., Childers, N. F., Brody, H. W. 1942. The effect of leafhopper injury on apparent

photosynthesis and transpiration of apple leaves. *J. Agr. Res.* 65 : 265–81

123. Meade, A. B., Peterson, A. G. 1964. Origin of populations of the six-spotted leafhopper, *Macrosteles fascifrons* in Anoka County, Minnesota. *J. Econ. Entomol.* 57 :885–88

124. Medler, J. T. 1941. The nature of injury to alfalfa, caused by *Empoasca fabae* (Harris) *Ann. Entomol. Soc. Am.* 34 :439–50

125. Medler, J. T. 1957. Migration of the potato leafhopper—a report on a cooperative study. *J. Econ. Entomol.* 50 :493–97

126. Medler, J. T. 1962. Long range displacement of Homoptera in Central United States, *Proc. Int. Congr. Entomol., 11th, 1960.* 3 :30–35

127. Medler, J. T., Pienkowski, R. L., Kieckhefer, R. W. 1966. Biological notes on *Empoasca fabae* in Wisconsin. *Ann. Entomol. Soc. Am.* 59 :178–80

128. Metcalf, Z. P. 1962–1968. General catalogue of the Homoptera *U.S. Dep. Agr. ARS Fasc. VI,* Parts 1–17 :5429 pp.

129. Metcalf, Z. P., Osborn, H. 1920. Some observations of insects of the between tide zone of the North Carolina Coast. *Ann. Entomol. Soc. Am.* 13 :108–19

130. Miller, L. A., Delyzer, A. J. 1960. A progress report on studies of biology and ecology of the six-spotted leafhopper, *Macrosteles fascifrons* (Stal) in southwestern Ontario. *Proc. Entomol. Soc. Ontario* 90 :7–13

131. Miller, R. L., Hibbs, E. T. 1963. Distribution of eggs of the potato leafhopper, *Empoasca fabae,* on solanum plants. *Ann. Entomol. Soc. Am.* 56 :737–40

132. Mulla, M. S., Madsen, H. F. 1955. A new leafhopper attacking prunes in California. *J. Econ. Entomol.* 48 :476

133. Mulla, M. S. 1956. Two mymarid egg parasites attacking *Typhlocyba* species in California. *J. Econ. Entomol.* 49 :438–41

134. Murphy, D. M. 1940. A great northern bean resistant to curly top and common bean mosaic. *Phytopathology* 30 :779–84

135. Newton, R. G., Barnes, D. K. 1965.

Factors affecting resistance of selected alfalfa clones to the potato leafhopper. *J. Econ. Entomol.* 58: 435–39

136. Nickiparick, W. 1965. The aerial migration of the six-spotted leafhopper and the spread of the virus disease, aster yellows. *Int. J. Bioclimatol. Biometeorol.* 9:219–27

137. Nielson, M. W., Toles, S. L. 1968. Observations on the biology of *Acinopterus angulatus* and *Aceratagallia curvata* in Arizona. *Ann. Entomol. Soc. Am.* 61:54–56

138. Oman, P. W. 1949. The Nearctic leafhoppers. A generic classification and check list. *Wash. Entomol. Soc. Mem.* 3:1–253

139. Ossiannilsson, F. 1946. On the sound-production and the sound-producing organ in Swedish Homoptera Auchenorrhyncha. (A preliminary note) *Opusc. Entomol.* 11:82–84

140. Ossiannilsson, F. 1948. On the sound-production of the females of certain auchenorrhynchous Homoptera. *Entomol. Tidskr. Stockholm*

141. Ossiannilsson, F. 1949. Insect drummers, *Opusc. Entomol. Suppl.* X: 1–145

142. Owen, F. V., et al. 1938. Curly top resistant sugar beet varieties in 1938. *U.S. Dep. Agr. Circ.* 513: 1–10

143. Painter, R. H. 1951. *Insect Resistance in Crop Plants.* New York: Macmillan. 520 pp.

144. Palmiter, D. H., Coxeter, W. J., Adams, J. A. 1960. Seasonal history and rearing of *Scaphytopius acutus* (Say). *Ann. Entomol. Soc. Am.* 53:843–46

145. Parnell, F. R. 1935. Origin and development of U. 4 cotton. *Empire Cotton Growers Rev.* 12:177–82

146. Parnell, F. R., King, H. E., Rustin, D. F. 1949. Jassid resistance and hairiness of the cotton plant. *Bull. Entomol. Res.* 39:539–75

147. Perkes, R. R. 1969. Substrate vibrations as a possible means of sexual communication in the beet leafhopper. *Bull. Entomol. Soc. Am.* 15(3):215 (Abstract, Chicago meeting)

148. Piemeisel, R. L. 1932. Weedy abandoned lands and the weed hosts of the beet leafhopper. *U.S. Dep. Agr. Circ.* 229:1–23

149. Piemeisel, R. L. 1938. Changes in

weedy plant cover on cleared sagebrush land and their probable causes. *U.S. Dep. Agr. Tech. Bull.* 654:1–44

150. Pienkowski, R. L., Medler, J. T. 1962. Effects of alfalfa cuttings on the potato leafhopper, *Empoasca fabae. J. Econ. Entomol.* 55:973–78

151. Pienkowski, R. L., Medler, J. T. 1964. Synoptic weather conditions associated with long range movement of the potato leafhopper, *Empoasca fabae*, into Wisconsin. *Ann. Entomol. Soc. Am.* 57:588–91

152. Pienkowski, R. L., Medler, J. T. 1966. Potato leafhopper trapping studies to determine flight activity. *J. Econ. Entomol.* 59:837–45

153. Pierce, W. D. 1918. The comparative morphology of the order Strepsitera together with records and descriptions of insects. *Proc. U.S. Nat. Mus.* 54(2242):391–501

154. Pitre, H. N. 1966. Corn virus diseases of recent occurrence in southeastern United States with emphasis on the corn stunt virus-vector-plant interrelationship in Mississippi. *Proc. Centr. Br. Entomol. Soc. Am.* 21:43–47

155. Pitre, H. N. 1967. Green house studies of the host range of *Dalbulus maidis*, a vector of corn stunt virus. *J. Econ. Entomol.* 60: 417–21

156. Pitre, H. N., Douglas, W. A., Combs, R. L., Jr., Hepner, L. W. 1967. Annual movement of *Dalbulus maidis* into the southeastern United States and its role as vector of the corn stunt virus. *Ann. Entomol. Soc. Am.* 60:616–17

157. Pollard, H. N., Yonce, C. E. 1965. Significance of length of tibial spines relative to oviposition processes by some leafhoppers. *Ann. Entomol. Soc. Am.* 58:594–95

158. Pollard, H. N. 1965. Description of stages of *Homalodisca insolita*, a leafhopper vector of phony peach disease. *Ann. Entomol. Soc. Am.* 58:699–701

159. Pollard, H. N. 1965. Fecundity of *Homalodisca insolita*—a leafhopper vector of phony peach virus disease. *Ann. Entomol. Soc. Am.* 58:935–36

160. Poos, F. W. 1932. Biology of the

potato leafhopper, *Empoasca fabae*
(Harris) and some closely related
species of *Empoasca*. *J. Econ.
Entomol.* 25 :639–46

161. Poos, F. W., Johnson, H. W. 1936.
Injury to alfalfa and red clover
by the potato leafhopper, *J. Econ.
Entomol.* 29 :325–31

162. Poos, F. W., Smith, F. F. 1931. A
comparison of oviposition and
nymphal development of *Empoasca
fabae* (Harris) on different host
plants. *J. Econ. Entomol.* 24 :361–
71

163. Poos, F. W., Wheeler, N. H. 1943.
Studies of host plants of the leafhoppers of the genus *Empoasca*.
U.S. Dep. Agr. Tech. Bull. 850 :
1–51

164. Putnam, W. L. 1941. The feeding
habits of certain leafhoppers. *Can.
Entomol.* 73 :39–53

165. Raatikainen, M. 1967. Bionomics,
enemies and population dynamics
of *Javesella pellucida* (F). *Ann.
Agr. Fenn 6 (Suppl. 2)* :149

166. Romney, V. E. 1943. The beet leafhopper and its control on beets
grown for seed in Arizona and
New Mexico, *U.S. Dep. Agr. Tech.
Bull.* 855 :1–24

167. Romney, V. E. 1945. The effect of
physical factors upon catch of the
beet leafhopper (*Eutettix tenellus*
(Bak) by a cylinder and two
sweep-net methods. *Ecology* 26 :
135–47

168. Ross, H. H. 1963. An evolutionary
outline of the leafhopper genus
Empoasca subgenus *Kybos* and a
key to the Nearctic fauna. *Ann.
Entomol. Soc. Am.* 56 :202–23

169. Saugstad, E. S., Bram, R. A.,
Nyquist, W. E. 1967. Factors influencing sweep-net sampling of
alfalfa. *J. Econ. Entomol.* 60 :421–
25

170. Saini, R. S. 1967. Low temperature
tolerance of the aster leafhopper.
J. Econ. Entomol. 60 :620–21

171. Schoene, W. J. 1938. Ecological
Studies of the white apple leafhopper. *J. Econ. Entomol.* 31 :229–
32

172. Searls, E. M. 1932. A preliminary
report on the resistance of certain
legumes to certain homopterous
insects. *J. Econ. Entomol.* 25 :
46–49

173. Severin, H. H. P. 1919. Notes on the

behavior of the beet leafhopper,
Eutettix tenellus (Baker). *J. Econ.
Entomol.* 12 :303–8

174. Severin, H. H. P. 1919. Investigations
of the beet leafhopper *Eutettix
tenella* (Baker) in California. *J.
Econ. Entomol.* 12 :312–26

175. Severin, H. H. P. 1921. Summary of
life history of beet leafhopper
(*Eutettix tenella* Baker). *J. Econ.
Entomol.* 14 :443–36

176. Severin, H. H. P. 1924. Natural
Enemies of beet leafhopper, *Eutet-
tix tenellus* (Baker). *J. Econ.
Entomol.* 17 :369–77

177. Severin, H. H. P. 1929. Additional
host plants of curly top. *Hilgardia*
3 :595–636

178. Severin, H. H. P. 1930. Life history
of beet leafhopper, *Eutettix tenellus* (Baker) in California. *Calif.
Univ. Publ. Entomol.* 5 :37–89

179. Severin, H. H. P. 1933. Field obser-
vations on the beet leafhopper,
Eutettix tenellus in California.
Hilgardia 7 :281–360

180. Severin, H. H. P. 1940. Potato
naturally infected with California
aster yellows. *Phytopathology* 30 :
1049–51

181. Severin, H. H. P. 1946. Longevity,
or life histories of leafhopper
species on virus-infected and on
healthy plants. *Hilgardia* 17 :121–
33

182. Severin, H. H. P. 1949. Life history
of the blue-green sharpshooter,
Neokolla circellata. *Hilgardia* 19 :
187–89

183. Severin, H. H. P. 1950. *Texananus
incurvatus*. III life history on
virus-infected and on healthy
plants. *Hilgardia* 19 :546–48

184. Severin, H. H. P., Henderson, C. F.
1928. Some host plants of curly
top. *Hilgardia* 3 :339–92

185. Severin, H. H. P., Klostermeyer,
E. C. 1950. *Colladonus geminatus*
and *C. montanus* life histories on
virus-infected and on healthy
plants. *Hilgardia* 19 :553–60

186. Severin, H. H. P., Severin, H. C.
1927. Curly top sugar beets in
South Dakota. *J. Econ. Entomol.*
20 :586–88

187. Severin, H. H. P., Thomas, W. W.
1918. Notes on the beet leafhopper,
Eutettix tenellus (Baker). *J. Econ.
Entomol.* 11 :308–12

188. Sleesman, J. P. 1936. Influence of

variety on potato leafhopper populations. *Ohio Agr. Exp. Sta. Bull.* 561 :43

189. Sleesman, J. P. 1938. Variety differences in potato leafhopper populations. *Ohio Agr. Exp. Sta. Bull.* 592 :48

190. Sleesman, J. P. 1940. Resistance in wild potatoes to attack by the potato leafhopper and the potato flea beetles. *Am. Potato J.* 17 :9–12

191. Sleesman, J. P. 1940. Developing potato resistance to leafhoppers. *Ohio Agr. Exp. Sta. Bull.* 617 : 26–27

192. Sleesman, J. P. 1945. Search for a leafhopper resistant potato. *Ohio Agr. Exp. Sta. Bull.* 659 :114–15

193. Sleesman, J. P., Stevenson, F. J. 1941. Breeding a potato resistant to potato leafhopper. *Am. Potato J.* 18 :280–98

194. Smith, F. F. 1933. The nature of the sheath material in the feeding punctures produced by the potato leafhopper and the three cornered alfalfa hopper. *J. Agr. Res.* 47 : 475–85

195. Smith, F. F., Poos, F. W. 1931. The feeding habits of some leafhoppers of the genus *Empoasca*. *J. Agr. Res.* 53 :267–85

196. Smith, J. W., Georghiou, G. P. 1969. The precopulatory and copulatory acoustics of the beet leafhopper and their associated behavior. *Bull. Entomol. Soc. Am.* 15 :215 (Abstract 191, Chicago meeting)

197. Smith, K. M. 1926. A comparative study of the feeding methods of certain Hemiptera and the resulting effects upon the plant tissue, with special reference to the potato plant. *Ann. Appl. Biol.* 13 :109–39

198. Smith, R. C. 1943. Common insects of Kansas. *Kans. St. Bd. Agr. Rep.* 62(255) :1–186

199. Stahl, C. F. 1920. Studies on the life history and habits of the beet leafhopper. *J. Agr. Res.* 20 :245–52

200. Stearns, L. A., MacCreary, D. 1938. Leafhopper migrations across Delaware Bay. *J. Econ. Entomol.* 31 : 226–29

201. Steiner, H. M. 1936. New nymphal-adult parasite of the white apple leafhopper. *J. Econ. Entomol.* 29 : 632–33

202. Steiner, H. M. 1938. Effects of orchard practices on natural enemies of white apple leafhopper. *J. Econ. Entomol.* 31 :232–40

203. Steinhaus, E. A. 1949. *Principles of Insect Pathology.* New York : McGraw Hill. 757 pp.

204. Stoner, W. N., Gustin, R. D. 1967. Biology of *Graminella nigrifrons* a vector of corn (maize) stunt virus. *Ann. Entomol. Soc. Am.* 60 :496–505

205. Storey, H. H. 1938. Investigations of the mechanism of the transmission of plant viruses by insect vectors. II. The part played by puncture in transmission. *Proc. Roy. Soc. Ser. B* 125 :455–77

206. Strickland, A. H. 1961. Sampling crop pests and their hosts. *Ann. Rev. Entomol.* 6 :201–20

207. Subramanian, T. V. 1922. Some natural enemies of the mango leafhoppers (*Idiocerus* spp) in India. *Bull. Entomol. Res.* 12 :465–67

208. Taylor, L. R. 1960. Mortality and viability of insect migrants high in the air. *Nature* 186 :410

209. Thorsteinson, A. J. 1960. Host selection in phytophagous insects. *Ann. Rev. Entomol.* 5 :193–218

210. Turian, G. 1960. Mycose a *Empusa apiculata* Thaxtcher les Cicadelles du genre *Deltocephalus*. *Mitt. Schweiz. Entomol. Ges.* 33 :88–90

211. Turner, W. F., Pollard, H. N. 1959. Life histories and behavior of five insect vectors of phony peach disease. *U.S. Dep. Agr. Tech. Bull.* 1188 :1–57

212. Vargo, A., Shaw, K. 1969. The acoustical behavior of *Empoasca fabae* (Harris). *Bull. Entomol. Soc. Am.* 15 :214 (Abstract 179, Chicago meeting)

213. Wallis, R. L. 1962. Spring migration of the six-spotted leafhopper in the western great plains. *J. Econ. Entomol.* 55 :871–74

214. Wallis, R. L. 1962. Host preference of the six-spotted leafhopper. *J. Econ. Entomol.* 55 :998–99

215. Wellington, W. G. 1945. Conditions governing the distribution of insects in the free atmosphere. *Can. Entomol.* 77 :7–15 ; II Surface and upper winds, 21–28 ; III Thermal convection 44–49

216. Westal, P. H., Barrett, C. F., Richardson, H. P. 1961. The six-spotted leafhopper, *Macrosteles*

fascifrons (Stal) and aster yellows in Manitoba. *Can. J. Plant Sci.* 41 :320-31

217. Wilcox, R. B. 1951. Tests of cranberry varieties and seedlings for resistance to the leafhopper vector of false blossom disease. *Phytopathology* 41 :722–35

218. Williams, F. X., Swezey, O. 1905. Leafhoppers and their natural enemies. *Hawaiian Sugar Planters Assoc. Exp. Sta. Bull.* Part 1 4 : 123–57

219. Wilson, H. F., Childs, L. 1915. The rose leafhopper as a fruit pest (In 2nd biennial crop pest rept) *Oreg. Agr. Coll. Exp. Sta. 1913– 1914* 189–94

220. Wilson, M. C., Davis, R. L., Williams, G. G. 1955. Multiple effects of leafhopper infestation on irrigated and non-irrigated alfalfa. *J. Econ. Entomol.* 48 :323–26

221. Wolfenbarger, D., Sleesman, J. P. 1961. Resistance in common bean lines to the potato leafhopper. *J. Econ. Entomol.* 54 :846–49

222. Wolfenberger, D., Sleesman, J. P. 1961. Resistance to the potato leafhopper in lima bean lines, interspecific *Phaseolus* crosses, *Phaseolus* spp., the cowpeas and the Bonavist bean. *J. Econ. Entomol.* 54 :1077–79

223. Wolfenbarger, D., Sleesman, J. P. 1963. Variation in susceptibility of soybean pubescent types, broad bean, and runner bean varieties and plant introductions to the potato leafhopper. *J. Econ. Entomol.* 56 :895–97

224. Wressell, H. B. 1960. Resistance of alfalfa varieties to potato leafhopper. *Forage Notes* 6(1) :13

225. Yadava, C., Shaw, F. R. 1968. The preferences of certain coccinellids for pea aphids, leafhoppers and alfalfa weevil larvae. *J. Econ. Entomol.* 61 :1104–5

226. Yasumatsu, K., Masuda, H. 1932. On a new hunting wasp from Japan. *Ann. Soc. Nat. Hist. Fukwokensis* 1 :35–65

LIFE HISTORY OF THE CHIRONOMIDAE   6008

D. R. OLIVER
Entomology Research Institute, Canada Department of
Agriculture, Ottawa, Ontario, Canada

Information on the life history of the Chironomidae is scattered through a wide variety of papers in many languages. The present article is based mainly on papers that have been published since 1950 as the earlier papers were reviewed thoroughly in *Chironomus* by Thienemann (101). Additional references and information on most of the topics covered here may be found in this excellent book. Even with this restriction a high degree of selection is necessary and most of the papers on genetics, physiology, productivity and biomass, and fish food are not considered. Many of the papers cited have been chosen because they are most likely to be useful for further reference.

CLASSIFICATION

It seems necessary to comment briefly on the reasons for the confusion which exists in the classification of the Chironomidae. Part of the confusion arises from the dual use of two family names. The name *Tendipes* Meigen 1800 was used rather than *Chironomus* Meigen 1803 by many taxonomists as the type genus of the family. A recent ruling of the International Commission on Zoological Nomenclature has suppressed the Meigen 1800 names in favor of his 1803 names (see Fittkau 39). As a result, the name *Tendipes* was suppressed in favor of *Chironomus,* and similarly *Pelopia* was suppressed in favor of *Tanypus.* This action means that the correct name of the family is Chironomidae not Tendipedidae. Furthermore, it also establishes that the subfamily names Chironominae and Tanypodinae are correct. Although not covered by a ruling it is generally accepted that Orthocladiinae be used in place of Hydrobaeninae. *Tanytarsus* has been used as a genus in the tribe Chironomini, but another ruling (*Bull. Zool. Nom. 18,* Opinion 616) established it as a genus in the tribe Tanytarsini.

The classification of holometabolous insects presents special difficulties because the characteristics of all the life stages must be considered. Different ecological requirements of each stage often result in greater diversity, both ecological and morphological, in one stage than in another. The adults of the Chironomidae are usually more uniform in structure than are the immature stages, especially the larvae. The adult stage is somewhat ephem-

211

eral, completing the reproductive aspects of the life cycle in a fairly short time. Adults require only minimum shelter for mating, maturing eggs, and ovipositing. In contrast, the largest part of the life cycle is spent in the larval stage and the range of habitats occupied is perhaps unparalleled among other insect groups. All the energy required to complete the life cycle is built up in the larval stage, because the adults, with few exceptions, do not feed. This ecological diversity between the life stages is shown in the two systems of classification that have evolved; the larval and pupal system of Thienemann and his associates and the adult system of Edwards and Goetghebuer (in 13). The generic limits in the system based on the immature stages are frequently narrower than those based on adults. Fortunately, many recent studies have been based on all three stages (e.g., Brundin 13, 14; Fittkau 37), and a more stable system of classification is beginning to evolve.

The general classification used as a basis for discussion in this publication follows that of Brundin (14). Seven subfamilies, as follows, are recognized: Tanypodinae, Podonominae, Aphroteniinae, Telmatogetoninae, Diamesinae, Orthocladiinae, and Chironominae. Each subfamily, except the Telmatogetoninae, has several tribes, but only the tribes Chironomini and Tanytarsini belonging to the Chironominae are used here.

DISTRIBUTION AND HABITAT DIVERSITY

The distribution of the family is world-wide. The two species found in Antarctica are the southernmost free-living holometabolous insects known (106). Chironomids extend to the northern limits of land, and they make up one-fifth to one-half of the total number of species in the arctic insect fauna (80). Between these geographical extremes they have radiated into nearly every habitat that is aquatic or wet, including peripheral areas of the oceans (105). Some of these habitats have a very large number of species, e.g., 140 species live in Lake Innaren (11) and 168 in Lake Constance (86). There is no reliable estimate of the total number of species in the family: over 5000 species have been described to date and the species living in large areas such as Asia are almost unknown.

The distribution of each of the subfamilies, within their geographical range, is primarily governed by the availability of water ecologically suited to the requirements of the larvae. The Aphroteniinae, the smallest subfamily, with eight species, is strictly rheophilic; it is more or less confined to swift mountain streams in southern South America, South Africa, and Australia. The Podonominae, primarily rheophilic and cold-adapted, is much more common in the Southern Hemisphere, particularly the southern part where 130 species have been found, than in the Northern Hemisphere, where only 20 species have been recorded. One genus, *Lasiodiamesa,* has become adapted to warm pools in sphagnum bogs. The Diamesinae is also a rheophilic and cold-adapted group, though a few genera occur in lentic habitats; they inhabit the colder parts of the circumpolar lands and the mountain ranges throughout the world. All these subfamilies are absent in the tropical lowlands (14).

Most of the species in the family belong to one of the subfamilies Tanypodinae, Chironominae or Orthocladiinae, which are more or less worldwide in distribution. Most of the Tanypodinae and Chironominae are essentially thermophilous and adapted to living in standing water, though species of each do occur in cool habitats and in running water. Both subfamilies occur in all geographical regions except Antarctica. They are very abundant in the warmer parts of the Holoarctic and decrease in numbers with increasing latitude, or its climatological equivalent. Eighty percent of the chironomids living in forest streams of the central Amazon region are Chironominae (38). The Tanypodinae also are common in the tropics but they are probably more boreal (37). The Orthocladiinae occupy the widest range of habitats of all chironomids. It is the dominant subfamily in the arctic region (80) and, in contrast with the Chironominae and the Tanypodinae, decreases in numbers in increasingly warmer regions, though they are not uncommon in many warm habitats (e.g., 38). The larvae of this primarily cold-adapted subfamily live in all types of lentic and lotic habitats. It is the only subfamily with terrestrial species (13, 14) that live not only in the wet margins of water bodies but also in quite nonaquatic habitats such as cow dung (61).

The Telmatogetoninae, the largest marine group, is associated with rocky peripheral areas that are usually subjected to tidal action. The larvae are euryhaline and are often found in areas that receive freshwater runoff; one species of *Telmatogeton* lives in swift mountain streams in Hawaii (105). These peripheral marine habitats, as well as brackish rock pools, also have been invaded by larvae of the Orthocladiinae and Chironominae (46, 47, 83).

Brundin (14) believes that the primitive habitat in which chironomid evolution began was the upper reaches of mountain streams that arose from cool springs and ran through hygrophilous, temperate forests. It was a rather stable habitat, rich in dissolved oxygen and diatoms, with moderate annual variation in temperature and water level. In his account of the evolution of the Chironomidae within the framework of Hennig's sister group theory, Brundin (14) shows that the "pleisomorph lines" have remained rheobiontic, whereas the "apomorph lines" adapted to the different types of lentic water. As he points out, there are exceptions, e.g., the Tanypodinae are primarily lentic but the occurrence of the genera *Rheopelopia, Conchapelopia,* and *Macropelopia* in running water is a secondary adaptation. The Tanypodinae is considered to be the "apomorph" sister group of the Aphroteniinae and the Podonominae. Together these three subfamilies form the sister group of the remaining four subfamilies. Within this latter group the Chironominae is considered to be the "apomorph" sister group of the Telmatogetoninae, Diamesinae, and Orthocladiinae. As previously mentioned, many of the larvae in these two "apomorphic" subfamilies (Tanypodinae

California Baptist College
Riverside, California

California Baptist College
Riverside, California

and Chironominae) within each major sister group are lentic and thermophilous. Accepting the thesis that the larvae of the Chironomidae were initially cool-adapted, the success of the family in the colder regions is not
surprising. In the arctic region and in cold mountain streams larvae develop
at temperatures close to the limit of life and no major special adaptations
appear to have evolved (14, 80), except the ability of some larvae to withstand freezing (2). In contrast, a number of ecological and physiological
specializations have evolved to accommodate conditions found in warm
standing water, such as the free-living pupae of the Tanypodinae (14), the
presence of hemoglobin in the larvae of the Chironominae (14, 102), and
the most unusual ability of *Polypedilum vanderplankii* larvae to completely
dehydrate and remain viable under dry conditions for months (51).

THE LIFE CYCLE

The four life stages, egg, larva, pupa, and adult, are treated separately.
Unless stated otherwise, the discussion is about the aquatic groups as little
is known about the life cycle of the terrestrial Orthocladiinae, other than
that reported in Thienemann (101).

Egg

Egg mass.—The eggs are laid in a protective gelatinous matrix, which
expands in water to several times its original size. The shape of the mass
and the arrangement of the eggs vary, and each subfamily appears to have a
characteristic type of egg mass. The two main shapes are a linear type and
a more compact type. The linear type is characteristic of the Orthocladiinae and Diamesinae (9, 23, 106); in it the eggs are usually obliquely
arranged along the central axis of the matrix. The second type is characteristic of the Tanypodinae and Chironominae; it is usually spherical but
sometimes it is cylindrical, tear-shaped, or fig-shaped (3, 9, 23, 34, 48), and
occasionally attached to a solid object by a gelatinous stalk. The eggs are
peripheral or scattered throughout the matrix. The eggs of some Tantarsini
are arranged obliquely in a tube of viscous material which is embedded in a
transparent spherical matrix (24, 63).

Duration of egg stage.—This aspect of the life cycle has been generally
neglected in the literature, because it can be conveniently determined only
in the laboratory, as in the case of *Chironomus plumosus* (48). However,
from the little that is known, it may be inferred that the duration is temper-
ature-dependent and related to the overall length of the life cycle.

Hatching.—This process has rarely been described. In *Calopsectra neo-
flavellus,* by ingesting water the prolarva swells until the chorion ruptures
just posterior to the anterior proleg (24). In another species, unidentified,
the rupture occurs over the egg burster on the dorsal surface of the head
capsule. Davis (24) questions the observation by Thienemann (101) that

the larvae of two terrestrial Orthocladiinae break the chorion with their mandibles.

LARVA

The occurrence of four larval instars is probably universal in the Chironomidae; as this number has been reported for the Tanypodinae, Diamesinae, Orthocladiinae, and Chironominae (40, 101). However, five instars have been reported in the Tanypodinae (96). The instars can be easily separated by the width or length of the head capsule (8, 21). All the larvae of the family, except *Archaeochlus drakenbergensis,* which have two open spiracles in the eighth segment (14), are apneustic.

First instar larva.—Newly hatched larvae leave the egg mass and swim around in the water (1, 48, 69). The larvae are photopositive (62) and remain planktonic until a suitable habitat is found. The larvae of *C. plumosus* become less photopositive as soon as they have settled on the substrate (48). The energy required for swimming is obtained by feeding on suspended algae and detritus, but some nourishment may be derived from the yolk that remains from the egg (1). Larvae of *C. dorsalis* do not seize food particles directly from the water. They stop swimming, bend in the form of a horseshoe and remove particles adhering to the anal brush and claws of the posterior prolegs with their mouthparts and then continue swimming. The sticky viscous material on the anal appendages is apparently not derived from salivary gland secretion (1).

Kalugina (55) states that first instar larvae of different taxonomic and ecological groups are more alike morphologically than are mature larvae belonging to the same species. This similarity is apparently quite evident in the species which have narrow specialized habits, such as *Stenochironomus gibbus* that bore in submerged rotten wood. Perhaps the original suggestion by Dorier (26) that there are two morphological larval types—larvule (first instar) and larva (the remaining instars)—has merit when the degree of morphological similarity, the photopositive planktonic habit, and the peculiar method of feeding are considered. However, very few species have been investigated thoroughly, and none live in streams. If the first instar larvae of the rheophilic groups were also photopositive and planktonic, they would be easily swept away by the current.

Mode of life.—In a suitable habitat the first or second instar establishes a mode of life that usually continues throughout the larval stages. Chironomids divide into two groups, depending on whether or not a larval case is constructed. All Chironominae, except some of the predaceous species, build a larval case on or within the substrate in which they live. The case consists of particles from the substrate, and is lined and held together with silk-like threads secreted by the salivary glands (103). The structure of the case varies with the different genera (23, 93, 103) and may even vary within a species, in response to different environmental conditions (53). The extent of tube building in the Orthocladiinae is not fully known. Many species live in close association with some type of substrate, and the somewhat unorganized case built by *Cardiocladius* (19) may be more common than has been reported, because it is so difficult to observe. The larvae of this genus live in tunnels within a loose mass of particles cemented together by salivary secretion. In contrast, *Abiskomyia* and *Heterotanytarsus* build distinct cases similar to those built by Trichoptera (101). The tunnels mined in *Potomogeton* and *Nostoc* by *Cricotopus* (8, 10) substitute for cases; these tunnels are not lined with silk even though other members of this genus that are not miners spin silken cocoons (23). *Lasiodiamesa,* and the Corynoneurini and the Tanypodinae are free-living and very good swimmers.

Larvae usually penetrate only a few centimeters of the substrate. In both lotic and lentic habitats with soft bottoms about 95 percent of the larvae occur in the upper 10 cm of the substrate (6, 16, 33, 41, 73), with very few below 40 cm. Most of the larvae that penetrate below 10 cm are tube-building Chironominae; Tanypodinae have been found at depths of 7 cm (16). Brundin (12) summarized the evidence for a relationship between depth penetration in lake substrata and oxygen concentration in the overlying water. Deep penetration is possible only when the concentration of oxygen is high. Overwintering larvae may penetrate to greater depths as suggested by the following observations on *C. plumosus.* In a lake in Poland, Kajak (54) found that 50 percent and 20 percent of the larvae were at 8 cm in the spring and summer, respectively. In the laboratory at 6.0°C, feeding is sporadic and nonfeeding larvae burrow in the substrate down to 51 cm (48).

The free-swimming larvae of the Tanypodinae and Corynoneurini move freely within the confines of their habitat. However, the extent of movement by benthic larvae, especially the tube-builders, is not fully appreciated. The bottom fauna of lakes and streams move freely over the substrate (68) but this movement often extends into the water over the substrate (42, 74). Little is known about why the larvae are stimulated to leave the bottom, how they travel, or how they orientate. Much of this movement is nocturnal so the larvae are less vulnerable to fish predation. Sometimes the movement is a true migration and the larvae return to their original location, e.g., the diel vertical migrations of *Sergentia* larvae, which reaches a peak at about 2200 hr in Marion Lake (42). More often the movement results in a horizontal displacement from one part of the habitat to another, as reported for *C. plumosus* (4, 49). In Titisee, *Sergentia coracina* larvae gradually migrate deeper into the profundal zone as they become older (107).

Chironomid larvae are a common component of stream-drift and opinions differ on whether this drift is passive or active (25, 31, 104). Elliot (31) suggested that only a small proportion of the animals are swept away when they lose their grip or are jostled by other animals and they re-attach themselves to the bottom as soon as possible. Dimond (25) concluded that drift is density-related and most likely to occur when there is a surplus population. Probably the horizontal movement in lakes is, in part, density-dependent, related to such conditions as crowding, reduced resources, and unfavorable environment. Bay et al (5) report that when the oxygen concentration at the mud-water interface is depleted, some chironomid larvae leave the bottom and are carried about by water currents, and they return to the bottom only when the oxygen supply has been restored.

Environmental requirements.—The broad environmental requirements of the various subfamilies are relatively well known and have been outlined above. Certain groups are characteristic of lentic or lotic environment, cold or warm water, oxygen-rich or oxygen-poor water, etc. On the basis of this type of data, chironomid larvae frequently are used as indicators of different types of lakes. One example is the system of lake classification initiated by Thienemann and developed by Lenz and Lundbeck (in 11). However, further detailed studies at the specific level are needed to understand the various physical, chemical, and biological factors that interact to produce a suitable habitat for larval growth. Curry (20) provides a very useful list of the range of some environmental factors including temperature, oxygen concentration, pH, ionic concentration, depth, water type, and substrate for about 80 species in 4 subfamilies. Hilsenhoff & Narf (50) correlated the distribution of 10 species of chironomid larvae that occur in lakes in Wisconsin with 14 physical and chemical factors.

Chironomid larvae show specific differences in their resistance to low oxygen concentration, and this factor frequently determines their presence or absence in a particular habitat. Most of the species that are able to tolerate very low oxygen conditions belong to the Chironominae and Tanypodinae, and some of the tube-building Chironominae are reported to withstand prolonged periods of anaerobic conditions in nature (20). The critical factor for the tube-building species is the concentration of oxygen at the mud-water interface (12) because the mobile Tanypodinae can swim to higher strata of water where the concentration of oxygen is probably higher. The possession of hemoglobin is correlated with the ability of Chironominae larvae to function in low oxygen concentrations (60, 102). However, as Walshe (102) pointed out, the possession of hemoglobin is not the only respiratory modification because other larvae can remain active without hemoglobin.

The temperatures in which most species complete their development are between 0°C and 32°C (20). In arctic lakes some species (e.g., *Heterotrissocladius subpilosus* and *Pseudodiamesa arctica*) complete larval development in temperatures below 4°C (80). In contrast, several Tanypodinae, Chironominae, and some *Cricotopus* are able to live in temperatures between 32°C and 40°C (7, 20).

Most species live in water with a pH range of 6.0 to 8.0 (20), the normal range for surface waters, and a few are unaffected by a range from 5.0 to 9.0. But a very low pH can be a limiting factor because few species live in water below 4.0 (20). *C. plumosus* can live in water with a pH of 2.3 to 3.2, but it is apparently unable to complete development (44) and the abundance of this species in some lakes in Wisconsin is positively correlated with an increased pH of the mud (50).

The reasons for the distribution of the larvae of a species must be found in a combination of factors, rather than in one factor such as concentration of oxygen or the temperature of the water. For example, Harnisch (43) noted that the most important ecological factors in the tropics, anaerobic conditions and high temperatures, produce a special type of "oxybiosis" that differs from the normal diffusion type in temperate areas. In one woodland stream the distribution of larvae was primarily limited by food and temperature (66), whereas in another it was limited by current and substrate (15).

Food and feeding habits.—It has often been recorded that the larvae of the Tanypodinae are carnivorous (e.g., 70) but, whereas many species do attack and consume aquatic invertebrates, probably very few are obligate carnivores. Small prey are swallowed whole; larger ones are seized by the rear end and engulfed to the level of the thoracic segments. This process of engulfment is repeated several times until finally the head and thorax are bitten off and rejected (59). The larvae also feed by piercing the cuticle and sucking out the body contents (70). Many species that are known to be carnivorous also ingest algae and detritus (15, 23, 42), and some small species, e.g., *Larsia acrocinata,* feed exclusively on diatoms and detritus (42). A few genera in the Chironominae and Diamesinae are also carnivorous (15, 23, 42, 85, 87).

Except for these carnivorous groups and a few species that feed on plant tissue or are parasitic (22), most of the chironomid larvae are microphagous, feeding on small plants and animals and detritus. Detritus is any nonliving organic matter, plant or animal, that has begun to be attacked by microconsumers (15). Most of the microphagous species can be placed in the three noncarnivorous groups—algal, detrital, and algal-detrital feeders—used by Coffman (15) in his study of the species living in a stream. These groups were based on the percentage ingestion of calories that were derived from each type of food. Most of the Orthocladiinae and Chironominae are either algal or algal-detrital feeders, and only one *Cryptochironomus* is classed as a detrital feeder. Among the algal feeders, most fed on diatoms. All the Aphroteniinae and Podonominae, and many Diamesinae feed on diatoms (14). Throughout the family, diatoms are one of the most important sources of food. Brundin (14) has suggested that the primitive habitat was rich in diatoms and that early chironomid evolution could have proceeded independently of other types of plants.

Little is known about the feeding mechanism of the Orthocladiinae, but they probably seek out and ingest their food directly from the substrate on which they live. The feeding mechanisms of the Chironominae have been described in detail by Walshe (103). The larva of *Microtendipes pedellus* extends from its case, spreading salivary secretion with its anterior proleg and then withdraws into the tube dragging the secretion and adhering particles with it. Other species feed directly on the mud as they extend their cases on or below the surface. Three types of filter feeding have evolved in the Chironominae (103). In the first type, *Chironomus plumosus* spins a net across the lumen of its tube. Water is drawn through the tube by anteroposterior undulations, then the net and entrapped material is eaten. In the second type, a similar mechanism for straining food particles out of the water is used by the leaf-mining species of *Glyptotendipes, Polypedilum* and *Endochironomus*. In these species, the larvae turn around and spin the net across the lumen so that it is posterior to the body during the undulating movements. In the third type, the larvae of *Rheotanytarsus* spin a net between the arms of their cases so that it faces into the current and traps drifting particles of food. Periodically the larvae reach out of the tube and free a section of the net and roll it into a ball which is eaten or added to the case. The net is then repaired. One interesting species, *Kiefferulus martini,* has "feeding-brushes" extending from the head capsule (67), which look similar to the mouth brushes of the Simuliidae and are presumably used for straining food particles out of the water.

The leaf-miners cited previously ingest plant tissue whenever they extend their tubes, but others such as several species of *Cricotopus* feed entirely on the leaves of *Potomageton* (8). Some species of Chironominae and Orthocladiinae are parasitic on other chironomids, Gastropods, Ephemeroptera, and other aquatic invertebrates. This and other types or associations with aquatic invertebrates were reviewed by Steffan (95).

Growth and development.—The time required for growth and development varies under the influence of a number of factors, especially temperature (34, 56, 57). Within certain limits, as the temperature increases development proceeds at a faster rate than growth. Therefore, a mature larva that developed at a high temperature would be smaller than if it developed at a lower temperature (57). However, the situation is more complex: the size at which a larva is physiologically mature and able to pupate is also influenced by sex. Ford (40) observed that the bimodal distribution of the size of several structures in the head capsule of the final instar is related to sex; the female structures being larger. Mature larvae of *C. plumosus* weighing over 60 mg are female, and under 60 mg are male (48). In *C. nuditarsus* the male and female larvae have different growth rates, 1.90 for the male and 1.98 for the female (36) which, if occuring in other species, would account for the observed size differences between the sexes.

In uniform conditions found in the tropics, the potential for uninterrupted growth and development exists. At 30°C the life cycle of the Amazonian species, *C. strenzkei* is completed in 10 to 12 days (98). Although the natural habitat of this species is unknown a large number of generations can occur within one year. In Lake Victoria, the life cycle of *Tanypus guttatipennis* and *Procladius umbrosus* is completed in two lunar months (65). These two species are very interesting, because at any one time two separate populations (of each species) are present and they are separated by one lunar month.

Continuous growth and development rarely occur in temperate and polar regions as the larvae must adjust to the less favorable conditions during the winter months. Commonly, the effect of the lower temperature in the winter is to bring the larvae into a state of inactivity, and sometimes all growth and development must cease when the larvae are enclosed by frozen substrate (2, 18). The larvae of overwintering generations of multivoltine and bivoltine species take longer to develop than the summer generations. This can be inferred from the numerous emergence curves that have been published (72, 86, 91). Similarly, the period of rapid growth of the larvae of univoltine species occurs during the warm period of the season, which depends to some extent on whether the species is warm- or cold-adapted. In Marion Lake, the warm-adapted Chironominae generally grow faster in the summer and fall, whereas most of the cold-adapted Orthocladiinae grow faster in the winter (42). In the arctic, as far as is known, all the life cycles are one year or longer (80).

From the above, it is possible to conclude that changes in the duration of the larval period occur over a wide range of latitude. With increasing latitude or its climatological equivalent, the duration of the larval period becomes longer and one of the main factors responsible for this increase is temperature. However, it is clear that the duration of the larval period is not entirely determined by temperature. *C. anthracinus* has a two-year life cycle and growth is reduced both in summer (related to low oxygen concentration) and in winter (related to low temperature) (52, 53). Two-year life cycles have also been reported for other species (42, 90, 107) and it is possible that these are normally one-year species living in suboptimal conditions. *Sergentia coracina* has a two-year life cycle in Titisee (an alpine lake) and a one-year life cycle in more northern lakes. Wülker (107) suggested that the two-year life cycle occurs in response to the conditions at the southern range of the species, but it also could be a response to the conditions at high altitudes.

It is also possible to conclude that the larvae of many species have the ability to grow and develop as conditions permit. But the processes involved in the slowing down or cessation of growth and development have received little attention. In some instances it is probably a facultative response to unfavorable conditions, but there have been two reports of apparent genuine diapause (32, 84). In the species *Metriocnemus knabi* and *C. tentans,* the diapause is controlled by photoperiod; short-day conditions induce and maintain diapause, whereas long-day conditions favor the termination of diapause.

Little is known about the state of overwintering larvae in temperatures near or below freezing. The overwintering larvae of *Limnochironomus* and *Endochironomus,* in temperatures near freezing, build a special type of cocoon which differs from the normal larval cases (89, 94). The larvae are folded head to tail in completely closed semicircular cocoons. In one study at Point Barrow on chironomid larvae that overwinter in frozen substrate, it was found that the larvae did not supercool, but begin to freeze as the temperature falls below 0°C. They prepared for freezing by a slight dehydration, which increased as the freezing proceeded (92). This experiment has an interesting parallel with observations on species that live in an entirely opposite environment, i.e., in shallow unshaded rock pools that are filled with water or are wet for only short periods (51). The larvae that live in these pools withstand the dry and very hot periods by dehydration.

Pupa

The pupal stage, except in taxonomic studies, has not been much investigated. Brundin (14) distinguished seven groups, based on the arrangement of the leg sheaths.

As for larvae, there are two basic types: free living and sedentary. The Tanypodinae and two genera of the Podonominae, *Podonopsis* and *Lasiodiamesa,* have free-living pupae, which are among the strongest swimmers in the Diptera Nematocera (14). Most of the pupae of the rest of the family are sedentary and are protected by some sort of a case, although the pupa of *Cryptochironomus fulvus,* lies free of the substrate (23). Pupation occurs within the case of the last larval instar which is sometimes rebuilt, or in a case built by the last instar of free-living species (23, 48, 87, 93).

Compared with the larval stage, the duration of the pupal stage is very brief, ranging from a few hours to a few days (e.g., 10, 23). In *C. plumosus,* the duration depends upon the temperature of the substrate (48).

When the pupa is mature (pharate adult stage) and upon receiving the proper stimulus, it moves to the surface of the water and adult eclosion occurs. The swimming ascent of the sedentary pupa is aided by air that has accumulated within the pupal skin around the thorax of the adult (72). This is probably one of the most critical times in the life cycle of the chironomids, as not only are the pupae exposed to predation, but it precedes, and in a sense determines, the emergence of the adult into a very different set of environmental conditions. The factors that guide this event are not well understood, but some of them are referred to in the section on emergence.

Adult

The adult stage lasts, at the most, several weeks during which time the reproductive aspects of the life cycle are carried out. The adults have reduced biting mouthparts and do not feed. However, two species of *Smittia* are known to imbibe nectar from arctic flowers (80), and an Australian species has been found to have fully developed biting mouthparts (28).

Emergence.—The adult emerges from a dorsal split in the thorax of the pupal skin. In aquatic habitats, it is a very rapid event, occurring in 10 to 30 sec, at most several minutes, and the adult is able to fly almost immediately (14, 48, 72, 78). The males of some *Clunio* aid the eclosion of the females by stripping off the pupal skin (45, 79).

The emergence of adults from any given habitat, except in very uniform environments, is not a random occurrence but has a very definite phenological sequence. Each species has a more or less restricted time of emergence with less intensive emergence before and after a peak (11, 23, 42, 73, 86, 91). The emergence period is related to the duration of the larval period and the different types, univoltine to multivoltine, are outlined in the section on the larva. The position of the seasonal emergence period or periods for each species is controlled by the prevailing environmental conditions and can very from year to year. In warm habitats, emergence can occur throughout the year (17, 65). Toward higher latitudes, the larvae spend more time overwintering, and the emergence period becomes shorter as shown by the emergence curves published by Mundie (73), Reiss (86), and Sandberg (91). In the arctic region it is very short and the seasonal emergence of each species is highly synchronized (80). A full discussion of the "restriction of the emergence period as being progressively related to latitude, or its climatological equivalent" and of the degrees of synchronization, was presented by Corbet (17), in a review of the temporal emergence of aquatic insects, much of which is applicable to the Chironomidae.

Many species exhibit a diel periodicity of emergence (71, 81, 82). The emergence peak usually occurs after sunset, or during the day in late morning or early afternoon, however, it can also occur in the early morning (78). Some species, e.g., *Ablabesmyia cingulata,* have no diel periodicity, and they emerge at a more or less constant rate throughout the diel (71). In general, summer species of the temperate regions emerge after sunset (71, 81, 82), and spring and arctic species emerge at about solar noon (71, 80). According to Palmén (82), the emergence after sunset is correlated with a change in light intensity (light to dark) and is independent of temperature. In some bivoltine species that emerge in June and August there is a shift in the emergence peak between generations that matches the shift in the time of sunset (71). In contrast, the diurnal species emerge during the brightest part of the diel, and temperature may play a more important role.

Probably both seasonal and diel emergence are under the dual control of water temperature and light intensity, and their relative importance as exogenous factors varies from place to place. A recent experiment by Fischer & Rosin (35) showed how the dual control affects the diel emergence pattern of *C. nuditarsis;* at 18°C peak emergence occurs after the onset of darkness, whereas at 13°C the peak shifts to the morning. However, other factors can be involved: *C. plumosus* requires a feeding stimulus provided by diatom blooms in order to pupate and emerge (48), and profundal chironomids emerge and oviposit in relation to spring and autumn overturn, when oxygen concentration is at its maximum (12).

Lunar emergence rhythms have been recognized in several marine species (17, 47, 75, 76). Along the coast of the North Sea, the emergence of *Clunio marinus* is restricted to two periods each month, occurring within a few days of the new and the full moon. Within each period of semilunar emergence, the diel emergence occurs at the time of low tide. In relation to local variation in tides, the emergence pattern varies from place to place and the diel emergence pattern is under gene control (76). In the Baltic Sea where the tidal influence is negligible, *C. marinus* has only one emergence period per year and the diel emergence occurs after sunset (79). Several chironomids living in Lake Victoria also have lunar emergence rhythms (17).

Flight and other locomotion.—The various types of flight of aquatic chironomids, for purposes of discussion, can be grouped into three categories: flight after eclosion to a resting site and the various adjustments to environmental conditions; swarming flight; and oviposition flight of the females. Nielsen (78) presents a very detailed description of the first category for *Glyptotendipes paripes* in Florida. In this species, and probably in many others, the flight to shore is governed by the direction of the wind. In fact, chironomids seem to be specifically adapted for prolonged flight on wind after emergence which contrasts with later behavior when wind easily inhibits flight. The selection of the resting site is related to temperature and wind, according to Konstantinov (58), humidity is the main factor as it is very important for the adult to prevent water loss, particularly soon after emergence when the body is still soft. This suggests a possible correlation with the time of diel emergence: the species that emerge after sunset in the summer would find night conditions more favorable for preventing loss of water than conditions during the day (81). This situation would not be so critical for the spring species during the cooler days of the spring. Hashimoto (46) divided the marine midges into three types based on locomotion; the flying type with well-developed wings; the walking type; and the gliding type (only *Clunio* and *Pontomyia*), in which the male is specialized for gliding on the water and the reduced, apterous female resembles a pupa.

One of the most frequently noted features of the adult midge is the ability to form swarms. Syrjämäki (97) grouped chironomids into daylight swarmers and twilight swarmers. Twilight swarming occurs at dusk or dawn, and the beginning and end of the swarming period is primarily governed by light intensity (97), but can be modified by temperature (99). The factors influencing daylight swarming need further investigation (97). In the arctic, *Smittia extrema* exhibits a diel periodicity in continuous daylight, and, irrespective of the weather, activity is minimal at midnight (100).

Swarms can be composed entirely of males (64), males of several species (23), or a mixture of males and females (78). As far as is known, chironomids form their swarms in relation to a marker that is visually recognized by the swarming adults. The selection of different markers and the reaction to them are probably major factors in determining the composition of the swarms (see Downes, 27, for a full discussion of swarm markers). There is no evidence that the sound produced by swarming males entices other males to join the swarm. Males are attracted to the flight sound of females (88, 100) and in *C. plumosus* both the male response and the frequency of sound produced by the female vary with temperature (88).

There are a number of reports of directional flight of the female prior to oviposition, but how a female is able to select the correct site is not understood. Because the first instar larvae are very mobile, the selection of the site may not be that critical. Few observations have been reported on the time of oviposition, but most of these suggested the commonest times for oviposition are dusk and dawn (23, 73, 78), and oviposition may be influenced by the same factors that induce twilight swarming. It would be interesting to know if the species that emerge and swarm at twilight also oviposit at this time and similarly for the species that emerge in the daylight.

Mating.—Characteristically, chironomids initiate copulation in a swarm and there appears to be little doubt that the main function of the swarm is to bring the sexes together (27). Mating is often completed in a few seconds and the mated pair may never reach the ground (99). In some Orthocladiinae the pair drifts to the ground and may remain in copula for several hours (80). Mating on the ground without swarming occurs in a number of species (78, 97). In some of these species swarming also occurs but Syrjämäkii (97) suggested that because most of the mating takes place on the ground swarming is "only a relict fraction of its behaviour."

The position of the two sexes during the mating process has been described in detail for *Allochironomus crassiforceps* (97), and for *Pontomyia* and *Clunio* (45), but otherwise has not received much attention. The male of *A. crassiforceps* mounts from above ("end-to-face position") and when they switch to an "end-to-end position" a rotation occurs between the 7th and 8th abdominal segments, thereby allowing the male to remain upright. In the marine species, the hypopygium rotates 180 degrees before copulation, therefore the mating begins in the "side-by-side position" switching to the "end-to-end position." This is the only report (45) of hypopygial rotation before copulation in the Chironomidae.

Sperm is transferred by means of an intermediary spermatophore in some species (77, 80), but it is not known if this is universal in all species. In *Glyptotendipes paripes* the spermatophore has two cavities and its opening is applied to the opening of the spermathecal ducts (77).

Egg maturation.—In most of the species that have been studied the females emerge with follicles that are one-third to one-half developed (3, 34, 80). Whereas some Tanytarsini and *Clunio* emerge with fully developed eggs (45, 80), *Pontomyia* completes oviposition within 20 min from the time of emergence (45). A parthenogenetic Tanytarsini also lays eggs soon after emergence (29). Parthenogenesis, which is not very common in the Chironomidae, occurs mainly in the Tanytarsini and the Orthocladiinae (30, 63).

In some species a second egg batch is produced after the oviposition of the first batch (34, 80). Following the oviposition of the second batch by *C. nuditarsis,* a few of the third series of follicles begin to develop; however, no third oviposition was observed (34).

The number of nurse cells in the developing follicle appears to vary in the different groups. The Chironominae have one nurse cell as illustrated by Fischer (34), whereas the Orthocladiinae and the Diamesinae have seven (personal observation).

CONCLUSION

This very diverse family has radiated into many different types of habitats. Although we are a long way from understanding the ecological diversity involved, we are now beginning to understand the natural or phyletic relationships of the various groups that make up the family, and a framework for further progress is established. Much of this review is based on the Chironominae because it is better known than other subfamilies. This is particularly true of the adults. Above all, further detailed studies on species in other subfamilies are needed.

ACKNOWLEDGEMENTS

I wish to thank Mr. J. A. Downes and Dr. J. F. McAlpine for their constructive criticism of the manuscript.

LITERATURE CITED

1. Alekseyev, N. K. 1965. [On the nutrition of Chironomidae during the planktonic period of life.] *Nauch. Dokl. Vyssh. Shk. Biol. Nauki* 1: 19–21
2. Andersen, F. S. 1946. East Greenland lakes as habitats for chironomid larvae. *Medd. Groenland* 100 :5–65
3. Anderson, J. F., Hitchcock, S. W. 1968. Biology of *Chironomus atrella* in a tidal cove. *Ann. Entomol. Soc. Am.* 61 :1597–1603
4. Barthelmes, D. 1961. Über die horizontale Wanderung der Teichbodenfauna. *Z. Fisch.* 11 :183–87
5. Bay, E. C., Ingram, A. A., Anderson, L. D. 1966. Physical factors influencing chironomid infestation of water-spreading basins. *Ann. Entomol. Soc. Am.* 59 :714–17
6. Beatty, K. W. 1968. *An ecological study of the benthos of Castle Lake, California.* PhD thesis, Univ. California, Davis
7. Berczik, Á. 1964. Angaben über das Vorkommen von Chironomidenlarven Gewässer. *Opusc. Zool., Budapest* 5 :43–47
8. Berg, C. O. 1950. Biology of certain Chironomidae reared from *Potamogeton. Ecol. Monogr.* 20 : 83–101
9. Branch, H. E. 1928. Description and identification of some chironomid egg masses. *Ann. Entomol. Soc. Am.* 21 :566–70
10. Brock, E. M. 1960. Mutualism between the midge *Cricotopus* and the alga *Nostoc. Ecology* 41 :474–83
11. Brundin, L. 1949. Chironomiden und andere Bodentiere der südschwedischen Urgebirgsseen. *Rep. Inst. Freshwater Res. Drottningholm* 30 :1–914
12. Brundin, L. 1951. The relation of O_2-microstratification at the mud surface to the ecology of the profundal bottom fauna. *Rep. Inst. Freshwater Res. Drottningholm* 32 :32–42
13. Brundin, L. 1956. Zur Systematik der Orthocladiinae. *Rep. Inst. Freshwater Res. Drottningholm* 37 :1–185
14. Brundin, L. 1966. Transantarctic relationships and their significance, as evidenced by chironomid midges with a monograph of the subfamilies Podonominae and Aphroteniinae and the Austral Heptagyiae. *Kgl. Sv. Vetenskapsakad. Handl.* 11 :7–472
15. Coffman, W. P. 1967. *Community structure and tropic relations in a small woodland stream, Linesville Creek, Crawford County, Pennsylvania.* PhD thesis, Univ. Pittsburgh, Pittsburgh, Pa.
16. Cole, G. A. 1953. Notes on the vertical distribution of organisms in the profundal sediments of Douglas Lake, Michigan. *Am. Midl. Nat.* 49 :252–56
17. Corbet, P. S. 1964. Temporal patterns of emergence in aquatic insects. *Can. Entomol.* 96 :264–79
18. Crisp, G., Lloyd, L. 1954. The community of insects in a patch of woodland mud. *Trans. Roy. Entomol. Soc. London* 105 :269–313
19. Curry, L. L. 1954. Notes on the ecology of the midge fauna of Hunt Creek, Montmorency County, Michigan. *Ecology* 35 :541–50
20. Curry, L. L. 1965. A survey of environmental requirements for the midge. In *Biological Problems in Water Pollution,* 127–41. Cincinnati : U.S. Pub. Health Serv. Publ. 999-WP-25, 376 pp.
21. Czeczuga, B., Bobiatyńska-Ksok, E., Niedzwiecki, E. 1968. On the determination of age structure of the larvae of Tendipedidae. *Zool. Pol.* 18 :317–28
22. Czeczuga, B., Bobiatyńska-Ksok, E. 1968. Ecological-Biological aspects of the parasitic larvae *Cryptochironomus* ex. gr. *pararostratus* Harn. *Int. Rev. Ges. Hydrobiol.* 53 :549–61
23. Darby, R. E. 1962. Midges associated with California rice fields, with special reference to their ecology. *Hilgardia* 32 :1–206
24. Davis, C. C. 1966. A study of the hatching process in aquatic invertebrates. *Hydrobiologia* 27 :196–207
25. Dimond, J. B. 1967. Evidence that drift of stream benthos is density related. *Ecology* 48 :855–57
26. Dorier, A. 1933. Sur la biologie et les metamorphosis de *Psectro-*

cladius obvius Walk. *Trav. Lab. Hydrobiol. Piscicult. Grenoble* 25: 205–15

27. Downes, J. A. 1969. The swarming and mating flight of Diptera. *Ann. Rev. Entomol.* 14 :271–98

28. Downes, J. A., Colless, D. H. 1967. Mouthparts of the biting and bloodsucking type in Tanyderidae and Chironomidae *Nature* 214: 1355–56

29. Edward, D. H. D. 1963. The biology of a parthenogenetic species of *Lundstroemia* with descriptions of the immature stages. *Proc. Roy. Entomol. Soc. London* 38 :165–70

30. Edward, D. H. D., Colless, D. H. 1968. Some Australian parthenogenetic Chironomidae. *J. Aust. Entomol. Soc.* 7 :158–62

31. Elliot, J. M. 1967. Invertebrate drift in a Dartmoor stream. *Arch. Hydrobiol.* 63 :202–37

32. Englemann, W., Shappirio, D. G. 1965. Photoperiodic control of the maintenance and termination of larval diapause in *Chironomus tentans. Nature* 207 :548–49

33. Ferencz, M. 1968. Vorstudium über die vertikale verteilung des zoobenthos der Theiss. *Tiscia (Szeged)* 4 :53–58

34. Fischer, J. 1969. Zur Fortpflanzungsbiologie von *Chironomus nuditarsis* Str. *Rev. Suisse Zool.* 76 :23–55

35. Fischer, J., Rosin, S. 1968. Einfluss von Licht und Temperatur auf die Schlüpf-Aktivität von *Chironomus nuditarsis* Str. *Rev. Suisse Zool.* 75 :538–49

36. Fischer, J., Rosin, S. 1969. Das larvale Wachstum von *Chironomus nuditarsis* Str. *Rev. Suisse Zool.* 76 :727–34

37. Fittkau, E. J. 1962. Die Tanypodinae. *Abh. Lar.-sys. Insekt.* 6 :1–453

38. Fittkau, E. J. 1964. Remarks on limnology of central-Amazon rainforest streams. *Verh. Int. Verin. Limnol.* 15 :1092–96

39. Fittkau, E. J. 1966. *Chironomus,* nicht *Tendipes. Arch. Hydrobiol.* 62 :269–71

40. Ford, J. B. 1959. A study of larval growth, the number of instars and sexual differentiation in the Chironomidae. *Proc. Roy. Entomol. Soc. London* 34 :151–60

41. Ford, J. B. 1962. The vertical distribution of larval Chironomidae in the mud of a stream. *Hydrobiologia* 19 :262–72

42. Hamilton, A. L. 1965. *An analysis of a freshwater benthic community with special reference to the Chironomidae.* PhD thesis, Univ. British Columbia, Vancouver

43. Harnisch, O. 1960. Untersuchunger am O₂-Verbrauch der larven von *Chir. anthracinus* and *Chironomus plumosus. Arch. Hydrobiol.* 57: 179–86

44. Harp, G. L., Campbell, R. S. 1967. The distribution of *Tendipes plumosus* (Linné) in mineral acid water. *Limnol. Oceanogr.* 12 :260–63

45. Hashimoto, H. 1957. Peculiar mode of emergence in the marine chironomid *Clunio. Sci. Rep. Tokyo Kyoiku Daigaku, Sect. A* 8 :177–226

46. Hashimoto, H. 1962. Ecological significance of the sexual dimorphism in marine chironomids. *Sci. Rep. Tokyo Kyoika Daigaku, Sect. A* 10 :221–52

47. Hashimoto, H. 1965. Discovery of *Clunio takahashii* Tokunaga from Japan. *Jap. J. Zool.* 14 :13–29

48. Hilsenhoff, W. L. 1966. The biology of *Chironomus plumosus* in Lake Winnebago, Wisconsin. *Ann. Entomol. Soc. Am.* 59 :465–73

49. Hilsenhoff, W. L. 1967. Ecology and population dynamics of *Chironomus plumosus* in Lake Winnebago, Wisconsin. *Ann. Entomol. Soc. Am.* 60 :1183–94

50. Hilsenhoff, W. L., Narf, R. P. 1968. Ecology of Chironomidae, Chaoboridae, and other benthos in fourteen Wisconsin Lakes. *Ann. Entomol. Soc. Am.* 61 :1173–81

51. Hinton, H. E. 1960. A fly larva that tolerates dehydration and temperatures of −270° to +102° C. *Nature* 188 :336–37

52. Jónasson, P. M. 1965. Factors determining population size of *Chironomus anthracinus* in Lake Esrom. *Mitt. Int. Verein. Theoret. Angew. Limnol.* 13 :139–62

53. Jonasson, P. M., Kristiansen, J. 1967. Primary and secondary production in Lake Esrom. Growth of *Chironomus anthracinus* in relation to seasonal cycles of phytoplankton

and dissolved oxygen. *Int. Revue Ges. Hydrobiol.* 52 :163–217

54. Kajak, Z. 1958. [An attempt at interpreting the quantitative dynamics of benthic fauna in a chosen environment in the "Konfederatha" pool adjoining the Vistula.] *Ekol. Pol.* 6 :205–91

55. Kalugina, N. S. 1959. [Changes in morphology and biology of chironomid larvae in relation to growth.] *Trudy Vses. Gidrobiol. Obshchest.* 9 :85–107

56. Konstantinov, A. S. 1958. [The effect of temperature on growth rate and development of chironomid larvae.] *Dokl. Akad. Nauk SSSR* 120 :1362–65

57. Konstantinov, A. S. 1958. [On the growth type of chironomid larvae.] *Dokl. Akad. Nauk SSSR* 120 : 1151–54

58. Konstantinov, A. S. 1961. [On the biology of the midges of the family Chironomidae.] *Nauch. Dokl. Vyssh. Shk. Biol. Nauki* 4 :20–23

59. Kurazhkovskaya, T. N. 1966. [Anatomy of intestine and salivary glands of chironomid (Diptera) larvae.] *Tr. Inst. Biol. Vnutr. Vod. Akad. Nauk SSSR* 14 :286–96

60. Lavrovsky, V. A. 1966. [On the participation of hemoglobin in the oxygen transport in chironomid larvae.] *Zh. Obshch. Biol.* 27 :128–30

61. Lawrence, B. R. 1954. The larval inhabitants of cow pats. *J. Anim. Ecol.* 23 :234–60

62. Lellák, J. 1968. Positive Phototaxis der Chironomiden—Larvulae als regulierender Faktor ihrer Verteilung in stehenden Gerwässern. *Ann. Zool. Fenn.* 5 :84–87

63. Lindeberg, B. 1958. A parthenogenetic race of *Monotanytarsus boreoalpinus.* Th. from Finland. *Ann. Zool. Fenn.* 25 :35–38

64. Lindeberg, B. 1964. The swarm of males as a unit for taxonomic recognition in chironomids. *Ann. Zool. Fenn.* 1 :72–76

65. MacDonald, W. W. 1956. Observations on the biology of chaoborids and chironomids in Lake Victoria and on the feeding habits of the "elephant-snout" fish (*Mormyrus kannume* Forsk.). *J. Anim. Ecol.* 25 :36–53

66. MacKay, R. J. 1969. Aquatic insect communities of a small stream on Mont St. Hilaire, Quebec. *J. Fish. Res. Bd. Can.* 26 :1157–83

67. Martin, J. 1963. The cytology and larval morphology of the Victorian representatives of the subgenus *Kiefferulus* of the genus *Chironomus. Aust. J. Zool.* 11 :301–22

68. Moon, P. P. 1940. An investigation of the movements of fresh-water invertebrate faunas. *J. Anim. Ecol.* 9 :76–83

69. Mordukhai-Boltovskoy, F. D., Shilova, A. I. 1955. [On the temporary planktonic modus vivendi of the larvae of *Glyptotendipes.*] *Dokl. Akad. Nauk SSSR* 105 :163–65

70. Morgan, M. J. 1949. The metamorphosis and ecology of some species of Tanypodinae. *Entomol. Mon. Mag.* 85 :119–26

71. Morgan, N. C., Waddell, A. B. 1961. Diurnal variation in the emergence of some aquatic insects. *Trans. Roy. Entomol. Soc. London* 113 : 123–37

72. Mundie, J. H. 1956. The biology of flies associated with water supply. *J. Inst. Pub. Health Eng.,* 55 : 178–93

73. Mundie, J. H. 1957. The ecology of Chironomidae in storage reservoirs. *Trans. Roy. Entomol. Soc. London* 109 :149–232

74. Mundie, J. H. 1959. The diurnal activity of the larger invertebrates at the surface of Lac la Ronge, Saskatchewan. *Can. J. Zool.* 37 : 845–956

75. Neumann, D. 1966. Die lunare und Tägliche schlüpfperiodik der mücke *Clunio.* Steuerung und abstimmung auf die gezeitenperiodik. *Z. Vergl. Physiol.* 53 :1–61

76. Neumann, D. 1967. Genetic adaptation in emergence time of *Clunio* populations to different tidal conditions. *Helgolaender Wiss. Meeresunters.* 15 :163–71

77. Nielsen, E. T. 1959. Copulation of *Glyptotendipes* (*Phytotendipes*) *paripes* Edwards. *Nature* 184 : 1252–53

78. Nielsen, E. T. 1962. Contributions to the ethology of *Glyptotendipes*

(*Phytotendipes*) *paripes* Edwards. *Oikos* 13 :48–75

79. Olander, R., Palmén, E. 1968. Taxonomy, ecology and behaviour of the Northern Baltic *Clunio marinus* Halid. *Ann. Zool. Fenn.* 5 :97–110

80. Oliver, D. R. 1968. Adaptations of Arctic Chironomidae. *Ann. Zool. Fenn.* 5 :111–18

81. Palmén, E. 1955. Diel periodicity of pupal emergence in natural populations of some chironomids. *Ann. Zool. Soc. Vanamo* 17 :1–30

82. Palmén, E. 1956 (1958). Diel periodicity of pupal emergence in some North European chironomids. *Proc. Int. Congr. Entomol., 10th,* 2 :219–24

83. Palmén, E., Aho, L. 1966. Studies on the ecology and phenology of the Chironomidae of the Northern Baltic. *Ann. Zool. Fenn.* 3 :217–44

84. Paris, O. H., Jenner, C. E. 1959. Photoperiodic control of diapause in the pitcher-plant midge, *Metriocnemus knabi. In Photoperiodism and Related Phenomena in Plants and Animals,* ed. R. B. Withrow, 601–24. Am. Assoc. Advan. Sci., 55, Washington. 903 pp.

85. Pringle, G. 1960. Invasion of the egg masses of the mollusc *Bulinus* (*Physopsis*) *globosus,* by larval chironomids. *Parasitology* 50 :497–99

86. Reiss, F. 1968. Ökologische und systematische Untersuchungen an Chironomiden des Bodensees. *Arch. Hydrobiol.* 64 :176–323

87. Roback, S. S. 1957. The immature tendipedids of the Philadelphia area. *Monogr. Acad. Nat. Sci. Philadelphia* 9 :1–152

88. Römer, F., Rosin, S. 1969. Untersuchungen über die Bedeutung der Flugtöne beim Schwarmen von *Chironomus plumosus* L. *Rev. Suisse Zool.* 76 :734–40

89. Saether, O. A. 1962. Larval overwintering cocoons in *Endochironomus tendens* Fabricius. *Hydrobiologia* 20 :377–81

90. Saether, O. A. 1968. Chironomids of the Finse Area, Norway, with special reference to their distribution in a glacier brook. *Arch. Hydrobiol.* 64 :426–83

91. Sandberg, G. 1969. A quantitative study of chironomid distribution and emergence in Lake Erken. *Arch. Hydrobiol. Suppl.* 25 :119–201

92. Scholander, P. F., Flagg, W., Hock, R. J., Irving, L. 1953. Studies on the physiology of frozen plants and animals in the arctic. *J. Cell. Comp. Physiol.* Suppl. 1, 42 : 56 pp.

93. Scott, K. M. F. 1967. The larval and pupal stages of the midge *Tanytarsus* (*Rheotanytarsus*) *fuscus* Freeman, *J. Entomol. Soc. S. Afr.* 30 :174–84

94. Sokolova, G. A. 1966. [On the hibernation mode of larvae of *Limnochironomus* ex. gr. *nervosus* Staeg.] *Zool. Zh.* 45 :140

95. Steffan, A. W. 1968. Zur Evolution und Bedeutung epizoischer Lebensweise bei Chironomiden-Larven. *Ann. Zool. Fenn.* 5 :144–50

96. Styczynski, B., Rakusa-Suszczwski, S. 1963. Tendipedidae of selected water habitats of Hornsund region (Spitzbergen). *Polsk. Arch. Hydrobiol.* 11 :327–41

97. Syrjämäki, J. 1964. Swarming and mating behaviour of *Allochironomus crassiforceps* Kieff. *Ann. Zool. Fenn.* 1 :125–45

98. Syrjämäki, J. 1965. Laboratory studies on the swarming behaviour of *Chironomus strenzkei* Fittkau in litt. *Ann. Zool. Fenn.* 2 :145–52

99. Syrjämäki, J. 1966. Dusk swarming of *Chironomus pseudothummi* Strenzke. *Ann. Zool. Fenn.* 3 :20–28

100. Syrjämäki, J. 1968. A peculiar swarming mechanism of an Arctic Chironomid at Spitsbergen. *Ann. Zool. Fenn.* 5 :151–52

101. Thienemann, A. 1954. Chironomus. Leben, Verbreitung und wirtschaftliche Bedeutung der Chironomiden. *Binnengewässer* 20 :1–834

102. Walshe, B. M. 1951. The function of haemoglobin in relation to filter feeding in leaf-mining chironomid larvae. *J. Exp. Biol.* 28 :57–61

103. Walshe, B. M. 1951. The feeding habits of certain chironomid larvae (subfamily Tendipedinae). *Proc. Zool. Soc. London* 121 :63–79

104. Waters, T. F. 1965. Interpretation of stream drift in streams. *Ecology* 46 :327–34

105. Wirth, W. W. 1949. A revision of

the Clunionine midges with descriptions of a new genus and four new species. *Univ. Calif. Publ. Entomol.* 8:151–82

106. Wirth, W. W., Gressitt, J. L. 1967. Diptera: Chironomidae (Midges).

Antarctic Res. Ser. 10:197–203

107. Wülker, W. 1961. Lebenszyklus und Vertikalverteilung der Chironomide *Sergentia coracina* Zett. im Titisee. *Verh. Int. Verein Limnol.* 14: 962–67

BIONOMICS OF AUTOGENOUS MOSQUITOES[1]   6009

ANDREW SPIELMAN

*Department of Tropical Public Health, Harvard School
of Public Health, Boston, Massachusetts*

The role of blood-feeding in the life cycle of mosquitoes was a controversial subject at the turn of the century. Theobald (101) wrote "Why
[mosquitoes] should draw blood at all and what causes predominate this san-
guinary habit are unknown. [Ross's statement that they require blood before
insemination can occur] must not, I think, be taken too much into account. I
feel bound to state that blood is not an essential for the majority of species
and in this Professor Howard agrees. . . ." Similarly, Mitchell (71) stated,
"[Goeldi] seems to consider blood necessary to the development of fertile
eggs, in contrast to Dr. Dupree's tests which show that it is not always so."
The consensus was expressed by Smith (93), "On the whole the balance of
evidence is perhaps against the idea that blood is a necessity for egg development." Each of these early workers recognized that blood-feeding was
generally followed by the deposition of eggs, but they seem to have considered this a facultative relationship. Indeed, each had independently reared
successive generations of *Culex pipiens* without blood-feeding, and had observed large populations of mosquitoes in localities that appeared to be devoid of suitable vertebrate hosts: Mosquitoes were thought to be primarily
phytophagous. Even Christophers (11), in his pioneering study of ovarian
development in *Anopheles,* was ambiguous in specifying the stage at which
starved mosquitoes cease oogenesis.

The concept of facultative blood-feeding was discarded during the following decade, and the earlier reports forgotten. It became apparent that
adult *Aedes aegypti* do not produce eggs on a carbohydrate diet. In one
study, eggs were obtained solely after feeding whole blood; cell or serum
fractions alone did not suffice (31). Other workers found that a meal of
peptone permitted egg production while sugar alone did not (27, 91). The
classical studies on the ovaries of *Anopheles* (75) and of *Culex* (73) established that oogenesis ceases at a specific development stage and that a meal
of vertebrate blood promotes further ovarian growth.

Sporadic reports recalled the earlier observations of facultative blood-

[1] This investigation was supported in part by Public Health Service Grant
AI-00046.

feeding (46, 74), but they had little impact. In 1929, three independent studies (5, 41, 86) reported the "original" observation that certain blood-starved *C. pipiens* produce viable eggs. This facultative blood-feeding has been the subject of many subsequent reports and it is the purpose of this review to bring together some of this literature.

DEFINITION

Roubaud (86) coined the term "autogeny" to denote egg production by mosquitoes that had not fed upon blood, while "anautogeny" was applied to the more common obligate requirement for vertebrate hosts. It seems unfortunate that the prefix "an-" was attached to a term describing a process complicated by a special requirement, i.e., blood-feeding. However, this terminology is universally accepted and it is not expedient to change it. The original definition has been expanded (99) to include other insects, even nonhematophagous species. Thus, "anautogenous" denotes a situation wherein there is a specific food-mediated ovarian arrest, and "autogenous," the absence of such a diapause. The term has been employed to describe populations of psychodids (22), ceratopogonids (65, 66), tabanids (84), and muscids (54). Facultative blood-feeding is common among simuliids (23).

Arrested oocytes are characterized by the virtual absence of microvillae and pinocytotic vesicles (1, 85). However, within 4 hr of a blood meal, the oolemma is covered by micropinocytotic elements. This condition, involving active uptake of yolk precursor, defines the nondiapausing oocyte. Roth & Porter (85) concluded that the appearance of the follicular envelope serves as a second point of distinction between the two physiological states, and that the follicle cells of *A. aegypti* become widely separated after the blood meal. This would result in free flow of hemolymph between the hemocoel and the oocyte. However, no such separation of follicular cells was observed by Anderson & Spielman (1) and the follicular epithelium proved to be no more permeable after blood-feeding than during ovarian arrest.

A special kind of facultative autogeny was proposed for certain arctic mosquitoes (14). Female *Aedes impiger* and *A. nigripes* feed upon vertebrate hosts within the first week of adult life and, several days later, produce eggs. However, females that did not feed upon blood commence oogenesis at approximately 10 days of age. Paradoxically, unfed female *A. nigripes* produce almost as many eggs as do blood-gorged mosquitoes. On the other hand, fecundity of *A. impiger* is reduced sixfold when blood is withheld. An earlier report suggested that a meal of sugar stimulated autogenous development in these mosquitoes (13), but this theory has been abandoned.

A special problem in definition is posed by mosquitoes that require two meals of blood before eggs mature. Certain *Anopheles gambiae* (16, 30) and *A. elutus* (70) have this requirement but it may be that only malnourished larvae give rise to such "super-anautogenous" adults.

GENETICS

Initially, there was confusion concerning the genetic control of autogeny. Boissezon (5) argued that any female *C. pipiens* would produce eggs if the larvae had been reared on a sufficiently rich diet. However, Roubaud (86, 87) disagreed, and it is now generally accepted that certain strains of *C. pipiens* are characterized by autogeny (67, 100, 110). Unfortunately, Roubaud was not consistent and coined the term "spanogyny" to describe a supposed loss of autogenous potential after several generations without blood-feeding. The concept that fecundity must be renewed periodically by some property of the blood meal was astonishingly persistent.

While genetically anautogenous mosquitoes cannot be made autogenous by superabundant larval feeding, autogenous development can be suppressed by the starving or overcrowding of genetically autogenous larvae (48, 55, 59, 95). Even overcrowding of pupae results in partial suppression of autogeny (95). Such pupae seem to exhaust nutrient reserves by keeping each other in motion. Adult mosquitoes that are malnourished as larvae begin autogenous development but frequently fail to produce mature eggs. Their ovaries contain a mosaic of ova in different stages of development (12, 106). Autogeny in certain mosquitoes has special requirements. Autogenous *A. taeniorhynchus* require a meal of sugar before eggs develop (59). Among *C. tarsalis* the expression of autogeny is reduced when light is less than 12 hr per day (36). Fewer egg rafts than normal are produced by an autogenous population subject to these conditions of light, and the adult itself appears to be sensitive to this photoperiodic effect. This suppression of autogeny is surprising and requires confirmation.

Autogeny is inherited as a simple Mendelian dominant in *Aedes atropalpus* (80). Reproduction without blood-feeding was first recognized in this species in 1912 (40) and it was assumed that this was the general rule. However, O'Meara and Craig found three anautogenous populations and interbred laboratory strains derived from them with autogenous strains from other locations. F_2 and back-crosses approached the classical ratios, thus demonstrating that autogeny is conferred by a simple dominant autosomal gene. This genetic factor has complete penetrance; virtually all female mosquitoes that possess one allele from an autogenous parent produce eggs autogenously. Other alleles affect the number of eggs produced.

Among members of the *Culex pipiens* complex, however, autogeny is controlled by genes on at least two of the three pairs of chromosomes (56, 95). It is not clear how these genes interact. One interpretation, based on a study of hybrid populations, holds that each of the two or more alleles functions independently (95). Each allele is semidominant and the effects are additive. The other interpretation is more complex. Laven (56) analyzed the work of Aslamkhan, who marked each of the three chromosomes at one or more known loci. He concluded that Chromosome III carries a pair of major genes regulating autogeny while Chromosome I, the sex chromosome, carries a series of four alleles that modify the penetrance of the alleles on Chromosome III. The effects of the various alleles, then, would not be additive. Aslamkhan found that 32.4 percent of mosquitoes that inherit one Chromosome I from an autogenous parent are themselves autogenous; 12.5 percent with one such Chromosome II are autogenous; and 43.7 percent of those with one Chromosome III are autogenous. Laven discounted autogeny associated with Chromosome I, attributing it to an artifact of cross-over. He then reasoned that any population with one each of autogeny-bearing Chromosomes I + III would be 76.1 percent autogenous if the effects of the different alleles were additive. Aslamkhan produced five populations with this chromosomal make-up and these contained 48.0 percent, 72.3 percent, 72.4 percent, 80.0 percent, and 81.5 percent autogenous females, respectively. Although he did not present detailed statistical analyses, Laven concluded that the variance of these results is so great that simple additive effects are excluded.

Eight studies of the inheritance of autogeny in the *C. pipiens* complex have already been tabulated (95). In addition to Aslamkhan's work, five studies have been published since 1957 (25, 48, 96, 103, 107). Of these, only the study employing Egyptian strains (25) included a complete set of backcross populations. El Dine's results were somewhat anomalous, and she made no attempt to explain them. Her hybrid populations were all highly autogenous; even those resulting from the back-cross to the anautogenous parent. She reported that the strains did not breed true and it may be that the anomalous results were due to contamination. Vinogradova (107) worked with Russian strains and Umino (103) with Japanese strains. Both workers experienced difficulty in obtaining hybrids and their F_1 and F_2 populations were predominantly anautogenous. They too, may not have employed homozygous parent stocks. Although fragmentary, results of the remaining studies were consistent with the concept of multiple, semidominant alleles.

PHYSIOLOGY

Since autogenous mosquitoes produce eggs without having fed, nutrient ingested during larval life must provide the protein from which yolk is formed. Indeed, autogenous mosquitoes contain larger stores of nutrient than do anautogenous mosquitoes (12, 50, 88, 102). Roubaud and Toumanoff described the extent of the fat body in the two biotypes of *C. pipiens* and concluded that the autogenous mosquitoes amass much the greater stores of fat during larval life. Clements confirmed these observations histochemically. He found that fat bodies of recently emerged mosquitoes of both types contain fat, glycogen, and protein but that these components disappear during the first few days of adult life, concurrently with the disappearance of vestigial larval musculature. Although these reserves in anautogenous females are relatively small, Clements hypothesized that they might be sufficient for the production of a few eggs. Twohy & Rozeboom (102) and Lang (50) established analytically that autogenous females are larger and contain more lipid, glycogen, and nitrogen than do anautogenous females. However, at variance with Clements, they found that starved anautogenous females do not possess sufficient energy reserves to survive five days, much less produce eggs at the end of that time. Anautogenous mosquitoes exhaust their glycogen stores about a day before autogenous eggs would have been produced. One point of internal contradiction in Twohy & Rozeboom's work lies in the finding that the concentration of nitrogen is greatest in the tissues of anautogenous mosquitoes. Only the large size of autogenous females permits the conclusion that each contains more protein.

The blood of mature larvae of an autogenous *C. pipiens* population contains a protein fraction that is not present in larvae of an anautogenous population (10). This finding is all the more remarkable because the fraction is not found in other larval tissues, in adult mosquitoes, or in mature eggs. It is apparent that this fraction does not represent specific yolk precursors that are deposited during the larval stage, and Chen (10) suggests that the fraction may be involved in the endocrine control of autogeny. It is unfortunate that this study was based upon a comparison of old laboratory colonies of diverse origin. The strains even differed morphologically. Therefore, Chen's finding may not be related to autogeny.

Flight muscles contain a potential source of nutrient that may support the autogenous production of yolk, provided that the ability to fly is not essential. Indeed, one nonhematophagous mosquito, *Aedes communis,* may utilize its flight muscles for this purpose (37). A contrary report holds that mosquitoes of this species retain the flight muscles after eggs have been produced (3). Both of these studies present convincing evidence yielding opposite conclusions that are based upon anatomical observations of mosquitoes from the same region. However, it may be that different populations were observed. Furthermore, Hocking studied mosquitoes derived directly from the field while Bckel based his work on laboratory colonies.

Like *A. communis, A. rempeli* is a nonhematophagus, arctic mosquito (94). Females of this species retain their flight muscles after deposition of eggs while fat body and vestigial larval muscles are resorbed during oogenesis.

Transplantation of endocrine organs may result in autogenous production of eggs. Blood-starved, anautogenous *C. pipiens* and *A. aegypti* produced eggs when supplied either with transplanted corpora allata from gonoactive mosquitoes or with extracts of roach corpora allata (52). Apparently, oogenesis was not due to some nutritional property of the oily extracts because metabolizable oils themselves do not induce deposition of yolk (51). A recent study of this problem produced an entirely different result (62). No eggs were formed following transplantation of corpora allata from blood-fed to blood-starved anautogenous females. On the other hand, eggs developed normally in blood-fed females allatectomized at three–five days of age. Deposition of yolk was prevented when allatectomy was performed immediately after emergence, i.e., long before the time of ovarian arrest (58). Implantation of active glands restored the ability to develop eggs following later blood meals. It follows from this that the corpora allata are required solely during early oogenesis, permitting ovaries to reach the resting stage. The function of the corpora allata is the same in autogenous as in anautogenous mosquitoes. On the other hand, medial neurosecretory cells of the brain do influence ovarian arrest (61). Blood-starved females develop mature eggs after neurosecretory cells from gonoactive females are implanted in their abdomens.

Another attempt to induce yolk deposition in blood-starved, anautogenous females involved transfusion of hemolymph (29). Minute quantities were transfused from gonoactive females to others that had not yet become gonoactive. Transfused mosquitoes all had previously fed upon blood and were decapitated. Gillet concluded that appropriate transfusion induced female *A. aegypti* to develop eggs. However, this study is not convincing because no blood-starved mosquitoes were studied and the results were somewhat equivocal.

A third approach at inducing oogenesis in blood-starved anautogenous mosquitoes was successful. Eggs developed after beta-ecdysone was injected into the hemolymph of female *A. aegypti* (1). Lesser quantities of inoculum caused follicles to degenerate. It is interesting that a synthetic juvenile hormone preparation failed to affect yolk deposition, thus confirming that the role of the corpora allata may be limited to the prearrest stages of oogenesis. Ecdysone, itself, has not been demonstrated in adult insects and its role in normal oogenesis remains problematic. Prothoracic glands are absent from adult *A. aegypti* (7).

Experiments involving ligation of the abdomen provided the earliest evidence suggesting endocrine control of ovarian diapause. Detinova (17) found that eggs do not develop when a ligature is placed anterior to the abdomen within 3 hr after the blood meal. On the other hand, Clements (12) concluded that ligatures applied immediately after blood-feeding do not prevent oogenesis. Here, again, the literature is confused by contradictory results. Gillett (28) reported that decapitation of female *A. aegypti* at 4 hr after blood-feeding prevents oogenesis. Larsen & Bodenstein (52) concurred but found that the lag period is less than 1 hr.

There is similar confusion on yet another approach toward identifying the endocrine control of autogeny. Burgess & Rempel (7) provided convincing histological evidence that the corpora allata of *A. aegypti* become increasingly active soon after emergence but not after blood meals. Earlier, Clements (12) had reported that changes in the appearance of the corpus allata are not correlated with ovarian activity of *C. pipiens*. On the other hand, Detinova (17) and Mednikova (69) concluded that the activity of *Anopheles* corpora allata is correlated with blood-feeding. Working with *C. pipiens*, Larsen & Bodenstein (52) described a similar correlation.

The medial neurosecretory cells of the brain were the subject of a similar study (53). Neurosecretory material is apparently produced in these cells after the blood meal and is transported to the corpora cardiaca. The corpora cardiaca, then, would presumably store this neurosecretory material and be involved in its release into the blood stream. It should be noted that the corpora cardiaca and corpora allata of adult mosquitoes are adjacent and difficult to separate (7). This factor is surely at the root of some of the conflicting results described above because the manner of manipulation of the allatum complex may determine subsequent effects. Trauma to surrounding tissues may limit the quantity of neurosecretory material released from the severed nerve trunks.

An enigmatic aspect of autogeny is that only the first clutch of eggs of an autogenous mosquito is produced without benefit of a blood meal. Hematophagous species are anautogenous in subsequent ovarian cycles. Kal'chenko (44) reported multiple clutch formation by blood-starved *C. pipiens,* but this requires confirmation.

It is probable that some property of the blood meal induces the brain to secrete a hormone which is transported via nerves to the corpora cardiaca, and that ovarian arrest is disrupted following release of this material into the hemocoel. Three crucial but unanswered problems remain: What prevents ovarian diapause in autogenous mosquitoes? Are the ovaries stimulated directly by secretions of the medial cells of the brain or is another hormone (ecdysone?) released secondarily? What factors control the number of ova that mature?

A related question that has been subject to much speculation is, can ovarian diapause be interrupted by natural stimuli other than vertebrate blood? Three kinds of feeding stimuli have been considered: blood of invertebrate animals, high protein foods such as milk, and hormono-mimetic material of plant origin. Adenosine triphosphate is the most important constituent of vertebrate blood that induces filling of the midgut (39) and presumably permits completion of oogenesis. By itself, ATP may not suffice but the critical experiment has not been performed. In any case, it seems unlikely that ATP is available to mosquitoes in nature in any food other than vertebrate blood.

Some mosquitoes (*A. aegypti* and *C. pipiens*) feed readily in the laboratory upon other insects and subsequently produce eggs (35). The physiological effects of this food and the extent to which this feeding pattern occurs in nature are unknown.

Anautogenous *A. aegypti* will feed upon milk in the laboratory and produce small batches of eggs (18). After an *A. aegypti* colony was reared upon milk for a number of generations, autogenous females appeared and became predominant (63).

Phytoecdysones are ubiquitous in nature and comprise as much as 10 percent of the dry weight of certain common plants (78). When imbibed with sugar, beta-ecdysone permits anautogenous *A. aegypti* and *C. pipiens* females to produce a few eggs (1). Mosquitoes imbibe a variety of materials from plants, including substances that do not sustain life (82). It may be that phytoecdysones are frequently ingested and, if so, the definition of autogeny needs to be re-examined.

DISTRIBUTION

It seems that autogeny is found wherever it is sought. After one such search, eight new records were cited (9). Ten of 17 species contained autogenous members. In 1951, 12 autogenous mosquito species were known (2). Five years later, the total rose to 20 (95) and, by 1965, 36 species were listed (107). That number could now be augmented, but there seems to be little point in compiling another list since it would include species in most genera and will soon be out of date. In some populations, as *Aedes dorsalis* (32) and *Culex peus* (109), autogeny is rare, while in others, as *Deinocerites cancer* (33) or *Opifex fuscus* (34), it is universal.

Culex pipiens.—Culex pipiens occurs throughout the middle latitudes of the world (68), but the distribution of autogenous populations is limited. In the northern hemisphere, autogenous populations are broadly distributed throughout the range of the temperate zone subspecies, *C. p. pipiens*. With the exception of North Africa and Israel (47) and southeastern Australia (24), autogenous populations range north of the 33rd parallel. Many records of autogenous *C. pipiens* exist, mostly from urban loci.

Autogenous *C. pipiens* appear to comprise distinct populations. Twenty percent of female *C. pipiens* collected in an urban site in Boston were homozygous for autogeny while less than 3 percent were presumed hybrids. Clearly, these autogenous and anautogenous populations are genetically isolated. There are a combination of temporal, spatial, and behavioral mating barriers (98).

Mosquitoes of the two populations in the Boston study were all similar morphologically, and male genitalia were typical of the temperate zone subspecies *C. pipiens*. This pair of populations is interfertile but may be genetically incompatible with certain *C. pipiens* populations located elsewhere (57). Many males produce sperm that cannot fertilize the eggs of apparently similar females from distant populations. Although autogenous and anautogenous populations in Boston are of the same mating type, this relationship has not been rigorously studied in other locations.

A rather different situation exists in Japan (90). Here, sympatric autogenous and anautogenous populations can be distinguished by the shape of the aedeagus and by other differences. The autogenous form is typical for *C. p. pipiens* (the *"molestus"* biotype) while anautogenous populations resemble the tropical subspecies, *C. p. quinquefasciatus* or are intermediate (designated *C. p. pallens*). *Molestus* and pallens populations are distributed over the main islands of Japan, occurring in most cities. Populations of the Japanese biotypes are genetically isolated and two barriers to hybridization have been demonstrated. There is some evidence suggesting a barrier to effective insemination in that cross-mated females frequently have empty spermathecae. This may result from a mechanical barrier to sperm transfer (97). The second barrier results in nonviable eggs which contain fully formed larvae that cannot hatch. Populations of similar type intermate readily even though they may originate in different parts of Japan. However, attempts to intermate *molestus* and *pallens* populations, even from the same city, generally result in failure.

It is tempting to conclude from Sasa's observation that autogenous and anautogenous populations in Japan are isolated in nature. However, the genetic structure of Japanese *molestus* populations has not been established and it may be that *molestus* itself is comprised of autogenous and anautogenous components. Solution of this problem awaits study of individual female *C. pipiens* collected directly from the field.

California *C. pipiens* populations are structured like those in Japan. In an unpublished thesis, Iltis (42) recorded precisely the geographical distribution of many local populations of the complex. He found a general north-south cline between the *pipiens* and *quinquefasciatus* morphological extremes. However, the cline is highly modified by local climatic conditions, and in certain locations it is narrow and reversed. In the study area, autogeny is common where the *pipiens* morphological type is found and is absent from nearby *quinquefasciatus* areas. The prevalence of autogeny varies from location to location and between seasons. These data suggest that autogenous *pipiens* populations are genetically isolated from contiguous *quinquefasciatus* populations but not necessarily from anautogenous *pipiens* populations.

C. pipiens populations in Australia are generally similar to those of Japan and California (21). Again, the prevalence of autogeny varies from place to place but rarely approaches unity (20). It follows, therefore, that anautogenous *"molestus"* are present and it may be that they are not panmictic with the autogenous variety.

Most studies of Palearctic autogenous populations involve analysis of the behavior of laboratory colonies, each representing the population of a fairly large region. These colonies were generally started from samples containing many mosquitoes. In one such study, 40 colonies were established from the area of Rome and each of them contained autogenous members (49). Two additional biological properties generally characterize such autogenous populations, including the ability to mate without prior flight and inability to hibernate (87). Various combinations of these and other properties have been summarized by Rioux (83) and Vinogradova (108). Some traits are inherited independently rather than as pleiotropic effects of the genes controlling autogeny. Their frequent association suggests that autogenous populations comprise genetic entities that are distinct from anautogenous populations.

In general, students of the problem agree that autogenous *C. pipiens* populations are genetically isolated from nearby anautogenous populations. Studies in the laboratory suggest that populations in nature may be homozygous for autogeny because approximately ⅓ of a hybrid population begins autogenous formation of eggs without completing the process (98). This condition would prevent synchronous development of an anautogenous clutch of eggs and is surely pathological.

The temporal and geographic distribution of autogenous *C. pipiens* is largely determined by factors permitting winter survival. Anautogenous females develop large stores of fat when reared under conditions of temperature and light that approach those that occur naturally during late summer (15, 26, 38, 77). Such adult mosquitoes, but not autogenous *C. pipiens* (87), are capable of hibernation. Indeed, all hibernating adults collected in Boston were homozygous for anautogeny (96). However, ability to hibernate is a recessive character so that it is theoretically possible for hybrid forms to survive the winter in the adult instar (105). Autogenous *C. pipiens* must breed continuously throughout the winter which limits them to urban locations where sheltered breeding sites are available. They are found breeding throughout the winter in Boston (98), and *"molestus"* populations breed in Tokyo in mid-winter (76).

Temporal distribution of the two forms may also be influenced by relative fecundity. In nature, autogenous females apparently do not feed upon blood and this limits them to a single clutch of approximately 65 eggs (89, 96). In contrast, anautogenous females produce more than 200 eggs in each clutch and do so repeatedly.

There is controversy concerning the feeding habits of autogenous *C. pipiens*. Originally, the term *"molestus"* was applied to the autogenous race because it seemed to be avidly anthropophilic (43). Birds are the principle hosts for anautogenous mosquitoes. Shute (92) provided additional observations in support of this concept but he did not analyze the genetic make-up of those *C. pipiens* that fed upon man. It may be that episodes of man-biting by temperate zone *C. pipiens* result from hybridization between autogenous and anautogenous populations (96, 107). In any case, autogenous females do not feed until after their autogenous eggs are deposited, and subsequent clutches are generally small (19, 45, 98). The rate of maturation strongly influences reproductive potential and, again, anautogenous populations have the advantages (50, 102). During the breeding season, Japanese *pallens* populations increase to maximum abundance approximately a month earlier than do autogenous populations (76).

The breeding sites in which autogenous populations are found are generally enclosed while relatively open sites support anautogenous breeding. It is apparent that anautogenous mosquitoes must have access to suitable hosts and this requires that there be free access and egress from the breeding site. In situations in which hosts are abundant, anautogenous populations would be favored because the full reproductive potential of the species could be realized. When hosts are scarce, of course, autogenous populations would predominate, provided that sheltered winter breeding sites are available. It may be that restricted egress prevents dispersion from the breeding site and thereby favors autogenous populations.

The apparent difference in flight activity and dispersion suggests that autogenous populations may be more inbred than are anautogenous populations. Indeed, the genetic load of one autogenous population is surprisingly small (104). Anautogenous *C. pipiens* might carry larger reserves of genetic variability which would be shared with sympatric autogenous populations from time to time. This load would quickly be revealed in the autogenous population when new restricted breeding sites are founded.

Culex tarsalis.—Laboratory colonies of *C. tarsalis* may be either autogenous or anautogenous, thereby indicating that there is genetic control of this property (4, 8). The prevalence of autogeny varies with season; ranging from approximately 80 percent in early summer to 20 percent by fall (72). This is the reverse of the distribution described for *C. pipiens*. It is even more surprising that the fecundity of autogenous females declined concurrently with their prevalence. Monthly average fecundity ranged from 116 eggs in May to 30 in October. The maximum was 220, far more eggs than are produced by autogenous *C. pipiens*.

Aedes atropalpus.—In North America, the range of *A. atropalpus* is divided into four zones, each occupied by a separate subspecies (80). Only the type subspecies contains autogenous members and this population occupies the Appalachian Mountains and the hilly areas around the northern Great Lakes. The remaining anautogenous populations are distributed to the south and west. Apparently, all female *A. a. atropalpus* are autogenous and they are isolated by distance from other populations with which they can interbreed. Only one collection, from Texas, contained both autogenous and anautogenous mosquitoes (79).

Autogenous *A. atropalpus* are remarkably fecund, producing approximately 200 eggs without food (79). They rarely take blood. Anautogenous females deposit a similar number of eggs, but mature somewhat more rapidly. O'Meara suggested that the distribution of the autogenous subspecies may be limited, in part, by the availability to the larvae of food. *A. a. atropalpus* larvae require more food and feed for a longer time than do larvae of other subspecies.

Other Aedes.—*A. albonotaus* is distributed in the northern Bahamas and is common on Grand Bahama Island (99). Larvae of this species occupy a niche that seems to be identical to that of *A. aegypti* on this island, breeding in automobile tires and other small water containers in and around houses. In contrast to *A. aegypti,* all *A. albonotatus* appear to be autogenous and rarely feed upon human hosts. It is interesting that populations of the two species do not occupy the same towns. They intermate but are not interfertile and this may serve to prevent coexistence. Females of the two species appear to have equal access to hosts.

A. taeniorhynchus is distributed along the Atlantic coast of much of the United States and is numerous in Florida. Occasionally, eggs recovered from one sample of sod give rise to both autogenous and anautogenous strains (64). Strains interbreed freely but it is not known whether they do so in nature. Nor has the geographical or temporal distribution of autogeny been described.

Partially autogenous strains of *A. togoi* are known from Formosa (55) and Nagasaki (81), but other strains from Japan are anautogenous. It may be that autogeny arose through laboratory selection as has been demonstrated for *A. aegypti* (60). An alternative hypothesis is that the prevalence of autogenous genes varies geographically. Such genes appear to be rare within many *Aedes* populations (9).

Conclusions

The development of blood-feeding is the central problem in considering the evolution of mosquitoes. The ancestral Nematocera probably fed upon blood (23) and autogeny seems to have accompanied this feeding pattern as "insurance" against an unfavorable adult environment. Mosquitoes in such genera as *Toxorhynchites* and *Malaya* exploit a rich larval environment and possess mouthparts that cannot pierce vertebrate skin. However, most autogenous populations have retained the ability to take blood and can exploit the adult environment to enhance fecundity. Obligate blood-feeding by *C. pipiens* permits maximum fecundity and rate of maturation. In contrast, some *Aedes atropalpus* populations appear to employ obligate blood-feeding as compensation for a deficient larval environment. Such mosquitoes need not accumulate yolk protein during the larval stage and therefore can breed successfully when food is scarce.

The stimuli that induce mosquitoes to produce eggs in nature remain to be defined. Autogenous mosquitoes seem to be common, recalling the theories of earlier workers. In addition, anautogenous mosquitoes that have not fed upon vertebrate blood may produce eggs. The blood of other insects or perhaps material derived from certain plants may provide the necessary stimuli. Thus, the assumption is no longer tenable that mosquitoes become numerous solely where vertebrate blood is available.

The hormonal control of oogenesis is particularly difficult to unravel. As with many insects, corpora allatum activity is essential for oogenesis, but only in the newly emerged adult. Ovaries in the resting stage seem to develop after brain hormone is released from storage in the corpora cardiaca. However, it may be that this hormone does not act directly upon the oocyte. Perhaps another hormone, such as ecdysone, is released secondarily. Nor have the stimuli been identified that cause the corpora cardiaca to release stored hormone. Knowledge of the physiology of autogeny depends upon an understanding of the control of this ovarian arrest.

Vectorial capacity depends, in part, upon survival of a vector insect after it has become infected and upon its choice of hosts. Autogenous mosquitoes take blood only after the first clutch of eggs has been deposited, thus limiting the potential period of infectivity. On the other hand, host preference may be influenced by genes associated with autogeny (25, 92, 96, 107). It may be that human hosts more readily attract certain autogenous than anautogenous mosquitoes and in this manner, autogeny would enhance vectorial capacity. The public health significance of autogeny remains to be defined.

LITERATURE CITED

1. Anderson, W. A., Spielman, A. The mode of entry of yolk precursors into ovarian follicles of mosquitoes. In preparation
2. Report of the advisory committee. 1953. International cooperation to the study of the *Culex pipiens* complex. *Trans. Int. Congr. Entomol., 9th* 2:299–300
3. Beckel, W. E. 1954. The lack of autolysis of the flight muscles of *Aedes communis* (DeGeer) in the laboratory. *Mosquito News* 14:124–27
4. Bellamy, R. E., Kardos, E. H. 1958. A strain of *Culex tarsalis* Coq. reproducing without blood meals. *Mosquito News* 18:132–34
5. Boissezon, P. de 1929. Experiences au sujet de la maturation des oeufs chez les Culicides. *Bull. Soc. Pathol. Exot.* 22:683–89
6. Burgess, L. 1967. Pycnosis of the nuclei of ecdysial gland cells in *Aedes aegypti* (L.). *Can. J. Zool.* 45:1294–95
7. Burgess, L., Rempel, J. G. 1966. The stomodaeal nervous system, the neurosecretory system, and the gland complex in *Aedes aegypti* (L.). *Can. J. Zool.* 44:731–65
8. Chao, J. 1958. An autogenous strain of *Culex tarsalis* Coq. *Mosquito News* 18:134–36
9. Chapman, H. C. 1962. A survey for autogeny in some Nevada mosquitoes. *Mosquito News* 22:134–36
10. Chen, P. S. 1967. Electrophoretic patterns of larval haemolymph proteins in autogenous and anautogenous forms of *Culex pipiens* L. *Nature* 215:316–17
11. Christophers, S. R. 1911. The development of the egg follicle in Anophelines. *Paludism* No. 2:73–88
12. Clements, A. N. 1956. Hormonal control of ovary development in mosquitoes. *J. Exp. Biol.* 33:211–23
13. Corbet, P. S. 1964. Autogeny and oviposition in Arctic mosquitoes. *Nature* 203:669
14. Corbet, P. S. 1967. Facultative autogeny in Arctic mosquitoes. *Nature* 215:662–63
15. Danilevskii, A. S., Glinyanaya, E. N. 1958. Variation of the gonotropic cycle and the imago diapause with the length of day in the blood-sucking mosquitoes. *Uchen. Zap. Leningr. Gosud. Univ. (Ser. Biol. Nauk)* No. 46:34–51. In Russian
16. Davidson, G. 1954. Estimation of the survival-rate of anopheline mosquitoes in nature. *Nature* 174:792–93
17. Detinova, T. S. 1945. On the influence of glands of internal secretion upon the ripening of the gonads and the imaginal diapause in *Anopheles maculipennis*. *Zool. Zh.* 24:291–98. In Russian
18. Dimond, J. B., Lea, A. O., Brooks, R. F., DeLong, D. M. 1955. A preliminary note on some nutritional requirements for reproduction in female *Aedes aegypti*. *Ohio J. Sci.* 55:209–11
19. Dobrotworsky, N. V. 1954. The *Culex pipiens* group in South-Eastern Australia. III. Autogeny in *Culex pipiens* form *molestus*. *Proc. Linn. Soc. N.S.W.* 79:193–95
20. Dobrotworsky, N. V. 1955. The *Culex pipiens* group in South-Eastern Australia. IV. Crossbreeding experiments within the *Culex pipiens* group. *Proc. Linn. Soc. N.S.W.* 80:33–43
21. Dobrotworsky, N. V. 1967. The problem of the *Culex pipiens* complex in the South Pacific (including Australia). *Bull. W.H.O.* 37:251–55
22. Dolmatova, A. V. 1946. The autogenous development of eggs in *Phlebotomus papatasii* Scop. *Medskaya Parazit.* 15:58–62
23. Downes, J. A. 1958. The feeding habits of biting flies and their significance in classification. *Ann. Rev. Entomol.* 3:249–66
24. Drummond, F. H. 1951. The *Culex pipiens* complex in Australia. *Trans. Roy. Entomol. Soc. London* 102:369–71
25. El-Dine, K. Z. 1966. Preliminary studies on autogeny in *Culex pipiens* complex. *J. Egypt. Pub. Health Assoc.* 41:115–26
26. Eldridge, B. F. 1968. The effect of temperature and photoperiod on

blood-feeding and ovarian development in mosquitoes of the *Culex pipiens* complex. *Am. J. Trop. Med. Hyg.* 17 :133–40

27. Fielding, I. W. 1919. Notes on the bionomics of *Stegomyia fasciata,* Fabr. *Ann. Trop. Med. Parasitol.* 13 :259–96

28. Gillett, J. D. 1956. Initiation and promotion of ovarian development in the mosquito *Aedes (Stegomyia) aegypti* (Linnaeus). *Ann. Trop. Med. Parasitol.* 50 :375–80

29. Gillett, J. D. 1958. Induced ovarian development in decapitated mosquitoes by transfusion of haemolymph. *J. Exp. Biol.* 35 :685–93

30. Gillies, M. T. 1954. The recognition of age groups within populations of *Anopheles gambiae* by the pregravid rate and the sporozoite rate. *Ann. Trop. Med. Parasitol.* 48 :58–74

31. Gordon, R. M. 1922. Notes on the bionomics of *Stegomyia calopus* Meigen, in Brazil. Part II. *Ann. Trop. Med. Parasitol.* 16 :425–39

32. Grimstad, P. R., Wendell, K. G., Hill, W. B., Martin, H. J., Jr. 1969. Autogeny in *Aedes dorsalis* (Meigen). *Calif. Vector Views* 16 : 71–72

33. Haeger, J. S., Phinizee, J. 1959. The biology of the crabhole mosquito, *Deinocerites cancer* Theobald. *Rep. Florida Antimosquito Assoc.* 30 : 34–37

34. Haeger, J. S., Provost, M. W. 1965. Colonization and biology of *Opifex fuscus. Trans. Roy. Soc. N.Z.* 6 : 21–31

35. Harris, P., Cooke, D. 1969. Survival and fecundity of mosquitoes fed on insect haemolymph. *Nature* 222 : 1264–68

36. Harwood, R. F. 1966. The relationship between photoperiod and autogeny in *Culex tarsalis. Entomol. Exp. Appl.* 9 :327–31

37. Hocking, B. 1954. Flight muscle autolysis in *Aedes communis* (DeGeer). *Mosquito News* 14 : 121–23

38. Hosoi, T. 1954. Egg production in *Culex pipiens pallens* Coquillett. II. Influence of light and temperature on activity of females. *Jap. J. Med. Sci. Biol.* 7 :75–81

39. Hosoi, T. 1959. Identification of blood components which induce gorging of the mosquito. *J. Insect Physiol.* 3 :191–218

40. Howard, L. O., Dyar, H. G., Knab, F. 1912. *The mosquitoes of North and Central America and the West Indies. Carnegie Inst. Wash. Publ. 1.* 520 pp.

41. Huff, C. G. 1929. Ovulation requirements of *Culex pipiens* Linn. *Biol. Bull.* 56 :347–50

42. Iltis, W. G. 1966. *Biosystematics of the Culex pipiens Complex in Northern California.* PhD thesis, Univ. of California, Davis

43. Jobling, B. 1938. On two subspecies of *Culex pipiens* L. *Trans. Roy. Entomol. Soc. London* 87 :193–216

44. Kal'chenko, Y. E. 1962. On the biology of the mosquito *Culex pipiens molestus* Forsk. *Entomol. Rev.* 41 :52–54

45. Kanda, T. 1965. Studies on the ovarian development in *Culex pipiens* s. 1. following artificial ingestion of blood components and other nutrients. *Jap. J. Sanit. Zool.* 16 :49–56. In Japanese

46. Kirkpatrick, T. W. 1925. *The mosquitoes of Egypt.* Cairo : Government Press. 224 pp.

47. Knight, K. L. 1951. A review of the *Culex pipiens* complex in the Mediterranean subregion. *Trans. Roy. Entomol. Soc. London* 102 : 354–64

48. Krishnamurthy, B. S., Laven, H. 1961. A note on inheritance of autogeny in *Culex* mosquitoes. *Bull. W.H.O.* 24 :675–77

49. La Face, L. 1961. L'autogenia nella popolazioni del complesso "*Culex pipiens*" nella provincia di Latina. *Rend. Ist. Super. Sanita, Ital. Ed.* 24 :693–98

50. Lang, C. A. 1963. The effect of temperature on the growth and chemical composition of the mosquito. *J. Insect Physiol.* 9 :279–86

51. Larsen, J. R. 1965. Inability of metabolizable oils to mimic corpus allatum hormone stimulation of ovarian development in mosquitoes. *Nature* 206 :428

52. Larsen, J. R., Bodenstein, D. 1959. The humoral control of egg maturation in the mosquito. *J. Exp. Zool.* 140 :343–81

53. Larsen, J. R., Broadbent, A. 1968. The neurosecretary cells of the brain of *Aedes aegypti* in relation

to larval molt, metamorphosis and ovarian development. *Trans. Am. Microscop. Soc.* 87 :395–410

54. Larsen, J. R., Pfadt, R. E., Peterson, L. G. 1966. Olfactory and oviposition responses of the house fly to domestic manures, with notes on an autogenous strain. *J. Econ. Entomol.* 59 :610–15

55. Laurence, B. R. 1964. Autogeny in *Aedes (Finlaya) togoi* Theobold. *J. Insect Physiol.* 10 :319–31

56. Laven, H. 1967. Formal genetics of *Culex pipiens.* In *Genetics of Insect Vectors of Disease,* Chap. 2, 17–64. Amsterdam, London and New York : Elsevier. 794 pp.

57. Laven, H. 1967. Speciation and evolution in *Culex pipiens.* In *Genetics of Insect Vectors of Disease,* Chap. 7, 251–75. Amsterdam, London and New York : Elsevier. 794 pp.

58. Lea, A. O. 1963. Some relationships between environment, corpora allata, and egg maturation in aedine mosquitoes. *J. Insect Physiol.* 9 :793–809

59. Lea, A. O. 1964. Studies on the dietary and endocrine regulation of autogenous reproduction in *Aedes taeniorhynchus* (Wied.). *J. Med. Entomol.* 1 :40–44

60. Lea, A. O. 1964. Selection for autogeny in *Aedes aegypti. Ann. Entomol. Soc. Am.* 57 :656–57

61. Lea, A. O. 1967. The medial neurosecretory cells and egg maturation in mosquitoes. *J. Insect Physiol.* 13 :419–29

62. Lea, A. O. 1969. Egg maturation in mosquitoes not regulated by the corpora allata. *J. Insect Physiol.* 15 :537–41

63. Lea, A. O., Knierim, J. A., Dimond, J. B., DeLong, D. M. 1955. A preliminary note on egg production from milk-fed mosquitoes. *Ohio J. Sci.* 55 :21

64. Lea, A. O., Lum, P. T. M. 1959. Autogeny in *Aedes taeniorhynchus* (Wied.). *J. Econ. Entomol.* 52 : 356–57

65. Lee, V. R. 1968. Parthenogenesis and autogeny in *Culicoides bambusicola* Lutz. *J. Med. Entomol.* 5 :91–93

66. Linley, J. R. 1968. Autogeny and polymorphism for wing length in *Leptoconops becquaerti* (Kieff.). *J. Med. Entomol.* 5 :53–66

67. MacGregor, M. E. 1932. The occurrence of Roubaud's "race autogène" in a German strain of *Culex pipiens* in England : with notes on rearing and bionomics. *Trans. Roy. Soc. Trop. Med. Hyg.* 26 :307–14

68. Mattingly, P. F. 1951. The *Culex pipiens* complex. Introduction. *Trans. Roy. Entomol. Soc. London* 102 :331–42

69. Mednikova, M. V. 1952. Endocrine glands corpora allata and corpora cardiaca of mosquitoes. *Zool. Zh.* 31 :676–85. In Russian

70. Mer, G. G. 1936. Experimental study on the development of the ovary in *Anopheles elutus. Bull. Entomol. Res.* 27 :351–59

71. Mitchell, E. G. 1907. *Mosquito Life.* New York and London : Knickerbocker Press. 281 pp.

72. Moore, C. G. 1963. Seasonal variation in autogeny in *Culex tarsalis* Coq. in northern California. *Mosquito News* 23 :238–41

73. Nath, V. 1924. Egg-follicle of *Culex. Quart. J. Microscop. Sci.* 69 :151–76

74. Neumann, R. O. 1912. Brauchen die Steckmücken zur Reifung der Eir Blut als Nahrung. *Arch. Schiffs Tropenhyg.* 16 :27–30

75. Nicholson, A. J. 1921. The development of the ovary and ovarian egg of a mosquito, *Anopheles maculipennis,* Meig. *Quart. J. Microscop. Sci.* 65 :395–448

76. Noguchi, Y., Ogata, K., Kazama, T., Imai, S. 1965. Seasonal prevalence of *Culex molestus* breeding in septic tanks. *Jap. J. Sanit. Zool.* 16 : 133–37. In Japanese

77. Oda, T. 1969. Studies on the follicular development and overwintering of the house mosquito, *Culex pipiens pallens* in Nagasaki Area. *Trop. Med.* 10 :195–216

78. Ohtaki, T., Milkman, R. D., Williams, C. M. 1967. Ecdysone and ecdysone analogues : their assay on the fleshfly *Sarcophaga peregrina. Proc. Nat. Acad. Sci., U.S.* 58 : 981–84

79. O'Meara, G. F. 1969. *Analysis of Geographical Variation in* Aedes atropalpus *with Emphasis on Autogeny.* PhD thesis, Univ. of Notre Dame, Notre Dame, Indiana

80. O'Meara, G. F., Craig, G. B., Jr. 1969. Monofactorial inheritance of

autogeny in *Aedes atropalpus*. *Mosquito News* 29 :14–22

81. Omori, N. 1962. A review of the role of mosquitoes in the transmission of Malayan and Bancroftian filariasis in Japan. *Bull. W.H.O.* 27 :585–94

82. Patterson, R. S., Smittle, B. J., DeNeve, R. T. 1969. Feeding habits of male southern house mosquitoes on ³²P-labeled and unlabeled plants. *J. Econ. Entomol.* 62 :1455–58

83. Rioux, J. A. 1958. *Les Culicides du "Midi" Méditerranéan. Étude systématique et écologique. Encyclo. Entomol.* 35 :303 pp.

84. Rockel, E. G. 1969. Autogeny in the deerfly, *Chrysops fuliginosus. J. Med. Entomol.* 6 :140–42

85. Roth, T. F., Porter, K. R. 1964. Yolk protein uptake in the oocyte of the mosquito *Aedes aegypti. J. Cell Biol.* 20 :313–32

86. Roubaud, E. 1929. Cycle autogène d'ttente et générations hivernales suractives inapparentes chez le moustique commun, *Culex pipiens* L. *C. R. Acad. Sci. Paris* 188 :735–38

87. Roubaud, E. 1933. Essai synthétique sur la vie du moustique commun (*Culex pipiens*). L'évolution humaine et les adaptations biologiques du moustique. *Ann. Sci. Nat. (Zool.)* 16 :5–168

88. Roubaud, E., Toumanoff, C. 1930. Sur une race physiologique suractive du moustique commun, *Culex pipiens* L. *Bull. Soc. Pathol. Exot.* 23 :196–204

89. Saito, K., Hayashi, S. 1966. Studies on the age of mosquitoes. VII. A survey on the composition of a population of *Culex pipiens molestus* in a well closed habitat. *Jap. J. Sanit. Zool.* 17 :180–83. In Japanese

90. Sasa, M., Shirasaka, A., Kurihara, T. 1966. Crossing experiments between *fatigans, pallens* and *moleshus* colonies of the mosquito *Culex pipiens* s. 1. from Japan and southern Asia, with special reference to hatchability of hybrid eggs. *Jap. J. Exp. Med.* 36 :187–210

91. Sen, S. K. 1917. A preliminary note on the role of blood in ovulation in Culicidae. *Indian J. Med. Res.* 4 :729–53

92. Shute, P. G. 1953. Contribution to the *Culex pipiens* complex. *Trans. Intern. Congr. Entomol. 9th* 2 : 289–91

93. Smith, J. B. 1904. *Report of the New Jersey State Agricultural Experiment Station upon the mosquitoes occurring within the state, their habits, life history, etc.* Trenton : MacCrellish and Quigley. 472 pp.

94. Smith, S. M., Brust, R. A. 1970. Autogeny and stenogamy of *Aedes rempeli* in arctic Canada. *Can. Entomol.* 102 :253–56

95. Spielman, A. 1957. The inheritance of autogeny in the *Culex pipiens* complex of mosquitoes. *Am. J. Hyg.* 65 :404–25

96. Spielman, A. 1964. Studies on autogeny in *Culex pipiens* populations in nature. I. Reproductive isolation between autogenous and anautogenous populations. *Am. J. Hyg.* 80 :175–83

97. Spielman, A. 1966. The functional anatomy of the copulatory apparatus of male *Culex pipiens. Ann. Entomol. Soc. Am.* 59 :309–14

98. Spielman, A. 1967. Population structure in the *Culex pipiens* complex of mosquitoes. *Bull. W.H.O.* 37 : 271–76

99. Spielman, A., Weyer, A. E. 1965. Description of *Aedes (Howardina) albonotatus,* a common domestic mosquito from the Bahamas. *Mosquito News* 25 :339–43

100. Tate, P., Vincent, M. 1936. The biology of autogenous and anautogenous races of *Culex pipiens* L. *Parasitology* 28 :115–45

101. Theobald, F. V. 1901. *A monograph of the Culicidae or Mosquitoes.* London : Brit. Mus. (Nat. Hist.), I, 424 pp.

102. Twohy, D. W., Rozeboom, L. E. 1957. A comparison of food reserves in autogenous and anautogenous *Culex pipiens* populations. *Am. J. Hyg.* 65 :316–24

103. Umino, T. 1965. Some genetical studies on *Culex pipiens* complex (III) Experimental studies on the fertilization, larval colour phenotypes and autogeny of the hybrids between autogenous and anautogenous colony. *Jap. J. Sanit. Zool.* 16 :282–87. In Japanese

104. VandeHey, R. C. 1969. Incidence of

genetic mutations in *Culex pipiens.*
Mosquito News 29 :183–89

105. Vinogradova, Ye. B. 1961. The biological isolation of subspecies of *Culex pipiens* L. *Entomol. Rev.* 40 :29–35

106. Vinogradova, Ye. B. 1965. The morphological and biological characteristics of some natural populations of the blood-sucking mosquitoes *Culex pipiens* L. in the territory of the Soviet Union. In *The ecology of injurious and entomophagous insects,* ed. V. A. Zaslavskii, 31–57. *Trudy Zool. Inst. Leningrad* 36, 208 pp. Cited in *Rev.*

Appl. Entomol. 56 :83–84

107. Vinogradova, Ye. B. 1965. Autogenous development of the ovaries in blood-sucking mosquitoes. *Zool. Zh.* 44 :210–19. In Russian

108. Vinogradova, Ye. B. 1966. Bloodsucking mosquitoes of the *Culex pipiens* L. complex, their practical importance, taxonomy and biology. *Entomol. Rev.* 45 :131–40

109. Washino, R. K., Shad-Del, F. 1969. Autogeny in *Culex peus* Speiser. *Mosquito News* 29 :493–95

110. Weyer, F. 1935. Die Rassenfrage bei *Culex pipiens* in Deutschland. *Z. Parasitenk.* 8 :104–15

BIOECOLOGY OF EDAPHIC COLLEMBOLA AND ACARINA[1]

6010

JAMES W. BUTCHER, RENATE SNIDER, AND RICHARD J. SNIDER[2]

*Department of Entomology, Michigan State University,
East Lansing, Michigan*

INTRODUCTION

In the United States, biologists have been preoccupied with studies on those soil arthropods which directly affect yield and quality of crop plants. With the exception of individual workers who are largely taxonomists, little attention has been given to noneconomic forms here. It is becoming increasingly evident that these animals are important contributors to fundamental fertility humification processes and that agronomic or plant protection practices may affect them adversely.

An awareness in this country of the significance of soil arthropods other than those which attack crops directly is coupled with a realization that the European workers have contributed the major body of literature in this field. As yet, practically no comprehensive treatment of this subject matter has been supplied by American authors. Notable exceptions are the review by Christiansen (19) which deals with the bionomics of Collembola, and the introduction to soil fauna by Kevan (78). Some of the more recent basic European reference columns which represent sources for students of this subject matter are the works of Burges & Raw (12); Kühnelt (82); Murphy (113); Jackson & Raw (69); Palissa (126); Sheals (149); and in the last decade the valuable Proceedings of the Colloquia of the Soil Zoology Committee of the International Society of Soil Sciences: Doeksen & Van der Drift (28) and Graff & Satchell (43). The existence of periodicals such as the *Revue d' Ecologie et de Biologie du Sol* and *Pedobiologia,* both of which came into existence after 1960, testify to the burgeoning literature in this field abroad.

In this review, no exhaustive attempt has been made to cover literature which is available in the previously mentioned compendia. The reader is

[1] Support for this review was provided by the Food and Drug Administration (Grant No. 00223) and the Michigan Agricultural Experiment Station (Manuscript No. 5098).

[2] Nature Way Association, 5539 Lansing Road, Lansing, Michigan.

249

referred to those sources for information dealing with the broad aspects of soil biology. Because of the large volume of literature, the authors' own research interests in specific microarthropods, and the limitations of space, it has been necessary to restrict this review to the bioecology of soil Collembola and Acarina. In the sections dealing with bionomics it was prudent to further limit coverage to Collembola and Oribatei. In the portion dealing with microarthropod significance it was not always possible to separate Oribatei from other Acarina. To have done so would necessitate eliminating many papers which provide an overview of significant research, and observations relevant to the important role of microarthropods in fertility-humification processes.

The review is broken down into two major sections: 1. Biology and Ecology of Collembola and Oribatei; and 2. Significance of soil Collembola and Acarina. It is based almost entirely upon studies reported since 1960, with emphasis upon the period 1965-1969.

BIOLOGY AND ECOLOGY OF COLLEMBOLA

EMBRYONIC DEVELOPMENT

Oviposition.—The eggs of Collembola are laid in batches or singly (Christiansen, 19). Recent work has extended the information on oviposition for a rather diverse number of species (Green, 44; Hale, 50; Strebel, 155; Sharma, 146; Wallace, 176). In our laboratory studies we have been able to demonstrate that *Proisotoma minuta, Hypogastrura nivicola,* and *Onychiurus justi* lay their eggs in batches or clumps, while *Tullbergia granulata, Lepidocyrtus violaceus, Pseudosinella violenta,* and *Entomobryoides purpurescens* tend to lay their eggs singly. The number of eggs and the time required to lay them is highly variable. Most investigators agree that Collembola will oviposit in crevices, depressions, under loose material and generally in sheltered places (Hale, 50; Strebel, 155). Our work (to be published) with many species in culture has revealed that Collembola will oviposit just about anywhere; in the food, on each other, the container sides and top, crevices, holes, on exuviae, and on fungi. This variation in oviposition site was pointed out by Pedigo (127) working with *Lepidocyrtus cinereus.* However, that author may have forced his specimens to those alternatives by saturation of the culture substrate.

Development.—When first laid, the eggs are generally spherical and smooth. They may be white or colored as in the case of *Neanura muscorum;* creamy-pink in *Isotoma viridis;* and orange-red in *Tomocerus minor.* Exposure to light can produce coloring of the egg, apparently due to the formation of pigment in the embryo. Exposure for varied periods of time produces various intensities of color (Hale, 51).

After being laid, the egg increases in size through water uptake (Hale,

51; Pedigo, 127; Sharma, 146, 147). Hale (51) states that the initial swelling is completed after a few hours, whereas Sharma (146) states that swelling is completed on the third or fourth day after oviposition. There is really no discrepancy between authors when we consider that Hale was measuring size increase due to water uptake and Sharma was observing size increase due to cell division, ultimately resulting in optimum size for the egg.

Further enlargement takes place when the embryo grows (approximately after 20 percent of the total development time) and the egg loses its spherical shape. Increased growth splits the chorion, exposing the underlying serosal cuticle (Hale, 51; Sharma, 146, 147). There are exceptions to this condition, in which the chorion remains closely attached to the egg, as in the case of some *Onychiurus* species (Hale, 51) and *Lepidocyrtus cinereus* (Pedigo, 127).

Sharma (147) reports the appearance of a hairy outgrowth on the chorion one day after oviposition in *Tomocerus vulgaris,* and cites Sharma & Kevan (148) and others who have observed the same phenomenon. Hale (51) maintains that there is no connection between hairs and chorion (12 species studied); and that the chorion is without consistent appendages, transparent and unpigmented. He also states that the serosal cuticle develops hairs and spines, but only where it is exposed to air. These structures are probably acting as anchors after the oviposition or dispersion of the eggs, or both.

Developmental period and temperature.—The time needed for development varies with temperature. Each species has an optimum for development, which can be shown by testing a wide range of temperatures in laboratory experiments.

In general, lower temperatures increase the length of development time, while higher temperatures decrease it. Detailed data are given by Thibaud (157, 158) on lethal temperatures for the eggs of six Hypogastruridae: $-10°C$ and $+21$ to $+28°C$ ($28°C$ for *Ceratophysella bengtssoni* is lethal). The thermal optimum is the temperature at which the development time is the shortest and mortality lowest. The thermal optimum for the six species of Hypogastruridae used by Thibaud (158) was $10°C$. Table 1 lists comparative data on the effects of temperature on hatching time as observed by different investigators.

Hale (51) observed 14 species of Collembola at constant temperature (2, 4, 7, 12, 16, and $23°C$ were the temperatures generally used). In cases in which development was proportional to time, he calculated the "thermal constant" (Wigglesworth, 179). Using calculated thermal constants obtained in the laboratory, he then made estimates of developmental times in the field. The developmental data of Ashraf (3) is unreliable because his experimental temperatures varied from 13–$27°C$.

Diapause.—Eggs are relatively resistant to low temperatures and tempo

TABLE 1.

Author	Species	Temp. °C	Days	% Hatched
Green (44)	*Folsomia candida*	25	7 (5–15)	—
Pedigo (127)	*Lepidocyrtus cinereus*	23.4 ±3	7.1 (6–9)	—
Sharma (146)	*Isotoma olivacea*	24	9.8 (9–11)	86.2
Sharma (147)	*Tomocerus vulgaris*	24	7.9 (7–9)	86.6
		17	12.2 (11–14)	—
		11	23.7 (21–27)	73.3
Strebel (155)	*Hypogastrura boldorii*	9	36	—
		13	20	66.6
Thibaud (158)[a]	*Typhlogastrura balazuci*	27.5	lethal	
		22.5	lethal	
		18.5	22–24	40
		14.5	28–30	75
		10.5	44	95
		6.5	68	95
		0	8–10 mo	60

[a] Also examined by Thibaud were the following species: *Schaefferia caeca, S. willemi, Mesachorutes quadriocellatus, Mesogastrura ojcoviensis,* and *Ceratophysella bengtssoni.*

rary freezing which can, however, considerably retard development. It has been suggested by Hale (51) and Joosse (74) and clearly shown by Wallace (176) that other means of surviving long periods under extreme conditions do exist.

Dicyrtoma spp. have two kinds of eggs, summer and overwintering; the latter seemingly require more time to develop (Hale, 51). *Anurida maritima,* living in "nests" in vertical claywalls of creeks at the tidal zone of seashores, lays eggs in November which do not hatch until the following April. This condition is considered to be diapause (Joosse, 74). The production of diapausing eggs by *Sminthurus viridis* is apparently influenced by the increasing maturity of food plants in the spring. At the beginning of the summer they lay egg batches which not only withstand Australian summer conditions, but must be exposed to such conditions before the eggs can hatch (Wallace, 176).

Post-Embryonic Development

Hatching.—In pigmented species, various authors have observed the increased pigmentation of the eye patches as the time of eclosion approaches (Hale, 51; Sharma, 146, 147; Pedigo, 127). The hatching process seems to occur most commonly as a result of the pumping action of the embryo which splits open the equatorial line of the egg between the head and the tip of the abdomen. The most common method of emergence is head first (Hale, 51). However, a single observation was made by Pedigo (127) of *Lepidocyrtus cinereus* emerging abdomen first. Hale (51) also mentions that *Dicyrtoma* sp. escapes the egg by eating its way out through the egg shell, which is covered with fecal material. This would seem to be only logical for the species that deposit fecal coverings on their eggs as they lay them. In our laboratory we have observed *Ptenothrix unicolor* performing such tactics during egg laying. Later, when the covering had dried, it formed a rather rigid cement (data unpublished).

Growth.—Some species aggregate in one place where joint moulting takes place; usually several age groups together. Body contact seems to be important (Strebel, 155; Joosse, 74).

The number of moults and time required to reach sexual maturity varies according to the species and other factors, mainly temperature and humidity. Higher temperatures reduce instar duration, and lower temperatures increase it to a point where no further moulting can take place ($-1°C$ for Hypogastruridae which live for months with no moults). Juveniles seem to be more resistant to lower and less resistant to higher temperatures (Thibaud, 159).

Size increase with each instar has been measured by various authors, using body and head capsule length, or the ratio of various parts of the body as indicative criteria (Hale, 52; Sharma, 147; Janetschek, 71). It is assumed that growth stops with attainment of maturity. A gradual increase in instar duration has been noted up to a maximum size, but no information is available on whether this increase continues through the adult stage (Thibaud, 158; Hale, 52). It is interesting to note that Green (44) found the first instar to be of longer duration than succeeding ones, although with maturity the instar durations began to increase again. We should mention here that many contradictions can arise from the extreme diversity of measurement techniques and the absence of replications by investigators over the last forty years. Observations on the growth of individuals have been numerous in the laboratory, but only a few attempts have been made to combine laboratory and field observations. We cannot overemphasize the necessity of such syntheses as verification.

Valuable information was obtained by Hale (52), using data from the lab to ascertain the development of a population of *Onychiurus latus* in the field. Although instar duration in the field varies with the temperature at different times of the year, head capsule size stays constant with a given instar. Thus, it was possible to determine various instars at random dates.

A slightly different approach was made by Healey (65), using weight classes instead of head measurements. While the procedure was not accurate enough for studying the detailed age structure of a population, it certainly served the purpose of biomass estimations.

Mating.—Probably all Collembola accomplish fertilization by the transfer of a sperm packet (Christiansen, 19; Hale, 50; Chang, 18). The process is varied and recent work has shown that both males and females can take an active part by performing species-specific movements prior to the deposition and then uptake of the spermatophore (Bretfeld, 11), but in species with random spermatophore deposition, we still have no explanation of how the species select the correct spermatophore.

Parthenogenesis.—This is a proven fact in *Folsomia candida* (Green, 44) but there have been suggestions that other species are at least capable of facultative parthenogenesis (Hale, 49, 54; Christiansen, 19).

Fecundity.—If we compare studies on fecundity, it becomes abundantly clear that many estimates are untrustworthy and not comparable with each other. This condition is due to the variety of methods employed and the lack of observations in a wide enough range of varied environmental factors.

There are four basic methods for investigating fecundity (Hale, 51): 1. estimation of average egg batch size in mass culture; 2. isolation of gravid females until oviposition; 3. dissection of gravid females; 4. culturing of "pairs."

Culture conditions may have an effect on oviposition and fecundity. Hence, low figures for egg production are misleading because females obviously deposit eggs more than once in their lifetime (Hale, 51). It is probable that in many cases the potential fecundity was never attained (Green, 44).

According to our own and observations of others, some general trends can be deduced. There is an optimal temperature for fecundity. The influence of the food quality is becoming more obvious. A wide variety of foods have been used in laboratory rearing of many species. Yeast is one of the most common food sources for collembolan culture. If precautions are not taken, molds and bacteria may soon infest the yeast. It has been ascertained that different kinds of food and even different strains of Actinomycetes can considerably alter the reproductive rate of a species. Thus, using one kind of food for any given species may introduce serious underestimates of actual fecundity and mask its potential under natural conditions (Müller & Beyer, 112; Naglitsch & Grabert, 117; Törne, 161, 164, 165; Vail, 169).

The influence of competition between species and between whole groups of animals living under similar conditions has yet to be studied extensively. Christiansen (20) gives detailed information on the possibility of competitive elimination. Whatever the type of competition, there are probably several mechanisms involved: destruction of spermatophores, egg destruction, toxification of the substrate, and interference with oviposition, the latter probably being the most common factor. Density of competitor species has a stimulating effect if numbers are low, but will lower productivity in all species if increased. Somewhat comparable results were obtained by Green (45) with crowded *Folsomia candida* cultures. The formation of eggs was not reduced, but oviposition was inhibited.

The structure and pH of the substrate may promote or inhibit oviposition of certain species (Ashraf, 3; Wallace, 175). Again, it is clear that integration of laboratory and field observations is necessary for better understanding of the various phenomena, such as patterns of population fluctuations, aggregation, migration. So far, where only single species have been studied, generations and seasonal peaks have been clearly indicated (Hale, 51, 53, 54; Healey, 65; Janetschek, 71; Joosse, 74; Knight & Read, 81; Cassagnau, 15).

EDAPHIC FACTORS

When dealing with the factors of the soil which impinge upon Collembola and mite species, it is difficult to distinguish any one factor from all others (Nosek, 120).

Humidity.—Requirements vary between troglomorphic forms, edaphic forms, hemiedaphic forms, epigionistic forms and xeromorphic forms. For instance, species from caves, lakeshores, seashores, and brackish pond shores need very humid conditions, even though there may be slight differences between adults and juveniles (Poinsot, 128, 131; Vannier & Thibaud, 173; Thibaud, 159; Moeller, 109). It is interesting to note that the ability to choose the higher humidity in a humidity gradient may be lost in true cave forms (Mais, 104).

The effects of low humidity result in migration, lower reproduction, and higher mortality (Vannier & Thibaud, 173; Stebaeva, 151; Vannier, 171, 172; Strebel, 155; Wallace, 175). Other changes result in striking behavioral patterns such as construction of protective cells observed in *Subisotoma variabilis* and *Isotomurus* sp. (Poinsot, 129, 130, 132), and in *Onychiurus fimatus* (Massoud et al, 107). Also observed are patterns of inactivity (Janetschek, 71; Strebel, 155), ecomorphosis (Poinsot, 130), and drastic changes in phenotype such as neutralization of males and females, and retardation in sexual development of juveniles. These latter phenomena, however, may be produced by a sudden drop in temperature as well as lowering of the humidity (Cassagnau, 15).

Temperature.—The influence of temperature can have a marked effect on the activity of a species. It affects every part of the life cycle of Collembola, as previously discussed. Many authors feel that laboratory studies could

help clarify the role of temperature in the dynamics of a collembolan population through its influence on reproductive potential (Healey, 66; Moeller,
108, 109; Strenzke, 156; Poinsot, 131). Population peaks may occur at any
time of the year. But the fact that most species are abundant in early summer and fall seems to demonstrate an optimal combination of temperature
and humidity. The number of generations per year would thus seem to be
limited in the field (Hale, 51). Resistance to high and low temperatures
varies with the species and age of the animal (Thibaud, 159; Moeller, 108;
Healey, 66; Janetschek, 72).

Light.—Since Christiansen (19) reviewed the existing literature dealing
with light and its effects on Collembola, little work has been done on this
factor in recent years. There is a close correlation between life forms and
vertical distribution. Those forms with well-developed eyes and pigment occur in the upper (0-3 cm) layer, while pigmentless and eyeless forms
occur in the lower (3-6 cm) layer in grassland (Hale, 53). Individuals of
Hypogastrura boldorii were seen to aggregate on walls and other surfaces
in direct sunlight, even in July and August. Normally they seem to be active
in diffuse daylight (Strebel, 155). Recently Uchida & Fujita (167) have
been able to show that in the well-known swarming of various species on
snow it is not phototaxis that plays the primary role, but temperature (on
nights with high air temperature they remain on the snow).

Salinity.—During faunistic studies in Antarctica, Wise et al (180) found
microarthropods in soils with 0.12–1.8 percent salinity. It seems that salinity is a factor that may be tolerated by Collembola to a degree, and it may in
some species prevent the invasion of certain habitats by being too high
(Moeller, 108; Janetschek, 72; Strenzke, 156; Poinsot, 128).

pH.—Although there are species of Collembola definitely related to soils
of specific acidity (Nosek, 120; Hale, 54; Marcuzzi, 106), the general tolerance range lies between 6.0 and 7.8 (Davis, 26). Ranges between 6.85–9.6
were reported by Wise et al (180). Ashraf (3) reports that reproduction
was highest at pH 7.2–7.5 with no eggs at pH 8.5–9.7. Other authors have
also mentioned this condition (Davis, 26; Marcuzzi, 106).

Food and Microflora

It is clear that Collembola in culture will accept various diets, both artificial and natural, which are different from their natural selection (habits).
Even though many foods have been enumerated (Christiansen, 19), yeast
seems to be the most universal and successful diet for the widest variety of
species. We suspect that there is an optimal diet for any given species,
which would probably be the food chosen under natural conditions. So far,
the only way to monitor the natural diet is by analysis of the gut contents,
although this may not be entirely satisfactory. Various substances, taken in

through the mouth, can influence the gut flora and thus the metabolism of the collembolan (Törne, 164, 166).

There have been numerous reports on food preferences; some species can feed on both fungal hyphae and spores, others on hyphae or spores only, and finally some on neither. The characteristic affinity of each species for fungi, or even a certain fungus (Farahat, 38; Müller & Beyer, 112; Knight & Angel, 80; Törne, 164), and for bacteria (Törne, 161, 166) seems to be reflected in population dynamics, and species composition. Also there is a distinct affinity, species-specific, for a certain soil type or habitat based on microflora composition (Zivadinovic & Cvijovic, 190; Knight & Read, 81; Stebaeva, 152; Nosek & Ambroz, 121; Nosek, 120; Loub & Haybach, 88). By using experimental models, it will be possible to determine the factors influencing the natural distribution of some species of soil animals (Törne, 164).

Törne (160) indicated that the role of the microflora is a major factor acting as a mediator on the effects of the physical factors. This outlook has since been confirmed, together with the role of Collembola in the distribution and control of microorganisms, plus the humification of organic matter in the soil (Kurceva, 84; Naglitsch, 115; Wood, 182; Coleman, 21; Ghilarov, 42; Törne, 165, 166; Ambroz & Nosek, 2). But what is still needed "is to devise a series of experiments to bridge the gap between laboratory and field conditions in which the 'catalytic' effects of one group of organisms on the other can be measured" ... (Macfadyen, 95).

BIOLOGY AND ECOLOGY OF ORIBATIDS

THE EGG

Murphy & Jalil (114) reported round, small eggs in *Tectocepheus velatus;* in general, however, it is probably more common for the egg to be cylindrical, elongate, and sometimes colored or textured (Hartenstein, 56–60; Woodring & Cook, 188). When first deposited, they are much smaller than more mature eggs (188). The eggs are laid singly or in small clumps, the highest number reported being 16 (Berthet & Gérard, 7). Usually, oviposition occurs in old exuviae, the axils of moss bracts, under detritus, and in pores of the substrate (Hartenstein, 56–60; Block, 9; Berthet & Gérard, 7; Rockett & Woodring, 139; Luxton, 91; Woodring & Cook, 188; Lebrun, 85). As a rule, a number of eggs are kept inside the body of the female until all of them are mature. They are then deposited all at once (Woodring & Cook, 187, 188).

EMBRYONIC DEVELOPMENT

The oribatid mites, like the Collembola, are influenced in their development by temperature. Table 2 summarizes the findings of several authors.

TABLE 2. Embryonic Development of Oribatei

Author	Species	Temp. °C	Duration (days)
Jalil (70)	*Hermannia scabra*	20	15
Rockett & Woodring (139)	*Pergalumna omniphagous*	25	11 (9–13)
	Ceratozetes jeweli	25	12 (10–15)
Woodring (186)	*Galumna confusa*	23	(7–14)
	Galumna parva	23	(7– 8)
	Rostrozetes flavus	23	(8– 9)
	Scheloribates parabilis	23	(8–11)
	Scheloribates nudus	23	(8–11)
Shaldybina (145)	*Melanozetes mollicomus*	?	10
Murphy & Jalil (114)	*Tectocepheus velatus*	25	13
	Tectocepheus sarekensis	25	13
Bhattacharyya (8)	*Damaeus onustus*	15	17.4
Hartenstein (58)	*Ceratozetes gracilis*	20	32
Hartenstein (56)	*Belba kingi*	20	14
Hartenstein (57)	*Eremobelba nervosa*	20	15
Hartenstein (60)	*Protoribates lophotrichus*	20	23

Post-embryonic Development

The length of time required to reach the adult stage depends on temperature. Lebrun (85) summarizes this information and presents a detailed table. Table 3 summarizes the data not included in Lebrun's table.

It is interesting to note that the smaller species have a shorter development time and life, as far as is known, than the larger species (Hartenstein, 58; Lebrun, 85; Woodring & Cook, 188). The rate of development is also influenced by food. If the diet is adequate, development tends to proceed more rapidly (Hartenstein, 56; Woodring & Cook, 188); this same trend was observed by Woodring (186) in crowded cultures versus cultures with only a few individuals. Rockett & Woodring (139), rearing 5 to 20 individuals per culture tube, could not record any effect in the average length of the life cycle. They suspected that the reason for this was that the species they were working with were good fungivores and kept the fungal growth in the cultures to a minimum. The earlier experiments by Woodring were with species that were poor fungivores and therefore fungal growth impeded movements and feeding, resulting in slower growth. In most cases, but not necessarily, the duration of the stages increases with each stage, the pre-ecdysial resting stage or premoulting period increasing in the same proportion (Woodring, 186; Woodring & Cook, 188; Jalil, 70; Rockett & Woodring, 139; Murphy & Jalil, 114; Lebrun, 86).

Life cycles in the field take longer than under constant controlled condi-

TABLE 3. Life Cycle in Days from Egg Deposition to Adult Emergence for Certain Oribatei

Author	Species	Temp. °C	Duration (days)
Woodring (186)	*Galumna confusa*	23	58
	Galumna parva	23	44
	Rostrozetes flavus	23	46
	Scheloribates parabilis	23	25–28
	Scheloribates nudus	23	25–28
Jalil (70)	*Hermannia scabra*	20	185
Rockett & Woodring (139)	*Pergalumna omniphagous*	25	42
	Ceratozetes jeweli	25	53
Shaldybina (144)	*Chamobates spinosus*	?	27–35
Murphy & Jalil (114)	*Tectocepheus velatus*	25	102
	Tectocepheus sarekensis	25	102

tions in the laboratory. In general, small species may have several generations a year, whereas larger species have one single generation per year (Lebrun, 86; Hartenstein, 58; Woodring & Cook, 188; Block, 9). We lack information on the duration of the adult stage. Woodring & Cook (188) recorded 10–12 months for *Ceratozetes cisalpinus,* and Block (9) suggested that if adults are found in samples throughout the year, then their life span must be a year or more. Lebrun's method (86) of combining observations of field populations and laboratory-reared populations will, we feel, prove to be extremely valuable.

Moulting.—Before moulting, oribatid mites become quiescent (premoult period), then the old cuticle splits, and the animal remains quiescent again after partly emerging (Woodring & Cook, 187). This period of time, the hardening period, is of definite length and varies according to the species. Woodring (186) observed that the hardening period lasted 4–8 hr in *Galumna parva;* in most other observed species it seems to lie between 10 and 24 hr (Woodring, 186; Woodring & Cook, 188; Murphy & Jalil, 114).

Mating.—Fertilization occurs through spermatophore deposition by the male and uptake by the female. The spermatophores are stalked; the stalks are of various lengths and the packet of semen usually is supported by one finger (Rockett & Woodring, 139; Woodring, 186; Woodring & Cook, 188). According to rare observations, the females pick up the spermatophore by moving over it and settling down on top of it; after which the sperm packet is ruptured by unknown means (Woodring & Cook, 187, 188). In numerous observations of spermatophore deposition, it appeared that their number is fairly large, much more so in cultures in which females are present. Commonly, the spermatophores are deposited in groups, preferably on a smooth, dry surface. A good number of them are destroyed by individuals brushing against them or eating them (Woodring & Cook, 188; Woodring, 186; Rockett & Woodring, 139). See also review by Schaller on indirect sperm transfer, pp. 407–46.

Parthenogenesis.—Copulation never having been observed in oribatids, the spermatophore transfer method seems to be the most common, together with parthenogenesis, which is thought to be common especially in the Camisiidae and Eremaeidae (Woodring & Cook, 188). Records on parthenogenesis have also been made by Murphy & Jalil (114).

Fecundity.—Dissection and microscopic observation of gravid females have revealed a wide range in the number of eggs matured and laid at one single oviposition. Woodring & Cook (187) give an average figure of 12, with as many as 18 maximum. Most other observations seem to lie between 2 and 12. As a rule, the number of matured eggs will vary with the temperature, month of the year, and probably with the diet (Block, 9; Woodring & Cook, 188; Rockett & Woodring, 139). Except for the observation that belbid eggs are deposited throughout the life of the adult (Woodring & Cook, 188), there is no recent information available on the number of ovipositions or total number of eggs produced in a lifetime of any species.

EDAPHIC FACTORS

Temperature.—There seems to be a definite tendency in oribatids to choose a certain optimal temperature, or at least to avoid unfavorable temperatures. In temperature gradient experiments, Madge (101) showed that they invariably aggregated in a narrow temperature range. The range was slightly higher for adults than for juveniles. Choice appears to be unrelated to other prevailing environmental conditions. Tolerance for extreme temperatures varies according to the species. In general, most of the oribatids are inactivated by cold temperatures, although some reproduce and prosper during winter months (Woodring & Cook, 188). High temperatures apparently have no influence on their general activity except to increase movement slightly (Madge, 98), and may reduce or prevent egg laying or kill the sperm (Woodring & Cook, 188). Extremely high temperatures are finally lethal, each species having a characteristic thermal death point dependent upon relative humidity (Madge, 100).

Humidity.—Experiments have shown the humidity requirements to be species-specific. Some species are able to survive for days at 13 percent RH, whereas others need 50 percent to survive for three days in the adult stage and 70–80 percent to survive in the nymphal or larval stages (Atalla &

Hobart, 4; Hayes, 64). Species taken from various habitats clearly show different reactions in constant and alternating humidities. No marked responses were noted in a species from permanently wet sphagnum; others had characteristic initial response which could be altered, intensified, or reversed by changing additional environmental factors (Madge, 98, 99).

If attempts are made to correlate reactions to a combination of environmental stimuli, including humidity, the relative importance of each factor to the animal becomes a little clearer. Humidity and temperature are most important, influencing each other and triggering various behavioral responses (desiccation lowers the preferred temperature; and generally speaking, the higher the temperature, the greater the choice for the higher humidity). The effects of substrate structure and texture, isolation, and starvation seem to be secondary (Madge, 98, 99, 101, 102). The fact that there is a waterproof cover on the cuticle of the animals makes it probable that water loss occurs through the oral, anal, and genital and spiracular apertures. Due to its waxy nature, the cover is susceptible to melting at a species-specific temperature (Madge, 97).

Salinity.—Salinity is of great importance in salt-marsh biotopes. Mites living under these conditions must be able to withstand high osmotic pressures, especially in the egg and immature stages, and the sudden changes in osmotic pressure brought about by tidal inundation. Laboratory experiments have shown that being enveloped by salt water does not affect most Acarina (viviparous forms even reproduce under these conditions), and that oribatids do not seem to have any preference for saline or nonsaline substrate. Extensive studies on salinity and other factors related to the same habitat have been carried out by Luxton (90-93).

pH.—It would be misleading to correlate density of mites and pH, especially in summer, when temperature, humidity, and animal respiration combined acidify the substrate (Lebrun, 85). This is especially so in soils rich in organic matter, which is preferred by oribatids and which determines their abundance (Loots & Ryke, 87; Frank, 39). On the other hand, the distribution of certain species is clearly related to pH, as some of them have a distinct perference for a certain range (Moritz, 110).

Food

In his exhaustive review on the food of Oribatei, Berthet (6) follows the classification by Schuster (143), which may roughly be confirmed by three general methods (Hartenstein, 55; Luxton, 49): (a) examination of the gut contents (a very accurate, but difficult method); (b) studying the feeding of a species on known foods in the laboratory (where the conditions are artificial and the choice of foods is limited); and (c) studying the biochemical compounds needed for growth. The classification comprises:

1. Microphytophagous mites which feed on pollen, small pieces of algae, lichens, soil, and highly decomposed humus, but mainly fungal hyphae.

2. Macrophytophagous mites which feed on dead plant tissues, parenchyma of needles, or directly on dead wood (especially Phthiracaridae).

3. Nonspecialized mites which feed on leaf tissue and mycelia (Berthet, 6; Lebrun, 85; Schuster, 143).

It is interesting to note that the more "primitive" forms are generally microphytophagous or nonspecialized, whereas the higher-placed species are macrophytophagous. Moreover, the species large in size are generally macrophytophagous or nonspecialized, and therefore play the main role in the process of litter decomposition.

Essentially, oribatids are herbivores, feeding on all types of rotting vegetable tissues, fungi, algae, lichens (Woolley, 189), but acceptance of and even preference for animal food has been observed. Cricket powder alone proved to be inadequate for survival (Woodring, 186), whereas saprophytic and nonsaprophytic nematodes were eaten with vigor and in great quantity (Rockett & Woodring, 139). The preferences are extremely varied, and experiments have shown that response to a certain kind of food offered reaches from "rejected" to "accepted," with all degrees of effect on the rate of feeding, development, and reproduction (Luxton, 91; Berthet, 6; Woodring & Cook, 188).

In the lab, various foods have been used for successful rearing. The most common are ground dried mushroom; chopped lichen, and the resulting fungal growth; decomposing leaves; and an artificial diet composed mainly of dextrose and casein (Woodring & Cook, 188; Rockett & Woodring, 139; Woodring, 186). Bhattacharyya (8) used brewer's yeast, and we, too, have successfully reared several species on brewer's yeast. In the case of species specialized in boring into needles, it has been shown that not the tree species, but rather the state of decomposition was the determining factor of preference. It is felt that this may have been due to lab conditions, since, in the field, the same mite species showed consistently lower populations under one of the tree species (Hayes, 62, 63).

SIGNIFICANCE OF SOIL COLLEMBOLA AND ACARINA

In attempting to assess the significance of soil invertebrates, two major kinds of studies have been made. One of these is the pedological (soil morphology) approach, and the second has to do with nutrient cycling.

Only a few invertebrates exercise a direct and major role in altering soil morphology, and the role of the Acarina and Collembola must be regarded as minor (Wood, 182). Furthermore, these smaller animals, in combination with enchytraeids and nematodes, may account for less than 10 percent of the soil metabolism (Macfadyen, 94), or as low as 5 percent (van der Drift, 170). Their significance should not be discounted, however, since they may exercise indirect effects through stimulation of microbial activity, spore distribution, inhibition of mycostasis and bacteriostasis, and encouragement of microbial growth as a consequence of grazing on senescent colonies. If this is a reasonable assumption, it follows that pesticides may have an effect on nutrient cycling; negatively by permitting energy to remain bound up in undecomposed plant debris, or positively by killing predators of saprophagous arthropods; thus releasing detritus-feeding microarthropod populations and permitting an increase in liberation of nutrients. What follows is a summary of pertinent literature which relates to these phenomena. Since the two groups (Acarina and Collembola) dealt with there appear to exercise a minor function in pedological processes, emphasis is placed upon literature pertaining to (a) their role in breakdown, decomposition, and nutrient cycling; (b) distribution with respect to some problems of accurate censusing and their relationship to relevant trophic and environmental factors; (c) studies on secondary production and energy flow processes; and (d) effects of insecticides on their presence and abundance.

Role of Acarina and Collembola in Nutrient Cycling

The soil is considered to be a biological system subjected to both biological and abiotic influences (Nosek & Ambroz, 121). Fertility is a dynamic fact which cannot be equated with soil quality. It is an expression of its creative energy and vitality. In soil development, microflora, soil fauna and vegetation have influence which can be ascertained only through study of all of these living components. In Rendzina soils which are shallow and dry, the dominant species are *Onychiurus armatus, Folsomia alpina,* and *Folsomia nana.* Humic substances are stable because they are bound as calcium-humate to the C-horizon (limestone or dolomite). There is intensive bacterial activity but few fungi. In podzol soils there are few bacteria but numerous fungi. *O. armatus, O. subarmatus, F. nana,* and *Tetracanthella arctica* are prominent species. Ambroz & Nosek (2) describe the role of Collembola in embryo soil development as (a) reducing growth of molds; (b) contributing to humus fraction by their excrement; (c) creating a favorable balance between bacteria and fungi; and (d) reducing vegetable detritus and producing enzymes.

Based on a comparison of three north England moorland soils occurring within 250 yards of each other, Wood (180) suggests that when other invertebrates are abundant, microarthropods and enchytraeids have little pedological significance. In the absence of comminution and incorporation, litter accumulation leads to raw humus formation (L layer, gleyed podzolic brown earth). Accumulation and comminution without incorporation leads to moder formation (F and H layers, gleyed podzolic brown earth); comminution and mechanical incorporation produce mull-like moder fabrics, (mull-like rendzina and A horizon of the brown earth); and chemical incorporation leads to mull formation. Only a few invertebrates are regarded as important to the soil. Activities which may be considered in relation to the different processes which influence the fate of plant material are accumulation of litter; comminution of litter; and incorporation with mineral soil, mechanically or chemically.

Ghilarov (41, 42) recommends increasing populations of beneficial soil invertebrates in order to increase fertility and soil formation. This can be done by introducing species which can survive if mulching or liming is carried out. Invertebrates are essential for soil formation from the standpoint of mineral breakdown and depth of humus horizon. Natural reforestation and natural succession of vegetation usually depend upon the activity of soil animals, which are an obligatory link in rate of mineral turnover. Disturbance of tropical soil biocoenoses can result in rapid nutrient impoverishment, and soil mulching increases the level of humus formers. Oribatei are thought to be especially important as decomposers of plant debris. On mulched plots they made up 60 percent of total arthropods compared to 8.5 percent on newly mowed or fallowed plots. The succession of Collembola and mite species in compost piles is described, and by the time it ripens, compost is enriched by forms that may be active in the soil as well. When ripe, the upper 10 cm of leaf compost contains over 30,000 invertebrates per M^2, and over 1.5 million per M^2 in composted turf with manure.

According to Ghilarov (41), composting during autumn encourages invertebrate development during the winter. Also, composting of urban wastes is promising, and various organic wastes with the addition of soil and sewage mud have almost the same life forms as can be observed in leaf composts. Finally, soil invertebrates are credited with contributing to the activity of microorganisms in an indirect way, by accelerating processes of decomposition and mineralization of plant remnants. Without microarthropods, leaching of compost nutrients can be very rapid (Stöckli, 154).

Törne (162, 163, 165, 166) reported that *Folsomia candida* decomposed cellulose (filter paper). The rate and intensity of decomposition was determined in part by microbial influences and could be evaluated as a time-dependent phenomenon. Population dynamics of detritus feeding soil animals are essentially regulated by microbial processes, and the reproduction rate of a species could be regulated by manipulating the intestinal flora with various strains of Actinomycetes. Further, dispersion of microorganisms through soil animals is a functional necessity for extraintestinal microbial decomposition of indigestible substances, and decomposition intensity is increased by the addition of their own microbial flora. They also can decrease the decomposition intensity by direct feeding on the decomposers. It follows that the rate of decomposition by Collembola species may be related to qualitative differences in the gut flora. Tendency to aggregation may be related to qualitative differences in the gut flora and may be due to a need to create more favorable microbial conditions in competition with the microbial flora of the environment.

Farahat (38), stimulated by Kühnelt's (83) observation that litter rich in mycelia is poor both in species and individuals of Collembola due to a shifting of pH towards acidity, found that certain species of Collembola can feed on both fungal hyphae and spores, while others can feed only on hyphae or on spores. Others cannot feed on any fungal structure. He concluded that every species has its own characteristic affinity for the fungi.

Heath et al (67) attempted to assess the breakdown of leaves of plant species by comparing disappearance from small and large mesh bags buried 2.5 cm deep in woodland or fallow soil. Three different disappearance rates were noted, depending upon plant species. Fallowing generally lessened the numbers of soil animals, possibly because animal species normally feeding on tree leaves are not adapted to fallow conditions; or microbial activity differed on the two sites. Lettuce, beet, kale, and bean leaves disappeared in a few weeks at rates independent of mesh size, suggesting microbial breakdown. Oak discs in large mesh bags in woodland soil lost 40 percent of their lamina in 4.5 months, but studies on effects of animal exclusion were not reported. Madge (103), using leaf discs in mesh bags, showed that tropical leaf litter disappeared most rapidly during the wet season due mainly to the activity of mites and Collembola.

Knight & Angel (80) exposed *Tomocerus* to four different diets: litter and fungi, fungi, humus and fungi, and humus, litter and fungi. Employing dissection and controlled feeding, it was reported that fungal spores were ingested in greater quantity than any other single food material, but specimens collected from the field had ingested more litter and fungal hyphae than fungal spores.

Tools for use in detailed litter-soil ecosystem studies are offered by Coleman (21). He studied the recolonization of sterile, fungal-inoculated cores of litter and soil by species of microarthropods. Soil microarthropods were observed to recolonize gamma-irradiated inoculated samples in patterns varying with different species of fungi. Respiration of cores after irradiation was at first lower, and later higher than that of unirradiated cores.

Reichle & Crossley (137) inoculated *Liriodendron* trees with ^{137}Cs to study forest macrocryptozoa. Animals were assayed on a Packard single channel autogamma spectrometer system. By correcting to time zero (initial inoculation) comparisons were made of relative levels of activity in the trophic structure over successive years. Gregers-Hansen (46) studied decomposition of diethylstilboestrol in soil by the $^{14}CO_2$ produced from the decomposition of ^{14}C-labeled DES.

Kurceva (84) suppressed invertebrate activity in oak litter with napthalene and concluded that microbial decomposition was extremely slow as a consequence; three times slower in dry years and six times in wet years. The necessity for combined arthropod-microbial action was emphasized. It appears to Loub & Haybach (88) that mites and Collembola can eat anything, and Collembola could not have population maxima in winter if they were specialized feeders on species of microbes, since the latter are at a minimum at that time. No obligatory connection was observed between microbes and the presence of particular animal species.

Marcuzzi (106) concluded that Parasitiformes, Sarcoptiformes, Hypogastruridae, and Entomobryidae are indifferent to humus content. Isotomidae and Onychiuridae prefer the highest values while the lowest are preferred by the Trombidiformes.

Laboratory studies have shown that a large variety of food materials may be accepted, but that the amount and quality of the food have decisive influence on growth and reproduction of both Collembola and Acari (Müller & Beyer, 112; Luxton, 91; Woodring & Cook, 188). In general they feed on decaying plant material, deriving nourishment from fungal and bacterial growth on those remains. They are important to the soil by mechanically breaking up and comminuting organic matter, thus making it more accessible to microorganisms and to constant microbial inoculation by feces (Luxton, 91; Naglitsch, 115).

MICROARTHROPOD DISTRIBUTION AND ABUNDANCE

It is difficult to generalize about the factors that affect soil microarthropod distribution. Conclusions in the literature are based upon experimental evidence, extensive or casual observations, and unevenly based speculation or inference.

In this review, summaries are provided of available pertinent information on the assumption that interested readers may wish to consult the original source. A brief summary of the factors discussed is presented at the conclusion of this section.

Plant cover and flora.—Naglitsch & Steinbrenner (116) point out that the effects of crops on soil fauna can persist for three years. Abundance of fauna for crops in decreasing order was meadow, alfalfa, potatoes, beets, and oats. Abundance and variety of fauna were greatest the year following alfalfa, and spring-sampled Collembola showed a more homogeneous dispersion under all one-year crops. The influence of clear cutting on oribatid woodland communities in two beech and three mixed pine stands was described by Moritz (111) as more quantitative than qualitative; characterized by a shifting in dominance. In general, changes in oribatid fauna were relatively small in the first years after felling and for the first three years, the communities remain "mesophil forest stand synusiae."

Ryke (141) described marked abundance fluctuations in mite and Collembola populations on two adjacent plots covered with the grasses, *Pennisetum clandestinum* and *Eragrostis curvula*, respectively. On the *Pennisetum* plot a minimum of 18,157 mites/M² were observed during a hot, dry period (October, 1962) to a maximum of 56,876/M² during a relatively wet, cool period in July 1965. In the *Eragrostis* plot, fluctuations between 3511/ M² (October) and 33,144/M² (July) were observed. Over 100 species of Acari and Collembola are found in most grassland soils according to Wood (182). The variety of Collembola (16-32 species) is fairly uniform; that of Acari (10–161 species) is not. (Only 5 of 63 species of Collembola were found in 7 or more of 10 grasslands studied by Dhillon & Gibson (27) and 40 species were found in only one.) Wood employed Sorenson's Quotient of Similarity

$$QS = \frac{2C}{a + b} \times 100$$

where, for any two habitats, C is the number of species common to both, and a and b are the total numbers of species for each habitat. A trellis diagram was used in which habitats with the most similar fauna (highest QS values), are placed close together for purposes of comparison. Lists of species from grassland habitats tabulated by various authors were compared. Communities of Collembola in different soils in the same locality are more similar (QS values 52–90) than communities from different localities (QS 10–68). Communities of different grasslands in continental Europe show reasonable similarities to each other while British grasslands show less. Acari communities from different grassland soils from the same locality are very similar (QS 60–83); but soils from different localities show very few similarities. The same is true for comparisons of oribatid mites in a variety of heathlands and forest soils in Finland, Sweden, and central Europe. Of 18 different grasslands few of the 500 species occurred in more than 12 sites. Among these were *Tullbergia krausbaueri; Folsomia quadrioculata; Isotoma viridis* (Collembola), *Rhodacaris roseus* (Mesostigmata), *Tectocepheus velatus; Minunthozetes semirufus* and *Punctoribates punctum* (Cryptostigmata). Most (90 percent) appear to be distributed by chance, substantiating Haarlov's (48) observation that communities may reflect surrounding species complexes rather than being peculiar to habitat. Davis (26) feels that area species complexes mean more than that each of them happens to be there separately.

Alejnikova (1) found that each type of soil in forest, meadow, and fields has a typical soil invertebrate community abundance and composition, dependent mainly upon humus content and plant association. Deciduous forests had the most abundant and varied fauna. In an oak-hickory community in Missouri, mites averages 296/ft² and Collembola 162/ft² to a 10-in depth (Dowdy, 29). Most of the mites were taken from the litter and the 0–2" layer of soil. In a blue grass community, only 26 mite and 48 collembolan specimens were collected for each square foot.

Fucus seaweed wrack was not populated by *Hypogastrura viatica* at 100 M from shore, but this collembolan reproduced well when implanted, according to Moeller (109).

Stebajeva (150) recorded 75 Collembola species in the Tannu-Ola mountains and adjacent valley, and the richest fauna was present in the deciduous forests (45 species). At the transition zone of mountains to the plains there were only 17 species, but 13 new steppe species. In the steppe, species composition and abundance are lower in litter layers and the typical soil forms are largely replaced.

In studies of Collembola diversity and frequency in the alpine region of Mt. Tatyama (2700 M), Niijima (119) noted a decrease according to vegetation in the following orders: *Pinus, Phyllodoce,* and *Carex.* Isotomids (78–93 percent) and podurids (3–17 percent) were the principal inhabitants of the *Pinus* community, while the podurids and onychiurids constituted a majority in the *Phyllodoce* community.

Hayes (63) found that with the exception of *Hoploderma spinosum* numerical abundance of coniferous woodland phthiracarid mites appeared related to the tree species, and Ibarra et al (68) found that in sheep and cattle pasture oribatid distribution appeared to be related to areas of choice food and favorable microclimatic conditions. Sheep and cattle cestode host species showed the same distribution. In assessing the importance of a number of factors, Schalk (142) found that species composition of Oribatei was more strongly influenced by plant coverage than by soil type.

Soil type and organic matter.—Moritz (110) inferred "phenological groupings "(coincident habits and abundance peaks) for groups of oribatids from four different types of deciduous forest. By comparisons of vertical distribution, 50 percent of the individuals were "structure followers" e.g., restricted to the decomposition layer; less to the upper layer. They were small-to-medium size with a tendency toward vertical migration. "Surface forms" were mostly large, with many species but few individuals per species. "Limited structure followers" are found in the surface layer with a limited tendency to migrate downward. "Depth forms" are very small with a low number of individuals per species. Of the 20 families found, the most rich in species are the Hypochthoniidae, Eremaeidae, Carabodidae, Ceratozetidae, and Phthiracaridae. The most constantly represented family (28 percent) is the Eremaeidae (*Oppia* and *Suctobelba* principally). Poole (134) observed that Collembola distribution in a coniferous forest in North Wales was influenced by arrangement of the surrounding trees, and the influence may be subterranean. Physical factors could not explain the distribution patterns, but numbers were related to depth and moisture of the organic layers. In soil samples from a *Picea abies* plantation in Denmark, Poole (135) found Collembola averaging 1.1 to 1.7/ccm in the organic layer with wide variation between samples. Three common species showed strong positive correlation in a small area but not in cores from a large area. Influence of the rhizosphere was inferred on the basis of larger numbers of Collembola taken away from the tree base.

Loots & Ryke (87) found a significant correlation between the ratio of Oribatei and Trombidiformes and the percentage of organic matter in different soils. Oribatei dominate in soils with a high organic matter content, whereas Trombidiformes are more abundant in soils where the organic matter content is relatively low. The abundance of oribatids may be related to organic matter as a source of food, soil pore volume, moisture content, and living space. Lutz et al (89) see a significant parallel in the communities of plants, microbes, and soil microfauna as a result of studies on soil microfaunal synusiae. Oribatids are dominant in all biotopes, and even though the species vary, their abundance appears to depend directly on the humus content of the environment.

Oribatei numbers and species were greater in sandy soils according to Schalk (142). He also found four species of Oribatei which were characteristic of deep layers although the 0–5 cm layer showed highest density for Collembola, Oribatei, and other Acari. Wood (183) found Acari and Collembola populations to be similar in cushion protorendzinas under mosses on limestone boulders ($277,000–281,000/M^2$); a mull-like rendzina on cliff ledges ($178,000/M^2$); and a brown earth of low base status under *Nardus* ($167,000/M^2$). Prostigmata and Cryptostigmata dominated in all but the brown earth sites, where Collembola were more abundant than Prostigmata. Few Mesostigmata were found in mosses, and Astigmata were scarce in mosses and the mull-like rendzina. The greatest number of species occurred in mull-like rendzina (128) and a total of 200 species were recorded from 11 sites. The logarithmic relationship between species and individuals did not hold; species-individual curves showed brown earth populations to be most homogeneous, while mull-like rendzina populations were more heterogeneous. Assuming reliability of sampling techniques, density comparisons depend upon extraction efficiency, and population density of Acari and Collembola cannot be correlated with humus form, soil, or vegetation type. Stronger correlations may exist between feeding habits. In a rather unique situation, Moeller (109) found that the numbers of *Hypogastrura viatica* feeding on *Fucus* seaweed wrack were higher in sand where decomposing organic matter was present.

Cultivation and disturbance.—Distribution of microarthropods was found to be irregular under one-year crops, in part due to cultivation (Naglitsch & Steinbrenner, 116). In a blue grass community, only 26 mite and 48 collembolan specimens were collected for each square foot and most were from the 0–2" depth (Alejnikova, 1). Man and animal disturbances of natural communities is cited as a possible explanation for the small numbers of individuals in grass as opposed to forest communities. Schalk (142) found that cultivation had no apparent effect upon Collembola, but it did affect Oribatei and Acari.

Weather and microclimate.—Loub & Haybach (88) studied annual cycles of soil microfauna and microflora in the Vienna forest. Dry plots (Pannonic climate) showed microbial and mite abundance peaks in spring and fall when soil moisture content is highest, while Collembola exhibited three maxima in May, September-October, and December. In moderate climate, a first maximum follows soon after a microbial maximum and again in December. Changes in faunistic composition, seasonal abundance, and population density of Acari and Collembola in central Thailand were reported by Ogino et al (124). Decrease in litter brought about minimal populations during the dry season. There was evidence that the litter forms moved into the surface soil in December, and the number of individuals in the surface soil, as well as total numbers per plot, gradually increased from August to March, with an abrupt rise in May. Striking seasonal changes in relative abundance of Collembola and Acarina were noted.

Cassagnau (16) correlated the distribution of Collembola species with plant sociological and climatic environment of the "Montagne Noire." "Mediterranean" and "Atlantic" climatic zones are confirmed in each of six habitat types: litter, moss on soil, humus, deep soil, moss on trees, and moss on rocks. Combinations of individual distribution led to his formation of nine different synusiae.

Healey (65, 66) discusses winter changes in soil arthropod populations of *Pteridium* moorland in south Wales. Few adverse effects were observed when soil was frozen for six weeks, and *Onychiurus procampatus* Gisin continued to breed and grow. Alejnikova (1) discussed seasonal and numerical abundance with respect to vertical stratification, noting that highest mite populations occurred in September and peaks of Collembola populations occurred in February and December. Populations fluctuated between 2600 and 13,000 in *Pteridium* moorland plots.

Effects of other fauna.—There is not a great deal of literature on the relationship of other animals in the soil to microarthropod abundance. That this might be a rewarding area of research is demonstrated by the work of Karg (75) who studied the numerical response of microarthropods to nematode infestations in potato fields. Parasitiformes feed on nematodes and reach high densities in infested soil, but their numbers are reduced by low, unstable temperatures. Acaridiae (root mites) may be six times higher in nematode-infested plots, while Oribatei and Trombidiformes are reduced. The Collembola *Hypogastrura* and *Tullbergia* reach higher densities in infested plots while *Onychiurus armatus, Isotoma notabilis,* and *Sminthuridae* were reduced. Wallace (175) reported that although numerical variation of *Sminthurus viridus* in west Australia was affected by weather, soil type, botanical composition of the pasture, predation by *Bdellodes lapidaria,* and scavenging on dead bodies by newly hatched nymphs, population collapse may occur because of the latter due to high uric acid accumulation. Storage of excessive waste material is suggested as a cause of mature *Sminthurus* deaths.

Nature of the distributions.—An increasingly important body of literature is being developed which hopefully may permit more reliable population estimates. Some of the technics, while not necessarily new, have validity in soil microarthropod research. They enable us to better understand distribution but, more important, permit greater reliance on abundance estimates.

Employing a variance/mean ratio or relative variance method of Nef (118) four species of mixed coniferous woodland Pthiracarid mites (*Hoploderma magnum, Phthiracarus piger, Oribotritia loricata,* and *H. spinosum*) were shown by Hayes (62) to exhibit aggregated populations at all seasons of the year. Again, highly clumped distribution of *Onychiurus procampatus* gave index of aggregation values between 2.5 and 4.0 (Healey, 65).

Ibarra et al (68) fitted sheep and cattle pasture oribatid population data to Poisson, Neyman type A, and negative binomial distributions. Judging the goodness of theoretical approximation by a chi-square test, the authors found that the negative binomial gave the best fit in all but one case. An analysis of variance of the transformed mite data confirmed the contagious distribution. A logarithmic transformation was found to normalize oribatid sampling frequency distribution, but it did not homogenize heterogeneity of variances sufficiently where density variation was very great according to Gérard & Berthet (40). These authors studied the choice of transformation and its efficiency, using data from two forest communities. They concluded that interpretations of results must be made with care, especially when densities are low.

Attempts to classify sites and communities have been made by Wood (184, 185) who found the greatest densities of Acari and Collembola in the upper 4 cm of mull-like rendzina, brown earth, and two gleyed podzolic brown earths. Surface concentrations were greater in the podzolic brown earths (87–91 percent of the total) than in the other two (76–79 percent), and vertical distribution of individual species related closely to fested plots while *Onychiurus armatus, Isotoma notabilis,* and *Sminthuridae* of feeding habits enable closely related species to exploit the same habitat. Trellis diagrams and species analyses (200 recorded mite species) were not adequate to classify sites faunistically. Differential species groupings were described, along with characteristic species identification. Tests for homogeneity and Sorenson's index were used in the comparisons. In Wood's judgement, Poores (136) "multidimension net of variation" ideas may be valid for soil arthropods. The usefulness of a community approach to soil microarthropod studies is thought to be limited because of possibly obscure common environmental features and existence of ecological races. Challet & Bohnsack (17) attempted to assess differences between transects near Point Barrow, Alaska. Faunal similarities were sought using Jacardes Coefficient of Community (Whittaker 178) as follows:

$$C \text{ of } C = \frac{c}{a - b - c}$$

where C of $C = \%$ of similarity; $a =$ no. of genera (species) in inhabitat "a"; $b =$ the same for habitat "b" and $c =$ no. of genera both have in common. The authors conclude that numbers of individual species, differences between habitats and differences between transects can be summarized partially by a two-way classification of species on the basis of frequency and relative abundance, using modifications of tests devised by Davis (26) and Weis-Fogh (177).

Again, Hale (54) found distribution of Collembola on four vegetation sites to be aggregated, probably in areas of food concentrations, and vertical distribution was correlated with life forms. Rarer species were better indicators of two mineral soils. Biomass measurements are given for instars of three species along with average population densities for the study sites. Higher population densities reported here are attributed to improved extraction methods. Random collection compromised evaluation of the aggregation tendency. The data did not follow a Poisson distribution and aggregation is thought to occur at some point after hatching. There was a close correlation between life forms and vertical distribution, and differences in vertical distribution were noted on the different sites and between species. Six species dominated the four main sampling sites and "indicator" species were described.

Lebrun (85) carried out a statistical study of 48 litter and humus species of Oribatei in an oak forest in relation to microclimatic and edaphic conditions. Extreme annual density fluctuations ranged from 70,000 (December) to 180,000 (October) individuals/M^2. Some 65 percent of the species showed pronounced aggregating tendencies. Water influence was manifested by competition factors; e.g., suppression of vital space available, and oxygen and water saturation deficits. Reaction of oribatids to temperature variations led to classifying the species in thermophil, mesotherm, and oligotherm groups.

An approach to the causes of aggregation was used by Jensen & Corbin (73) who employed sampling boards of different size and color to test the effect of different microclimates on aggregation of various cryptozoa. The data suggest that time of day, weather, and underboard microclimate influence collembolan aggregations. Tomocerinae in the litter-fermentation and humus-soil microstrata were sampled in open field, ecotone, and pine forest communities by Knight & Read (81). Humus-soil populations were significantly larger and more uniform in each area. Also, *Tomocerus* population size was positively related to fungi as a food supply, negatively to soil mites (predation and competition) and flooding (suppression of microflora). Similarly, Poole (133) showed that Collembola populations averaged 46,700/M² in a New South Wales Douglas-fir plantation; populations were aggregated and more than one species was aggregated in the same place. Vertical distribution showed family differences and was related to soil compaction. Aggregation may be due to a tendency to remain near egg clusters, tendency to aggregate, or patchily distributed environmental factors. Highly aggregated *Tomocerus krausbaueri* populations, on the other hand, lay eggs singly or in pairs, and physical factors show no variations on the same order as Collembola populations. Species increase and decrease in abundance in the whole community, and marked differences exist on apparently similar sites. Families show characteristic vertical distributions with the greatest numbers in the fermentation layer. Drouth and soil texture are important in vertical movement.

Perhaps because of inadequacy of accepted sampling techniques, Rusek (140), in comparing Collembola abundance between two woodlot plots, concluded that differences up to 30 percent were insufficient to characterize a community. Absolute constance was considered the most significant factor in community characterization. Species showing an absolute constance of 50 percent or more, or a relative constance of 6.2 percent or more were considered characteristic. They are usually the dominant species. He established characteristic species for collembolan fauna in an Acereto-Fraxinetum association. In habitat "A" abundance was 50,000/M² while in "B" it was 93,400/M². The greatest difference is attributable to absolute dominance of *Onychiurus armatus, Isotomiella minor, Folsomia quadrioculata,* and *Folsomia multiseta.* Dominance difference for species on the same or on two different localities can reach 28 percent.

Finally, Usher (168) found the distribution of Collembola within a block of samples to be of three types: (*a*) uniform, (*b*) random, and (*c*) aggregated. Using a co-ordinate technique for specifying locations, numbers and distance between aggregations, it was concluded that aggregation could result from some fixed attribute of the species such as size of the egg cluster or availability of a food niche.

It appears that microarthropod distribution can be influenced by water, pore space, oxygen saturation deficits, temperature variations, microclimate, fungi, soil mites (in the case of Collembola), flooding, cropping, cultivation, organic matter, ground cover, altitude, litter, nematode populations, man and animal disturbance, soil compaction, soil type and texture, predation, and feeding habits, among others. Aggregation tendencies are attributed to water, temperature, time of day, microclimate, season, food source, microflora, vegetation, clustering of eggs, patchily distributed environmental factors, tendency to aggregate, subterranean influence of surrounding trees, depth and moisture of organic layers, rhizosphere, and feeding habits. Reliability of each of these influences depends in part upon the extent to which individual investigators have sought to document their speculations, inferences or conclusions. It is reassuring that greater attention is being paid to adapting and developing statistical devices to soil arthropod populations. Reliability of data is undoubtedly compromised by as yet inadequate sampling and extraction techniques. Imperfect as the best of these may appear, they represent considerable advances over earlier work.

SECONDARY PRODUCTION AND ENERGY FLOW

The paper of Macfadyen (96) provides a comprehensive point of departure for a review of significant literature on energy flow as it relates to microarthropods. In his view, "energy flow" studies can provide basic information for comparison and analysis of the functions of ecosystems. By considering the total energy flow passing through the invertebrate populations, he thought it possible to determine the relative importance of species which are in competition on the same trophic level, or on successive trophic levels of the same ecosystem. Also, it can permit a quantitative evaluation of the interaction between organisms. The author defines "production" as the quantity of energy accumulated in animal tissue (by growth and reproduction). The term "productivity" is used in only a general way, and "energy flow" is broken down into (a) "consumption" (amount of energy and food ingested) and (b) "assimilation" (the amount of energy converted to biomass or respired). "Biomass" refers to the total quantity of organic matter represented by the organisms in question. It is a value at a particular point in time. It may increase through production and immigration and decrease through mortality, size reduction, and emigration. "Production" is the net result of a series of energy flows, conveniently expressed in calories. Energy flow studies may try to establish an "energy budget" for one species, or for various species, for trophic levels, or communities.

Measurement of secondary production involves (a) analysis of the age structure of the natural population; (b) determination of at least two of the three criteria (assimilation, production, respiration) for the various age classes under the changing factors of the environment; and (c) calculation of weight, growth rate, reproduction rate, assimilation, consumption, and elimination rate; i.e., the total energy corresponding to respiration and, ultimately, production.

Most data obtained under laboratory conditions are open to question, and micro-methods are inadequate to measure most environmental conditions with precision. In general, a four-stage approach is followed: the population is inventoried, and respiratory activity must be measured. (It cannot be assumed that a simple relationship exists between metabolism and temperature.) Ecological factors must be measured, since actual temperatures to which they are exposed under natural conditions are difficult to establish, and lab and field data must be integrated. This is done most simply by multiplying the number of individuals by a mean respiration rate, thus calculating energy flow passing through various links of a food chain. (Usually arbitrary temperature estimates are used.) Alternatively, biomass of the population may be calculated and multiplied by a respiratory constant which is dependent on respiration and body mass of the species in question. Macfadyen concludes that energy flow studies can produce important information on food chains, liberation of nutritive elements, and causes of characteristic soil properties.

Cragg (22) compared collembolan biomass and respiration with other groups of moorland soil animals. A biomass of $0.6g/M^2$ accounted for 0.3 percent of the total biomass on limestone grassland, and $0.1g/M^2$ accounted for 0.1 percent of the total biomass on *Juncus* grassland. Using Bornebusch's (10) respiratory determinations on the two areas, 1.1 mg and 0.3 mg of O_2 are respired per M^2 per hour at 13°C by Collembola—2.0 percent and 3.0 percent, respectively, of the total. The usefulness of earlier attempts at volume calculations as indicators of biomass was questioned by Edwards (31). In his studies, body lengths of small animals were plotted against cube root of weights to obtain straight line graphs useful for making accurate estimates of live weights based upon bulk measurements of specimens. Specific gravity was obtained by immersing animals in solutions of increasing density until they remained suspended halfway down the tube. Weight estimates were obtained from a regression fitted to a double log plot.

Berthet (6), in dealing with metabolism of oribatids, determined O_2 consumption with the Cartesian diver technique in order to obtain data at different temperatures (15, 10, 5°C—sometimes also at 25 and 0°C). Sixteen species were studied, and a strong relationship was shown to exist between temperature increments and respiratory activity. The average Q_{10} measured between 0° and 15°C was approximately 4 for the 16 species studied. Relationship between weight and metabolic activity was investigated, and the exponent of weight was found to be equal to 0.70. A general relation was elucidated:

$$Y = 18.059 + 0.70W - 0.487Z$$

where $Y = \log_{10} O_2$ consumed, $\mu 1 \times 10^{-3}$ per individual per day;

$$W = \log_{10} \text{ of the weight in } \mu g;$$

$$Z = \frac{1}{T \ abs} 10^4$$

Results thus obtained were discussed in relation to factors that can modify metabolic activity, along with data from the literature applicable to these mites and other edaphic animals. Using the data of Lebrun (85), the author attempted to apply the general formula to a natural oak forest population. Total annual O_2 consumption was calculated for the oribatid population. It was concluded that variations in biomass did not coincide with the variations in metabolic activity. Importance of each species is highly variable from one month to another. Activity is dominated throughout the year by two species, *Steganacarus magnus* and *Platynothrus peltifer* which are responsible for 60 percent of the O_2 consumed by adult Oribatids in litter and 50 percent of that observed in humus. Total annual O_2 consumption of the adult oribatids was 4.5 L/M² per year. If nymphs were taken into account, it appears that this figure should be multiplied by two or three and the O_2 consumption would likely be in the area of 10 to 15 L. The author compares these figures with extrapolations from the work of Bornebusch (10), Engelmann (36), and O'Connor (122) (for enchytraeids). In general, discrepancies are attributed to differing sampling techniques, different extraction methods, arbitrary interpretations of Krogh's curve, and temperature fluctuations in natural situations, among other things. The author accepts the respiration quotient figure on the order of 0.82 of O'Connor (122) and Macfadyen (94) for a liter of oxygen consumed, as corresponding to the liberation of 4,775 kcal, and an adult utilization of 4.51 of O_2 per M² per year, or, 21.5 kcal per M² per year. (This is 1/10th of that produced by the human population of Belgium (biomass of 15g/M²). It is also near the prairie soil figures of Macfadyen (94). Based on annual litter accumulations reported by Witkamp & van der Drift (181), Hartman (61), and Ovington (125), a figure of 1.225 kcal per square meter is arrived at. Adult oribatids can be considered to dissipate around 1.8 percent of this, or double, if the juveniles are included. This is only an estimate of energy dissipated and thus might be regarded as negligible. On the other hand, the quantity of material that passes through the total oribatid population in one year is about equal to 50 percent of the annual leaf fall. By breaking up and mixing the litter, the oribatids in effect act as a catalyst of microbial metabolism and in such a context they play an important role in the community.

Healey (65) notes that attention has been concentrated on population metabolism on the assumption that in populations of small animals this may amount to 70–90 percent of total energy flow. Although Acarina and Collembola are numerically significant, in terms of biomass and metabolism, they may be relatively unimportant in comparison with nematodes and oligochaetes. By focusing on *Onychiurus procampatus* (70 percent of the total Collembola biomass) it was thought that estimates could be extended to the balance of the population. In trying to make a preliminary assessment of the significance of Collembola in energy flow, stratified random sampling, a Macfadyen high gradient extractor, and a Cartesian diver respirometer were employed. Error of each estimate was assessed separately, although the need for multivariate analysis was acknowledged. Population estimates introduced by far the largest single source of error, with fluctuations between 2600 and 13,000 (mean 6000) per M². Confidence limits of 35 percent were reduced to ± 19 percent for log transformed data. Biomass was calculated by head capsule lengths and live weight relationships.

For the year, biomass varied between 98 and 459 mg, with an annual figure of 217 mg/M². Respiration measurements at 15°C over the whole weight range gave rates of 10×10^3 μl O_2/ hr for an animal weighing 10 mg, and 32×10^3 μl O_2/hr for an animal weighing 100 mg. A regression of respiration on live weight was fitted on a double log plot. A high Q_{10} of 3.5–4.0 was estimated.

For each month in this study, mean weight of animals in each size class was calculated; the rate of O_2 consumption estimated from the regression, and this value converted to that of the monthly mean soil temperature by Krogh's curve. By multiplying the number of animals in the size class in the field population, O_2 consumption of the monthly size was obtained. These ranged from 10–129 ml O_2 per M², with a total June 1962–May 1963 figure of 548 ml O_2 per M². Of this total, 49.6 percent was due to immatures; 38.4 percent to adult females and 12.0 percent to adult males. Estimates of production were obtained by summing the numerical estimates for each size class for each of the 12 months of the year. Since 80 percent of the egg output died, the egg mortality fraction of production amounted to 49 mg. A respiration value of 86 ml was assigned to this bringing the total up to 634 ml.

Estimates of assimilation efficiency for *O. procampatus* (a fungal feeder) range between 40–70 percent. Using a 40 percent figure, estimates of minimum assimilation amount to 2.30 g for the year. Conservatively estimating calorific content of arthropod tissues at 6000/g, and the dry weight of adults at 45 percent, the calorific conversion of 2.5 cal per mg of live weight was obtained. Annual population energy flow in *O. procampatus* is estimated at 5.65 kcal per M² per annum. Since published calorific equivalents of fungal tissue are about 2000 cal per gram of dry weight (commercial mushrooms) a value of 2.30 g applied to fungal tissue gave an estimate of assimilation by *O. procampatus* of 4.60 kcal per M² per annum.

Even though there is still a great deal of disagreement over the reliability of described energy budgets, parameters such as biomass, rates of assimilation, respiration measurements, and the influence of temperature on respiratory activity, especially when field observations are analyzed, it is apparent that some provocative beginnings have been made. The reader is referred to the work of Odum (123), Engelmann (37), and a forthcoming review by Crossley (23) for more general discussions of the energy flow concept. Only a few of the more comprehensive attempts to analyze the role of oribatids and Collembola are reported here.

Effects of Insecticides on Microarthropods

Edwards (32) has synthesized most of the more recent literature of pesticides into some useful generalizations based in considerable measure upon his own research (Edwards et al, 30, 33–35). The author discusses the significance of application methods, differential susceptibility, persistence, dosage rates, and ecological factors in relation to insecticide effects on soil arthropods. In terms of its effect upon predatory mites, pauropods, and springtails, he concluded that insecticide persistence was more important than toxicity; amount of insecticide applied and number of soil animals killed is more logarithmic than proportional; repeated annual applications are more effective than a single dose; and fall and spring applications are most effective since populations are then at their peak. Fumigants are nonselective and treated areas tend to be repopulated more rapidly, with first invaders building up rapidly and temporarily. Selective action on predators can result in spectacular Collembola population increase, and this may also be true for many detritus feeders. Edwards concludes that effects of contaminations do not persist for more than a few years after its last traces have disappeared and even the most persistent insecticides are gradually degraded. A more comprehensive review of this subject is presented by Butcher (13). What follows is not intended to be exhaustive, but rather to suggest the complex nature of the problem and the benefits that might derive from a more imaginative selection of parameters.

Studies of the effects of pesticides on more than one component of an ecosystem are rare, possibly because of the difficulty in obtaining adequate understanding of the relationships involved and in identifying complex perturbations. Barrett (5) attempted such an approach in order to ascertain the effects of carbaryl stress on a grassland system. He analyzed two major trophic levels (phytophagous and predaceous) and attempted to document a "nontarget" effect on litter microarthropods on the basis of a decreased "litter bag" weight after treatment. Citing similar studies by Crossley & Hoglund (24) and Crossley et al (25), the author concludes that pesticide-decomposer relationships conceivably could adversely affect mineral recycling. Griffiths et al (47) states that an insecticide effect is dependent upon (a) the toxicity of the chemical; (b) the extent to which the species contacts the chemical; and (c) the insects' power of recolonization. It was further stated that interdependence of faunal elements may produce repercussions which can be measured, as increases in Collembola readily produce increases in Collembola-feeding ground beetles. The difficulty in isolating the effect of insecticides from other influences was described by Knight & Chesson (79) who believed that low moisture values were probably more important in bringing about population decreases than were the DDT treatments, even though a comparison of post-treatment and pretreatment results showed statistically significant values for Sminthuridae in the LF-level, and for populations of Poduridae in the upper stratum.

Although short-term effects can be shown it is often difficult to document long-term changes which can be attributed to insecticide action. A case in point is the work of Stegeman (153) who applied carbaryl to a 25-year-old red pine plantation and a mixed hardwood stand. Neither mites (oribatids, mesostigmatids, and juveniles) nor Collembola were exterminated. Reduction in population was relatively proportional to the severity of the treatment up to 10 lbs/A and there was little additional effect at 50 lbs/A. Five months after treatment mite populations were far greater on treated plots than in controls, and Collembola populations were reduced by all treatments. Little direct effect was shown by any of the treatments, although long-term effects were unknown. In another study, Voronova (174) found that oribatid mites were lower in carbaryl (0.5 g/M^2) treated than in check plots one year after treatment, largely due to decrease in numbers of a few species. The effects on Collembola were mixed when results of two years' data were compared, and could not be separated from weather influences.

Kelsey & Arlidge (77) showed that serious side effects of Isobenzan (1, 3, 4, 5, 6, 7, 8, 8-Octachloro-3a, 4, 7, 7a-tetrahydro-4, 7-methanonaphthalan) applications include population reduction of all recorded faunal members (except nematodes) and soil bulk density was significantly increased after application of 2 lbs active ingredient per acre. Absorptive roots developed within a mat of slowly decomposing plant debris on the surface in treated plots. Plant growth was retarded and capacity of soil to absorb and retain moisture was reduced. Although pasture mat formation was not evident, soil organic carbon was considerably increased. Through its effect on soil fauna Isobenzan causes severe and persistent changes in soil properties which affect plant growth. Different formulations of the same insecticide may produce different levels of microarthropod reduction. Karg (75, 76) found the highest mortality of forest soil microarthropods to be caused by technical lindane when compared with BHC (hexachlorocyclohexane) (gamma content 80 percent) and technical BHC (20 percent gamma) all applied at the rate of 1 kg/hectare gamma content in all cases. Maximum persistence was observed in grassland when it was treated with an 80 percent gamma formulation. The duration of persistence depended upon the initial concentration as well as humus particles. Sminthuridae were significantly reduced three days after treatment with lindane-DDT spray to oak stands (2 percent lindane - 10 percent DDT); mortality of Parasitiformes was high, and Sarcoptiformes nymphs in the litter layer were reduced. In a similar experiment after three weeks, animals in the soil were not significantly affected: whereas in the litter, mortality had increased to 30 percent for Parasitiformes, 50 percent for Sarcoptiformes, and up to 90 percent for Collembola. One year after treatment most of the microarthropods had increased in numbers. After one year the pesticide had penetrated to the soil and Parasitiformes were 50 percent reduced, while the number of species was reduced from 16 to 10. In large scale investigations, 96 mite and Collembola species were present one year after treatment as compared with 91 species before treatment, possibly indicating a short-term effect. In other studies, Collembola had increased in numbers in treated plots while declining in control plots, probably due to reduced numbers of Gamasina (predatory mites). Tarsonemini showed a slight increase, but there were no significant changes in Acaridae and Oribatei. In pea fields changes in species composition were noted after dimethoate treatment. A 36 percent overall reduction of microarthropods was observed after two DDT-lindane sprays. In similar treatment against potato beetles, after six weeks the number of predatory mites underneath the plants had increased, while *Folsomia* (Collembola) had decreased. Mites had decreased between the plants while Collembola had increased.

The effects of pesticides on target and nontarget microarthropods; predator prey relationships; and on soil fertility and mineral re-cycling have not been investigated extensively. The possible influence of soil microarthropods on insecticides is even less well documented. In a recent paper, Butcher, Kirknel & Zabik (14) fed *Folsomia candida* (Willem) upon pp' DDT (10–20 ppm)-agar-mycelium substrates. The animals were subsequently macerated and analyzed for DDT and its metabolites. Within 24 hr after they began feeding, quantities of DDE were recovered from the extracts as contrasted with that recovered from the Agar surface and subsurface. The percentage of DDE recovered increased consistently during observation periods of from 12 to 69 days. This study suggests that the role of soil microarthropods in the breakdown of persistent pollutants deserves considerably more attention than it has received.

Insecticides may act to inhibit or promote nutrient cycling, depending upon whether the effects are broad spectrum or selective. A good deal of the literature on pesticides does not address itself to indirect or subtle effects of perturbations. The summary here of some significant papers on this subject indicates that it is feasible to document influences on more than one ecosystem component. It also points up a need to analyze long-term effects in ecosystems of varying extent in order to illuminate rate and nature of recolonization and equilibrium establishment. Although there may well be a shift in use away from persistent to less persistent insecticides, the production and use of pesticides is currently increasing at a 15 percent annual rate, and it has been predicted that by 1975 insecticides will more than double in use (HEW Pesticides Report, 138). There is increasing apprehension in the United States over the side effects of insecticides, commercial fertilizers, and other production agriculture practices. This is not likely to diminish interest in the significance of soil arthropods, particularly in view of strong evidence attesting to the ubiquity of insecticides and their metabolites. The studies of Manley (105) working at Michigan State University showed how annual applications of persistent and short-lived insecticides could dramatically alter the ecology of *Melampyrum lineare* through its direct and indirect effects upon primary terrestrial arthropod consumers. The implications of these and similar studies highlight the dilemma that faces agriculturists and ecologists alike, and necessitates a continuing re-examination of the conditions under which pesticides may be authorized for use.

CONCLUSION

Soil must be considered as a complex of living and nonliving components which are present in different combinations and which possess identifiable gross characteristics. Microarthropods contribute directly to the humus fraction and also police the soil, permitting necessary complexes of soil organisms to exist. Even though their role in comminution and mixing may be small in comparison with that of larger invertebrates, microarthropods exercise an important function in mineral turnover, vegetation succession, and as decomposers of organic matter. In combination with microflora which they may disperse, they assist in decomposition of matter which they themselves cannot digest.

As concern about man's influence on the soil ecosystem intensifies, there is likely to be increasing curiosity about the ramification of his actions and their consequences. Our prior commitment to this area of research in North America has been almost negligible. We have been almost completely dependent upon European workers for our understanding of beneficial soil arthropod taxonomy, biology, ecology, and their role in nutrient cycling and fertility-humification processes. The contributions of viable research groups such as the Institute of Ecology at the University of Georgia, Oak Ridge National Laboratory, the Research Institute of the Canadian Department of Agriculture at London, Ontario, the Departments of Entomology at the University of Wisconsin, Michigan State University, and MacDonald College, Quebec, are increasingly supplementing the more specialized efforts of individuals at Ohio State University, Colorado State University, Grinnell College, and elsewhere. While these are encouraging beginnings, a great deal must be done if we are to implement successfully the ambitious and complex ecosystems analyses currently being undertaken, or which are being proposed under United States International Biological Program involvement.

LITERATURE CITED

1. Alejnikova, M. M. 1965. Die Bodenfauna des Mittleren Wolgalandes und ihre regionalen Besonderheiten. *Pedobiologia* 5:17–49
2. Ambroz, von Z., Nosek, J. 1967. Mikrobielle Aktivität und Apterygotenbesatz in initialen Böden der Niederen Tatra. *Pedobiologia* 7: 1–10
3. Ashraf, M. 1969. Studies on the biology of Collembola. *Rev. Ecol. Biol. Sol* 6(3):337–47
4. Atalla, E. A. R., Hobart, J. 1964. The survival of some soil mites at different humidities and their reaction to humidity gradients. *Entomol. Exp. Appl.* 7:215–28
5. Barrett, G. W. 1968. The effects of an acute insecticide stress on a semi-enclosed grassland ecosystem. *Ecology* 49(6):1019–35
6. Berthet, P. 1964. L'activité des Oribatides d'une chénaie. *Mem. Inst. Roy. Nat. Bel.* 152:1–152
7. Berthet, P., Gérard, G. 1965. A statistical study of microdistribution of Oribatei (Acari). I. The distribution pattern. *Oikos* 16:214–27
8. Bhattacharyya, S. K. 1962. Laboratory studies on the feeding habits and life-cycles of soil-inhabiting mites. *Pedobiologia* 1:291–98
9. Block, W. C. 1965. The life histories of *Platynothrus peltifer* (Koch 1839) and *Damaeus clavipes* (Hermann 1804) in soils of Pennine moorland. *Acarologia* 7:735–43
10. Bornebusch, C. H. 1930. The fauna of the forest soil. *Forstl. Forsogsv. Dan.* 11:1–224
11. Bretfeld, G. 1969. Neuer Paarbildungstyp bei der indirekten Spermatophorenübertragung der Collembolen. *Naturwissenschaften* 8: 425
12. Burges, A., Raw, F. 1967. *Soil biology.* London and New York: Academic. 532 pp.
13. Butcher, J. W. Insecticide influence on soil arthropods. *Symp. on Pesticides in the Soil, Michigan State Univ.* In press
14. Butcher, J. W., Kirknel, E., Zabik, M. 1969. Conversion of DDT to DDE by *Folsomia candida. Rev. Ecol. Biol. Sol* 6(3):291–98
15. Cassagnau, P. 1964. Ecologie et biologie des Symphypléones epigés de la haute vallée d'Aure (Hautes-Pyrénées). *Rev. Ecol. Biol. Sol.* 1:451–500
16. Cassagnau, P. 1965. Ecologie édaphique de la Montagne Noire basée sur les groupements de Colemboles. *Rev. Ecol. Biol. Sol* 2:339–75
17. Challet, G. L., Bohnsack, K. K. 1968. The distribution and abundance of Collembola at Pt. Barrow, Alaska. *Pedobiologia* 8:214–22
18. Chang, S. L. 1966. Some physiological observations on two aquatic Collembola. *Trans. Am. Microsc. Soc.* 85(3):359–71
19. Christiansen, K. 1964. Bionomics of Collembola. *Ann. Rev. Entomol.* 9:147–78
20. Christiansen, K. 1967. Competition between collembolan species in culture jars. *Rev. Ecol. Biol. Sol* 4(3):439–62
21. Coleman, D. C. 1966. The recolonization of gamma-irradiated soil by small arthropods. A preliminary study. *Oikos* 17:62–70
22. Cragg, J. B. 1961. Some aspects of the ecology of moorland animals. *J. Ecol.* 49:447–506
23. Crossley, D. A. Role of microflora and fauna in soil systems. *Symp. on Pesticides in the Soil, Michigan State Univ.* In press
24. Crossley, D. A., Jr., Hoglund, M. P. 1962. A litter bag method for the study of microarthropods inhabiting leaf litter. *Ecology* 43:571–73
25. Crossley, D. A., Jr., Witkamp, M., Dobson, G. J. 1963. Gross effects of arthropods and microflora on rates of leaf litter break down. *Physics Div. ann. progr. rep. for period ending June 1963. ORNL-3492,* 98–99
26. Davis, B. N. K. 1963. A study of microarthropod communities in mineral soils near Corby, Northants. *J. Anim. Ecol.* 32:49–71
27. Dhillon, B. S., Gibson, N. H. E. 1962. A study of the Acarina and Collembola of agricultural soils. *Pedobiologia* 1:189–209
28. Doeksen, J., Van der Drift, J., Eds. 1963. Soil Organisms. *Proc. Colloq. on Soil Fauna, Soil Microflora and their Relationships.* Amsterdam: North-Holland Publ. Co. 453 pp.
29. Dowdy, W. W. 1965. Studies on the ecology of mites and Collembola.

Am. Midl. Nat. 74:196–210
30. Edwards, C. A. 1965. Some side-effects resulting from the use of persistent insecticides. *Ann. Appl. Biol.* 55:329–31
31. Edwards, C. A. 1967. Relationship between weights, volumes and numbers of soil animals. See Reference 43
32. Edwards, C. A. 1969. Soil pollutants and soil animals. *Sci. Am.* 220: 92–99
33. Edwards, C. A., Dennis, E. B., Empson, D. W. 1967. Pesticides and soil fauna: effects of aldrin and DDT in an arable field. *Ann. App. Biol.* 60:11–22
34. Edwards, C. A., Thompson, A. R., Lofty, J. R. 1967. Changes in soil invertebrate populations caused by some organophosphorous insecticides. *Proc. Brit. Insecticides and Fungic. Conf., 4th,* 48–55
35. Edwards, C. A., Thompson, A. R., Beynon, K. I. 1968. Some effects of chlorfenvinphos, an organophosphorous insecticide, on populations of soil animals. *Rev. Ecol. Biol. Sol* 5(2):199–224
36. Engelmann, M. D. 1961. The role of soil arthropods in the energetics of an old field community. *Ecol. Monogr.* 31:221–38
37. Engelmann, M. D. 1968. The role of soil arthropods in community energetics. *Am. Zool.* 8:61 69
38. Farahat, A. Z. 1966. Studies on the influence of some fungi on Collembola and Acari. *Pedobiologia* 6:258–68
39. Frank, F. 1965. A contribution to the knowledge of seasonal fluctuation of Oribatids (Oribatei, Acari) of the forest soil on the mountain Ingman. *Veterinaria, Saraj.* 14: 19–24 (Engl. summary)
40. Gérard, G., Berthet, P. 1966. A statistical study of microdistribution of Oribatei (Acari). II. The transformation of the data. *Oikos* 17: 142–49
41. Ghilarov, M. S. 1965. Some practical problems in Soil Zoology. *Pedobiologia* 5:189–205
42. Ghilarov, M. S. 1968. Soil stratum of terrestrial biocenoses. *Pedobiologia* 8:82–96
43. Graff, O., Satchell, J., Eds. 1967. Progress in Soil Biology. *Proc. Colloq. on dynamics of soil communities, 1966, Braunschweig, Amsterdam.* 656 pp.

44. Green, C. D. 1964. The life history and fecundity of *Folsomia candida* (Willem) var. *distincta. Proc. Roy. Entomol. Soc. London Ser. A* 39:125–28
45. Green, C. D. 1964. The effect of crowding upon the fecundity of *Folsomia candida* (Willem) var. *distincta* (Bagnall). *Entomol. Exp. Appl.* 7:62–70
46. Gregers-Hansen, B. 1964. Decomposition of diethylstilboestrol in soil. *Plant Soil* 20(1):50–58
47. Griffiths, D. C., Raw, F., Lofty, J. R. 1967. The effects on soil fauna of insecticides tested against wireworms (*Agriotes* spp.) in wheat. *Ann. Appl. Biol.* 60:479–90
48. Haarlov, N. 1960. Microarthropods from Danish soils: ecology, phenology. *Oikos, Suppl.* 3, 176 pp.
49. Hale, W. G. 1964. Experimental studies on the taxonomic status of some members of the *Onychiurus armatus* species group. *Rev. Ecol. Biol. Sol* 1(3):501–10
50. Hale, W. G. 1965. Observations on the breeding biology of Collembola. I. *Pedobiologia* 5:146–52
51. Hale, W. G. 1965. Observations on the breeding biology of Collembola. II. *Pedobiologia* 5:161–77
52. Hale, W. G. 1965. Postembryonic development in some species of Collembola. *Pedobiologia* 5:228–43
53. Hale, W. G. 1966. A population study of moorland Collembola. *Pedobiologia* 6:65–99
54. Hale, W. G. 1966. The Collembola of the Moor House National Nature Reserve, Westmorland: a moorland habitat. *Rev. Ecol. Biol. Sol* 3:97–122
55. Hartenstein, R. 1962. Soil Oribatei. I. Feeding specificity among forest soil Oribatei (Acarina). *Ann. Entomol. Soc. Am.* 55:202–6
56. Hartenstein, R. 1962. Soil Oribatei. II. *Belba kingi,* n. sp., and a study of its life history. *Ann. Entomol. Soc. Am.* 55(4):357–61
57. Hartenstein, R. 1962. Soil Oribatei III. Studies on the development, biology and ecology of *Metabelba montana* and *Eremobelba nervosa* n. sp. *Ann. Entomol. Soc. Am.* 55(4):361–67
58. Hartenstein, R. 1962. Soil Oribatei IV. Observations on *Ceratozetes gracilis. Ann. Entomol. Soc. Am.* 55(5):583–86
59. Hartenstein, R. 1962. Soil Oribatei

V. Investigations on *Platynothrus peltifer*. *Ann. Entomol. Soc. Am.* 55(6):709–13

60. Hartenstein, R. 1962. Soil Oribatei VI. *Protoribates lophotrichus* and its association with microorganisms. *Ann. Entomol. Soc. Am.* 55(5):587–91

61. Hartmann, F. 1952. *Forstökologie.* Wien: G. Fromme. 461 pp.

62. Hayes, A. J. 1963. Studies on the feeding preferences of some pthiracarid mites. *Entomol. Exp. Appl.* 6:241–56

63. Hayes, A. J. 1965. Studies on the distribution of some phthiracarid mites in a coniferous forest soil. *Pedobiologia* 5:252–61

64. Hayes, A. J. 1966. Studies on the activity and survival of some phthiracarid mites at different relative humidities. *Pedobiologia* 6:281–87

65. Healey, I. N. 1967. The energy flow through a population of soil Collembola. *Secondary productivity of terrestrial ecosystems,* ed. K. Petrusewicz, Warszawa-Kraków, 695–708

66. Healey, I. N. 1967. An ecological study of temperatures in a Welsh moorland soil, 1962–63. *J. Anim. Ecol.* 36:425–34

67. Heath, G. W., Arnold, M. K., Edwards, C. A. 1966. Studies in leaflitter breakdown I. Breakdown rates of leaves of different species. *Pedobiologia* 6:1–12

68. Ibarra, E. L., Wallwork, J. A., Rodriguez, J. 1965. Ecological studies on mites found in sheep and cattle pastures. I. Distribution pattern of oribatid mites. *Ann. Entomol. Soc. Am.* 58:153–59

69. Jackson, R. M., Raw, F. 1966. *Life in the Soil.* Studies in Biology No. 2. New York: St. Martin's Press. 59 pp.

70. Jalil, M. 1965. The life cycle of *Hermannia scabra* (C. L. Koch 1879). *Oikos* 16:16–19

71. Janetschek, H. 1967. Growth and maturity of the springtail *Gomphiocephalus hodgsoni* Carpenter from South Victoria Land and Ross Island. *Antarctic Res. Ser.* 10:295–305

72. Janetschek, H. 1967. Arthropod ecology of South Victoria Land. *Antarctic Res. Ser.* 10:205–93

73. Jensen, P., Corbin, K. W. 1966. Some factors affecting aggregation of *Isotoma viridis* Bourlet and *Arion fasciatus* Nilsson. *Ecology* 2:332–34

74. Joosse, E. N. G. 1966. Some observations on the biology of *Anurida maritima. Z. Morphol. Oekol. Tiere* 57:320–28

75. Karg, W. 1962. Untersuchungen über die Veränderungen und Wechselbeziehungen der Mikroarthropoden in kartoffelnematodenverseuchten Flächen. *Nachrichtenbl. Deut. Pflanzenschutzdienstes* NF. 16(9):187–95

76. Karg, W. 1967. Beeinflussung der Bodenbiozönose in Forst und auf landwirtschaftlich genutzten Flächen durch Insektizide für den Flugzeugeinsatz. *Nachrichtenbl. Deut. Pflanzenschutzdienstes* 21:167–75

77. Kelsey, J. M., Arlidge, E. Z. 1968. Effects of isobenzan on soil fauna and soil structure. *N. Z. J. Agr. Res.* 11(2):245–59

78. Kevan, D. K. McE. 1962. *Soil Animals.* London: H. F. & G. Witherby Ltd. 237 pp.

79. Knight, C. B., Chesson, J. P. 1966. The effect of DDT on the forest floor Collembola of a loblolly pine stand. *Rev. Ecol. Biol. Sol* 3(1):129–39

80. Knight, C. B., Angel, R. A. 1967. A preliminary study of the dietary requirement of *Tomocerus. Am. Midl. Nat.* 77:510–17

81. Knight, C. B., Read, V. 1969. Microstratification of *Tomocerus* in a pine-open field continuum. *Rev. Ecol. Biol. Sol* 6:221–34

82. Kühnelt, W. 1961. *Soil Biology with Special Reference to the Animal Kingdom.* London: Faber and Faber. 397 pp.

83. Kühnelt, W. 1963. Über den Einfluss des Mycels von *Clitocybe infundibuliformis* auf die Streufauna. *Soil organisms,* eds. J. Doeksen, J. Van der Drift, 281–88. Amsterdam: North Holland Publ.

84. Kurceva, G. F. 1964. Wirbellose Tiere als Faktor der Zersetzung von Waldstreu. *Pedobiologia* 4:8–30

85. Lebrun, P. 1965. Contribution a l'étude écologique des Oribatides de la litière dans une forêt de Moyenne-Belgique. *Mem. Inst. Sci. Nat. Belg.* 153:1–96

86. Lebrun, P. 1968. Ecologie et biologie de *Nothrus silvestris* C. L. Koch

1839. *Pedobiologia* 8 :223–38
87. Loots, G. C., Ryke, P. A. J. 1967. The ratio Oribatei : Trombidiformes with reference to organic matter content in soils. *Pedobiologia* 7 :121–24
88. Loub, W., Haybach, G. 1967. Jahreszyklische Beobachtungen der Mikroflora und Mikrofauna von Böden im südlichen Wienerwald. *Rev. Ecol. Biol. Sol* 4 :59–80
89. Lutz, J. L., Traitteur-Ronde, G. 1965. Über Zusammenhänge in Artenbestand von Pflanzen, Bodenkleintieren und Mikroben des Hochmoores nebst ökologischen Ausblicken :bodenzool. Teil. *Biosoziologie*, ed. R. Tuxen, 215–29. Den Haag
90. Luxton, M. 1963. Some aspects of the biology of salt-marsh Acarina. *Acarology*, fasc. h.s. *C. R. Ier Congr. Int. Acarol.*, 172–82
91. Luxton, M. 1966. Laboratory studies on the feeding habits of salt-marsh Acarina, with notes on their behaviour. *Acarologia* 8 :163–75
92. Luxton, M. 1967. The zonation of salt-marsh Acarina. *Pedobiologia* 7 :55–66
93. Luxton, M. 1967. The ecology of salt-marsh Acarina. *J. Anim. Ecol.* 36 :257–75
94. Macfadyen, A. 1963. *Animal ecology. Aims and methods*, 2nd ed. London :Pitman. 344 pp.
95. Macfadyen, A. 1965. Heterotrophic productivity in the detritus food chain in the soil. *Proc. Int. Congr. Zool., 16th*, 4 :318–23
96. Macfadyen, A. 1966. Les méthodes d'étude de la productivité des invertebrés dans les écosystèmes terrestres. *Terre Vie* 4 :361–92
97. Madge, D. S. 1964. The water relations of *Belba geniculosa* Oudmns. and other species of oribatid mites. *Acarologia* 6 :199–223
98. Madge, D. S. 1964. The humidity reactions of oribatid mites. *Acarologia* 6 :566–91
99. Madge, D. S. 1964. The longevity of fasting mites. *Acarologia* 6 :818–29
100. Madge, D. S. 1965. The effects of lethal temperatures on oribatid mites. *Acarologia* 7 :121–30
101. Madge, D. S. 1965. The behaviour of *Belba geniculosa* Oudms. and certain other species of oribatid mites in controlled temperature gradients. *Acarologia* 7 :389–406

102. Madge, D. S. 1965. Further studies on the behaviour of *Belba geniculosa* Oudms. in relation to various environmental stimuli. *Acarologia* 7 :744–57
103. Madge, D. S. 1965. Leaf fall and litter disappearance in a tropical forest. *Pedobiologia* 5 :273–88
104. Mais, K. 1969. Zur Kenntnis der ökologischen Valenz von *Onychiurus cavernicolus* und *O. vornatscheri*. Über Temperatur-, Feuchtigkeits-und Lichtreaktionen. *Pedobiologia* 9 :282–87
105. Manley, G. 1968. Insect predators of the jack pine parasite *Melampyrum lineare* Desr. MS thesis, Michigan State Univ., East Lansing, Michigan
106. Marcuzzi, G. 1967. Osservazioni ecologice sulla fauna del suolo con particulare riferimento al Veneto e alla regione Trentino-alto Adige. *Riv. Biol.* 60 :433–99
107. Massoud, Z., Poinsot, N., Poivre, C. 1968. Contribution a l'étude du comportement constructeur chez les collemboles. *Rev. Ecol. Biol. Sol* 5(2) :283–86
108. Moeller, J. 1965. Oekologische Untersuchungen über die terrestrische Arthropodenfauna im Anwurf mariner Algen. *Z. Morphol. Oekol. Tiere* 55 :530–86
109. Moeller, J. 1966. Oekologische Beobachtungen an *Hypogastrura viatica* (Tullberg 1872). *Veröff. Inst. Meeresforschg. Bremerhaven* (Sonderband) 11 :329–36
110. Moritz, M. 1963. Über Oribatidengemeinschaften norddeutscher Laubwaldböden, unter besonderer Berücksichtigung der die Verteilung regelnden Milieubedingungen. *Pedobiologia* 3 :142–243
111. Moritz, M. 1965. Untersuchungen über den Einfluss von Kahlschlagmasznahmen auf die Zusammensetzung von Hornmilbengemeinschaften norddeutscher Laub-und Kiefernmischwälder. *Pedobiologia* 5 :65–101
112. Müller, G., Beyer, R. 1965. Über Wechselbeziehungen zwischen mikroskopischen Bodenpilzen und fungiphagen Bodentieren. *Zentralbl. Bakteriol. Parasitenk. Infektionskr. Hyg. Abt. 2* 119(11) : 133–47
113. Murphy, P. W., Ed. 1962. *Progress in Soil Zoology.* London : Butterworths. 398 pp.

114. Murphy, P. W., Jalil, M. 1963. Some observations on the genus *Tectocepheus Acarologia* fasc. h.s. 1964. *C. R. I. Congr. Int. Acarol.*
115. Naglitsch, F. 1965. Methodische Untersuchungen über den Einfluss von Bodenarthropoden auf die Humifizierung organischer Substanzen. *Pedobiologia* 5:50–64
116. Naglitsch, F., Steinbrenner, K. 1963. Untersuchungen über die bodenbiologischen Verhältnisse in einem Futterfruchtfolge-Versuch unter spezieller Berücksichtigung der Collembolen. *Pedobiologia* 2:252–64
117. Naglitsch, F., Grabert, D. 1967. Zu Fragen des biogenen Abbaues von Stroh unter kontrollierten Versuchsbedingungen. *Pedobiologia* 7: 353–61
118. Nef, L. 1962. The distribution of Acarina in the soil, 56–59. *Progress in Soil Zoology,* ed. P. W. Murphy, London: Butterworths
119. Niijima, K. 1966. Diversity and frequency of Collembola found in 3 different vegetations of the alpine region of Mt. Tateyama, Central Japan. (Engl. summary) *Kontyu* 34(4):339–46
120. Nosek, T. 1967. The investigation on the apterygotan fauna of the Lower Tatras. *Acta Univ. Carol. Biologica* 349–528
121. Nosek, J., Ambroz, Z. 1964. Apterygotenbesatz und mikrobielle Aktivität in Böden der Niederen Tatra. *Pedobiologia* 4:222–40
122. O'Connor, F. B. 1962. Oxygen consumption and population metabolism of some populations of Enchytraeidae from North Wales. *Soil organisms. Proc. Colloq. Soil Fauna Soil Microflora Their Relationships 1962.* Amsterdam:North Holland Publ., 1963, 32–48
123. Odum, E. P. 1968. Energy flow in ecosystems: A historical review. *Am. Zool.* 8:11–18
124. Ogino, K., Saichuae, P., Imadate, G. 1965. Seasonal changes of soil microarthropod population in central Thailand. *Nature Life SE Asia* 4:303–15
125. Ovington, J. D. 1962. Quantitative ecology and the woodland ecosystem concept. *Advances in Ecological Research,* ed. Y. B. Cragg, 1:103–93. London and New York: Academic
126. Palissa, A. 1964. *Bodenzoologie in*

Wissenschaft, Naturhaushalt und Wirtschaft. Berlin: Akademie-Verlag. 180 pp.
127. Pedigo, L. P. 1967. Selected life history phenomena of *Lepidocyrtus cyaneus f. cinereus* Folsom with reference to grooming and the role of the collophore. *Entomol. News* 78(10):263–67
128. Poinsot, N. 1965. Sur la biologie des Collemboles liés aux eaux saumâtres en Camargue. *Ann. Fac. Sci. Marseille* 38:109–20
129. Poinsot, N. 1966. Existence d'un comportement constructeur chez un Collembole Isotomidae *Subisotoma variabilis* (Gisin 1949). *Rev. Ecol. Biol. Sol* 3(1):173–78
130. Poinsot, N. 1966. Sur un comportement constructeur chez le Collembole *Isotomurus* sp. Relation entre ce comportement et le phénomène de l'écomorphose. *Rev. Ecol. Biol. Sol* 3(4):585–88
131. Poinsot, N. 1966. Etude écologique des Collemboles des dunes de Beynes (Haute Camargue). *Rev. Ecol. Biol. Sol* 3:483–93
132. Poinsot, N. 1968. Cas d'anhydrobiose chez le collembole *Subisotoma variabilis* Gisin. *Rev. Ecol. Biol. Sol* 5:585–86
133. Poole, T. B. 1961. An ecological study of the Collembola in a coniferous forest soil. *Pedobiologia* 1:113–37
134. Poole, T. B. 1962. The effect of environmental factors on the pattern of distribution of soil Collembola in a coniferous woodland. *Pedobiologia* 2:169–82
135. Poole, T. B. 1964. A study of the distribution of soil Collembola in three small areas in a coniferous woodland. *Pedobiologia* 4:35–42
136. Poore, M. E. D. 1956. The use of phytosociological methods in ecological investigations. IV. General discussion of phytosociological problems. *J. Ecol.* 44:28–50
137. Reichle, D. E., Crossley, D. H. 1965. Radiocesium dispersion in a cryptozoan food web. *Health Phys.* 11: 1375–84
138. Report of the Secretary's Commission on Pesticides and their Relationship to Environmental Health Part I and II, 1969. U.S. Department of Health, Education and Welfare, U.S. Government Printing Office, 1969
139. Rockett, C. L., Woodring, J. P. 1966.

Biological investigations on a new species of *Ceratozetes* and *Pergalumna*. *Acarologia* 8 :511–20

140. Rusek, J. 1968. Die Apterygotengemeinschaft der Acereto-fraxinetum-Waldassoziation des Mährischen Karstes. *Acta Soc. Zool. Bohemislov.* 32(3) :237–61

141. Ryke, P. A. J. 1965. Numerical fluctuation in mite populations of grass covered soils. (Engl. summary). *Tydskr. Natuurwetensk.* 5 :32–47

142. Schalk, V. 1968. Zur Bodenfauna von Wiesen und Luzernebeständen unterschiedlicher Standorte unter besonderer Berücksichtigung der Oribatiden. *Pedobiologia* 8 :424–506

143. Schuster, R. 1956. Der Anteil der Oribatiden an den Zersetzungsvorgängen im Boden. *Z. Morphol. Oekol. Tiere* 45 :1–33

144. Shaldybina, E. S. 1966. Postembryonic development of *Chamobates spinosus* Sellnick, 1928 (In Russian, Engl. summary). *Zool. Z.* 45(5) :661–66

145. Shaldybina, E. A. 1967. Biology of *Melanozetes mollicomus* (Koch) (In Russian, Engl. summary). *Zool. Z.* 46(2) :1659–67

146. Sharma, G. D. 1967. Observations on the biology of *Isotoma olivacea* Tullberg 1871. *Pedobiologia* 7 : 153–55

147. Sharma, G. D. 1967. Bionomics of *Tomocerus vulgaris*. *Proc. Roy. Entomol. Soc. London* 42 :30–34

148. Sharma, G. D., Kevan, D. K. McE. 1963. Observations on *Pseudosinella petterseni* and *Pseudosinella alba* in eastern Canada. *Pedobiologia* 3 :62–74

149. Sheals, J. G. 1969. *The Soil Ecosystem*. London : The Systematics Assoc. 270 pp.

150. Stebajeva, S. K. 1963. Oekologische Verteilung der Collembolen in Wäldern und Steppen Süd-Tuvas. (In Russian, German summary). *Pedobiologia* 3 :75–85

151. Stebaeva, S. K. 1966. The ecological characteristics of Collembola dwelling in soils of the northern Baraba steppe. (In Russian, Engl. summary). *Zool. Zh.* 45(8) :1144–58

152. Stebaeva, S. K. 1967. Pedobiologische Experimente mit ausgetauschten Bodenblöcken im süd-östlichen Altai-Gebirge und der Severnaja Baraba. *Pedobiologia* 7 :172–91

153. Stegeman, C. 1964. The effects of the carbamate insecticide carbaryl upon forest soil mites and Collembola. *J. Econ. Entomol.* 57 :803–8

154. Stöckli, A. 1946. Die biologischen Komponenten der Vererdung der Gare und der Nährstoffpufferung. *Schweiz. Landwirt. Monatsh.* 24 : 3–19

155. Strebel, O. 1965. Beobachtungen über die Biologie und das Verhalten von *Hypogastrura boldorii* Denis 1931. *Mitt. Pollichia (Dürkheim)* 12(3) :178–200

156. Strenzke, K. 1963. Die Arthropodensukzession im Strandanwurf mariner Algen unter experimentell kontrollierten Versuchsbedingungen. *Pedobiologia* 3 :95–141

157. Thibaud, J. M. 1967. Action de différentes températures sur la durée du développement de 6 espèces de Collemboles Hypogastruridae épigés et cavernicoles. *C. R. Acad. Sci. Paris* 265 :2074–76

158. Thibaud, J. M. 1968. Contribution a l'étude de l'action des facteurs température et humidité sur la durée du développement embryonnaire des Hypogastruridae. *Rev. Ecol. Biol. Sol* 5(1) :55–62

159. Thibaud, J. M. 1968. Contribution a l'étude de l'action des facteurs température et humidité sur la durée du développement postembryonnaire et de l'intermue de l'adulte chez les Hypogastruridae. *Rev. Ecol. Biol. Sol* 5(2) :265–81

160. Törne, E. von. 1961. Ökologische Experimente mit *Folsomia candida*. *Pedobiologia* 1 :146–49

161. Törne, E. von. 1964. Ueber die Anzucht und Haltung individuenreicher Collembolenpopulationen. *Pedobiologia* 4 :256–64

162. Törne, E. von. 1965. Experimentelle Untersuchungen über den Einfluss der Lebenstätigkeit von Mikroorganismen und Bodentieren auf den Abbau von Zellulose, 1. *Pedobiologia* 5 :211–27

163. Törne, E. von. 1966. Ueber den Verlauf der Zelluloserotte unter biotisch verschiedenen Versuchsbedingungen. *Pedobiologia* 6 :226–37

164. Törne, E. von. 1967. Beispiele für mikrobiogene Einflüsse auf den Massenwechsel von Bodentieren. *Pedobiologia* 7 :296–305

165. Törne, E. von. 1967. Beispiele für indirekte Einflüsse von Bodentieren

auf die Rotte von Zellulose. *Pedobiologia* 7 :220–27

166. Törne, E. von. 1968. Beispiele für mikrobiogene Einflüsse auf den Massenwechsel von Bodentieren, II. *Pedobiologia* 8 :526–35

167. Uchida, H., Fujita, K. 1968. Mass occurrence and diurnal activity of *Dicyrtoma rufescens* in winter. *Sci. Rep. Hirosaki Univ.* 15(1, 2) : 36–48

168. Usher, M. B. 1969. Some properties of the aggregations of soil Arthropods : Collembola. *J. Anim. Ecol.* 38 :606–22

169. Vail, P. V. 1965. Colonization of *Hypogastrura manubrialis,* with notes on its biology. *Ann. Entomol. Soc. Am.* 58 :555–61

170. Van der Drift, J. 1970. The need for a coordinated soil biology research effort. *Symp. Pesticides in the Soil, Michigan State Univ.* In press

171. Vannier, G. 1967. Etude in situ des réactions de la microfaune au dessèchement progressif d'un type de sol donné. *C. R. Acad. Sci. Paris* 265 :2090–92

172. Vannier, G. 1967. Definition des rapports entre les microarthropodes et l'état hydrique des sols. *C. R. Acad. Sci. Paris* 267 :1741–44

173. Vannier, G., Thibaud, J. M. 1968. Le concept de disponibilité en eau appliqué a une population de collemboles Hypogastruridae vivant dans le guano de grotte. *C. R. Acad. Sci. Paris* 267 :278–81

174. Voronova, L. D. 1968. The effect of some pesticides on the soil invertebrate fauna in the South Taiga zone in the Perm region (USSR). *Pedobiologia* 8 :507–25

175. Wallace, M. M. H. 1967. The ecology of *Sminthurus viridis* (L.) I. Processes influencing numbers in pastures in Western Australia. *Aust. J. Zool.* 15 :1173–1206

176. Wallace, M. M. H. 1968. The ecology of *Sminthurus viridis,* II. Diapause in the aestivating egg. *Aust. J. Zool.* 16 :871–83

177. Weis-Fogh, T. 1947. Ecological investigations of mites and collemboles in the soil. *Nat. Jutlandica* 1 :139–270

178. Whittaker, R. H., Fairbanks, C. W.

1958. A study of plankton copepod communities in the Columbia-basin in southeastern Washington. *Ecology* 39 :46–65

179. Wigglesworth, V. B. 1953. *Principles of Insect Physiology,* 5th ed. London :Methuen

180. Wise, K. A. J., Fearon, C. E., Wilkes, O. R. 1964. Entomological Investigations in Antarctica, 1962–63 season. *Pacific Insects* 6 :541–70

181. Witkamp, M., Van der Drift, J. 1961. Breakdown of forest litter in relation to environmental factors. *Plant Soil* 15 :295–311

182. Wood, T. G. 1966. The fauna of grassland soils with special reference to Acari and Collembola. *Proc. N. Z. Ecol. Soc.* 13 :79–85

183. Wood, T. G. 1967. Acari and Collembola of moorland soils from Yorkshire, Engld. I. Description of the sites and their population. *Oikos* 18 :102–17

184. Wood, T. G. 1967. Acari and Collembola of Moorland soils from Yorkshire, Engld. 11. Vertical distribution in 4 grassland soils. *Oikos* 18 :137–40

185. Wood, T. G. 1967. Acari and Collembola of moorland soils from Yorkshire, Engld. III. The microarthropod communities. *Oikos* 18 : 227–92

186. Woodring, J. P. 1965. The biology of 5 new species of oribatids from Louisiana. *Acarologia* 7 :564–756

187. Woodring, J. P., Cook, E. F. 1962. The internal anatomy, reproductive physiology, and molting process of *Ceratozetes cisalpinus. Ann. Entomol. Soc. Am.* 55 :164–81

188. Woodring, J. P., Cook, E. F. 1962. The Biology of *Ceratozetes cisalpinus* (Berl.), *Scheloribates laevigatus* (Koch) and *Oppia neerlandica* (Ouds.) with a description of all stages. *Acarologia* 6 :101–37

189. Woolley, T. A. 1960. Some interesting aspects of oribatid ecology. *Ann. Entomol. Soc. Am.* 53 :251–53

190. Zivadinovic, J., Cvijovic, M. 1969. Afinitet Collembola I. Drugih Organizama Tla Prema Tipovima Tla. *Ekologija* 4(1) :13–22

EFFECTS OF INSECT DEFOLIATION ON GROWTH 6011
AND MORTALITY OF TREES[1]

H. M. KULMAN

Department of Entomology, Fisheries, and Wildlife, University of Minnesota,
St. Paul, Minnesota

The most important ensuing effects from insect defoliation are mortality, growth loss, rotation delays, and increased susceptibility to secondary insects and disease. These damages are called, collectively, growth impact. In an extensive report on damage to forests in 1952, it was estimated that 15% of the insect-related growth impact in the United States was caused by defoliation (66). The projections of growth loss and the role of defoliation in predisposing trees to secondary insects are probably underestimated since comprehensive reviews on these subjects were not available. Except for brief summaries in forest entomology texts, I found only three small reviews of the effects of defoliation on tree growth and mortality (27, 90, 136). The totality of the interactions of insects with the forest has recently been reviewed (136).

Equally scarce are papers on economic analysis of potential damage to forest stands from various intensities of defoliation. The first published formula for a financial analysis of costs relative to losses from defoliation was released in 1933 (159). Even though it excluded interest charges, it was a good attempt to put defoliator control on a basis of costs versus projected loss. Recent papers review analysis needs and formulae using site value, present and future value of the final crop and thinnings, rotation age, rotation and thinning delays, risk of additional defoliation, interest charges and other factors (3, 75, 76). In a few of the studies covered in my review, control costs are related to projected volume losses. These are mentioned in the sections with the specific defoliator. Effects of defoliation are included in a recent life table approach to the analysis of insect impact (175).

My review is limited to quantitative studies of tree mortality and increment reduction related to measured amounts of insect and artificial defoliation in which the foliage is removed at the time of treatment. The effects of defoliation on seed production, epicormic branching, wood quality, attack by secondary organisms, and foliation timing are often mentioned, but not ex-

---

[1] The survey of the literature ended in January 1970 and with the 1968 index of Forestry Abstracts. I would appreciate receiving information on literature omitted.

tensively reviewed. Pruning and other damage to leaf-bearing twigs and reduced photosynthetic efficiency of foliage and early leaf fall caused by mites, sucking insects, algae, fungi, microorganisms, weather, fire, air pollution, chemical defoliants, and residues on foliage are not included since they are difficult to quantify and/or their simultaneous influence on other parts of trees confounds foliage loss-related damage. Surveys made to determine losses resulting from defoliation of unspecified duration or amounts of foliage are also omitted since they are of little value in guiding future activities in pest management. They are, of course, of considerable short-term interest to forest managers for salvage and management operations.

TREE PHYSIOLOGICAL FACTORS AFFECTING DEFOLIATION, GROWTH, AND MORTALITY

Defoliation measurement.—The methodology of estimating the quantity of foliage loss will not be reviewed. However, foliage age, location of foliage in crowns, time of defoliation, and stage of leaf development modify the influence of defoliation on tree growth and mortality. Therefore, it is important that the units used for quantitative statements of foilage loss relate to these factors.

Studies and reviews of simulated insect defoliation of conifers (93, 100, 125) and hardwoods (101) show that the importance of foliage in promoting shoot and diameter growth decreases with age of the foliage and varies with season of the year. These effects have been clearly shown for white spruce, balsam fir, and pines (29). Needles and leaves in the tops of trees weigh more per unit area than foliage at lower levels. Hardwood leaves are smaller in the tops of trees than at the bottom, but the reverse relationship occurs with some coniferous foliage in which the longest needles occur higher in crowns. The differences in weight and size of foliage between the top and bottom of trees are greater in closed stands than in open stands (133). These studies illustrate the importance of indicating crown position and foliage age in descriptive statements of defoliation.

Both photosynthetic efficiency and production of foliage sometimes are influenced by site quality (142). Recent studies also show that growth rate can affect rate of photosynthesis (161).

In conifers and hardwoods of the temperate zone with predetermined shoots, early shoot growth and springwood utilize photosynthate stored from the previous season. Summerwood growth and, in conifers, needle elongation are largely dependent on current photosynthate. Trees with heterophyllous shoots and the summer flushes of growth in southern pines are dependent on current photosynthate for much of their shoot and foliage growth (85, 87, 93, 101, 125). The effects of leaf development at the time of defoliation, refoliation, and size of leaves in years following refoliation are covered in later sections with the defoliator (also in 7, 55, 92, 101). The quantity of foliage of different ages varies with tree species, level in crown, tree vigor, and tree age (hardwoods, 87, 101, 133) ; conifers, (93, 133, 152). Retention of foliage also varies between and within species (152). In some cases, differences in retention have been attributed to potassium deficiency (105).

Buds and new foliage produce hormones which affect the production and utilization of photosynthate. There is increasing evidence that growth is more closely correlated with utilization than production since stored photosynthate is often present but not utilized in the absence of hormones (86, 88; also see 173). Needles, stems, and roots of evergreen conifers also serve as food storage organs (89). In summary, defoliation of evergreen conifers could cause a loss of stored food, a loss of a place to store food, and affect the production of growth regulators and photosynthate.

Stored photosynthate in foliage of various ages and locations within crowns is used at different times to satisfy different photosynthate needs (30, 101). In artificial defoliation studies, removal of new foliage decreased growth of the upper stem more than in the lower stem, whereas the reverse was true when the old foliage was removed (34).

Late season defoliation leaves twigs unlignified and subject to winter damage (79, 91, 140, 160, 177). Specific effects of bud and shoot damage by defoliating insects have received little attention (see sections on spruce and jack pine budworms, sawflies, and 94, 124). Flowers and fruit often reduce foliage production and change the relationship between quantity of foliage and growth (see following section on growth measurement). Photosynthesis occurs in the bark of aspen in both summer and winter (129, 130). Possibly photosynthesis in bark would increase when trees are defoliated.

Genetic differences between clones and provenances must be considered for many of the physiological factors (88).

Most of the above mentioned items should be accounted for in developing statements of defoliation loss that will relate to tree growth losses and mortality.

Growth measurement.—The equipment and methodology used for growth measurement are complex. Workers should consult a standard forest mensuration text (i.e. 71).

In several conifers (118, 162) and hardwoods (63, 106), flowering is associated with reduced growth and foliage production (88). Therefore, the related (as with budworms) or random association of defoliation with flower production can be confounded with the effects of defoliation. However, in Douglas-fir (152), a heavy flower crop had no apparent effect on foliage production. The effects of flowering on tree growth have recently been reviewed (88).

Vertical distribution of diameter growth is also an important consideration (41, 84, 99, 153, 173). In conifers holding a dominant position in stands, ring width is greatest in the living crown at the point of maximum leaf surface and decreases regularly above and below this point except at the base of the trunk where it again increases. In suppressed trees the point of maximum ring width is higher than in dominant trees and there is little or no growth at the base (84). The many variations related to tree species, spacing, age, and suppression are covered in all of the articles (especially 84, 99, 173). Several studies have shown that defoliation influences diameter increment differently at various heights in the stem (34, 99). The effects of defoliation are usually greatest and first observable in the living crown (43, 112, 119, 183). However, in most studies cited here, the diameter increment was measured only at 4.5 feet above the ground. The statement "Measurements at breast height can account for only a small portion of the total variation in radial growth of Douglas-fir trees." (153), points up a serious limitation of many of the defoliation studies covered in this review.

Defoliation often causes incomplete and missing growth rings which complicates the measurement of increment losses (33, 65, 93, 97, 110, 126, 137). In studies of the effects of defoliation of jack pine and *Eucalyptus,* missing rings were composed of the summerwood of the defoliation year and the springwood of the following year (97, 110, 137). From studies on Douglas-fir it was concluded that spring- and summerwood are distributed differently between levels in the stem and controlled by different factors. Therefore, spring- and summerwood should be studied separately rather than together as a single annual growth ring (153).

In making growth measurements, the time of measurement is important since tree stems expand and contract according to changes in hydration which differ between tree species, seasons, humidity levels, and different parts of the same stem (84).

Seasonal measurements with dendrometers are often useful for showing the exact time of defoliation effects (7, 9, 26, 36), although the user should be cautious of hydration-caused diameter fluctuations. Increment cores and x-sectional discs are most frequently used to make diameter increment measurements. The cores or x-sections should not be taken until trees recover from the effects of defoliation. Studies on pines partially defoliated for one year show that recovery of shoot elongation occurs when the trees regain their full age-complement of foliage (93). However, variation in recovery time is rampant as shown by the recovery times discussed in the main body of this review.

Effective measures of increment reduction can be obtained by comparing growth rates at several heights in the stem of defoliated and nondefoliated trees. Trees randomly selected and protected with insecticides or microorganisms would be the most desirable controls since defoliation may be coincident with flowering and hydration. Recent Russian studies indicate that some defoliators, previously considered "primary insects," require "physiologically weakened stands" in order to develop large populations (60). It is generally known that fertilized trees are less favorable for the mass development of some defoliating insects (155). In Germany, *Diprion pini* pupae were larger and more numerous in nonfertilized pine stands than in fertilized stands (145). In later studies in which the sawflies were caged on trees growing on good and poor sites, survival rate and size of larvae was greater on trees from poor sites. The author suggests that unfavorable water balance in poor sites increased the sugar content of needles which is favorable for sawfly growth and survival (146). These studies strongly support the need for randomly assigned control plots protected from defoliation since naturally occurring nondefoliated areas may be free of insect damage by the action of factors that have adverse effects on the insects and beneficial or detrimental effects on tree growth.

Increment reduction of defoliated trees can be determined by comparison with the growth rate of damage-free nonhost trees in the same stands by using the ratio of pre- and postdefoliation growth rates (17, 20, 33, 38, 69, 119, 139). Although host and nonhost trees are exposed to similar environmental conditions, the method is subject to error since the nonhost tree species could have different growth patterns and thresholds of response to environmental influences (88, 122). In some cases, nondefoliated trees in infested stands show a positive response to the defoliation of their competitors (42, 72, 132).

When the nondefoliated controls are trees within the stand that escaped defoliation for unknown reasons, their suitability as controls is often determined by comparing the previous growth rate patterns of defoliated and control trees. Several studies have revealed a relationship between pre- and postdefoliation growth rate (33, 70, and section on spruce budworm). When nondefoliated trees are not available for comparison, postdefoliation growth rate is compared to predefoliation growth rate (33, 42, 70, 112). This method may be effective for comparing various intensities of defoliation. However, as a measure of volume loss it may be inaccurate because of growth pattern (see below).

Cambium-age-related growth pattern probably influences most of the above methods of growth measurement. From a detailed study of radial increments at all internodes in red pine (41), three ways of demonstrating radial growth (known as the Duff-Nolan method) are defined in Figure 1. The horizontal sequence does not accurately portray the influence of defoliation at lower levels in trees since it is the point of slowest growth and least sensitivity to defoliation. At higher stem levels, defoliation effects can be confused with intrinsic growth patterns. Theoretically, vertical sequence is most useful for measurement of effects of defoliation because it is free from age-of-cambium related growth pattern. From studies of growth reduction in three conifers, the oblique pattern was most useful for portraying the effects of defoliation of conifers in which defoliation is concentrated in the tops of crowns. However, the first effects of defoliation are detected most easily on the vertical sequence. Unfortunately, no mention was made for using the Duff-Nolan method to quantify growth loss (171). In studies on defoliation of lodgepole pine, a short-cut procedure for the Duff-Nolan

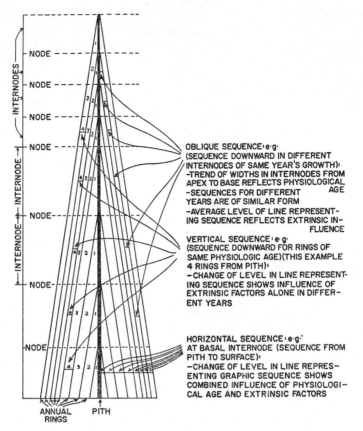

FIGURE 1. Diagrammatic explanation of the Duff-Nolan growth sequences using the simplified sequence terminology of Mott et al (119). Reprinted from *Concepts of Forest Entomology* by Kenneth Graham, Copyright (c) 1963 by Reinhold Publ. Corp., by permission of Van Nostrand Reinhold Company.

method was developed which also provided a way to determine the percentage of growth loss (156). A modification of this method was used to show growth loss in ponderosa pine from defoliation (31), but quantitative growth loss data were not developed from about ten other studies employing the Duff-Nolan method. Possibly temporal models in triangular coordinates such as (5) should be studied for their application in quantifying growth losses from defoliation.

Most applied work on tree growth related to effects of defoliation has been related only to photosynthate production by foliage. In order to alert readers to the importance of hormones and other factors, reviews on physiological considerations of growth have sometimes all but ignored the importance of photosynthate production. The vast number of studies on defoliation and growth covered in this review, attest to the close relationship between quantity of foilage lost and reduction in growth. This indicates that loss of foliage probably has a proportional effect on many of the above mentioned factors that influence growth. However, growth loss predictions are probably safe only for the particular tree species grown under a limited range of conditions. Since practical growth loss predictions are needed for only a relatively small number of tree species and defoliators, studies which use percentage of defoliation as a predictive index for growth loss will continue to be useful. However, an appreciation of the other physiological factors should improve the sensitivity of defoliation and growth measurements.

Mortality measurement.—The vast accumulation of circumstantial evidence of tree mortality directly or indirectly resulting from defoliation is covered in the remainder of the paper. Although the final killing agents are often secondary bark beetles, borers, and fungi; several studies show that secondary insects build up in dead and weakened trees and attack and kill trees that otherwise would have survived (see sections on spruce budworm and gypsy moth). As pointed out in the growth measurement section, some defoliators may need physiologically weakened trees in order to produce large populations. Another factor of concern is the life expectancy of the tree if it had escaped defoliation. Surely, mortality of a tree that would have died in 2 years from other causes is not as serious a loss as defoliation-related mortality of a tree that had a life expectancy of 20 years. Although it is beyond the scope of this paper to cover the nature of tree mortality in the absence of secondary agents, defoliation studies revealed that, "Death of trees after repeated insect defoliation could be due to the exhaustion of starch reserves to a level which, in the absence of a photosynthetic organ, does not support respiration or growth" (6). However, some studies suggest that even when food is available, growth is inhibited, and probably mortality caused, by various internal blocks to food conversion. In uninjured trees, deficiency of growth regulators and internal water stress are involved in growth control and probably mortality, when food is present (86).

Even if it could be possible to isolate mortality caused only by defoliation, the complex of "secondary" agents and pre- and postdefoliation weather conditions would still complicate direct cause-effect relationships. Therefore, it appears that mortality estimates in defoliated stands will be dependent on a comparison with similar stands that were not defoliated, or preferably with randomly selected plots protected with insecticides or microorganisms. Predictive losses will need to be tempered by pre- and postdefoliation considerations. When check plots are not available, some of the difficulties can be partially accounted for by determining the natural mortality rate before defoliation and, as with spruce budworm studies, by checking the relationship between predefoliation growth rate and postdefoliation mortality (33).

ARTIFICIAL DEFOLIATION

Simulated insect defoliation and studies conducted for other purposes in which the defoliation is similar to insect defoliation are valuable because they permit randomization and exact measurement of defoliation intensity. Simulated and natural insect defoliation differs because man usually pulls or cuts off foliage, whereas caterpillars often leave leaf petioles, midribs of leaves, some of the leaf blade, needle length contained in the fascicle sheath, and damage shoots (see 2, 21, 35, 93, 140).

White pine.—In Ontario, defoliation treatments in May involving new foliage reduced height growth 80–90% during the first year and 10–40% during the second year. Diameter growth reductions were 40–90 and 30–70%. Losses from defoliation of older foliage were generally lower (102). In Ohio, treatments that removed all new, one-year and two-year foliage were applied to different trees in July, September, and April. Shoot elongation was reduced about 75% from defoliation of new foliage, but removal of all old foliage had little effect. Complete defoliation killed 12 of 15 trees (37).

Red pine.—In Wisconsin, complete defoliation in early spring reduced dry weight of shoots by about 87%, new needle length by 65%, and terminal growth by 80%. The role of stored food in needles, stems, and roots is reviewed (89). In West Virginia, defoliation of new or old needles, or both, reduced shoot growth, but in the second and third years after defoliation, shoot growth reductions were evident only in defoliation treatments that involved new foliage. Needle length was significantly reduced after the first year of new foliage removal, significantly increased during the second year, and normal during the third year (93).

Scots pine.—In France, one- and two-year foliage was removed three weeks earlier than new foliage. Treatments that included new foliage reduced height growth by 50–60% and diameter growth by 50–72%. Defoliation of one and two-year foliage reduced height growth 15–36% and diameter growth by 15% (98). In German studies, needles were stripped leaving only the two most recent years of needle growth for a period of ten years. Height growth was reduced by 2–3%, volume by 15–45%, and needle production by 40% (22). In West Virginia, new foliage defoliation in July reduced shoot growth by 48 and 41% during the first and second years after defoliation. Removal of old foliage had little effect (93). In Michigan, removal of old foliage before new foliage developed reduced radial increment by 71% and terminal shoot elongation by 63% (184). In Virginia, complete defoliation either 30–40 days after new growth started or in September, killed all trees. Removal of fully developed new foliage killed one of three treated trees and reduced diameter growth more in the top than at the base. Defoliation of new foliage in early spring caused similar, but less severe growth reduction and no tree mortality. The author stated that old needles were retained longer, and new needles and scarious bracts grew abnormally long. Removal of old foliage before and after buds opened caused reductions in diameter growth, especially at the base of trees. The latter treatment caused the greatest diameter growth reductions—up to 75% (34).

Jack pine.—The treatments in last mentioned Scots pine study were also used on jack pine giving similar results (34). In Quebec, August defoliation of new, 1 + 2 + 3–, 1–, 2 + 3–, and 2-year foliage reduced height growth about 84, 33, 22, 20, and 14% and diameter growth 40, 50, 20, 20, and 5%, respectively. Bud production was reduced by complete defoliation (125).

Longleaf pine.—Removal of 30, 60, and 90% of the needle length of longleaf pine seedlings in Mississippi caused 24, 34, and 51% reduction in height increment and 14, 23, and 39% reduction in diameter increment. The effects were progressively greater in the February, July, and November treatments. Two 90% defoliations in the same season reduced height increment 63% and diameter increment 51%. Defoliation affected periodicity as well as amount of growth. Mortality was insignificant (21). Because of damage to roots of longleaf pine during transplanting, the foliage is often cut back to reduce transpiration so that the roots will be able to supply the needed moisture (1). In Louisiana, removal of about 60% of the needle length just prior to planting reduced diameter increment 8% and height increment by 10% in measurements made at 2, 4.5, and 5.5 years after treatment. Treatments had no effect on survival (35). In Mississippi, 75, 50, 25, and 0% reduction in needle length of longleaf pine resulted in 52, 60, 43, and 35% survival of the seedlings at the end of the first season. Partially defoliated seedlings survived and grew better than nondefoliated seedlings. Data are also given on defoliation in various seasons (1). Papers review previous studies on the subject.

Other evergreen trees.—In Japan, defoliation of seedlings of Japanese red pine before the start of spring growth in January, March, and April reduced height growth by 78, 70, and 34%. Growth normally starts in late March (53, also see 52). Removal of old foliage from Japanese black pine greatly reduced radial growth, but when buds and old foliage were removed no radial growth occurred. The role of stored food and hormones is discussed (124). In Great Britain, late April defoliation of Austrian pine reduced height increment 67% (172). August defoliation of six species of evergreen shrubs in North Carolina caused winter mortality, whereas spring defoliation caused only reductions in diameter increment of 0–74% (91).

Hybrid Populus.—In the Netherlands, complete defoliation in June reduced growth by 50% on both a weight and x-sectional area basis (48). In France, complete defoliation for two years reduced height growth by 18%, radial growth by 44%, and volume by 60%. The average annual increment loss for the three years since the first defoliation was 40% for complete defoliation and 3.7% for 50% defoliation (78). In Yugoslavia, trees were defoliated wholly or partially at monthly intervals. Some trees were defoliated both in May and mid-July. The latter treatment reduced diameter increment by 60%, delayed spring flushing by almost a month and left many shoots unlignified and thus subject to winter kill. Single, complete, and partial defoliation treatments reduced diameter increment by about 40 and 20%, with May defoliation causing a greater reduction. August defoliation had little effect on growth, but delayed spring flushing of leaves and left shoots unlignified and subject to winter damage (79; also see 101).

Maples.—In West Virginia, complete disbudding of sugar maple just prior to full opening of leaves caused a reduction of 65% in terminal shoot diameter increment and 47% in bud production. Partial disbudding caused proportionately less damage. Red maple was much less sensitive to all treatments. Partial disbudding may have stimulated linear shoot growth (94). In Wisconsin, sugar maple defoliated at seven different dates from July 7 to August 26 showed complex time and defoliation interactions. Generally complete, upper two-thirds, and upper one-third defoliation of the crown in closed stands caused about 88, 72, and 39% mortality of terminal shoots after one year. Tree mortality was less than 10% after one year but rose to 43% in the following year. This extensive paper also covers epicormic branch sprouting, refoliation, size of leaves, and defoliation in open stands (55).

Elm.—In Connecticut, two-year old potted American elms were defoliated before and after the termination of shoot growth. One and two defoliations caused 12 and 30% twig die back, 22 and 31% less shoot growth, 75 and 69% less diameter increment, and a 29 and 66% reduction in the size of leaves. Refoliation and susceptibility to the Dutch elm disease, *Ceratocytis ulmi,* are discussed (172).

Plum.—In Italian studies, diameter increment was proportional to defoliation and flowering. Defoliated branches had shorter than normal fibers and there was a tendency for gummosis in the parenchymal cells (106). In Wisconsin, complete defoliation reduced spur elongation by 75%. Complex series of partial leaf defoliation were evaluated mostly for effects on bud production (140).

Larch.—In Virginia, eastern larch was defoliated for one or two years when the foliage was half, mostly, and completely developed. Growth reduction was about proportional to the severity of defoliation, greatest for early defoliation, and more equally distributed between top and base of tree than with similar studies on pines (34). Other studies are covered in the section on larch sawfly (56, 73). In Russian studies using a dial gauge dendrometer, four different defoliation treatments on European larch caused reductions in diameter increment proportional to defoliation intensity. Complete defoliation caused total and immediate cessation of diameter growth (26).

Other deciduous trees.—The treatments in the last mentioned larch study were also used on *Betula verrucosa* giving similar results (26). In Wisconsin, complete defoliation of yellow birch, ironwood, and basswood saplings caused 30, 20, and 0% tree mortality and 78, 76, and 8% bud mortality (55). Rumanian studies on *Quercus robur* showed that complete defoliation reduces diameter increment 40–50% and causes extensive new shoot mortality on 10–20% of the trees. Two years of defoliation caused total mortality of new shoots on 10–30% of the trees and partial mortality of new shoots on 20–30% of the trees (107). In Russia, complete and upper crown defoliation of oaks in spring reduced diameter growth by 75 and 38% and height growth 60 and 43%. Summer replications caused 60 and 42% diameter and 18 and 0% height growth reductions (164). In other Russian studies, *Fraxinus pennsylvanica* was defoliated at several different dates. Only summerwood growth was reduced by defoliation (127). See phasmatid section for *Eucalyptus* (108), p. 303.

DEFOLIATION BY SAWFLIES

Scots pine defoliated by the European pine sawfly, Neodiprion sertifer.—In Sweden, heavy defoliation of old foliage for one and two years reduced diameter growth 39 and 52% (44). In Hungary, light, medium, and heavy defoliation caused 10–20%, 20–30%, and 30–45% reductions in annual growth. Losses were determined by comparing Scots pine growth with growth of nonhost pines (83). In Michigan, 20, 65, 85, and 100% defoliation before new foliage developed caused 14, 23, 37, and 63% loss in terminal shoot elongation and 18, 47, 53, and 71% reduction in radial increment. The 100% defoliation was simulated two weeks before maximum sawfly defoliation occurred. Trees survived three years of complete defoliation. Data are also given for increment reduction associated with two years of defoliation and twig damage (184).

Scots and Austrian pine defoliated by Diprion pini.—In Holland, 20, 50, and 85% defoliation of Scots pine caused 21, 41, and 57% radial increment reductions at 1.3 meters and 26, 44, and 60% losses at a point one-half way up the tree. Mortality was only 2–3% (104). In Yugoslavia, Austrian pine that was heavily defoliated for two years produced 71 and 86% less foliage and 69 and 90% less height growth in the following two years. Mortality, diameter growth loss and lower cone yields were also noted (185). In Russia, there was no diameter growth on pines in the years following 85% defoliation (115).

Shortleaf and Virginia pine defoliated by the Virginia pine sawfly, Neodiprion pratti pratti.—In Virginia, two years of 55% defoliation caused 33 and 16% increment losses, but trees survived and regained the normal rate of growth one year later (116, 117).

Red pine defoliated by the red pine sawfly, Neodiprion nanulus nanulus.—In Wisconsin, height was reduced 21 and 64% after one and two years of complete defoliation. No mortality occurred (80).

Loblolly pine defoliated by Neodiprion taedae linearis.—In Arkansas, a loss in increment of 51 and 28% occurred in the first and second years following 75% defoliation (174)[2].

Jack pine defoliation by Neodiprion swainei.—In Quebec, missing and incomplete growth rings were common. The author questions the use of conventional ring counting procedures (126).

White and black spruce defoliated by the European spruce sawfly, Diprion hercyniae.—In Canada, loss of 20–50% of the old foliage caused measurable losses. Successive defoliations for 13, 7, and 4 years resulted in increment losses for 17, 12, and 10 years. In white spruce, 4–7 years of 90% old and 50% new foliage loss caused little mortality, but trees that lost 95% old and 75% new foliage frequently died. Black spruce died from lower rates of defoliation than did white spruce (139).

White fir defoliated by the white-fir sawfly, Neodiprion abietis.—In California, trees with more than 65% defoliation during the outbreak lost about 66% in radial increment. Recovery was not evident two years later, but mortality was rare and mostly limited to intermediate and suppressed trees (158).

Shortleaf and loblolly pine defoliated by the red headed pine sawfly, Neodiprion lecontei.—In North Carolina, 84% of the 1–5-foot loblolly and shortleaf pine in open areas died after 76–100% defoliation. In shaded areas, trees were killed with as little as 25% defoliation (12). Complete defoliation of shortleaf pine in Illinois rarely killed trees. Defoliation by the first

[2] U.S. Department of Agriculture, Southern Forest Experiment Station 1948 Report and Forest Research Note 60 (1949) give identical data for *Neodiprion banksianae.*

and second sawfly broods caused about 60 and 40% reductions in radial increment (15).

Jack pine defoliated by Neodiprion rugifrons.—In Wisconsin, first generation sawflies consumed only old foliage and caused little damage, but second generation sawflies consumed new foliage and killed 75% of completely defoliated trees. Trees with only 10% of the foliage left had dead tops and twigs (181).

Norway spruce defoliated by Lygaeonematus abietinus.—In Denmark, light, medium, heavy, and very heavy defoliation damage caused 24, 32, 32, and 32% reduction in basal area increment and 21, 50, 55, and 71% reduction in height growth. Since older larvae feed only on new foliage, branch and top killing was common (131).

White spruce defoliated by the yellow-headed spruce sawfly, Pikonema alaskensis.—In Minnesota, defoliation of 80% or more of new and one-year old foliage on 1.5- to 2.5-meter tall trees caused about 60% reduction in terminal shoot elongation in the first year and 50% in the following year. Two years of similar defoliation caused about 80% reduction. Shoot elongation of branches were more drastically reduced. Trees with 90–99% of all foliage consumed rarely died. However, old records of spruce mortality are cited (95).

Japanese larch defoliated by Cephalcia alpina.—In Holland, 85, 50, and 20% defoliation caused 50, 20, and 10% radial increment loss at 1.3 meters and 35, 20, and 8% radial increment loss at a point one-half way up the tree. Increment losses were less in trees growing in good loamy soils than in trees on sandy soils. Economic analyses are given (104).

Eastern larch defoliated by the larch sawfly, Pristophora erichsonii.—In Manitoba, simulated sawfly defoliation rates of 25, 50, and 70% for three years reduced volumetric increment during the third year by approximately 60, 75, and 88% of the predefoliation growth rate. Reductions in foliage weight followed similar patterns, with 70% defoliation reducing foliage weight by 50%. Measurable effects on root mortality were detected only after four years of 70% defoliation (73). Similar results were found in Michigan, but increased radial increment was reported in the base of trees during the year of defoliation (56). Also see larch in artificial defoliation section (26, 34).

In Minnesota, increment loss from natural defoliation was measured by comparing the growth of sawfly-defoliated larch with black spruce which is not defoliated by the sawfly. On this basis, larch suffered an 18–84% reduction of increment from two years of defoliation (38). Since black spruce and larch differ in their moisture requirements, these increment reductions could be partly confounded with tree responses to available moisture (122). In Canadian studies using historical records of defoliation, a relationship between intensity of defoliation and losses in increment was demonstrated at several locations. Reductions up to 30% were frequent (122). By using the Duff-Nolan method, reduced stem diameter increment in defoliated crowns was demonstrated with the oblique pattern (119). The same workers later found that both defoliation and flooding produced similar growth patterns (122). In Minnesota, a relationship was noted for previous defoliation, radial increment, needle length, and production, and length and number of new shoots (23).

Several papers make reference to accounts of trees surviving after six years of defoliation with increment losses of 80% or more (38, 59, 122). Preliminary studies on the effect of thinning on the growth of defoliated trees were inconclusive (120).

One to three years of artificial defoliation in Virginia and Michigan did not cause mortality (34, 56). In Manitoba, 50% of the trees died after two years and about 80% after three years. Three seasons of partial defoliation did not cause mortality (73).

In Manitoba, stands with natural defoliation histories of six years of heavy, four years of heavy plus two of medium, and two years of heavy plus two of medium were associated with mortality of 26–40, 21, and 5% of the trees in the two years following defoliation (166, 167). In Minnesota, defoliation rates of 75–100% in three or four of the last four years of an infestation resulted in 22–29% mortality. Lesser defoliation caused very little mortality (13). In Canada, three to four years of severe defoliation can cause tree mortality in poorly drained bogs, but trees on well-drained sites can withstand six to eight years of equal defoliation (122). In Ontario, trees survived nine years of heavy defoliation but, afterward, up to 30% of the living and dead trees showed signs of attack by *Dendroctonus simplex* (61). Outbreaks of this bark beetle are known to occur on trees previously attacked by the larch sawfly (40). In the reviews of the old and often poorly quantified records of larch mortality following defoliation, there is a general consensus that six or more years of heavy defoliation will frequently cause mortality (38, 59, 122). In English studies on defoliated European larch, growth rings were formed in the living crown every year until the trees died. However, growth rings were missing at the base of trees for one or more years before death (65).

Defoliation reduces the production of new shoots which are used by the sawfly for oviposition (23, 38, 39). Fewer eggs/shoot were found in old defoliated sites which have shorter shoots than new sites. The effects of these items, however, are tempered by the production of epicormic shoots and shoots arising from foliage short shoots after several years of defoliation (67, 121).

DEFOLIATION BY PHASMATIDS

Eucalyptus defoliated by Didymuria violescens.—In Australia, moderate defoliation of pole-size *Eucalyptus delegatensis* did not affect diameter growth during the defoliation period, but summerwood was reduced by 89%. In the following season crown recovery was good, but diameter increment was reduced by 50%. Damage from partial (top one-third of tree) and complete simulated phasmatid defoliation, presented in graphs, showed large growth losses with January defoliation causing greater losses than May treatments. Complete defoliation killed 83% of the trees defoliated in January, but all of the trees defoliated in May lived (108). The summerwood of trees lightly defoliated by phasmatids was reduced 13% in the defoliation year and increased 13% in the following year. The summerwood of the defoliation year and the springwood of the following year were sometimes almost absent (110). This growth pattern was used to date past phasmatid infestations (137). With *Eucalyptus regnans,* two years of defoliation killed 83% of the trees and reduced diameter growth for two years in the surviving trees (109). Effects of defoliation and refoliation on starch reserves in *Eucalyptus* has been extensively reviewed (6).

Oaks defoliated by the northern walkingstick, Diapheromera femorata.—In mixed oak stands in Michigan defoliated in seven to nine alternate years, 30% of the trees in the black oak group were decadent and 49% were dead. There was no mortality and insignificant decadence in white oaks and other species not commonly defoliated by the walkingstick. Two-, 4-, 6-, and 8-inch diameter trees suffered 72, 44, 23 and 6% mortality (58).

DEFOLIATION BY THE JACK PINE BUDWORM, *Choristoneura pinus*

Growth loss.—In Minnesota studies, new and old foliage losses were recorded separately, x-sections were cut at every 5-foot interval from the base of trees, summerwood and springwood were measured separately, and non-defoliated control trees were used. Springwood of the defoliation year, 1956, was not affected by defoliation, but 1956 summerwood ring and both springwood and summerwood rings in 1957 and 1958 were reduced proportionately to the severity of 1956 defoliation. Growth, expressed in area of the annual rings, was reduced in light, medium, heavy, and very heavy defoliation classes in 1956 summerwood by 32, 60, 83, and 99%; in 1957, springwood by 54, 76, 99, and 99%; and in 1957, summerwood by 27, 44, 73, and 91%. In 1958, springwood increment was reduced 52 and 86% from heavy and very heavy defoliation and summerwood increment by 20 and 86% (97). In other Minnesota studies, progressive and regressive trees with moderate defoliation produced an average of 55 and 61% less volume than lightly defoliated trees during the first two years after one year of defoliation (74).

In Ontario, radial growth of Scots pine was conspicuously reduced about

50 and 80% during the first and second year of heavy defoliation as shown on graphs developed from the Duff-Nolan method using the oblique sequence (135). In Minnesota, 1956 shoot growth of red pine was not affected by 1956 defoliation, but in 1957 there was a highly significant correlation between length of shoot growth and 1956 defoliation (96).

Mortality and top-killing.—In simulated budworm defoliation studies, complete defoliation killed both jack and Scots pines, but when only the new foliage was removed, all jack pines died and one of the four Scots pines died (34). Several surveys of top-killing and mortality of over- and understory jack pine from budworm defoliation have been made (16, 57). In Minnesota, top defoliation caused trees to develop rounded or flat tops. Medium and heavy defoliation in a young pole-sized stand caused 2–6% mortality in progressive and provisional trees and 9–13% in regressive trees during the first two years after defoliation. Twenty-nine to 44% of almost completely defoliated trees died (10). Red pine trees that were defoliated 70% or less did not suffer top killing, but trees that were defoliated 80% or more had crown killing correlated with the length of the crown defoliated (96).

Male cone production.—Budworm survival in the spring is benefited by male cones. Studies in Ontario (68) and Minnesota (97) have shown that defoliation reduces the production of male cones in subsequent years. In the latter study, six defoliation categories from very light to very heavy yielded cone productions of 32, 60, 40, 12, 0, and 0% in the third year after defoliation.

DEFOLIATION BY THE SPRUCE BUDWORM, *Choristoneura fumiferana*

Growth loss.—Spruce budworm outbreaks have occurred since the 1700's as shown by comparing growth ring patterns in spruce and fir with ring patterns in tree species not subject to budworm attack (extensively reviewed in 20). The vast number of studies on mortality and growth loss that are not directly correlated with defoliation intensity or duration, or both, are not reviewed. Projections of growth and mortality losses in spruce-fir forests from such studies are of little value since the causal variable, defoliation, is not accounted for in the analyses. Since heavy male cone production reduces tree growth and is often associated with heavy defoliation (118), care must be taken to avoid confounding the effects of cone production with the effects of defoliation.

As early as 1922, studies in pure coniferous stands in New Brunswick and Quebec demonstrated 33–50% less increment in red and white spruce and balsam fir in the ten-year period after the start of heavy defoliation as compared to the previous decade. Lower increment losses occurred on balsam fir and red spruce in a mixed northern hardwood stand. Pure stands with flat-top trees suffered a 50% growth loss. Defoliation estimates were not tied directly to increment losses. Two to three annual growth rings were often missing in defoliated balsam fir that survived. Therefore, the starting point for the ten-year postdefoliation period was identified by the larger-than-normal growth rings formed during the first year of defoliation at the base of the tree (33). Recent studies have shown that larger rings also occur at the base of nondefoliated trees, which suggests coincidence with environmental factors (17) or "typical" growth patterns (119). Missing rings were common in all tree species prior to death, but in trees that survived, missing rings were common only in balsam fir (33).

In Ontario studies it was demonstrated that growth suppression did not occur until two–four years after heavy defoliation of the new shoots. A "growth ratio" between nondefoliated jack pine, red pine, and balsam fir and defoliated balsam fir and white spruce was used. On the basis of these ratios and tables of defoliation, I calculated a 35, 57, 81, and 91% reduction in radial growth during the third to the sixth year of complete defoliation of the new needles in plots 5 and 6. Specific defoliation records for white spruce in the same plots were not available, but the reductions in growth rate were 31, 44, 44, 60, and 70% for the second to sixth years after the start of heavy defoliation on the associated balsam fir in the same stands (17).

In Quebec, growth reductions of 60–75% occurred in balsam fir during a five-year period of depressed growth compared to the previous five years. The growth reduction started three–six years after the start of heavy defoliation. All of the current foliage was consumed for about six years. The percentage of growth loss was greater on fast-growing than on slow-growing trees. From similar measurements of five-year ring groups made elsewhere on the bole where live branches were absent, it appeared that measurements at 4.5 feet were represntative except in the live crown where they were consistently 30% greater (112). In another study in the same infestation area, diameter growth was reduced in the tops of trees and increased at the base during the first year of defoliation. During the postdefoliation period, the growth rate in the tops of trees was greater whether or not the trees eventually died. Summerwood was greatly reduced or missing (160). In Ontario, defoliation and growth loss graphs indicated that during a four-year period of defoliation of 40, 80, 40, and 20% that a growth loss of 25% was observed during the third year followed by losses of 37, 45, and 55% in the three subsequent years. Other graphs in this study and elsewhere (14, 19, 163) on growth reductions in balsam fir and white spruce were not suitable for quantitative interpretations of growth loss or defoliation, or both.

Ontario studies, using a dendrometer to record accumulated diameter growth, showed that the effects of current foliage defoliation of balsam fir were expressed mostly after the first of July (14). With the Duff-Nolan method, the oblique sequence showed progressive reduction throughout the

entire stem and reflected growth reductions during the first year after 65%
defoliation of the new foliage occurred. Other growth sequences were not
sensitive to growth reductions until the second year. Quantitative measurements of growth loss were not presented (119). In a recent Minnesota
study, check plots were randomly selected and protected with insecticides.
Defoliation was recorded annually on each tree with new and old foliage
considered separately. The net periodic increment between defoliated and
check stands differed by 54 square feet of basal area and 1014 cubic feet per
acre of balsam fir. Significantly less height growth occurred on the defoliated plots, crown length and width were reduced, and crown position frequency was changed. Individual tree ring volume was not significantly affected until the second year of defoliation and production of new shoots
originating from normal as well as adventitious buds was stimulated by defoliation. (H. O. Batzer, to be published in 1971).

Tree mortality.—Most of the old and poorly quantified records of
mortality related to spruce budworm defoliation have been reviewed (14,
17, 20, 45, 160). In Quebec, heavy, medium, and light defoliation (⅔, ⅓–⅔,
⅓ of all foliage) yielded 89, 62 and 0% mortality within the next four
years. Although some defoliation was not recorded in some years, the correlation with defoliation intensity was convincing. Vigor, crown class,
crown ratio, and stem diameter at 4.5 feet were also recorded and related to mortality. Since they were not correlated with defoliation, the
relationship had little predictive value. In trees defoliated from 1944–1949
with 100% of the new foliage consumed in 1945, mortality started in 1948
at 1% and was 16, 26, 46, and 39% during the next four years (112). In
Ontario, 4–6, 7, 8, and 9+ years of complete defoliation of the current
foliage of balsam fir caused 2–14, 18–43, 26–66, and 51–100% tree mortality
(45). In other Canadian studies, trees with 100, 90, 75, 50, and 25% of
their foliage gone at the end of the infestation period suffered 100, 100, 82,
56, and 28% mortality. Trees with less than 25% of the foilage missing did
not die. The relationship between defoliation, winter killing, and secondary
insects is discussed (160). In New Brunswick and Quebec, growth in the
ten-year period before and after defoliation was recorded. Flat-top balsam
fir and white and red spruce killed by defoliation had a predefoliation
growth rate that was only 60, 75, and 48% of the average of all trees in
the plots. Complex relationships of mortality, growth rate, diameter at
4.5 feet, and stand composition are discussed in detail. Unfortunately, defoliation history was inadequately stated (33).

In Ontario, tree mortality, in stands with an intermediate production of
male cones, started in the fifth year after the start of heavy (75% +) defoliation of new foliage. The cumulative defoliation was 350%. Once
started, mortality increased rapidly with total mortality occurring in the
ninth year with a cumulative defoliation of 800%. In all stands it took a

maximum of two years for mortality to increase from 25–75%. Once the cumulative defoliation had been sufficient to cause some 4-inch diameter trees to die, subsequent mortality was, to some degree, independent of subsequent defoliation. The cumulative defoliation needed to cause specific percentages of tree mortality was less for poor sites and for trees producing many male cones. The number of years from the start of heavy defoliation to the start of tree mortality was not changed by the occurrence of one year of light or moderate defoliation. If trees had old and new foliage eaten, they died sooner than trees only missing new foliage (18).

In Minnesota, studies in which check plots were randomly selected and protected with insecticides, the cumulative defoliation during the infestation period was significantly correlated with the number of dead and damaged trees. The net mortality during the five-year postmortality period was 68% of the original stand, with the smaller trees dying first. Mortality in the check plots was largely restricted to suppressed trees. (H. O. Batzer, to be published in 1971; also see 19). In New Brunswick, four years of over 90% defoliation plus three years of 34–48% defoliation caused 30 and 40% mortality of dominant and codominant balsam fir. There was a decrease in average tree height from top mortality in the last two years of defoliation (8).

In Ontario, mortality of white spruce occurred during the years of defoliation or immediately after, but before bark beetles became prevalent. Later mortality of trees that would otherwise have survived the defoliation, was caused by bark beetles what reached an epidemic level in trees killed by defoliation. Outbreaks of *Dendroctonus piceaperda* may be a natural development in budworm-defoliated balsam fir-spruce stands (163). In Ontario, balsam fir started to die five years after all new foliage was gone and complete mortality occurred after eight years of defoliation. Extensive studies of bark beetles and borers indicated that they did not kill trees that might otherwise have survived. The literature on secondary insects is thoroughly covered (14).

Top and root mortality.—Since budworm defoliation is concentrated in the tops of trees, terminal mortality is common. Stands with 10% of the trees dead had 75–100% of the survivors with five or more feet of the terminal dead (160). Recent Minnesota surveys showed that up to 78% of the balsam fir had two or more feet of the top dead after three years of complete defoliation of new growth (143). In New Brunswick, topkill was often found in trees that survived the 1912–20 budworm outbreak. After 20 or more years, all trees with overgrown dead leaders one-half inch or more in diameter and with at least five annual rings were associated with heart rot (157).

The effect of defoliation on root mortality is well established (160). Seventy and 100% defoliation of new shoots of balsam fir caused >30 and >75% rootlet mortality. When defoliation was reduced, young trees immediately produced new rootlets, but mature and over-mature trees did not produce new rootlets even though the trees would partially refoliate before death (138).

DEFOLIATION BY THE LODGEPOLE NEEDLE MINER, *Evagora starki*

Canadian studies using a short-cut Duff-Nolan growth sequence analysis showed that increment losses in lodgepole pine occur only after 50% defoliation in young trees and after 40% in mature trees. Forty percent defoliation also marked the start of reduced lateral and terminal shoot elongation (156). In other studies with increment losses of 21–75%, an increase in percentage of increment losses was correlated with the length of time that defoliation exceeded 40% (32). Studies on the Duff-Nolan method showed that vertical sequences could show the earliest sign of increment reduction (119). Mortality was insignificant in all Canadian studies (32, 156). In California there was extensive mortality in needle miner-weakened trees that were attacked by *Dendroctonus monticolae* (128).

DEFOLIATION BY *Dendrolimus* SPP.

In Russian studies with *Dendrolimus sibiricus*, mortality of Siberian fir from 90–100% defoliation was high, requiring salvage within one to two years. In 70–80% of defoliated stands, salvage was needed in two to three years. Borer attack was usually less important in trees defoliated less than 60% (103). In Japan, simulated *Dendrolimus spectabilis* defoliation of Japanese red pine caused an immediate reduction in diameter increment. Height increment losses were not detected until the following season. Both height and diameter increment were reduced 50–60% from 90% defoliation and 10–20% after 60–70% defoliation. Detailed graphical data are given (52). In other Japanese studies, 40, 60, and 100% natural defoliation caused an average of 12, 20, and 50% reduction in diameter growth and 5, 22, and 42% reduction in height growth (51).

DEFOLIATION OF CONIFERS BY MISCELLANEOUS LEPIDOPTERA

Douglas-fir, Engelmann spruce, and grand fir defoliated by the western spruce budworm, Choristoneura occidentalis.—Graphs in British Columbia studies show about 33, 50, 50, and 60% reductions in growth starting after second-year defoliation. Total defoliation ranged from 20–60%. Relationships between defoliation and growth were complicated by changes in shoot defoliation and age composition of foliage as the infestation progressed (151). In Oregon it was shown that various damage sign categories were correlated with top killing and growth loss with the greatest growth reduction occurring in the upper crowns (182, 183).

Western hemlock, balsam fir and Sitka spruce defoliated by the black-headed budworm, Acleris variana.—In Alaskan studies with one year of heavy defoliation, top kill in dominant, codominant, and intermediate hemlocks was 85, 52, and 10%, and for spruce top kill was 93, 86, and 71%. Data are given on many related damage items (111, 168). Detection of previous outbreaks by ring analysis was studied in British Columbia (134). Two years of heavy defoliation of balsam fir in eastern Canada caused neither top killing nor marked increment reduction. Destruction of new shoots resulted in production of adventitious shoots on the two-year-old shoots (113).

Western hemlock, Douglas-fir, amabilis fir, and Sitka spruce defoliated by the hemlock looper, Lambdina fiscellaria lugubrosa.—In British Columbia 80–90% defoliation caused 65–78% tree mortality, but 50–75% defoliation caused only 10–25% mortality. Five to 10% of the hemlock and Douglas-fir died from less than 45% defoliation (82, also see 134). In Washington studies, over 60% of the hemlock died within three years after 70% defoliation (77) (see 76 for economic analysis).

Balsam fir defoliated by Lambdina fiscellaria fiscellaria.—Trees usually died in the year following complete defoliation. Diameter increment of surviving trees was reduced by more than 50% (176).

White fir defoliated by the Douglas-fir tussock moth, Hemerocampa pseudotsuga.—In California, defoliation had an immediate effect on radial growth, with the upper crown being most affected. Heavy, moderate, and light defoliation caused 74, 67, and 31% reduction in radial increment and 50, 15, and 0% mortality in saplings and small poles. In sawtimber, there was 20% mortality from heavy defoliation. Borers and bark beetles were associated with 95% of the sawtimber mortality. In poles and saplings they were associated with only about one-third of the trees. Top killing occurred in 12% of the heavily defoliated trees. The caterpillars feed on new foliage first (178, also see 180).

Spruce defoliated by Zieraphera diniana.—In Germany, 16% loss in total increment followed complete defoliation of May shoots for two years and partial defoliation for two years. The growth on defoliated trees was greater than on foliated trees during the first two years of defoliation but steadily decreased in the third and fourth years. In the first year after defoliation, the growth rose to 50% of the average predefoliation growth rate. Seed production occurred only on nondefoliated trees (24).

Spruce defoliated by the nun moth, Lymantria monacha.—In Rumania, heavy defoliation caused 50–80% loss in radial increment and 10% mortality. Defoliation was concentrated on dominant trees. In response to increased illumination, codominant and suppressed trees had a greater diameter increment than did dominant trees (132).

Mexican weeping pine defoliated by Euproctis terminalis, Buzura abruptaria, *and* Orgyia basalis.—In South Africa, *E. terminalis* reduced volume 22% during four years following the first year of defoliation. Complete defoliation occurred only during the third year after which no defoliation was observed (62). In Rhodesia, complete defoliation by *Buzura* and *Orgyia* caused a reduction in volume of 70 and 40% (in 3).

Ponderosa pine defoliated by the pine butterfly, Neophasia menapia.—In Idaho, with growth losses shown graphically, defoliation caused over 70% increment reduction and 26% mortality of mature trees. Many trees did not add basal increment for 1–11 years (47). In more recent studies using the Duff-Nolan principle, the reduction of annual increment was 39%. The mortality was insignificant (31).

European larch defoliated by Coleophora laricella *in Germany.*—Trees suffered 33–45% reduction in diameter increment from about 40% defoliation (148).

Red pine defoliated by the red pine midge, Thecodiplosis piniresinosae *in Wisconsin.*—Loss of 75% of the new foliage at the end of the season caused terminal and lateral shoot mortality on some trees. Losses in terminal shoot elongation are shown in graphs (81).

Western hemlock defoliated by Melanolphia imitata.—In Canada, 90% defoliation caused tree mortality or top killing. Some top killing occurred in trees with less than 50% defoliation (150).

DEFOLIATION BY THE GYPSY MOTH, *Porthetria dispar*

Growth loss.—The average defoliation rate of red, scarlet, black, and white oaks over a ten-year period was 37% and the average growth reduction was 34% compared to the predefoliation decade. In years of no defoliation there was a 24% growth reduction and in years of 100% defoliation, a 52% reduction. Using the ratio of 100/52, the 37% defoliation should yield a 34% growth decline—only 4% less than the measured 38%. Since the measurements were made on dominant trees that survived and still were in "good condition," the study underestimates the average growth reductions (114). In other long-term New England studies, 21–40, 41–60, 61–80, and 81–100% defoliation rates were related to 9, 20, 19, and 30% reductions in the radial growth of four species of oak and 0, 15, 27, and 13% reductions in white pine using the 0–20 defoliation class as controls. The relationship of defoliation, growth loss, and rainfall are presented graphically (4).

In a recent study, one year of complete defoliation reduced diameter growth in white pine, eastern hemlock, and oaks 14, 40, and 24% during the following year, and the yearly average losses for the five-year postdefoliation years were 16, 28, and 7% for each species. Reductions were based on the growth during the previous year or five years and adjusted for the growth decline recorded in trees that were not defoliated. Lighter defoliation rates gave proportional reductions. After five years, 31% of the white pine that was completely defoliated had a normal amount of foliage and 47% were missing less than 20%. Tables are given for foliage recovery at various defoliation levels for white pine and eastern hemlock (70).

In Hungarian studies on several species of oaks, three to four years of complete defoliation reduced annual increment by more than 50% and reduced wood quality (169). In Rumania, complete defoliation of oaks reduced increment by 30–40% (25). Similar reports of losses in Eastern Europe are reported in (164) and Forestry Abstracts 22, No. 814, 815; 23, No. 3971; 25, No. 996; 27, No. 2501; 28, No. 364, 4197, and 29, No. 5992. Also see artificial defoliation section (107, 164), pp. 296–99.

Tree mortality.—Oak mortality in the older studies is difficult to separate from normal mortality, but losses of 50% of the dominant white oaks in the year following complete defoliation were impressive. However, there were records of a white oak that survived five complete strippings. Over a ten-year period of unstated defoliation, 50% of the oaks died (114). *Agrilus bilineatus* was associated with the dead trees in another study with similar losses (4).

Young white pine losing 100% of the old foliage and 41–61, 61–80, and 81–100% of the new foliage, suffered 12, 11, and 32% mortality (114). In recent studies, defoliation of over 90% of the white pine foliage caused 11, 16, and 42% mortality of the dominant and codominant, intermediate and over-topped crown classes of white pine. With eastern hemlock, 68, 92, and 42% of the trees died in the three crown classes. Only 5% of the oaks died (70).

Similar losses on several tree species are described in a review of two unpublished and poorly quantified evaluations of losses related to gypsy moth defoliation (165).

DEFOLIATION BY THE FOREST TENT CATERPILLAR, *Malacosoma disstria*

Growth loss.—Small studies with a dendrometer in Michigan showed that complete defoliation of trembling aspen immediately reduced growth. Defoliation rates of 25 and 100% reduced growth 38 and 67% (36). In Ontario, frost and caterpillar defoliation gave similar growth responses. Partial and complete defoliation showed 40 and 80% growth reductions on published dendrographs (141). In New Brunswick, the growth rate of 60-year-old aspen was not affected during the first week after complete defoliation of fully developed leaves. However, growth ceased during the following week and resumed after refoliation started. About 30% of the growth occurred after refoliation. The leaves were one-half grown when defoliation occurred in the second year and most growth occurred after refoliation. In the third year, the first-year pattern was repeated. The growth reduction was 42, 52, and 77% for the three years and was below normal during the next year without defoliation. White birch was more severely affected with 23, 87, and 52% defoliation yielding 22, 68, and 86% growth loss. The difference is probably related to later defoliation and less refoliation (7). In other studies in which the growth pattern of associated white spruce was used as a basis for projecting aspen growth patterns, an 8.4% growth loss was measured in the second year of heavy defoliation. The growth rate returned to about a normal level in the following defoliation-free year (69). In Minnesota, first year light (L); first year moderate (M) plus second year L; first year heavy (H) plus second year M; and second plus third year H defoliation caused about 10, 47, 78, and 88% reduction in radial growth (9, 11). In other Minnesota studies on aspen 40 years old and younger, it was shown that L, H, H followed by H, and L followed by H and H caused 16, 72, 79, and 87% reduction in basal area growth. During the first year after H, first year after H and H, and first year after H, H, and L, the growth reductions were 22, 16, and 16%. During the second year after defoliation the losses were not significantly different from the predefoliation averages. Careful analysis showed that the reductions were due to defoliation and not coincident rainfall differences (43, 49, 50, all covered in detail in 42).

In Ontario, larval feeding was always completed when the foliage was about 50% developed and shoot elongation was about two to three weeks from completion. The last two to three weeks of shoot elongation accounted for continued foliation (apparent refoliation) of trees stripped of leaves. Refoliation occurred only when complete defoliation was accompanied by the destruction of developing shoot tips. Frost defoliation occurred before diameter growth started. Growth was therefore delayed until refoliation occurred (141, also see 101). In Minnesota, the foliage was fully developed by the end of the feeding period and true refoliation occurred. Refoliated leaves were smaller than normal and subnormal leaves were found in about 90 and 25% of the trees in the first and second years after three years of defoliation (42).

Tree and twig mortality.—In most studies tree mortality caused by defoliation was negligible (9, 11, 36, 42, 54, 69, 141). In Minnesota, fungus cankers and wood borers did not appear to become more abundant on defoliated trees (42). However, follow-up studies in these plots six to ten years after the end of defoliation showed that mortality caused by insects, *Nectria* and *Hypoxylon* cankers, and "unknown" causes, was greatest in stands that had three years of heavy defoliation. *Hypoxylon* was the only mortality agent that showed a strong graded relationship to defoliation intensity and longevity (28). In New Brunswick, with trees 20 or more years older than in the Minnesota stands, 29% of the aspen and 21% of the white birch died. *Argrilus* spp. were important contributing factors to the mortality (7).

In Minnesota, shoot mortality was less than 5% of the crown in 95% of the trees (42). In other studies, however, one, two, and three years of heavy defoliation caused 2, 5, and 14% crown die back on high vigor trees and 8, 9, and 25% in low vigor trees (11). Two years after two to three seasons of heavy defoliation, more than 80% of the trees had dead twigs (9).

DEFOLIATION OF HARDWOODS BY MISCELLANEOUS LEPIDOPTERA

Oaks defoliated by the winter moth, Operophtera brumata *and by* Tortrix viridana.—In Nova Scotia studies on winter moth defoliation of red oak, cumulative defoliation of 100, 200, 300, 400% caused about 1, 5, 15, and 40% tree mortality and 15, 33, 50, and (no data)% branch mortality. Cumulative defoliation of 50, 100, 200, and 250% reduced basal area growth by 3, 6, 11, and 14%. The relationship of flushing time and tree damage is discussed (46). In England, the combined defoliation activities of both insects caused about a 60% reduction in summerwood increment, but insignificant differences in springwood (170). The effects of defoliation by *T. viridana* (147) and both defoliators (164) has been discussed in recent Russian studies. Also see artificial defoliation section (107, 164), pp. 296–99.

European beech defoliated by Dasychira pudibunda *in Germany.*—Trees suffered 7 and 13% diameter increment losses after 90% defoliation for one and two years. Economic analysis indicated that chemical control was not justified (144).

Sugar maple and ironwood defoliated by saddled prominent Heterocampa guttivitta *in Wisconsin.*—Complete defoliation for three years caused 45 and 11% mortality in maple and ironwood trees, and 34 and 5% mortality in saplings. Other trees were less drastically damaged (55).

Poplars defoliated by Pygaera anastomosis *in Italy.*—Trees suffered 12 and 30% reduction in volumetric increment from one and two defoliations in the same year. With two defoliations, all twigs were killed and epicormic branches were common, but trees rarely died (2).

Wattle defoliated by the wattle bagworm, Kotochalia junodi *in South Africa.*—With light-to-heavy defoliation rated 1–7, cumulative defoliations of about 10–20 over a ten-year period reduced diameter growth by about 2–3 inches and bark yield about one ton/acre. Cumulative defoliations under 5 had little effect. The depressing effects of defoliation were independent of

California Baptist College
Riverside, California

tree age and number of years between defoliation and harvesting. Defoliation reduced diameter growth and bark yield more on poor sites than on good sites (149).

Sagebrush, *Artemisia* spp., defoliated by *Aroga websteri* suffered 20, 30, 40, and 95% mortality from 25, 50, 75, and 100% defoliation (64).

Heavy defoliation of 2 brush species by the Great Basin tent caterpillar, Malacosoma fragile.—Heavy defoliation caused very little mortality, but 12–40% of the twigs died (179).

DEFOLIATION RELATED TO HARDWOOD DECLINE CONDITIONS

Decline conditions of hardwood trees has been associated with defoliation and many other factors. Most of the decline symptoms, tree and twig mortality, reduced increment, and smaller and fewer leaves are common after effects of defoliation. See (63) and (88) for the effects of flowering.

The extensive world-wide literature on decline conditions in oak has been reviewed (154). The relationship between oak defoliation and tree and crown mortality and reduced increment has been correlated with defoliation during drought periods in extensive Pennsylvania studies. A complex of 5 major and 17 minor defoliators were involved (123).

A decline condition of sugar maple, maple blight, has been associated with a similar complex of factors. However, experimental and observational studies have shown that all symptoms can be duplicated with defoliation (55).

It appears that the symptom patterns of the decline conditions do not differ greatly from patterns observed in most studies on the effects of flowering and insect defoliation covered in this review. Possibly the refoliation phenomenon, the difficulty of detecting partial defoliation, and the subtle effects of late season defoliation in predisposing twigs to winter killing all tend to obscure the insect-related causes of the decline conditions. Russian (60) and German (145, 146) studies indicate that large populations of some defoliating insects build up only in stands of trees that are in a weakened condition. This should be considered in future studies on the various decline conditions.

SUMMARY

Defoliation and growth measurements relate to a complex of factors. Flower production and the size, abundance, food storage capacity, hormone production, photosynthetic efficiency, and respiration of foliage vary with tree age, species, season of the year, and the age, exposure, and crown location of the foliage. These must be considered when making quantitative statements of foliage loss because they affect foliage growth and production, tree survival, and the quantity and/or location of reduction in shoot and diameter growth. Variations in ring growth at different vertical levels in the trunk and the separate response of spring- and summerwood growth are especially important.

Trees that escaped defoliation could be poor experimental controls if the defoliator populations were affected by flowering, flushing time, or foliage qualities that independently affect tree growth or mortality. Increased illumination in defoliated stands or differences in the threshold of response to environmental factors by nonhost tree species or escapes, limit their use as controls. Pre- and postdefoliation growth measurements are confounded with cambium-age-related growth pattern. Protected control plots are recommended.

Almost all of the 174 studies relate growth loss proportionately to the quantity of foliage loss. The importance of foliage for growth and survival decreases directly with its age and inversely with its exposure. Evergreen conifers usually survive a single complete defoliation in spring before the elongation of the new foliage. Later defoliation that includes the new foliage often causes mortality. Pines and deciduous trees usually show an immediate response to defoliation if it occurs before shoot or ring growth is completed. Growth responses in spruce and firs are often delayed for one to several years. The first growth responses in the trunk are usually found within the living crown, but considerable variation is reported. Most deciduous trees survive several defoliations and produce new sets of leaves in the same season. Secondary agents, such as site, tree age, and climatic conditions influence the survival of defoliated trees.

Although extensive material was available for several insects, many of the studies suffer from insensitive defoliation and growth measurements.

SCIENTIFIC NAMES OF TREES

Aspen, trembling: *Populus tremuloides*
Basswood: *Tilia glabra*
Beech, European: *Fagus sylvatica*
Birch, white and yellow: *Betula, papyrifera* and *lutea*
Douglas-fir: *Pseudotsuga menziesii*
Elm, American: *Ulmus americana*
Firs, amabilis, balsam, grand, Siberian, white: *Abies, amabilis, balsamea, grandis, siberica,* and *concolor*
Hemlock, eastern and western: *Tsuga, canadensis* and *heterophylla*
Ironwood: *Ostrya virginiana*
Larch, eastern, European, and Japanese: *Larix, laricina, decidua,* and *leptolepis*
Maples, Norway, red, silver, and sugar: *Acer, platanoides, rubrum, saccharinum,* and *saccharum*
Oaks, red and white: *Quercus, rubra* and *alba* (red and white oak groups refer to subgenera *Erythrobalanus* and *Leucobalanus*)
Pines, Austrian, jack, Japanese black, Japanese red, loblolly, lodgepole, longleaf, Mexican weeping, red, Scots, shortleaf, Virginia, and white: *Pinus, nigra,*

banksiana, thunbergii, densifloria, taeda, contorta, palustris, patula, resinosa, sylvestris, echinata, virginiana, and *strobus*

Poplars: *Populus* spp.

Spruce, black, Engelmann, Norway, red, Sitka, and white: *Picea, mariana, engelmannii, abies, rubens, stichensis,* and *glauca*

Wattle: *Acacia mollissima*

ACKNOWLEDGMENT

I am indebted to T. T. Kozlowski, H. A. I. Madgwick, J. C. Gordon, E. I. Sucoff, I. Millers, L. D. Nairn, A. T. Drooz, H. O. Batzer, E. V. Bakuzis, and J. A. Witter for their help.

LITERATURE CITED

1. Allen, R. M. 1955. Foliage treatments improve survival of longleaf pine plantings. *J. Forest.* 53 :724–27
2. Arru, G. M. 1964–65. Studies on the morphology and biology of *Pygaera anastomosis* (L.). *Bull. Zool. Agrar. Bachicoltura* 2(6) :206–72
3. Austara, O. 1968. The economic consequences of insect defoliation of pines in East Africa. *East Afr. Agr. Forest. J.* 34 :203–6
4. Baker, W. L. 1941. Effect of gypsy moth defoliation on certain trees. *J. Forest.* 39 :1017–22
5. Bakuzis, E. V., Brown, R. M. 1962. Elements of model construction and the use of triangular models in forestry research. *Forest Sci.* 8 :119–31
6. Bamber, R. K., Humphres, F. R. 1965. Variations in sapwood starch levels in some Australian forest species. *Aust. Forest.* 29 :15–23
7. Barter, G. W., Cameron, D. G. 1955. Some effects of defoliation by the forest tent caterpillar. *Can. Dep. Agr. Forest Biol. Div. Bi-Mo. Progr. Rep.* 11(6) :1
8. Baskerville, G. L. 1960. Mortality in immature balsam fir following severe budworm defoliation. *Forest. Chron.* 36 :342–45
9. Batzer, H. O. 1955. Effects of defoliation by the forest tent caterpillar. *Entomol. Soc. Am. No. Central States Br. Proc.* 10 :27–28
10. Batzer, H. O., Bean, J. L. 1962. Spruce budworm defoliation causes continued top killing and tree mortality in northeastern Minnesota. *US Forest Serv. Lake States Forest Exp. Sta. Tech. Note 621.* 2 pp.
11. Batzer, H. O., Hodson, A. C., Schneider, A. E. 1954. Preliminary results of an inquiry into the effects of defoliation of aspen trees by the forest tent caterpillar. *Minn. Forest. Note 31.* 2 pp.
12. Beal, J. A. 1942. Mortality of reproduction defoliated by the red-headed pine sawfly (*Neodiprion lecontei* Fitch). *J. Forest.* 40 :562–63
13. Beckwith, L. C., Drooz, A. T. 1956. Tamarack mortality in Minnesota due to larch sawfly outbreak. *J. Forest.* 54 :268–69

14. Belyea, R. M. 1952. Death and deterioration of balsam fir weakened by spruce budworm defoliation in Ontario. *J. Forest.* 50 :729–38
15. Benjamin, D. M. 1955. The biology and ecology of the red-headed pine sawfly. *US Dep. Agr. Tech. Bull. 1118.* 57 pp.
16. Benjamin, D. M., Banash, S. E., Stewart, R. B. 1961. Losses attributable to the jack-pine worm during the 1955–1957 outbreak in Wisconsin. *Univ. Wisc. Forest Res. Notes 73.* 4 pp.
17. Blais, J. R. 1958. Effects of defoliation by spruce budworm (*Choristoneura fumiferana* Clem.) on radial growth at breast-height of balsam fir (*Abies balsamea* (L.) Mill.) and white spruce (*Picea glauca* (Moench) Voss.). *Forest. Chron.* 34 :39–47
18. Blais, J. R. 1958. The vulnerability of balsam fir to spruce budworm attack in northwestern Ontario, with special reference to the physiological age of the tree. *Forest. Chron.* 34 :405–22
19. Blais, J. R. 1962. Collection and analyses of radial growth data from trees for evidence of past spruce budworm outbreaks. *Forest. Chron.* 38 :474–84
20. Blais, J. R. 1968. Regional variation in susceptibility of eastern North American forests to budworm attack based on history of outbreaks. *Forest. Chron.* 44(3) :17–23
21. Bruce, D. 1956. Effect of defoliation on growth of longleaf pine seedlings. *Forest Sci.* 2 :31–35
22. Burger, H. 1951. Pruning, needle-stripping and increment of young spruce and pine. (German.) *Mitt. Forstl. VersAnstl, Mariabrunn No. 47* 8–16. (*Forest. Abstr. 13* No. 156)
23. Butcher, J. W. 1951. *Studies on larch sawfly populations, factors contributing to their fluctuations and early recognition of host decadence.* PhD thesis. Univ. Minn., St. Paul. 136 pp.
24. Capek, M. 1962. The effect of defoliation by *Zeiraphera diniana* Guen. on the annual increment of spruce. (German, French sum.) *Schweiz. Z. Forstw.* 113 :635–42

25. Cazacu, I., Fratian, A. 1966. The need for economic calculations in operations to control insect defoliators. (Rumanian, English sum.) *Rev. Padurilor* 81:466–72 (*Forest. Abstr. 28* No. 4138)

26. Chalupa, V. 1965. Influence of the reduction of leaves on the beginning and course of radial growth. *Commun. Inst. Forest. Czech.* 4: 61–73. (*Forest. Abstr. 27* No. 4580)

27. Church, T. W. 1949. Effects of defoliation on growth of conifers. *US Forest Serv. Northeast. Forest Exp. Sta. Pap. 22.* 12 pp.

28. Churchill, G. B., John, H. H., Duncan, D. P., Hodson, A. C. 1964. Long-term effects of defoliation of aspen by the forest tent caterpillar. *Ecology* 45:630–33

29. Clark, J. 1961. Photosynthesis and respiration in white spruce and balsam fir. *Syracuse Univ. State Coll. Forest. Tech. Publ. 85.* 72 pp.

30. Clausen, J. J., Kozlowski, T. T. 1967. Food sources for growth of *Pinus resinosa* shoots. *Advan. Frontiers Plant Sci.* 18:23–32

31. Cole, W. A. 1966. Effect of pine butterfly defoliation on ponderosa pine in southern Idaho. *US Forest Serv. Res. Note INT–46.* 7 pp.

32. Cook, J. A. 1961. Growth reduction in lodgepole pine defoliated by the needle miner, *Evagora (Recuvaria) starki* Freeman. *Forest. Chron.* 37:237–41

33. Craighead, F. C. 1925. Relation between mortality of trees attacked by the spruce budworm (*Cacoecia fumiferana* Clem.) and previous growth. *J. Agr. Res.* 6:541–55

34. Craighead, F. C. 1940. Some effects of artificial defoliation on pine and larch. *J. Forest.* 38:885–88

35. Derr, H. J. 1963. Needle clipping retards growth of planted longleaf pine. *Tree Planters, Notes* 57:31–33

36. Dils, R. E., Day, M. W. 1950. Effect of defoliation upon the growth of aspen. *Mich. Agr. Exp. Sta. Quart. Bull.* 33:111–13

37. Dochinger, L. S. 1963. Artificial defoliation of eastern white pine duplicates some effects of chlorotic dwarf disease. *US Forest Serv. Res. Note CS–16.* 6 pp.

38. Drooz, A. T. 1960. The larch sawfly, its biology and control. *US Dep.*

Agr. Tech. Bull. 1212. 52 pp.

39. Drooz, A. T., Meyer, D. 1955. Determination of the age of tamarack twigs and an indication of aborted twig elongation. *J. Forest.* 53: 454–55

40. Drouin, J. A., Turnock, W. J. 1967. Occurrence of the eastern larch beetle in Manitoba and Saskatchewan. *Manitoba Entomol.* 1:18–20

41. Duff, G. H., Nolan, N. J. 1953. Growth and morphogenesis in the Canadian forest species. I. The controls of cambial and apical activity in *Pinus resinosa* Ait. *Can. J. Bot.* 31:471–513

42. Duncan, D. P., Hodson, A. C. 1958. Influence of the forest tent caterpillar upon the aspen forests of Minnesota. *Forest Sci.* 4:72–93

43. Duncan, D. P., Hodson, A. C., Schneider, A. E., Batzer, H., Froelich, R., Meyer, D., Shiue, C. 1956. Influence of the forest tent caterpillar (*Malacosoma disstria* Hbn.) upon the aspen forest of Minnesota. *Minn. Off. Iron Range Res. Rehab. Publ.* 45 pp.

44. Eklund, B. 1964. On the effect of damage caused by the European sawfly as measured by the diameter growth as breast-height. (Swedish.) *Norrlands Skogs. Forb. Tidskr.* 3: 205–18. (1966. *Can. Dep. Forest. Transl.* No. 65)

45. Elliott, K. R. 1960. A history of recent infestations of the spruce budworm in northwestern Ontario, and an estimate of resultant timber losses. *Forest. Chron.* 36:61–84

46. Embree, D. G. 1967. Effects of the winter moth on growth and mortality of red oak in Nova Scotia. *Forest Sci.* 13:295–99

47. Evenden, J. C. 1940. Effects of defoliation by the pine butterfly upon ponderosa pine. *J. Forest.* 38:949–55

48. Fransen, J. J., Houtzagers, G. 1946. Loss of increment as a result of defoliation, and the seasonal growth of poplars. (Dutch.) *Ned. Boschb.-Tijdschr.* 18:36–39. (*Forest. Abstr. 11* No. 608)

49. Froelich, R., Hodson, A. C., Schneider, A. E., Duncan, D. P. 1955. Influence of aspen defoliation by the forest tent caterpillar in Minnesota on the radial growth of associated balsam fir. *Minn. Forest. Note 45.* 2 pp.

50. Froelich, R., Shiue, C., Duncan, D. P., Hodson, A. C. 1956. The effect of rainfall on the basal area growth of aspen as related to defoliation by the forest tent caterpillar. *Minn. Forest. Note 48.* 2 pp.

51. Furuno, T. 1964. The effects of leaf-eating insects upon the growth of forest trees. (Japanese, English sum.) *Bull. Kyoto Univ. Forests* 35 :177–206

52. Furuno, T. 1964. The effects of the feeding damage of pine caterpillar, *Dendrolimus spectabilis* Butler, upon red pine, *Pinus densiflora,* by artificial defoliation. (Japanese, English sum.) *J. Jap. Forest. Soc.* 46 :52–59

53. Furuno, T. 1965. The effects of artificial defoliation before growing period upon growth, especially height growth of Japanese red pine. (Japanese, English sum.) *Bull. Kyoto Univ. Forests* 36 :85–97

54. Ghent, A. W. 1958. Mortality of overstory trembling aspen in the Lake Nipigon forest area of Ontario in relation to recent outbreaks of the forest tent caterpillar and the spruce budworm. *Ecology* 39 :222–32

55. Giese, R. L., Kapler, J. E., Benjamin, D. M. 1964. Studies of maple blight, IV Defoliation and the genesis of maple blight. *Univ. Wisc. Res. Bull.* 250 :81–113

56. Graham, S. A. 1931. The effect of defoliation on tamarack. *J. Forest.* 29 :199–206

57. Graham, S. A. 1935. The spruce budworm on Michigan pine. *Univ. Mich. Sch. Forest. Conserv. Bull.* 6. 6 pp.

58. Graham, S. A. 1937. The walking stick as a forest defoliator. *Univ. Mich. Sch. Forest. Cir. 3.* 28 pp.

59. Graham, S. A. 1956. The larch sawfly in the Lake States. *Forest Sci.* 2 :132–60

60. Gremal'skii, V. I. 1961. The resistance of pine stands to defoliator pests. (Russian.) *Zool. Zh.* 40 : 1656–64. (US Dep. Commerce Transl. TT 65-50052)

61. Grisdale, D. G., MacLeod, L. S. 1962. Tamarack mortality associated with infestations of the larch sawfly and eastern larch beetle. *Can. Dep. Agr. Forest Biol. Div. Bi-Mo. Progr. Rep.* 18(5) :2

62. Grobler, J. H. 1956. Some aspects of the biology, ecology, and control of the pine brown tail moth. *Union S. Afr. Dep. Agr.* 186 pp.

63. Gross, H. L., Harden, A. A. 1968. Dieback and abnormal growth of yellow birch induced by heavy fruiting. *Can. Dep. Forest. Rural Develop. Forest. Br. Inform. Rep. O-X-79.* 7 pp.

64. Hail, R. C. 1965. Sagebrush defoliator in northern California. *US Forest Serv. Res. Note 75.* 12 pp.

65. Harper, A. G. 1913. Defoliation : Its effects upon the growth and structure of the wood of *Larix. Ann. Bot.* 27 :621–42

66. Hepting, G. H., Jemison, G. M. 1958. Forest Protection. In Timber Resources for America's Future. *US Forest Serv. Res. Rep.* 14 :183–220

67. Heron, R. J. 1953. Substitute short production by tamarack. *Can. Dep. Agr. Forest Biol. Div. Bi-Mo. Progr. Rep.* 9(3) :2

68. Heron, R. J. 1956. Jack pine staminate flower production. *Can. Dep. Agr. Forest Biol. Div. Bi-Mo. Progr. Rep.* 12(3) :2

69. Hildahl, V., Reeks, W. A. 1960. Outbreaks of the forest tent caterpillar, *Malacosoma disstria* Hbn., and their effects on stands of trembling aspen in Manitoba and Saskatchewan. *Can. Entomol.* 90 : 199–209

70. House, W. P. 1963. Gypsy moth-white pine damage study. In ed. N. Turner. Effect of defoliation by the gypsy moth, *Conn. Agr. Exp. Stu. Bull. 658.* 30 pp.

71. Husch, B. 1963. *Forest mensuration and statistics.* N.Y. Roland. 474 pp.

72. Ierusalimov, E. N. 1965. Changes in increment of a mixed oak forest after defoliation by insects. (Russian) *Arhangel'sk* 8(6) :52–55. (*Forest. Abstr.* 28 No. 2575)

73. Ives, W. G. H., Nairn, L. D. 1966. Effects of defoliation on young upland tamarack in Manitoba. *Forest. Chron.* 42 :137–42

74. Jacquith, P. H., Duncan, D. P., Kulman, H. M., Hodson, A. C. 1958. Preliminary study of growth losses in Minnesota jack pine following defoliation by the budworm. *Minn. Forest. Notes 74.* 2 pp.

75. Johnson, N. E. 1963. Some economic considerations in planning control

of insects affecting young forest trees. *J. Forest.* 61 :426–29

76. Johnson, N. E. 1967. The economic aspects of forest entomology and the forest manager. *Proc. Soc. Am. Foresters 1966.* 171–74

77. Johnson, N. E., Shea, K. R., Johnsey, R. L. 1970. Mortality and deterioration of looper-killed hemlock in western Washington. *J. Forest.* 68 : 162–63

78. Joly, R. 1959. The influence of forest defoliators on increment. (French.) *Rev. Forest. Fr.* 11 :775–84. (*Forest. Abstr. 21* No. 2046)

79. Kamilovski, M. 1966. Determining the best time to destroy defoliators of poplar. *God. Zborn. Zemi. Sum. Fak. Univ. Skopje No. 19* 157–86. (*Forest. Abstr. 29,* No. 2599)

80. Kapler, J. E., Benjamin, D. M. 1960. The biology and ecology of the red-pine sawfly in Wisconsin. *Forest Sci.* 6 :253–68

81. Kerby, W. H., Benjamin, D. M. 1964. The biology and ecology of the red-pine needle midge and its role in the fall browning of red-pine foliage. *Can. Entomol.* 96 :1313–22

82. Kinghorn, J. M. 1954. The influence of stand composition on the mortality of various conifers, caused by defoliation by the western hemlock looper on Vancouver Island, British Columbia. *Forest. Chron.* 30 :380–400

83. Kolonits, J. 1965. Life habits and damages caused by *Neodiprion sertifer* Geoffr. in Hungary. *Erdeszeti Kut.* 61 :225–39. (1968. *Can. Dep. Forest. Transl. No. 238*)

84. Kozlowski, T. T. 1963. Growth characteristics of forest trees. *J. Forest.* 61 :655–62

85. Kozlowski, T. T. 1964. Shoot growth in woody plants. *Bot. Rev.* 30, 335–92

86. Kozlowski, T. T. 1969. Tree physiology and tree pests. *J. Forest.* 67 : 118–22

87. Kozlowski, T. T., Clausen, J. J. 1966. Shoot characteristics of heterophyllous woody plants. *Can. J. Bot.* 44 :827–43

88. Kozlowski, T. T., Keller, T. 1966. Food relations of woody plants. *Bot. Rev.* 32 :293–382

89. Kozlowski, T. T., Winget, C. H. 1964. The role of reserves in leaves, branches, stems, and roots

on shoot growth of red pine. *Am. J. Bot.* 51 :522–29

90. Kramer, P. J., Kozlowski, T. T. 1960. *Physiology of trees.* New York: McGraw-Hill. 642 pp.

91. Kramer, P. J., Wetmore, T. H. 1943. Effects of defoliation on old resistance and diameter growth of broad-leaved evergreens. *Am. J. Bot.* 30 :428–31

92. Kulagin, J. A. 1966. Capacity for repeated foliation in woody plants. (Russian, English sum.) *Bot. Z.* 51(5) :723–26. (*Forest. Abstr. 28* No. 364)

93. Kulman, H. M. 1965. Effects of artificial defoliation of pine on subsequent shoot and needle growth. *Forest Sci.* 11 :90–98

94. Kulman, H. M. 1965. Effects of disbudding on the shoot mortality, growth, and bud production in red and sugar maples. *J. Econ. Entomol.* 58 :23–26

95. Kulman, H. M. 1971. Sawfly defoliation affects shoot growth of white spruce. *Minn. Forest. Note.* In press

96. Kulman, H. M., Hodson, A. C. 1961. The jack-pine budworm as a pest of other conifers with special reference to red pine. *J. Econ. Entomol.* 54 :1221–24

97. Kulman, H. M., Hodson, A. C., Duncan, D. P. 1963. Distribution and effects of jack-pine budworm defoliation. *Forest Sci.* 9 :146–57

98. Lanier, L. G. 1967. The influence of needles of different ages on the growth of *Pinus sylvestris.* (French.) *Proc. Congr. Int. Union Forest. Res. Organ., 14th, Munich* 24 : Part 5, 501–9

99. Larson, P. R. 1963. Stem form development of forest trees. *Forest Sci. Monogr. 5.* 42 pp.

100. Larson, P. R. 1964. Contribution of different-aged needles to growth and wood formation of young red pines. *Forest Sci.* 10 :224–38

101. Larson, P. R., Gordon, J. C. 1969. Leaf development, photosynthesis, and C^{14} distribution in *Populus deltoides* seedlings. *Am. J. Bot.* 56 :1058–66

102. Linzon, S. N. 1958. The effect of artificial defoliation of various ages of leaves upon white pine growth. *Forest. Chron.* 34 :50–56

103. Lonscakov, S. S., Mastov, A. D.,

Misel, Ju. A. 1958. The utilization of *Abies sibirica* stands in West Siberia, damaged by *Dendrolimus sibiricus*. (Russian.) *Lesn. Hoz.* 11,40–42. (*Forest. Abstr. 21* No. 2096)

104. Luitjes, J. 1958. On the economic significance of forest-insect pests (*Cephalcia alpina* Klug and *Diprion pini* L). (Dutch, English sum.) *Inst. Voor Toegepost Biol. Onderzoek Natuur (Itbon), Meded.* Nr. 40. 55 pp.

105. Madgwick, H. A. I. 1968. Some factors affecting the vertical distribution of foliage in pine canopies. *Symp. Primary Productivity Mineral Cycling Natural Ecosystems,* 231–45. Orono: Univ. Maine Press

106. Mancuso, B. 1955. Anatomical anormalies in the wood of defoliated branches of a cherry tree. (Italian) *Nuovo G. Bot. Ital.* 62:335–44. (*Forest. Abstr. 18* No. 2074)

107. Marcou, G., Catrina, I. 1962. Experimental research on the causes of dieback in oak in Rumania. (French.) *Proc. Congr. Int. Union Forest. Res. Organ., 13th,* Part 2, Sect. 24/13. 16 pp.

108. Mazanec, Z. 1966. The effect of defoliation by *Didymuria violescens* (Phasmatidae) on the growth of alpine ash Aust. *Forestry* 30:123–30

109. Mazanec, Z. 1967. Mortality and diameter growth in Mountain Ash defoliated by phasmatids. *Aust. Forest.* 31:221–23

110. Mazanec, Z. 1968. Influence of defoliation by the phasmatid *Didymuria violescens* on seasonal diameter growth and the pattern of growth rings in alpine ash. *Aust. Forest.* 32:3–14

111. McCambridge, W. F. 1956. Effects of black-headed budworm feeding on second-growth western hemlock and Sitka spruce. *Proc. Soc. Am. Foresters 1955,* 171–72

112. McLintock, T. F. 1955. How damage to balsam fir develops after a spruce budworm epidemic. *US Forest Serv., Northeast. Forest Exp. Sta. Pap.* 75. 17 pp.

113. Miller, C. A. 1966. The black-headed budworm in eastern Canada. *Can. Entomol.* 98:592–613

114. Minott, C. W., Guild, I. T. 1925. Some results of defoliation of trees. *J. Econ. Entomol.* 18:345–48

115. Moiseenko, F. P., Kozevnikov, A. M. 1963. Increment losses in Scots pine stands damaged by sawflies. (Russian.) *Lesn. Hoz.* 16: 8–11. (*Forest. Abstr. 25* No. 2508)

116. Morris, C. L., Schroeder, W. J., Robb, M. L. 1963. A pine sawfly, Neodiprion pratti pratti (Dyar) in Virginia. *Va. Div. Forest. Dep. Conserv. Econ. Devel.* 42 pp.

117. Morris, C. L., Schroeder, W. J., Knox, K. A. 1964. Growth loss in shortleaf and Virginia pines from sawfly defoliation. *J. Forest.* 62:500–1

118. Morris, R. F. 1951. The effects of flowering on the foliage production and growth of balsam fir. *Forest. Chron.* 27:40–57

119. Mott, D. G., Nairn, L. D., Cook, J. A. 1957. Radial growth in forest trees and the effects of insect defoliation. *Forest Sci.* 3:286–304

120. Nairn, L. D. 1956. Defoliation and thinning in relation to radial increment of tamarack. *Can. Dep. Agr. Forest Biol. Div. Bi-Mo. Progr. Rep.* 12(3):2

121. Nairn, L. D. 1958. The importance of adventitious growth in tamarack. *Can. Dep. Agr. Forest Biol. Div. Bi-Mo. Progr. Rep.* 14(6):3–4

122. Nairn, L. D., Reeks, W. A., Webb, F. E., Hildahl, V. 1962. History of larch sawfly outbreaks and their effect on tamarack stands in Manitoba and Saskatchewan. *Can. Entomol.* 94:242–55

123. Nichols, J. O. 1968. Oak mortality in Pennsylvania — a ten-year study. *J. Forest.* 66:681–94

124. Onaka, F. 1950. The effects of such treatments as defoliation, disbudding, girdling, and screening of light on growth and especially on radial growth of evergreen conifers. (Japanese English sum.) *Bull. Kyoto Univ. Forests* 18:55–95

125. O'Neil, L. C. 1962. Some effects of artificial defoliation on the growth of jack pine (*Pinus banksiana* Lamb.). *Can. J. Bot.* 40:273–80

126. O'Neil, L. C. 1963. The suppression of growth rings in jack pine in relation to defoliation by the Swaine jack-pine sawfly. *Can. J. Bot.* 41:227–35

127. Palandzjan, V. A., Hursudjan, P. A., Abramjan, B. M. 1960. Effect of defoliation on the formation of the annual ring in *Fraxinus pennsylvanica.* (Russian.) *Erevan* 13 :85–92. (*Forest. Abstr. 23* No. 4241)

128. Patterson, J. E. 1921. Life history of *Recurvaria milleri* Busck., the lodgepole needle miner, in the Yosemite National Park, California. *J. Agr. Res.* 21 :127–43

129. Pearson, L. C., Lawrence, D. B. 1958. Photosynthesis in aspen bark. *Am. J. Bot.* 45 :383–87

130. Pearson, L. C., Lawrence, D. B. 1958. Photosynthesis of aspen bark during winter months. *Proc. Minn. Acad. Sci.* 25 :101–7

131. Petersen, B. B. 1956. *Lygaeonematus abietinus* Christ as a pest on Norway spruce in South Jutland. (Danish, English sum.) *Soertryk af Det Forstlige Forsogsvoesen i Denmark* 22 :275–355

132. Popescu-Zeletin, I., Mocanu, V. G., Puiu, S. 1961. The development of spruce defoliated by *Lymantria monacha.* (German.) *Rev. Biol. Bucarist* 6 :119–37. (*Forest. Abstr. 24* No. 3954)

133. Potts, S. F. 1938. The weight of foliage from different crown levels of trees and its relation to insect control. *J. Econ. Entomol.* 31 : 631–32

134. Prebble, M. L., Graham, K. 1945. The current outbreak of defoliating insects in coastal hemlock forests of British Columbia. *British Columbia Lumberman* 29(2) :25–27, 42, 44, 46, 48

135. Prentice, R. M., Nairn, L. D. 1958. Increment reduction of Scotch pine following two years defoliation by the jack-pine budworm. *Can. Dep. Agr. Forest Biol. Div. Bi-Mo. Progr. Rep.* 14(2) :2

136. Rafes, P. M., Dinesman, L. G., Perel, T. S. 1964. Animal life as a component of a forest biogeocoenose. In V. Sukachev, N. Dylis, Ed. *Fundamentals of forest biogeocoenology,* Edinburgh and London: Oliver & Boyd. 672 pp. (English transl. 1968)

137. Readshaw, J. L., Mazanec, Z. 1969. Use of growth rings to determine past phasmatid defoliations of alpine ash forests. *Aust. Forest.* 33 :29–36

138. Redmond, D. R. 1959. Mortality of rootlets in balsam fir defoliated by the spruce budworm. *Forest Sci.* 5 :64–69

139. Reeks, W. A., Barter, G. W. 1951. Growth reduction and mortality of spruce caused by the European spruce sawfly, *Gilpinia hercyniae* (Htg.) *Forest. Chron.* 27 :140–56

140. Roberts, R. H. 1923. Effect of defoliation upon blossom bud formation. *Univ. Wisc. Agr. Exp. Sta. Res. Bull. 56.* 15 pp.

141. Rose, A. H. 1958. The effect of defoliation on foliage production and radial growth of quaking aspen. *Forest Sci.* 4 :335–42

142. Satee, T. 1967. Efficiency and quantity of leaves of closed stands of *Cryptomeria* joponica as influenced by site quality. *Proc. Congr. Int. Union Forest. Res. Organ., 14th,* Munich 21: Part 2, 395–404

143. Schmiege, D. C. 1961. Mortality and top killing of spruce-fir caused by repeated budworm defoliation. *US Forest Serv. Lake States Forest Exp. Sta. Tech. Note 597.* 2 pp.

144. Schneider, G. 1954. Is control of *Dasychira pudibunda* economically justifiable? (German.) *Forstu. Holzw.* 9 :378–79. (*Forest. Abstr. 17* No. 3041)

145. Schwenke, W. 1960. On the effect of forest fertilizers on the mass increase of *Diprion pini* in 1959 in central Franconia and the conclusions concerning outbreaks that may be deduced from it. (German, English, sum.) *Z. Angew. Entomol.* 46 :371–78

146. Schwenke, W. 1962. New knowledge on the origin and control of outbreaks of pests feeding on pine and spruce needles. (German, English sum.) *Z. Angew. Entomol.* 50 :134–42

147. Schwerdtfeger, F. 1960. The damage caused by *Tortrix viridana* L. *Holz-Zbl.* 86 :1450–52. (*Forest. Abstr. 22* No. 804)

148. Schwerdtfeger, F., Schneider, G. 1957. Defoliation of larch by *Coleophora laricella* and its effect on increment. (German.) *Forstarchiv.* 28 :113–17. (*Forest. Abstr. 19* No. 687)

149. Sherry, S. P., Ossowski, L. L. 1967.

The effect of defoliation by the wattle bagworm, *Kotochalia junodi* (Heyl.), on growth and bark yield of wattle plantations. *Wattle Res. Inst. Rep. 1966–67* 46–52

150. Silver, G. T., Collis, D. G., Alexander, N. E., Allen, S. J. 1962. The green-striped forest looper on Vancouver Island. *Proc. Entomol. Soc. Brit. Columbia* 59 :29–32

151. Silver, G. T. 1960. Notes on a spruce budworm infestation in British Columbia. *Forest. Chron.* 36 :362–74

152. Silver, G. T. 1962. The distribution of Douglas-fir foliage by age. *Forest. Chron.* 38 :433–38

153. Smith, J. H. G., Heger, L., Hejjas, J. 1966. Patterns in growth of earlywood, latewood, and percentage of latewood determined by complete analysis of 18 Douglas-fir trees. *Can. J. Bot.* 44 :453–66

154. Staley, J. M. 1965. Decline and mortality of red and scarlet oaks. *Forest Sci.* 11 :2–17

155. Stark, R. W. 1965. Recent trends in forest entomology. *Ann. Rev. Entomol.* 10 :303–24

156. Stark, R. W., Cook, J. A. 1957. The effects of defoliation by the lodgepole needle miner. *Forest Sci.* 3 : 376–400

157. Stillwell, M. A. 1956. Pathological aspects of severe spruce budworm attack. *Forest Sci.* 2 :174-80

158. Struble, G. R. 1957. Biology and control of the white-fir sawfly. *Forest Sci.* 3 :306–13

159. Summers, J. N., Burgess, A. F. 1933. A method of determining losses to forests caused by defoliation. *J. Econ. Entomol.* 26 :51–4

160. Swaine, J. M., Craighead, F. C., Bailey, I. W. 1924. Studies on the spruce budworm (*Cacoecia fumiferana* Clem.). *Dom. Dep. Agr. Bull. 37.* (Tech.) 91 pp.

161. Sweet, G. B., Wareing, P. F. 1966. Role of plant growth in regulating photosynthesis. *Nature* 210 :77–79

162. Tappeiner, J. C. 1969. Effect of cone production on branch, needle, and xylem ring growth of Sierra Nevada Douglas-fir. *Forest Sci.* 15 :171– 74

163. Thomas, J. B. 1958. Mortality of white spruce in the Lake Nipigon region of Ontario. *Forest. Chron.* 34 :393–404

164. Turcenskaja, I. A. 1963. Effect of *Lymantria dispar* and other defoliators on the growth of oaks. (Russian, English sum.) *Zool. Zh. Moskva* 42 :248–55. (*Rev. Appl. Entomol. Ser. B* 52 :594)

165. Turner, N., Ed. 1963. Effect of defoliation by the gypsy moth. *Conn. Agr. Exp. Sta. Bull. 658.* 30 pp.

166. Turnock, W. J. 1954. Tamarack killed by the larch sawfly. *Can. Dep. Agr. Forest Biol. Div. Bi-Mo. Progr. Rep.* 10(6) :2

167. Turnock, W. J. 1955. Additional mortality in tamarack plots. *Can. Dep. Agr. Forest Biol. Div. Bi-Mo. Progr. Rep.* 11(4) :2

168. US Dep. Agr. 1955. Annual report for 1954. *Alaska Forest Res. Center Sta. Pap.* 26–29

169. Varga, F. 1964. Loss in increment due to damage caused by *Lymantria dispar* in Turkey oak stand. (Hungarian, English sum.) *Sci. Pub. Univ. Forest. Ind.* 2 :219–26. (English transl. available)

170. Varley, G. C., Gradwell, G. R. 1962. The effect of partial defoliation by caterpillars on the timber production of oak trees in England. *Proc. Int. Congr. Entomol. Vienna, 11th, 1960,* 2 :211–14

171. Wallace, D. R. 1954. Some reactions of larch trees to sawfly defoliation. *Can. Dep. Agr. Forest Biol. Div. Bi-Mo. Prog. Rep.* 10(1) :2

172. Wallace, P. P. 1945. Certain effects of defoliation of deciduous trees. *Conn. Agr. Exp. Sta. Bull. 488* 358–73

173. Wareing, P. E. 1958. The physiology of cambial activity. *J. Inst. Wood Sci.* 1 :34–42

174. Warren, L. O., Coyne, J. F. 1958. The pine sawfly, *Neodiprion taedae linearis* Ross, in Arkansas. *Univ. Ark. Agr. Exp. Sta. Bull. 602.* 23 pp.

175. Waters, W. E. 1969. The life table approach to analysis of insect impact. *J. Forest.* 67 :300–4

176. Watson, E. B. 1934. An account of the eastern hemlock looper, *Ellopia fiscellaria* Gn., on balsam fir. *Sci. Agr.* 14 :669–78

177. Webster, H. V. 1943. Relation of defoliation by Japanese beetles and drought to frost injury of Amer-

ican elm. *Proc. Am. Hort. Sci.* 43 :
316–18

178. Wickman, B. E. 1958. Mortality of
white fir following defoliation by
the Douglas-fir tussock moth in
California, 1957. *US Forest Serv.
Calif. Forest Res. Sta. Note 137.*
4 pp.

179. Wickman, B. E. 1962. Effects of defoliation by the great basin tent
caterpillar on greenleaf manzanita
and mountain whitehorn ceanothus.
US Forest Serv. Res. Note 204.
3 pp.

180. Wickman, B. E. 1963. Mortality and
growth reduction of white fir following defoliation by the Douglas-
fir tussock moth. *US Forest Serv.
Res. Pap. PSW-7.* 15 pp.

181. Wilkinson, R. C., Becker, G. C.,
Benjamin, D. M. 1966. The biology
of *Neodiprion rugifrons,* a sawfly
infesting jack pine in Wisconsin.

Ann. Entomol. Soc. Am. 59 :786–
92

182. Williams, C. B. 1966. Differential
effects of the 1944–56 spruce budworm outbreak in eastern Oregon.
*US Forest Serv. Res. Pap. PNW-
33.* 16 pp.

183. Williams, C. B. 1967. Spruce budworm damage symptoms related to
radial growth of grand fir,
Douglas-fir, and Engelmann spruce.
Forest Sci. 13 :274–85

184. Wilson, L. F. 1966. Effects of different population levels of the European pine sawfly on young scotch
pine trees. *J. Econ. Entomol.* 59 :
1043–49

185. Zivojinovic, S. 1954. *Diprion pini* on
Mt. Maljen: history of the out-
break and consequences of the defoliation of *Pinus nigra.* (Croa-
tian.) *Zast. Bilja* 24 :3–19. (*Forest.
Abstr. 17* No. 656)

INTERACTIONS BETWEEN PESTICIDES AND WILDLIFE

6012

Oliver B. Cope

U. S. Department of the Interior, Bureau of Sport Fisheries and Wildlife, Washington, D.C.

The Sixties were a decade of enlightenment for fishery and wildlife science as these disciplines pondered the effects of pesticides upon fish, birds, and mammals. Before this period, knowledge of what happened when toxicants entered the environments of wildlife was largely confined to the benefits from pesticides as tools in the management of fish and wildlife and to limited measurements of acute toxicity of some economic poisons to a few species of birds, mammals, and fish. Up to 1960, the concept of pesticide-caused chronic effects in wildlife was accepted as a possibility, but little proof had been advanced, and only those with long-focus vision predicted the potential impact of pesticides on animal populations.

Wildlife science elected not to enjoy the luxury of ignorance but, rather, to accept its responsibility and intensify its investigations on pesticides and wildlife. Federal agencies, State agencies, and universities enlarged their funds, personnel, and scope of interest so that, nationwide, we supported a truly comprehensive effort directed to insecticide-wildlife relationships. The rationale providing the thrust for this campaign was similar to the expression that "We can never do merely one thing, because the world is a system of fantastic complexity." Wildlife-pesticide studies have been conditioned by the thought that man does more than one thing when he disperses pesticides. The studies, therefore, have probed many segments of the fish, bird, and mammal habitat, have considered acute, subacute, chronic, and indirect effects on wildlife, and have sought understanding at the cell, tissue, organ, organism, and population levels of organization.

The reader perceives progress, as from time to time reviews have appeared in the literature to provide updated summaries of the state of knowledge. The first of these (136) covered fish, wildlife, and pesticides up to 1956, and emphasized acute toxicity effects. Later reviews (39, 40, 93, 143), made it clear that the trends away from dominance of acute toxicity measurements led both inward, to measure physiological effects inside the bodies of the animals, and outward to gain an understanding of ecological effects. These shifts have been brought about through increased sophistication and sensitivity in chemical analytical instrumentation, added effort on pesticide problems by specialists not previously interested in pesticides and wildlife, improved knowledge of wildlife physiology, and inclinations in these directions as dictated by the course of the work. Newsom (124a) has discussed effects of pesticides on a variety of nontarget organisms.

Much of what we know about precise effects of pesticides on birds, mammals, and fish has been gained from experiments with laboratory animals, like white mice, and from wild or domesticated animals in the laboratory, like rainbow trout or *C. coturnix* quail. Extrapolation of these results to what might happen in nature is difficult and often impossible. However, some of the basic physiologic effects disclosed by laboratory investigations probably take place in the wild.

This paper is not intended as an exhaustive review of research on pesticides and wildlife. All studies are not mentioned here, even though many of those omitted have been excellent. Selected results are emphasized because they illustrate particular mechanisms.

ACUTE TOXICITY

Even though acute toxicity studies are usually undertaken mainly because of their relative simplicity, measurement of acute toxicity is a logical first step in pesticide studies because comparisons with other pesticides and with other species can quickly be established. With basic acute toxicity measurements at hand, we can plan more sophisticated studies and set priorities on a sensible basis.

Standard acute toxicity measurements differ for different vertebrate groups. Most fish acute toxicity data are expressed in terms of an amount of toxicant in the water and are reported as LC_{50}, TL_m, or TL_{50} values, the amount of toxicant in the water causing 50 percent mortality in a given time. For birds and mammals, standard tests feature an amount of pesticide in the diet, even though acute toxicity may also result from contact exposure or inhalation. The LD_{50} is the usual measure—the amount of ingested chemical resulting in 50 percent mortality in a given length of time.

Measurements of acute toxicity of pesticides to birds, fish, and mammals are scattered in the literature, and few long series of bioassays done with the same groups of tests animals at the same time are on record. One recent handbook (155) contains summaries of acute oral toxicities of most common pesticides to several bird and mammal species. There are also some other reliable reports (28, 29, 37, 38, 41, 79, 142, 144, 151, 157) that are useful for comparisons.

The contents of published bioassay reports illustrate several basic relationships which are useful to the resource manager concerned with the protection of birds, fish, and mammals, and to the researcher in predicting toxicities of other pesticides.

For fish, insecticides are generally more acutely toxic than are herbicides, fungicides, and algicides. Chlorinated hydrocarbons are usually more toxic than are organophosphates and methylcarbamates; compounds like endrin, endosulfan, and dieldrin, with acute contact 48-hr LC_{50} values below 5 μg per liter for the species we know about, are only slightly more toxic than highly toxic organophosphates like Dursban®, ethyl Guthion®, and Phosdrin®, but a more impressive difference between groups is seen by viewing them in the aggregate.

Organophosphates as a group are more toxic to white rats than chlorinated hydrocarbons. The methylcarbamate, Zectran®, is also more toxic to quail chicks, mallard ducklings, and pheasant chicks than is any chlorinated hydrocarbon insecticide that has been tested. Warm-blooded animals thus show a fundamentally different relative response to groups of insecticides than do the fish; our data do not show this for larvae of Fowler's toad, another cold-blooded vertebrate (137).

The generalities described above are valid with respect to broad chemical groups and higher animal categories, but specific predictions of toxicity are difficult. In a series of related insecticides, the one most toxic to most birds may not be the most toxic to most fishes or to most mammals. Endrin, for instance, is the most toxic chlorinated hydrocarbon insecticide to fish, birds, and mammals, but not to amphibians (38, 137, 142, 144, 157). DDT has a relatively high toxicity to fish, intermediate toxicity to mammals, frogs, and toads, and low toxicity to birds. Such inconsistency is also illustrated by some of the herbicides.

This unpredictable pattern is extended beyond animal groups to individual species. Chlorinated hydrocarbons, in general, are more toxic to rainbow trout than to bluegills, and more toxic to bluegills than to channel catfish. Yet, endrin is less toxic to rainbows than to the other species, and DDT more toxic to channel catfish than to the others. In review of the organophosphate data for quail and pheasant chicks and immature mallards, we find little evidence of toxicity dominance by individual insecticides or of consistent susceptibility by any species.

Despite the very general toxicity trends seen for basic chemical groups and for higher animal categories, and the great variation seen within these classes, there are fish family susceptibilities that might be used as a basis for classification in insecticide toxicity (112). In studies of 12 species of the families Ictaluridae, Cyprinidae, Percidae, Centrarchidae, and Salmonidae with nine insecticides representing the organophosphates, the chlorinated hydrocarbons, and the carbamates, susceptibility to the insecticides was generally similar within systematic groups. The ictalurids and cyprinids were the least susceptible and the salmonids the most susceptible to all the chemical groups. The variation in species susceptibility to the phosphorothionate and carbamate insecticides was greater than that for the chlorinated hydrocarbons, but not nearly as great as that observed for the phosphorodithionate insecticides. Differences in susceptibility seem to depend upon the tolerance of the species to cholinesterase inhibition, rather than upon accumulation of oxygen analogs to inhibit cholinesterase. For example, bluegills suffered 100 percent mortality with 84 percent brain cholinesterase inhibition from exposure to malathion, while bullheads exposed to the same dose showed similar inhibition, but no mortality.

FACTORS INFLUENCING ACUTE TOXICITY

When animal meets pesticide in nature, some factors are usually superimposed upon the general toxicity patterns described above to render the apparent acute toxicity different from that in the laboratory, where the measurements are made under a standard, controlled protocol. There are several of these factors, some relating to the animal itself, and some being characteristic of the environment. Reliable evidence has been garnered to give certain notions of the influence individual factors have on acute toxicity of some pesticides to some species of wildlife; many of the trends probably apply to other chemicals and other species.

Sex.—The sex of the animal often makes a difference in susceptibility to pesticides. The acute oral toxicity of aldrin, chlordane, DDT, and heptachlor is higher for male white rats than for females, while with endrin, endosulfan, mirex, toxaphene, and some others the females are more susceptible (157). The toxicity of most other chlorinated hydrocarbons is the same for both sexes. Among organophosphates, acute oral toxicity is higher to male white rats for Abate®, Azodrin®, methyl parathion, and some others, but higher to females for carbophenothion, coumaphos, Diazinon®, EPN, malathion, and parathion. Among carbamates, carbaryl has higher oral toxicity to females; for others, sex seems to make no difference. With dermal application both organophosphates and chlorinated hydrocarbons are more toxic to females than to males.

Most of the reported differences in sex susceptibility in rats are not large; with some, like endosulfan, endrin, carbophenothion, coumaphos, demeton, ethion, and phorate, there is a two- to three-fold difference in oral acute toxicity between sexes.

Sex and toxicity in fishes has not been adequately investigated. However, large differences between sexes should have been noticed in the multitudes of bioassay tests done in recent years. Even though most tests have featured immature fish, enough older fish have been used experimentally to have revealed important sex-connected differences in toxicity. Few differences have appeared.

Developmental stage and size.—We generally assume that susceptibility to pesticides decreases as the animal grows, and much of the evidence supports this concept. Feeding tests at Patuxent Wildlife Research Center (142) showed toxicity of a variety of insecticides and herbicides to immature bobwhite quail, *Coturnix* quail, ring-necked pheasant, and mallard duck immatures to be consistently higher than to adult birds. Young birds are quite vulnerable in the path of a pesticide spray, or spray drift. Without a protective feather cover the young bird is doubly jeopardized.

Among fish species on which considerable bioassay data are available, larger fish are less susceptible than smaller ones in the 0.5 to 5-g range. Some pesticides are highly toxic to fish eggs, while other are not. For example (138), antimycin, a piscicide, killed rainbow trout eggs at 19 ppb; tests with endosulfan and eggs from the same lot resulted in about the same egg mortality as in control lots, in 35- and 120-min exposures of up to 50 ppm of endosulfan. Northern pike show similar trends. We do not know whether selective permeability of the egg chorion or underdevelopment of endosulfan-sensitive structures is responsible. This seems important since endosulfan is one of our most toxic insecticides to rainbow and northern pike, but has little toxic effect on the eggs.

Verification of high susceptibility of small fish is given in the results of several studies in the wild in Canada (62). In connection with DDT spraying over New Brunswick forests where many young fish were killed after DDT was sprayed at .5-lb. per acre, underyearling Atlantic salmon were only 2-10 percent as abundant as in unsprayed situations, small parr only 30 percent, and large parr only 50 percent. After DDT spraying at .25-lb/acre, the underyearlings were about 50 percent as abundant, small parr about 80 percent as abundant, and large parr were hardly affected. Other studies of DDT forest spraying (97) gave similar results with caged and uncaged fish; in Coac Stream, 2–3 in parr exposed to .5-lb/acre of DDT had a 2 percent loss after eight days, while 1-in fry lost 56 percent. In McKenzie Brook, 1-in brook trout had 78 percent losses six days after the first spray at .25-lb/acre, while 5.25-in fingerlings suffered a 6 percent mortality in this time; after 20 days all the fry were dead, and no more fingerlings had died. These population changes were later reflected in sizes of adult populations returning to spawn.

Body size and toxicity from ingested DDT has been investigated (26) for coho salmon, and two factors were seen to be important. Laboratory-reared salmon of the same age were fed a diet containing DDT, and the median survival time was found to be directly related to body weight, perhaps because the smaller fish lack sufficient lipid to handle adequate storage detoxification of the DDT. The authors also accept the idea that the primary affect of body size is to control the rate of diet consumption by the fish, the small fish ingesting relatively large amounts of feed and, thus, DDT in proportion to their body weight.

Decreased toxicity with increased animal size, as discussed above, applies only after embryological stages. This was seen with endosulfan and rainbow trout, starting with the eggs. The eggs are resistant to 50 ppm of the insecticide over the 25-day incubation period, with continuous exposure (138). Mortality of 100 percent occurred in 250 ppb after 20 days and in

100 ppb after 30 days. The 48-hour LC_{50} is slightly more than 1 ppb for 2-in rainbow, but we assume that toxicity would decrease with increase in size beyond this, in agreement with other pesticides and other species. High toxicity is also seen in larval stages of Fowler's toad (137). Week-old larvae had a 48-hour DDT LC_{50} of 1.8 ppm, those 4 to 5 weeks old, 1.0 ppm, and 6-week-old animals, 0.41. From then on, toxicity decreased with age, with 7-week-old larvae showing a value of 0.75 ppm.

The toxicant.—Many common pesticides exist in a single chemical form, but there are many that occur as different isomers. Many other chemicals have pesticidal uses as various esters, salts, and amines. Aside from the occurrence of p,p'-DDT and o,p'-DDT,—the *gamma* and other isomers of benzene hexachloride, and variations in toxaphene, endosulfan, and chlordane, the common insecticides are generally free of varieties of the primary type. With many herbicides, however, it becomes important to specifically identify the salt, acid, amine, or ester form of the toxicant. Consider the phenoxy herbicides, where several salts, amines, and esters of 2,4-D, 2,4,5-T, and silvex, as well as different salts of endothall, azide, and others, are marketed.

The recognition of these different versions of toxicants is most important because the acute toxicities of the isomers, amines, esters, and salts of the basic chemicals to wildlife and fish species are usually different from each other. The *gamma* isomer of BHC is more toxic to insects than are the other isomers; this is also true with fish, and probably with birds and mammals. The propylene glycol butyl ether ester, the butyl ester, and the butoxyethanol ester of 2,4-D are more toxic to rainbow and to bluegills than are other esters (161). 2,4-D Acetamide is more orally acutely toxic to mallards, quail, and pheasants than are some other 2,4-D's. With white rats, in acute oral bioassays, the propylene glycol butyl ether ester of 2,4,5-T is more toxic than six other 2,4,5-T's tested (134).

In laboratory experiments with several kinds of 2,4-D, 2,4,5-T, and silvex compounds tested against bluegills, differences in toxicity were greater among the salts, esters, and amines of each basic chemical than among the acids of each chemical itself (86). Some of these are tenfold differences, so have some real significance.

Pesticide formulation.—The formulation in which a toxicant is distributed helps govern the toxicity to wildlife, just as it does to pest insects, weeds, undesirable fungi, or other target organisms.

The effect of formulation on toxicity to wildlife is best seen in the aquatic environment. Dusts, emulsifiable concentrates, oil solutions, wettable powders, granules, slurries, water solutions, and other formulations have their own chemical and physical characteristics (161). For fish, insecticide emulsions and oil solutions are the most toxic kinds of formulations (39,

119). Rainbow trout, red salmon, and three-spined stickleback in Alaska succumbed faster to emulsified and oil solutions of DDT, BHC, toxaphene, and chlordane than to acetone suspensions. Emulsifiable oil preparations of benzene hexachloride were 25 times more toxic to golden shiners than wettable powder formulations containing the same level of *gamma* isomer (119).

Granular formulations, wettable powders, and dusts, which usually release the active ingredients into the aquatic ecosystem at slow rates, have relatively low toxicities to fish (40). Thus, choosing formulations of these kinds can often afford protection to animals in the aquatic environment.

Temperature.—The temperature prevailing during exposure of wildlife to pesticides can also influence toxicity. Most combinations of toxicant and animal show impressive shifts in toxicity with increase in temperature. Poikilotherms show more pronounced toxicity variation with temperature than do warm-blooded vertebrates.

Rainbow trout (111) tested against 13 insecticides and 2 herbicides at 1.6°, 7.2°, and 12.7°C for 96 hr showed varying degrees of toxicity increase as temperatures rose. Dursban® had a relative increase in susceptibility of 7.18 from 1.6° to 12.7°C, trifluralin an increase of 5.00, and dieldrin an increase of 1.71. Bluegills tested at 12.7°, 18.3°, and 23.8°C also showed variation with different pesticides; trifluralin had a relative increase in susceptibility from low to high temperature of 4.04, malathion had 2.60, and diuron had 1.50. Some combinations indicated no immediate temperature effect after 24 hr, as with bluegills and lindane and azinphosmethyl over 24 hr, but at 96 hr these combinations showed increased toxicity as temperature rose.

We do not understand the entire mechanism involved here, but assume higher rates of toxicant uptake at higher temperatures. Conversely, higher temperatures should increase enzymatic activity to degrade pesticides in the animal body and lower toxicity. Rate of ventilation increases with increase of temperature, increasing the flow of water across the gills, and the uptake of toxicant. Indirect effects through increased animal metabolism that reduces oxygen in the water and increases waste products may also act to increase toxicity.

Entomologists have long known (118) of negative temperature coefficients with some insects and DDT and methoxychlor, with increased mortalities at lower temperatures. This is also seen in some fish; methoxychlor shows negative temperature coefficients with rainbow and bluegills, and other studies (38) show this to take place with DDT and the same species. This phenomenon, apparently related to chemical structure, is unusual with fish, but may be important under some circumstances. Widespread channel catfish mortality in the southeastern States in 1969 may be partially attributable to negative temperature coefficients. Fish containing body burdens of DDT and other chlorinated hydrocarbons developed a disease syndrome and died after fall temperatures declined. The phenomenon discussed here may have contributed to this complex, together with DDT mobilized from lipid storage after starvation.

Chemistry of the environment.—As with temperature, the chemistry of the fish environment generally has a greater impact on pesticide toxicity than that of birds and mammals. Some of the effects on fish are direct, influencing uptake of toxicant through the gills, some affect the availability of the toxicant after it enters the water, some change the chemistry and toxicity of the pesticide, and some accentuate the pesticide toxicity by adding to the stress on the fish.

The chemistry of the stream or lake substrate may control the dispersion of the pesticide in the water. Bottom muds of high organic content bind pesticides through adsorption, in contrast to sandy or rocky bottoms (40). Such adsorption often binds significant proportions of the amount entering the water, thus lowering the concentration in the water and reducing the hazard to fish. While the pesticide is attached to the substrate, it may degrade, may be slowly released to the water, or may persist unchanged for years, depending upon the chemical, the nature of the substrate, and the chemistry of the water. The dynamics of the binding and release of the toxicant can be of extreme importance to the fish population, and may be difficult to predict in many situations. It has been suggested that the concentration of toxicants in ponds can be reduced by the addition of silt or some other substrate to the water to adsorb the chemical and make it less available to fish.

The toxicity of pesticides may depend upon water chemistry. DDT in waters of high pH is not very toxic to salmonids (39). Alaska stream experiments in water of pH 8.4 showed low DDT toxicity, but after heavy rains that lowered the pH to 7.3, toxicity increased (42). Endosulfan toxicity to western white suckers also varies with pH. A pH of 8.4 or 9.4 reduces or eliminates its toxicity at 19°C within 24 to 96 hr (138), and reduction in toxicity at these pH's is less at lower temperatures. Malathion hydrolyzes rapidly in alkaline waters, but slowly in neutral or slightly acid waters; toxicity to fish varies accordingly.

Water chemistry may work in combination with other factors to affect toxicity. Effects may be important, as with the piscicide, antimycin, which has recommendations on the label for pH and temperature conditions (162). Studies on dieldrin and the sheepshead minnow (159) were conducted were the effects of dieldrin at three temperatures (19.4°, 25°, and 36.1°C), at three salinities (0, 15, and 30 ppt), and two hydrogen ion concentrations (7.2 and 8.2). Analysis of variance showed increases in toxicity to be attributable to temperature alone, to salinity alone, or to pH alone, as well as to interactions between pH $\times$ salinity, pH $\times$ temperature, salinity $\times$ temperature, and salinity $\times$ temperature $\times$ pH. In similar studies on the sailfin molly the main influence was temperature, with pH × salinity secondary and salinity alone of no statistical significance. In marine and estuarine animals it has been shown (98) that salinity and temperature should be considered together in their biological effects.

There is evidence that water hardness (with related pH) exerts some control over the toxicities of some pesticides to some aquatic animals, but no universal rule applies. In tests of 11 organophosphate insecticides against fathead minnows (79), Dipterex® was the only one to show different toxicity in hard than in soft water; it was higher in hard water (pH 8.2), perhaps because of rapid hydrolysis to more toxic substances in the more alkaline water. Other results with bluegills, largemouth bass, and fathead minnows and Hyamine 1622 and 2389 showed toxicity in soft water about twice that in hard water, but not with other herbicides, such as endothall, diquat, dalapon, or silvex (131). Another study showed the toxicity of silvex to black bullheads to be twice as great in hard as in soft water, but no significant differences occurred between hard and soft waters for bluegills, golden shiners, black bullheads, and goldfish tested against several other herbicides (38). Studies of Bayer 73 and rainbow trout, carp, channel catfish, and bluegills (113) generally showed greater toxicity in soft water than in hard.

Chemistry of the aquatic environment may, thus, be an important factor in toxicity of pesticides to aquatic animals. Unfortunately, we cannot always relate toxicant structure to these effects, so must rely upon developing a body of bioassay experience for predictions of toxicities in the wild.

Synergism and antagonism.—The toxicity of an individual pesticide may be enhanced or reduced by the presence in the environment of another pesticide. Knowledge of the occurrence and biochemical mechanisms in synergism and antagonism with insects has been extended to a few other invertebrates like *Daphnia* (126), to rats and mice, and to fish. Organophosphate synergism, antagonism, and additive effects in rats and mice (126) are more fully documented and better understood than in birds and fishes, but no combination approaching the impressive EPN-malathion case with dogs, where $1/40$ LD_{50} of malathion plus $1/50$ LD_{50} of EPN gave 100 percent mortality, has been measured for other vertebrates.

Laboratory tests with bluegills and 37 combinations of copper sulfate, DDT, carbaryl, parathion, malathion, and methyl parathion resulted in additive effects for 22 combinations, synergism for 13, and antagonism for 2. Most of the combinations showing synergism contained one or two organophosphates. Copper sulfate was synergistic with carbaryl and with parathion, but antagonistic with DDT and with malathion (110). Other fish tests showed EPN and malathion to be synergistic to fish (167), and some others showed synergism with DDT and 2,4-D (32). Some other combinations of piscicides and insecticides show synergism, and are the basis for patents.

Knowledge of synergistic effects on birds, mammals, and fishes in specific ecosystems is essential for effective resource management and protection. For one thing, many commercial herbicide, insecticide, and piscicide formulations contain more than one basic chemical. Another problem with synergism may come from the routine application in agricultural areas of more than one pesticide in quick succession in a single season.

Interactions of different insecticides when experimentally fed to mice have afforded some protection (154). One hour after aldrin, dieldrin, or chlordane were fed, the toxicity of parathion was enhanced in mice, but after four days it was decreased. Aldrin also protected mice four days later against paraoxon, EPN, Guthion®, TEPP, TOCP (tri-*o*-tolyl phosphate), DFP (di-isopropyl fluorophosphate), and physostigmine, but not against OMPA or neostigmine. Aldrin increased liver A-esterase and plasma B-esterase, decreased plasma A-esterase, and reduced inhibition of brain but not plasma cholinesterase by paraoxon. The authors suggest esterase increases probably account for only part of aldrin's protective action.

DDT is antagonistic to dieldrin storage in rat fat (150). Combinations of insecticides and drugs have variable effects. Phenobarbital decreases the rate of storage of dieldrin in mammal fat, but not in the amount of dieldrin in the liver (49). Phenothiazine-derived drugs potentiated poisoning from organophosphate insecticides in mammals (12), but chlorpromazine prolonged survival time of guinea pigs poisoned by malathion and parathion.

Another interesting effect has been noted with rotenone and cyclodiene insecticides (90). When low concentrations of rotenone were applied to plants, they catalyzed the photoisomerization of dieldrin, aldrin, endrin, and some other chlorinated hydrocarbons. From this finding may develop a method for controlling the persistence of residues in plants by using pesticide-photosensitizer combinations.

RESIDUES

In the laboratory and in the field, knowledge of the amount of pesticide in the animal and in the environment seems necessary for a grasp of the toxicology involved. Intensification of residue measurement has come about as the need for the data has become obvious and as the chemist has devised better analytical instruments with tighter specificity and higher sensitivity.

Wildlife scientists measure residues to trace pesticides through the environment, to relate the amount of pesticide to morbidity and mortality, to learn about rates of accumulation and loss in the animal, to see the chemical changes in the toxicant and in the animal after exposure, and to decide whether animals contain too much pesticide burden to be safely consumed by humans. These processes are interrelated in the dynamics of pesticide-wildlife interactions, emphasizing the utility of residue data in these studies.

Pesticides in the Environment

Pesticides sprayed from airplanes usually do not reach the ground or water surface without losses to the atmosphere through wind or drift (44, 169). They circulate in air currents and may later be deposited by rainfall in distant places. If the vegetative canopy is dense, only 20 percent of the spray may reach the ground or the water surface, but even low-level airplane spraying over open fields often results in losses of half of the toxicant and an uneven pattern of distribution.

Once on the ground or water surface, the toxicant may be subject to further dispersion. In running water the toxicant moves downstream in an ever-widening block, becoming less concentrated from adsorption and from dilution from tributary streams. Insecticides in oil solutions falling onto stream surfaces do not necessarily stay on the surface, but may penetrate to the bottom and affect fish and bottom invertebrates. In Trapper Creek in Montana, for example, stream bottom organisms were affected a few minutes after a DDT-oil spray fell on the surface at the rate of .19 pounds per acre (44). Dead and dying insects drifted downstream for a week, indicating that the DDT circulated to the bottom of the shallow stream in amounts sufficient to kill stonefly, mayfly, caddisfly, and dipteran immatures.

In lakes and ponds, pesticides also move within the system. Formulations like emulsifiable concentrates tend to disperse through the water, but do not always do so in a uniform way. In a toxaphene treatment of a New Mexico lake (94), the emulsified toxicant was distributed on the lake and appeared to spread quickly and evenly. Yet, after 24 hr, and for 2 weeks, water samples from the lee side of the lake at the 10-foot depth always contained more toxaphene than those from the windward side. Later, the concentration stabilized throughout the lake. Pesticides in oil solution tend to float on the surface and move at the surface in response to winds, but also penetrate beneath the surface.

A study worthy of attention (156) showed that pesticides concentrate in oceanic surface slicks and may move great distances on the surface. Chlorinated hydrocarbons were measured in slicks in amounts 10,000 times greater than in the surrounding water. The slicks, containing compacted organic and inorganic molecules that attract plankton and concentrate pesticides, may be several miles in length. This phenomenon may explain the occurrence of DDT in Antarctic penguins, seals, and fish.

Sprayed pesticide that does not immediately reach the ground or water surface may drift or be carried great distances, and in significant quantities. Atmospheric dust collected in the West Indies, after crossing 4000 miles of open ocean, contained appreciable amounts of chlorinated pesticides (156). There is much documentation of accidental damage to crops, domestic animals, and wildlife from this source (23, 107). Sprayed pesticides falling to the ground or applied to the soil may remain in place and intact for years (57), or may be subject to redistribution. Mathematical predictions can be made of accumulation of residues by the use of rate laws (74). Appreciable amounts of some pesticides may volatilize from soils (19, 76) or from crops (51). Movement in soils is usually minimal, but flushing by runoff from rains can leach pesticides considerable distances (105). It is not uncommon for pesticides to thus leach and cause mortality to aquatic animals, as in Alabama (172), where insecticides carried off cotton fields after heavy rains caused fish to die in 15 streams, and caused apparently complete eradication of fish in some of the streams.

Other losses of pesticides may take place after they reach the soil; these, removal by animals, degradation, and translocation by plants, are of varying importance in different situations, depending mostly on pesticide and formulation, quality of the soil, and the weather and climate.

From this, it is obvious that to understand the background of pesticide effects in a particular situation, we need to know more than the amount that was applied. The amount applied is subject to so many changes before and after it makes contact with the ground or water that there is only one way to know the amount when the pesticide contacts the wildlife—measure the residues.

Residues, morbidity, and mortality.—Knowledge of the magnitude of pesticide residues in animals, as it relates to mortality or morbidity, can be a useful diagnostic tool. There is need for reliable bases for deciding whether pesticides or something else, has caused sickness or death in wildlife, and research has aimed at measuring quantitative relationships between residues in the whole animal body, in some organ or tissue, or in some combination of organs or tissues, or both; and some vital, measurable disorder in the animal. Whole body residues are unreliable for this purpose because their variation from fish to fish is usually appreciable. In an exposed population in which some fish die, the surviving fish often contain higher whole body residues than do the fish that died (94). The surviving animals have a longer time to absorb the pesticides, and thus accumulate larger residues.

In spite of numerous studies on residues in fish, correlation has been demonstrated for only a few pesticide residue levels and morbidity or mortality. Pesticides, their metabolites, or resulting body chemical changes normally have been identified and measured in fish in association with distress or death, but the relationships between amount of chemical and acute effects are seldom consistent or predictable enough for diagnosis.

Research on parathion (121) has revealed that the concentration in the blood of brown bullheads has a close relationship to that in the water. It was demonstrated that parathion remains unaltered in the blood, and does not change to paraoxon. Unaltered parathion in the blood was associated with mortality only when present at 3 μg or more per gram of blood; only 7 percent of the fish had blood levels above 3 μg/g and still remained alive, though convulsing. All fish that died did so within 11 days, and there was no chronic mortality. Here, then, is a relationship between blood parathion and mortality reliable enough for practical diagnosis.

Lethality of endrin to fish has been related to residue levels in the blood of channel catfish (122) and gizzard shad (24) in laboratory exposures. In channel catfish a threshold of 0.30 μg endrin per gram of blood was established; when this amount was exceeded, death resulted regardless of time of exposure up to 25 days, or of concentration of endrin in the water. Overlap in range of blood concentration between suviving and dead fish was minimal; endrin in blood of survivers never exceeded 0.28 μg/g, and no fish died with less than 0.23 μg/g. With adult gizzard shad, the critical level (above which few shad survived) was 0.10 μg/gram of blood.

In birds, amounts of some insecticides in the brain are indicative of degree of damage to the animal. A study with DDT and cowbirds (146) showed differences between amounts of DDT + DDD in brains of dead and survivors exposed to the same amounts of DDT. In male cowbirds, DDT + DDD residues averaged 101 ppm in dead birds and 16 ppm in survivors. In dead birds, DDD generally exceeded DDT, while the two insecticides were about equal in survivors. The authors suggest storage without metabolism to DDD as a protective mechanism. Other studies with DDT and cowbirds showed that the amounts of DDT + DDD in the brains of dead birds were similar (145), regardless of the amount of time of exposure. Brain residues in dead or dying cowbirds were similar to those reported for robins, house sparrows, bald eagles, and white rats. Brain residue of DDT in birds seem to have some potential in diagnosis, but does not yet seem to have the precision of the few insecticide-fish blood relationships that have been studied.

Dieldrin in *Coturnix* quail brains correlated fairly well with death (148), although there was some overlap with amounts in surviving male birds in recent studies. Since brain weight and lipid content remain reasonably constant, dieldrin in the brain showed promise as an index of death. When these laboratory results were compared to brain levels in dead birds in dieldrin-treated areas, meadowlark, robin, starling, and wild woodcock levels were generally within the lower range for *Coturnix*, but below the means. There were differences among bird species, between sexes of *Coturnix*, and between *Coturnix* and the wild birds, but it appears that dieldrin poisoning can be judged from brain residues. Residues in other tissues are too variable for this purpose.

Mammal death in relation to pesticide residues has not been intensively studied, but there are data on white rats (77) and DDT that show constancy (35 to 52 ppm in the brain) and separability from levels in survivors. Cottontail rabbits and cotton rats dying in dieldrin-treated areas had brain residues of 13.8 and 7.9 ppm, respectively, within the range measured for wild birds in the area (7.1–21.7 ppm). The authors concluded that 4 or more ppm of dieldrin in the brain indicates that the birds or mammals were in the known danger zone and may have died from dieldrin.

Buildup of residues.—The uptake, accumulation, and loss of pesticide residues in the animal body has been investigated, and wildlife science has built a base of knowledge to help comprehend the mechanics and the chemistry of this important feature of animal-pesticide associations. In the aquatic environment we recognize pesticide solubility in the water as having a controlling influence on uptake through gills. Relatively water-insoluble compounds like chlorinated hydrocarbons tend to be absorbed through gill membranes at faster rates than more soluble ones, like organophosphate insecticides. Once inside the animal body, chlorinated hydrocarbons tend to be stored as residues more than do the organophosphates. The more acutely toxic chlorinated hydrocarbons tend to be stored as residues at slower rates than the less toxic ones. The composition of the animal influences residue accumulation, high lipid composition tending to encourage storage and to reduce toxicity.

In the aquatic environment, important factors controlling residue accumulation in the animals are solubility of the pesticide in the water, the lipid content of the animal, the species, and the time involved in the uptake. Pesticide chemicals vary widely in their water solubilities; mirex is practically insoluble; DDT is reported to be 0.0012–0.037 ppm; endrin, 0.23 ppm; dieldrin, 0.25 ppm; toxaphene, 3 ppm; lindane, 10 ppm; parathion, 20–25 ppm; malathion, 145 ppm; diuron, 40–42 ppm; dichlobenil, 20 ppm; 2,4,5-T acid, 280 ppm; and 2,4-D acid, 725 ppm (73).

As compounds are exchanged between the water environment and blood, across the gillls, and from blood to lipid materials in the body, the differences between solubilities in the media regulate the rates and extent to which pesticides enter the body and move to storage sites in the animal (75). Chlorinated hydrocarbon insecticides have solubilities in body fat of the order of grams per kilogram, so the trend is toward concentration in the fat. Compounds of low water solubility and high lipid solubility thus tend to accumulate rapidly and to high levels in the body. Fish concentrate lindane approximately 1×10^2 the amount in the water (66), toxaphene about 1×10^4 (152), dieldrin about 1×10^4 (35), and DDT about 1×10^5 (75) and 1×10^6 (131). These findings are in general agreement with amounts expected from the measured water solubilities and an assumed fat solubility of 10 g per liter for each compound.

The intimate relationship between residue buildup and lipid content of the animal has been emphasized in several fish and wildlife studies (82, 131, 146). The demonstrated increase in residues with increase in body fat of fish is sometimes the principal factor in rate of accumulation, as with dieldrin in Atlantic salmon (10), but sometimes an additive factor, as with DDD and DDE in Atlantic salmon, in which fish age and fat content are responsible for the whole body residue levels. DDT in these fish was significantly higher in high-lipid than in low-lipid fish at ages III+ and IV+, but not at age V+, showing a decrease with age among fish with high fat content and staying constant with age among those of low fat content. The condition of the fish, then, does not always bear a simple and direct relationship to insecticide levels (101), as illustrated by a variety of fish species from Wisconsin waters.

The strong relationship between residue buildup and time or age has been illustrated in studies on fish and chlorinated hydrocarbons. In cutthroat trout given various levels of DDT once a week in their diets (4) whole-body DDT-DDD-DDE residues increased in an exponential pattern, leveling off after a year in the low-fed groups and then maintaining the same levels as excretion and degradation balanced intake. The higher-fed groups did not reach equilibrium in this time. In the same experiment, trout given monthly DDT baths built up whole-body residues of the DDT complex in a similar pattern, but to lower levels.

Reticulate sculpins fed different amounts of dieldrin in *Tubifex* worms (35) also followed an exponential pattern in the buildup of residues. The retention of dieldrin was inversely related to the amount ingested. Sculpins exposed to constant concentrations of dieldrin in continuously renewed water accumulated the insecticide in patterns similar to those that were fed the insecticide. Curves from different exposures were nearly parallel to each other, and uptake and storage were functions of concentration and time.

Heptachlor fed daily to bluegills (11) at different rates reached peak levels at different times; those with the higher intakes had the earlier peaks. In concurrent contact experiments with one treatment per pond, peaks of whole-body residues were reached first in the low-concentration exposures.

Experiments with mirex and bluegills and goldfish (158) showed that residues increased in the exponential pattern seen in other cases; saturation was reached at the lowest feeding level in bluegills after 84 days, but not for the higher intake levels until 168 days, when some residues approached 100 ppm. In goldfish subjected to single treatments of mirex in ponds, residues reached nearly 200 ppm in fish from ponds treated at 1 ppm; equilibrium between intake and loss had not been reached in 308 days.

The herbicide, simazine, when fed to green sunfish at two levels, is rapidly lost from the fish. Whole-body residues, measured at 24-hr intervals after each of three feedings, showed sharp losses by 72 hr, and the fish had negligible residues at 168 hr. Simazine residues from contact baths at two levels increased until the end of the 21-day experiment, the concentrations in the water having remained near their concentrations for the full time (133).

Bluegills subjected to one-time wettable powder treatments of dichlobenil in earthen fish ponds developed very high whole-body residues in three days, but residue values declined rapidly thereafter (43). In contrast, the same chemical in a granular formulation in other ponds produced peak residues in largemouth bass, yellow perch, green sunfish, and bluegills, in 16 to 34 days.

Bluegills developed high three-day and seven-day levels when exposed to single treatments of diuron in earthen fish ponds and residues declined rapidly in the next week (115). In similar experiments with 2,4-D and bluegills (45), highest residues were found at 24 hr, with rapid disappearance from the fish thereafter.

Salmon and trout in Lake Michigan illustrate how time contributes to storage of DDT (130). Lake Michigan DDT levels were recently measured at 1–2 parts per trillion. Lake trout, 3–5 in long, contained about 1.1 ppm of DDT, while those 16 to 20 in long contained 6.6 ppm. Coho weighing 2–4 pounds after one year in the lake had 3–4 ppm DDT; a year later they were 11–12 pounds in weight and averaged 12–13 ppm DDT. Time alone was not responsible, according to the author; the percentage of fat increased as the fish grew, increasing the capacity of the fish to concentrate the insecticide.

These examples show that residues of chlorinated hydrocarbons in fish may be controlled by the nature of the chemical, the kind of animal, the level of treatment, the time of exposure, and the time during which loss of pesticide may occur. The same principles apply to birds and mammals; the pharmacodynamic changes in mammals have been reviewed (77) and discussed for birds (143).

Residue measurements in organs and tissues have helped in the search for indices of pesticide damage, in understanding buildup and loss of pesticides in the body, in ascertaining toxicity mechanisms, and in assessing the danger to human consumers. In trout, relatively high residues of DDT are found in liver, adipose tissue, gonads, and brain (4), organs with high lipid content. Salmon studies (130) revealed adult coho steaks to have 62.8 ppm of the DDT complex in dorsal median fat, 34.3 ppm in lateral line fat, 69.7 ppm in belly fat, 92.3 ppm in abdominal adipose, and 5.7 ppm in flesh; the average value for the whole steak was 14.9 ppm, indicating that a very high percentage of the fat and DDT could be trimmed and discarded before cooking and eating. Other chlorinated hydrocarbon pesticides have been demonstrated to accumulate in fat-bearing tissues in fish (93), birds, and mammals, but little information has been published on distribution of organophosphates in the animal body.

Resistance.—Residues play an important role in the development of vertebrate resistance to pesticides. Resistance of fish (64), frogs (23), laboratory mice (17), wild rats in the field (72), and pine mice in Virginia orchards (165) to pesticides have been reported, and resistant populations of some fishes are commonplace in some agricultural areas with long histories of insecticide exposure. Resistance in birds has not yet been reported. While mechanisms of resistance in insects have been explained, they are not understood for fish; authors have speculated that differential mortalities and selective pressures in nature have contributed to the development of resistance over long periods. A basis for high tolerance to DDT for laboratory mice has been discussed (17).

Some resistance in fish and mice has been found to be nonspecific, exposure to particular organochlorine insecticides leading to resistance to the exposed compounds and to cross-resistance involving other compounds, as well. With some fishes, resistance developed for some chlorinated hydrocarbons extends to resistance to certain organophosphate insecticides (63).

When resistance to insecticides appears in fish, there is a heavy accumulation of body residues. Since the tolerance of the animal has been increased beyond normal boundaries, heavy body burdens of pesticide are tolerated without causing the acute toxicity acting in nonresistant populations. For example, normal mosquito fish have a 48-hr TL_m of 1 ppb of endrin, while some resistant populations can survive in 1000 ppb for three weeks. Whole-body residues of endrin in some groups of resistant mosquito fish have exceeded 1000 ppm.

There is considerable ecological significance to the resistance-residue phenomenon. Resistance is genetically based; this has been established in fish through selective breeding, which has not only shown perpetuation of resistance through four generations but has resulted in increased resistance in successive generations. In nature, resistance may help exposed populations to survive, but the increased ability to store toxicants as residues can be detrimental to animals higher in the food chain. Experiments with feeding of high-residue mosquito fish to green sunfish resulted in death to the sunfish, even though most of the ingested mosquito fish were regurgitated by the sunfish (63).

Not all fish species develop resistance to all pesticides. In fact, increased sensitivity sometimes appears from generation to generation, as seen in exposures of DDT and endrin to sheepshead minnows (83).

Biological magnification.—Aquatic animals have the capacity to accumulate pesticides directly from the water, as well as from ingested organisms or detritus. One laboratory study with dieldrin and the reticulated sculpin (35) showed both these processes to occur but that they were not additive. Fish in water with dieldrin and dieldrin-treated worms accumulated no more residue than fish in similar water and uncontaminated worms. Massive residues may be found in the wild, as with oysters that contain 70,000 times the amount of DDT in the water (30), and with coho salmon in Lake Michigan that build up DDT residues over 1 million times the 1–2 parts per trillion in the water (130). We cannot assign proportions of these residues to water or food routes, but assume that both routes contributed to the totals.

Controlled studies on pesticides and food chains have shown some effects of magnification. One study (75) showed the following accumulations: algae, 5789 the amount in the water; invertebrates, 27, 750; and fish, 317,000. In the field, lake studies (88) showed that five years after the last application of DDD, no insecticide was found in the mud or water, but the plankton had 265 times the applied concentration; fat of carp and of frogs, 2000 times that applied; bluegills, up to 12,500; grebes, 80,000; largemouth bass, up to 85,000; white catfish, up to 118,750; and brown bullhead, up to 135,000.

Forest insect control studies (55) showed the following residues in the food chain heirarchy after DDT was sprayed at 1 pound per acre: herbivores, 0.03–0.1 ppm; predators with short food chains feeding mostly on herbivores, 0.1–5 ppm; predators with longer food chains feeding mostly on lower predators, 1–50 ppm. In a marsh whose soil averaged over 13 pounds of DDT per acre (170), measurements showed water to contain 0.00005 ppm; plankton, 0.04 ppm; shrimp, 0.16 ppm; eels, 0.28 ppm; mummichogs, 1.24 ppm; gulls, terns, and other sea birds, 3–26 ppm; and ring-billed gull, 75.5 ppm.

One group of researchers (75) rejects the usually accepted explanation that biological magnification depends upon the amounts of pesticides present in the various trophic levels as the pesticide moves stepwise up through the food chain. This step-by-step increase is supported by evidence for terrestrial animals (87) and is suggested by aquatic experience. These authors, however, accept the view that chlorinated hydrocarbon residue concentration in each trophic level of an aquatic food chain depends on the amount retained by the organisms through solubility and adsorption differences acting through systems of exchange equilibria, not on the amounts present at each level of the chain. Their experiments with DDT and food chains with fish show that there is merit in the hypothesis.

Biological magnification of pesticides in some birds and mammals has been important in large-scale mortalities among the vertebrates. Losses of robins receiving DDT from earthworms in Wisconsin, grebes receiving DDD from fish in California, and birds receiving toxaphene from fish in California have been attributed to buildup of extremely high residues in the food of these top carnivores. Thus, we may assume that as chlorinated hydrocarbons move up the rungs of the foodchain ladder and increase in residue concentrations, the higher the trophic level the more chance for acute or chronic toxicity in the animal population.

EFFECTS ON PHYSIOLOGY

Important alterations to physiological functions frequently result from pesticide contact with wildlife. This may be true for chronic exposures, as well as for acute situations. It is often said the chronic effects of pesticides on wildlife populations are more serious than acute effects. Chronic effects often go unnoticed so the problems may not even be recognized, in contrast to acute toxicity losses which attract attention and stimulate action. Fortunately, wildlife science has had its antennae tuned to the serious consequences of chronic pesticide intoxication, and an increasing proportion of wildlife-pesticide research in the past decade has been directed toward understanding chronic effects.

Reproduction

Of the chronic effects on wildlife after exposure to pesticides, reproductive change has received the greatest attention. This has happened partly because of the ease with which comparisons between untreated and treated groups of animals can be made in terms of quality and quantity of offspring produced, and partly because of its obvious importance to the welfare of animal populations. Most published material on pesticides and reproduction indicates reduced size of brood after contact, but yields little knowledge of the mechanism involved. Some studies go further, relating amount of exposure or residue content to reduced reproduction. A few studies have gone beyond and explain how the effect took place.

One important inquiry on lake trout and DDT (27) demonstrated the accumulation of DDT in lake trout flesh and eggs from several waters in New York. A DDT syndrome leading to mortality appeared in the developing trout fry. The weight of the fry at the time of development of the syndrome was related back to the amount of DDT in the ether-extracted oil of the eggs; 2.9 ppm DDT in the egg resulted in later appearance of the syndrome in the fry.

Insect chemosterilants have been tested against birds and fish. Exposures of *Coturnix* quail to single oral doses of 20 mg/kg and higher of apholate (141) resulted in effects on reproduction. Testis development was inhibited, with few spermatids and spermatozoa in the tubules. At high apholate levels, seminiferous tubules were small, and lumina were inconspicuous. Apholate delayed sexual maturity in the female quail which laid their first eggs at 49 days of age instead of at 41 days. Fifty percent production was attained at 50 days in the control group, and at 62 days in the group treated at 200 mg/kg.

Later studies on apholate and other alkylating agents and *Coturnix* quail (140) showed that reductions in egg production took place at 250 ppm apholate per day under conditions in which there was no effect on feed consumption. Egg fertilization and hatchability were negligible during the feeding period. After withdrawal of the sterilant, recovery was slow, with fertility at 42.1 percent in the 5th week, and near normal by the 7th week. At even higher levels of apholate, feed consumption was reduced, and egg production was completely inhibited.

The chemosterilant, TEPA, was tested against guppies, and repeated sublethal exposures showed effects on reproduction (149). The total young produced in high-level treatments was 7.8 percent of the total in the controls, and the low-treatment lots produced 55.7 percent of the control. Of six guppy pairs surviving the high treatment, only one pair reproduced, having four broods; in the controls, three pairs produced six broods, and two pairs produced five broods. The high-treatment lots produced an average of 2.8 young per parturition, the low-treatment lots 8.6, and the controls 13.6. The reproductive potential of the F_1 generation was apparently normal.

Another phase of the guppy-TEPA study stressed mechanism of the inhibitory influence. Single exposures of TEPA to guppies were made and fish were then paired in various combinations and reproduction observed. Results show no significant difference between reproduction in pairs having control males. Only 45 percent of the pairs having treated males reproduced, and there were significant delays in their broods. Histological changes took place in testes, but not in ovaries. The effect of TEPA is apparently specific to the spermatogenic cycle or male gamete, or both, in the guppy.

Unusual pathology was found in the testes of red-ear sunfish exposed to the herbicide Hydrothol® 191 (61). The testes of many fish contained large dark-staining cells resembling oogonia and primary oocytes and which displaced spermatogonia. Comparisons with normal ovarian cells show analagous structural details, but it is not verified that the Hydrothol actually caused a sex reversal.

Studies on DDT and cutthroat trout (4) showed that eggs from adults that had been exposed to selected amounts of DDT in the water for 30 minutes every 28 days for 15 months, and from adults fed 1 mg/kg once a week, were normal in their development; however, fry from these eggs suffered high mortality when the yolk-sacs were absorbed. The mortality was due to DDT derived from the parent fish.

In an estuary on the Texas coast, DDT residues in the ovaries of the speckled seatrout reached 8 ppm before spawning in 1968, and spawning was apparently unsuccessful later that year (31).

Brook trout experiments (108) showed that DDT fed at sublethal levels to maturing yearlings caused the females to produce fewer mature ova than the untreated fish, and the fish given the largest amounts of DDT had the fewest mature ova. In various matings from the experimental groups, eggs and fry suffered more mortality when at least one gamete came from a treated fish than when both gametes were from untreated fish. The greatest mortality among fry came during maximum utilization of the yolk-sac.

Walleye eggs and fry in Wisconsin (100) contained up to 9 ppm of DDT, but these residues were not associated with the success of the hatch or the median life span of the fry. Losses of fry took place after the fry had reached the feeding stage, but starvation was taken to be the cause of death. In an Alberta hatchery, rainbow, cutthroat, and brook trout eggs and fry were found to contain DDT, DDD, and DDE residues from contaminated feed given the brood stock. When more than 400 ppb were found in the eggs, there was a 30–90 percent mortality among the fry within 60 days following the swim-up stage (48).

Rainbow trout from New Zealand (84) lakes bore moderate amounts of the DDT complex in tissues, including testes and ovaries. Fry derived from these fish also contained DDT residues. In the hatchery, eggs and fry from adults with the highest residues suffered the highest mortality, with only 67.8 percent survival of fry.

The herbicide, diuron, in ponds (115) at initial concentrations of 0.5, 1.5, and 3.0 ppm, caused bluegill ovaries to be pale and nondescript at spawning time, and inhibition of reproduction took place. Depression of dissolved oxygen coincided with high diuron treatments; fish in treated ponds grew more slowly than unexposed fish, and low oxygen may have contributed to both abnormalities. Bluegills exposed to 2,4-D (45) spawned later than unexposed fish. The herbicide, dichlobenil, as a wettable powder, apparently depressed the numbers of bluegills produced after exposure to 40 ppm, but not to 10 ppm or in control ponds (43). Feeding and contact pond experiments with heptachlor and bluegills (11), and with mirex and bluegills and goldfish (158) did not affect reproduction.

Laboratory studies with guppies and dieldrin (33) revealed some effects on reproduction. In exposures near the "biologically safe concentration," exposed populations developed larger total numbers of young than the controls during the first two to three months; this is interpreted, not as a stimulus to reproduction, but as a change in feeding behavior of adults which normally prey on the immatures. This phenomenon disappeared after six months, and later in the experiment there was suggestive evidence that the dieldrin at the highest concentration had deleterious effects.

Reproduction in wild birds has been inhibited by pesticides, as deduced from population changes after exposure. Comparing pheasant eggs and chicks collected from DDT- and dieldrin-treated and untreated areas in California (89), researchers found more crippled and dead chicks from treated areas than from untreated areas. Heavy residues of DDT, DDE, and dieldrin prevailed in egg yolks and in the fat of hen pheasants from the treated territories. Similar events, with poor reproduction and large residues of the DDT complex in eggs and in fat of adults, are reported for herring gulls in Lake Michigan 81, 96). Unusual behavior, deserted nests, and very poor breeding success were associated with substantial dieldrin residues in eggs of red-winged blackbirds (70); and poor reproduction of woodcock in New Brunswick was found in areas treated with DDT (171).

Specific evidence of pesticide damage to reproduction in birds and mammals was found in experimental studies. Small numbers of offspring were found with house mice and laboratory mice fed DDT (20, 163), with rats fed chlordane (5) and heptachlor (117), and with white-tailed deer fed dieldrin (103). Survival effects have also been documented; dogs fed aldrin produced 12 pups, only 2 of which survived (99), rats fed chlordane had litters with low survival (5), rats fed aldrin suffered reduced oestrus frequency (15), deer fed dieldrin produced fawns with low survival (103), and rats fed heptachlor gave litters which survived poorly (117). The herbicide, 2,4,5–T, was teratogenic and fetocidal in strains of mice and rats. Cleft palate and cystic kidney were increased by the chemical (46).

Studies with birds have uncovered similar pesticide influence on reproduction. Diminished egg production has been induced by specific amounts of chlorinated hydrocarbon insecticides, such as DDT and Kepone,® in hens, pheasants, and quail (52, 53, 124, 135), and DDT and BHC in pheasants (13, 14). Several investigations have shown reduced egg production after the feeding of pesticides, but lowered food consumption and not the pesticides have been held responsible for the reduction. An effect more frequent than small clutch size has been poor survival of chicks. DDT, dieldrin, and toxaphene have had this effect on quail and pheasants (14, 51, 67) as have Kepone and mirex on chickens; several other studies have encountered poor hatchability of eggs attributable to insecticides. Other reproduction damage has been interference by DDT with spermatogenesis in chickens (3) and in bald eagles (106) and by heavy Kepone exposure in pheasants (53), delayed egg production with BHC and pheasants (13), nine-day delay of ovulation time with DDT and Bengalese finches (91), and unusual secondary sex characteristics, with Kepone inducing growth of female plumage on male pheasants and ducks (54). Residues of insecticides were measured in adults, eggs, or chicks in many of these studies; where analyses were made, high residues were found and help account for the effects.

A bizarre effect on bird reproduction has been found in many species in the past few years—decrease in eggshell thickness and strength. Raptorial birds were among the first to be studied with respect to disorders in calcium metabolism with consequent thin eggshells (80). The peregrine falcon in both Europe and parts of the United States has declined or been exterminated; thin eggshell has been recognized as a contributor, through parents breaking and eating their own eggs. Declining populations of bald eagles, ospreys, peregrines, and herring gulls have been associated with eggs with high residues of DDT and DDE in adults and with eggshells more than 19 percent thinner than normal. Experimental studies with sparrow hawks, DDT, and dieldrin have verified this mechanism of reproductive failure (129). Investigations of double-crested cormorants and white pelicans (6) have shown that DDE seems to be more responsible for thin eggshells than other insecticides or polychlorinated diphenyls. DDE has also been incriminated as the most severe cause of eggshell thinning, cracking, and embryo mortality in controlled studies on penned mallard ducks; DDD and DDT also impaired reproduction, but less seriously than did DDE (78).

The biochemical mechanisms in formation of thin eggshells in the presence of chlorinated hydrocarbon residues are suggested as being elevated steroid metabolism, abnormal vitamin metabolism, or inhibition of carbonic anhydrase and other enzyme systems. A study of ringdoves and DDT (128) showed that there was a decrease in blood estradiol early in the breeding cycle, a delay in egg-laying, a decrease in deposition of medullary calcium, and a decrease in eggshell weight. Injection of DDE reduced eggshell weight and inhibited carbonic anhydrase in the oviduct. In other work with Japanese quail, DDT and DDE (22) showed the insecticides to reduce carbonic anhydrase activity 16 to 19 percent in shell-forming glands. Other authors (92) relate reduction of eggshell thickness in pigeons to increases in thyroid weight and diminishing of colloid content of the follicles after feeding of DDT.

GROWTH

Pesticides affect growth in wildlife and may take several routes and have variable influences. Changes in growth of individuals have been measured in experimental work in the laboratory and in outdoor tests under controlled conditions, but reliable proof of effects on growth of birds, mammals, or fish in the wild is meager. Both direct and devious mechanisms are responsible for increases and decreases in sizes of animals.

Weight loss after pesticide ingestion or contact exposure has been demonstrated. *Coturnix* quail fed Azodrin® and Bidrin® sustained loss in weight (139), as did *Coturnix* and bobwhite quail fed Guthion® (69) and bluegills exposed to the detergent ABS (104) and bluntnose minnows and guppies treated with endrin (120).

In pond studies with bluegills and diuron, treated fish experienced retarded growth (115); the same was true with bluegills in pools treated monthly with the herbicide simazine and with sodium arsenite (68); with bluegills fed varied levels of endrin, the lots receiving the most endrin had the poorest weight increase. Silvex retarded growth of green sunfish and warmouth (85).

Some pesticide exposures have resulted in apparent growth acceleration, usually through indirect means. In a two-year study of cutthroat trout and DDT (4), the highest-treated fish were the largest at the end of the experiment. Here, it was theorized that selective mortality thinned out the weaker fish, leaving the most vigorous fish to survive and grow. Size effects were seen with redear sunfish in ponds treated with the herbicide Hydrothol 191; the highest-treated grew the fastest, after heavy mortality in the high-treated ponds left a few survivors to share the available food. A similar situation was observed in goldfish in ponds treated with 1 ppm of mirex (158). Treated fish were larger than the controls after 308 days; this was attributed to availability of food per fish, since mortality among the treated fish had reduced their numbers. This was also observed with 2,4-D and bluegills (45).

One author (160) has explained fish weight increase after endothall treatment in heavily weeded ponds as being related to increased availability of food after destruction of the aquatic weeds. Animal organisms became better available to the food chain and the fish, and thus increased fish production.

Weight increase effects through more direct means were measured in brook trout fed DDT (109) in experiments that featured constant feeding rates and no mortality. Trout fed 2 mg/kg of DDT per week for 31 weeks showed significantly greater weight gain than control fish, and the author considered the difference to be a direct effect of exposure to DDT. Males grew faster than females; the author explains this as storage of DDT in eggs developing in the adults, precluding the effect on body growth.

A common cause of weight loss in wild animal populations is starvation after pesticide application. Weight loss may be followed by death, as observed with sparrows, turkeys (7), woodcock (147), and Atlantic salmon (62). Studies of interactions of pesticides with other factors have emphasized the fact that starvation in mammals lowers liver microsomal activity

(56) and the protection provided by storage of pesticides in body fat is reduced or negated. These factors combine to help increase susceptibility to pesticide toxicity in starved animals. Suckers and whitefish in the Yellowstone River (36) died in large numbers three-four months after a DDT spray. The fish-food organisms in the stream suffered heavy mortality immediately after the spray, and the fish starved and became emaciated. When fat reserves were used by the fish, the DDT evidently became mobilized in the blood and contributed to fish death.

Growth may be affected in yet other ways. In experimental work, pesticide-contaminated feed is often rejected by laboratory animals; it is suggested that natural foods in the wild may be contaminated and have a repellant effect, causing the animal to seek other food, or starve and lose weight. Another mechanism that may be important in nature is an effect on food-seeking sensory organs, placing the animal at a disadvantage in competing for food. Detergents have this effect on chemoreceptors on the barbels of yellow bullheads, with erosion of taste buds and impairment of receptor function (16).

A spectrum of growth effects was illustrated in hard clam and American oyster bioassay studies (50). Embryos and larvae of these molluscs were exposed to a wide array of insecticides, herbicides, solvents, bactericides, fungicides, and algicides. Some pesticides caused slow growth in larvae but not in embryos, but most of the compounds affected embryonic development more than larval growth. Some compounds accelerated growth of larvae; it was suggested that these larvae grew fast because of bacteriostatic action of the pesticides or chelating effect. The idea has also been advanced that fast growth in pondfish after exposure to insecticides is due to reduction of parasite populations in the fish.

The reduction of adult oyster shell growth by pesticides is so positive and predictable that it is routinely used as an index to pesticide exposure (31). Shells of living oysters are filed smooth on their edges, measured, and the oysters exposed to pesticides in sublethal amounts. Shell growth of exposed oysters is then compared to the greater growth of control animals, giving an indication of amount of oyster exposure to pesticides.

Nervous System and Behavior

Some key roles of the nervous system in pesticide toxicity are immediately apparent and are well documented. The importance of the brain as a storage site was discussed earlier. The actions of organophosphates and carbamates as cholinesterase inhibitors will be treated later in enzyme discussions. The dramatic nervous symptoms of animals in acute distress from pesticide poisoning—nervousness, irritability, tremors, convulsions—have been repeatedly observed and described. To go beyond these, several careful studies have revealed pesticide-induced changes to bring about a variety of subtle, but important, functional disorders arising through the central or the peripheral nervous system.

Actual damage to the nervous system, with consequent behavioral aberrations, was measured in the barbels of yellow bullheads after exposure to 0.5 ppm of the detergents, LAS and ABS (16), a concentration far below any producing other sublethal damage in the past. This damaged the photoreceptors of the barbels, a serious effect on fish that find their food with these organs. Histological examination showed erosion of the taste buds, while electrophysiological measurements and observations of swimming and feeding showed impairment of receptor function. After six weeks in clean water, recovery was still only partially complete. Another scientist (102) found heptachlor and nicotine sulfate to curl and inactivate the barbels of another catfish, *Heteropneustes fossilis,* and described histopathological changes, but no behavioral disorders were reported from this study.

Pesticide influences on fish adaptations to temperature have been demonstrated. From a study with Atlantic salmon (127) came information that sublethal amounts of DDT influenced temperature selection by the fish. Low doses of the insecticide resulted in a downward shift in selected temperature, while higher doses made the shift go upward. The effect was more significant in salmon conditioned at 17°C than in those conditioned at 8°C; the authors suggest that DT may interfere with the normal acclimatization mechanism of the fish. Later investigation on the lateral line of brook trout (8) showed that response to pressure waves was influenced by 24-hr exposure to DDT. The lateral line nerve is rendered hypersensitive to the stimulus, especially at low temperatures, and the duration of response has a negative temperature coefficient. This increase in sensitivity might help explain the cold response described above. The central nervous system is apparently the site of pesticide sensitivity most responsible for changes in complex fish behavior (9). To support this, DDT (as well as acclimatization temperature) was shown to increase the cold-block temperature of the propeller tail in the brook trout; the cold blockage is situated in the spinal cord. Additional evidence on the central nervous system came from visual conditioning of a brook trout avoidance response formed in the optic tectum. DDT evidently interfered with the fishes' ability to form an association between an open doorway and escape from shock.

Other investigators reported on a sophisticated scheme for studying behavioral pathology in fish (164). Their work is an exploration of methodology capable of detecting and quantitatively recording a variety of responses linked with various processes in living organisms. Their apparatus, a continuous-flow system for exposing fish to water-pesticide mixtures, controls light and dark, and electric shock, in various combinations. In operation, the system tests the response of fish to a light stimulus, association of light with a shock stimulus, ability to unlearn the light-avoidance response, response to temperature changes, and response to vibration and sound stimuli. With this apparatus behavior patterns of untreated and toxaphene-treated fish were measured with four species. With 1.8μg/l of toxaphene, goldfish showed a heightened response to external stimuli in terms of total movement. For movement in relation to light stimulus, the toxaphene-exposed group was again significantly different from the control. A week later, memory was demonstrated, with the retention of the conditioned avoidance response, and the treated fish reacting more adversively than the control. With TEPP, experiments produced behavioral pathology at less than 1/200 of the 96-hr TL_m.

Pesticide-induced unusual feeding behavior in fish may deviate in several ways. An interesting experiment (33) revealed that in guppies exposed to dieldrin the exposed populations had more individuals than did the control populations. This was ascribed to dieldrin having lessened predatory and harassing behavior in the treated lots, with normal predation on immatures taking place in control lots.

Experiments with birds and mammals have also uncovered nervous system and behavior alterations. Brain waves have been altered; endrin injected into pigeons (132) increased brain-wave amplitudes in brain areas related to vision. Very low amounts of DDT reduced the ability of rats (56) to swim in cold water.

RESPIRATION

Pesticide interference with the vital respiration process has been inadequately studied in birds, mammals, and fish, yet conclusive evidence is at hand to show that oxygen and carbon dioxide transport often do not proceed normally after pesticide exposure. Some pesticides impair fish respiration by lowering oxygen levels in the water, as described earlier with diuron in bluegill ponds. More immediate interference comes about from pesticide effects on fish gills and blood and on bird and mammal lungs and blood.

Wood (unpublished data) described damage to fish gills showing engorgement of the vessels of the gill filaments and fusion of the lamellae in bluegills after exposure to the herbicide dichlobenil, and in trout after exposure to malathion, parathion, carbaryl, and heptachlor, to an aromatic solvent used for aquatic weed control, and to 4-(2,4-dichlorophenoxy) butyric acid, another aquatic herbicide. Others (115) found gill damage in bluegills exposed to the herbicide diuron, with the development of hemmorrhagic globes on lamellae, edema in the branchial cavity, fusion of lamellae, and later development of basal branchial epithelial hyperplasia in more than half of the lamellae. Some of these effects are specific for certain pesticides, and some are nonspecific; many are severe in terms of reducing the amount of exchange surface on the gills.

In preliminary work at the Fish Pesticide Research Laboratory, physiologists made in vitro exposures of carp blood to DDT, and concluded that the DDT decreased the plasma iron, plasma iron-binding capacity, hematocrit, total protein, and plasma osmolarity (71). Here, the total consequence to respiration in the whole animal is not known. In several long-term studies involving prolonged exposures of various pesticides and fish in ponds, tanks, and raceways, hematocrit measurements of the treated fish have usually not been different from those of the controls.

Another study (34) found significant increases in oxygen consumption by pumpkinseed sunfish in the presence of 1.68 ppb of dieldrin, and a significant reduction of cruising speed. TFM, a herbicide and lampricide, has been thought to act on sea lampreys by interference with oxygen transport, and produce either methemoglobin or high acidity in the blood to cause anoxia (2). This supposition was made because lethal doses of TFM cause vasodilation of gill filaments and liver, and edema of connective tissues, and because nitro-aromatic compounds oxidize hemoglobin to methemoglobin. Agris compared electrocardiograms of lampreys exposed to TFM and those in anoxia, and found the two patterns to be different from each other and different from the control. The anoxia pattern showed increased cardiac output and rate of blood passage through the gills, while the TFM pattern was indicative of drug action. Lampreys killed with TFM had red blood and no methemaglobin, and the vasodilation of gills and liver is considered to be consistent with effects of nitrite groups. So, although some pesticides do interfere with respiration in fish, TFM acts in a different way.

Respiratory failure is a common proximate cause of death from acute intoxication of birds and mammals from chlorinated hydrocarbon insecticides, often after hemorrhages in the lungs. This has been reported for woodcocks exposed to heptachlor (147). Rapid respiration is described in blue grouse dying from phosphamidon exposure (65).

Body Chemistry

Some pesticide-caused changes in body chemistry have been touched upon in previous parts of this article; let us now consider some others. Science has learned something of intermediary metabolism and its disruption when poisons enter the body, and some insight has been gained into changes in mineral imbalance of tissues and organs after exposure. In some cases the alterations in body chemistry have been understood in relation to basic aspects of physiology, but not in others.

Among the earliest-recognized and most-studied pesticide effects was the reduction of cholinesterase activity by the actions of organophosphates and carbamates (123). This topic is important because of the high toxicities of many of these agents to birds and mammals, and because of the promise it shows in using cholinesterase depression in brains and blood as an index of severity of exposure.

Weiss (168) pioneered fish studies on brain and blood cholinesterase inhibition and recovery in several species, and several workers have sought similar information for bird and mammal species. Their findings agree with the accepted view that organophosphates kill by inhibiting cholinesterase, with changes in nervous activity by the accumulation of acetylcholine at nerve endings.

More recent studies (1) on cholinesterase inactivation by carbamates in the shiner perch brain yielded comparative information on abilities of six insecticides to achieve 50 percent inhibition, and also showed that sodium, potassium, calcium, magnesium, and manganese ions activate acetylcholinesterase activity, and nickel, copper, lead, mercury, and iron ions inhibit AChE activity.

Reports in the literature claim that death occurs in several species of fresh-water fish whose brain cholinesterase is inhibited by 30–60 percent (166). This has been verified in several other studies, but in some cases greater reductions have taken place without death. Much remains to be learned about the kinds of cholinesterase that damage fish nervous systems, chronic effects, rates of recovery, and counteracting measures. Cholinesterase inhibition as we know it in fish brains shows too much variation to be reliably useful as a quantitative indicator of pesticide exposure, but can be helpful in a general qualitative way.

In mammals, carbamates and organophosphates, through phosphorylation and carbamylation, also inhibit acetylcholinesterase (95). Inhibition from organophosphates is slowly reversed by hydrolysis of the phosphorylated enzyme; inhibition from carbamates is also reversible. It has been shown that rat serum aliesterase activity rose within a few days after acute doses of aldrin, chlordane, DDT, heptachlor, and dieldrin, and that pretreatment of rats with aldrin reduced the toxicity of parathion by 7 times and TEPP by 5 times.

Goats, deer, coyotes, rabbits, pheasants, grouse, turkeys, chickens, ducks, eagles, hawks, prairie chickens, Hungarian partridges, and terns (47) were treated with malathion, phosphamidon, Baytex®, Dibrom®, DRC-714, parathion, and dimethoate. Marked depression of blood cholinesterase took place, but no two species reacted identically to any one insecticide. The cholinesterase activity in mammals was generally concentrated in the red blood cells except in rabbits and coyotes, which had the activity in the plasma; among birds it was in the plasma.

Another enzyme associated with the nervous system and which has been shown to be affected by insecticides is adenosine triphosphatase. Some of the adenosine triphosphatases take part in ion transport in the nervous system, and chlordane and other chlorinated hydrocarbon insecticides have been shown to inhibit these enzymes partially in rabbit brain (114). It has been demonstrated, also, that the action of Na, K, Mg-adenosine triphosphatase found in nerve endings of rat brain was inhibited by DDT and its analogs. The amount of inhibition was closely related to general toxicity in vivo of these compounds. Inhibition by DDT was 1000 times as great as that by DDE, and inhibition at low temperatures was higher than at high temperatures. This interference with ion transport is suggested as being involved in DDT poisoning, although it is perhaps not the principal cause, since nerve endings, not synapses, are known to be the main site for DDT attack.

Some chemical components of aquatic animals have been studied before and after exposure to insecticides to learn of any serious alterations. In northern puffers exposed in graded concentrations (0.05–1.0) of endrin (59) for 96 hr, mean hemoglobin content, relative liver size, serum chloride, gamma globulin, and uric acid remained essentially the same as those of the controls. Serum sodium, potassium, calcium, and cholesterol levels were higher than the controls, but levels of sodium, potassium, calcium, magnesium, and zinc in the liver were lower. Liver function was impaired in the exposed fish through transfer of these cations from liver to serum and by elevated cholesterol. Other investigations of the same species exposed to methoxychlor or methyl parathion (58) showed no changes in levels of sodium, potassium, calcium, magnesium, zinc, or iron in whole blood or serum after exposure to 30 ppb of methoxychlor for 45 days. But, fish exposed to 20,200 ppb of methyl parathion or to a combination of 15 ppb of methoxychlor and 10,100 ppb of methyl parathion underwent changes in feeding behavior and an increased rate of death after 96 hours. Survivors had low hemoglobin, low erythrocyte count, low hematocrit, complete inhibition of serum esterase enzymes, low liver magnesium, and low liver and gill zinc. It is suggested that these patterns offer a means of identifying unfavorable aquatic environments before more striking morphological or physiological changes occur.

Quahog clams exposed to graded concentrations of malathion and methoxychlor for 96 hr (60) also underwent tissue cation changes. Muscle and mantle calcium were reduced after exposure to each insecticide. Muscle and mantle potassium were increased by methoxychlor but reduced by lower concentrations of malathion. Mantle zinc and muscle sodium were increased by all concentrations of each insecticide, except one level of malathion. These patterns may also give an index of stresses of the environment, as with the puffers discussed above.

Cutthroat trout exposed to endrin in feed and by contact over a period of several months (125) sustained a variety of changes in tissue cations. Iron, manganese, copper, zinc, phosphorus, calcium, magnesium, dry matter, and ash were measured in several tissues, and the changes were influenced by kind and concentration of treatment, time, sex, and the acetone solvent used as a vehicle for the endrin in the contact exposures. In addition to mineral imbalance, endrin induced prolonged blood coagulation time and edema.

Pesticide effects on the thyroid have come to light from studies on several animal species. Hyperthyroidism appears to be induced in chickens and Bengalese finches (91), with accompanying weight loss, decreased testis weight, reduced ability to develop depot fat, and altered calcium metabolism resulting in thick eggshells. Since thin eggshells have been attributed to DDT-induced calcium metabolism disturbances in peregrine falcons and sparrow hawks, it appears that DDT affects calcium metabolism differently in different species. It is known that thyroxine has opposite effects on the concentration of calcium in the plasma of laying ducks and hens. Further research with pigeons fed DDT (92) revealed increases in liver weight and thyroid weight with increases of DDT residues in the liver. Moreover, affected thyroids suffered reduction in follicular size and colloid amount associated with hyperplasia of the epithelium. These conditions could indicate hypo- or hyperfunction of the thyroid, depending upon activity of the liver and pituitary, and stimulation or suppression of thyroxine formation or metabolism. The result, in nature, may be the induction of hypothyroidism in many wild birds by DDT, causing thin eggshells.

Studies of liver enzymes in coho salmon and rainbow trout were made (25) to compare responses with those known in mammals. Ingestion of DDT by salmon and trout failed to alter hepatic 6-phosphate dehydrogenase activity. Liver size decreased in small salmon but not in larger trout. These effects may differ from those in rats because DDT is unable to induce hepatic microsomal enzymes, possibly because the insecticide is so completely stored in lipids and does not reach sufficient concentrations in blood or the target organ, or because of the apparent absence of dietary or insulin control of this enzyme in fish.

Research on mammal and bird livers (56, 143) has revealed a variety of pesticide-induced stimuli of activity of hepatic microsomes, including oxidation of drugs and hydroxylation of steroids.

There is interest in the structural and chemical similarities of DDT and the estrogen, diethylstilbestrol, and the effects of the two have been studied in birds and mammals (21). Different results have emerged from similar studies, with both positive and negative evidence on DDT as an estrogen, but it seems clear that p,p'-DDT is only weakly estrogenic, while o,p'-DDT is more active. Chickens, *Coturnix* quail, and rats were injected with the two isomers of DDT and with 17 β-estradiol. The o,p'-DDT and the estradiol produced the same effects—100 percent increase in oviduct weight, increase in uterus net weight, increase in oviduct and uterus water content and RNA, and large increases in glycogen content of oviduct and uterus; little, if any, estrogenic activity was shown by p,p'-DDT. These effects are apparently not related to pesticide content in the oviduct, since residue measurements showed the two isomers to accumulate in the same amounts. In rats, p,p'-DDT residues were five times those of o,p'-DDT in the uterus. The significance of estrogenic properties of DDT in terms of fertility and reproduction in wild fish, birds, and mammals remains to be established.

PATHOLOGY

Few chronic studies on pesticides and wildlife have failed to reveal some form of pathology when histopathological examination has been made as part of the study. Several disorders described in the foregoing parts of this paper have been accompanied by pathological lesions, some showing morphological aberrations and some demonstrated by histochemical investigations.

In fish, gill lesions have been very common, and many have been nonspecific in nature. Gill pathology has appeared through irritating action by DDT, methoxychlor, malathion, methyl parathion, heptachlor, mirex, endrin, carbaryl, dieldrin, and the herbicides sodium arsenite, endothall, TFM, dichlobenil, diuron, and silvex in selected species. That the fish liver is an especially vulnerable site is shown by the frequency of liver pathology in several fish species; damage is described from exposure to silvex, sodium arsenite, toxaphene, lindane, parathion, malathion, methyl parathion, heptachlor, methoxychlor, endrin, diuron, dichlobenil, and 2,4-D, which produced the glycogen removal described earlier. Pathology in fish kidney has been caused by sodium arsenite, lindane, parathion, mirex, diuron, and dichlobenil. Other pesticides have induced morphological changes in ovaries, testes, intestines, and adrenals. Many of these pathological effects, especially those in the gills, run a benign course if the fish lives.

Histopathological studies have not been extensive in wild bird or mammal species, but pesticide involvement has caused pathology in laboratory rats. Carcinogenesis attributable to chlorinated hydrocarbon pesticides has been reported (18), and heptachlor administered over a long period has produced cataracts (117). Enlargement of liver, kidney, and brain, necrosis in kidney, degeneration of adrenals and brain, and hyperemia and edema in lungs were found in cats, monkeys, rats, and guinea pigs after exposure to endrin (153).

EFFECTS THROUGH THE ENVIRONMENT

Factors aside from toxicity through direct exposure frequently come into play to damage wildlife and fish by causing alterations of the environment. Among the obvious of these indirect effects is the suppression of food. With partial suppression, wildlife may select alternate foods and survive; or, they may ingest affected foods and take in the pesticide in greater amounts than those received directly through the gills, lungs, or skin. When the natural food supply is seriously depleted (36, 62), starvation may result in the birds, fish, or mammals, inducing changes in growth or productivity or leading to mobilization of fat-stored pesticide residues, acute toxicity, and death. Variations of food destruction effects may take several forms as different parts of complicated food webs are struck by pesticides; not only may a stratum in the food succession be lost, but behavioral changes may make some food organisms unusually vulnerable to predation or the action of water currents.

Herbicides often indirectly affect fish populations through destruction of aquatic vegetation. In a pond or lake without circulation, significant amounts of carbon dioxide may be released into the water when the plants fall to the bottom and decompose, and fish may die. In other cases, pesticides applied to ponds may deplete the oxygen in the water within a few hours, and thus place fish in jeopardy.

Other sorts of subtle, but significant, consequences occur as pesticides reach different ecosystems. One example is the inhibition of photosynthesis in marine phytoplankton by chlorinated hydrocarbon insecticides (116). Phytoplankters, *Skeletonema, Dunaliella, Coccolithus,* and *Cyclotella,* from different oceanic environments were tested against ^{14}C-labeled DDT, dieldrin, and endrin at various concentrations. Amounts of uptake varied with the kind of organism, the kind and concentration of insecticide, and time. So, also, were there differences in growth rates, cell division, and photosynthesis. Of the species studied, only *Dunaliella* remained insensitive to the insecticides. The fundamental importance of photosynthesis in natural waters leads us to suspect potential damage to fish when there is interference.

Just as chronic toxicity may frequently be more damaging to a population than is acute toxicity, indirect effects on wild animals can often have more serious consequences than sublethal direct influences.

EPILOGUE

What was learned in the 1960's about pesticides and wildlife was only a beginning, but it was a good beginning. As acute toxicity studies gave way to more sophisticated investigations of chronic toxicity, momentum increased, and the scope of interest broadened, penetrating on the one hand to effects at the cell level and on the other to population dynamics. Air, water, and terrestrial environments have come under inspection, involving soil physics and chemistry, interactions of microorganisms, the place of weather and climate in distribution of pesticides, the oceans, estuaries, and the big rivers.

With the start we have, with attention from new disciplines, with more precise and versatile measuring instruments, and with our focus on the wild as well as in the laboratory, fishery and wildlife science should go far toward a better understanding of the interactions among mammals, fish, birds, and economic poisons in the next decade.

LITERATURE CITED

1. Abou-Donia, M. B., Menzel, D. B. 1967. Fish brain cholinesterase: Its inhibition by carbamates and automatic assay. *Comp. Biochem. Physiol.* 21:99–108
2. Agris, P. F. 1967. Comparative effects of a lampricide and of anoxia on the sea lamprey. *J. Fish. Res. Bd. Can.* 24:1819–22
3. Albert, T. F. 1962. The effect of DDT on the sperm production of the domestic fowl. *Auk* 74:104–7
4. Allison, D. B., Kallman, B. J., Cope, O. B., Van Valin, C. C. 1964. Some chronic effects of DDT on cutthroat trout. *Bur. Sport Fish. Wildl. U.S., Res. Rep.* 64 30 pp.
5. Ambrose, A. M., Christensen, H. E., Robbins, D. J., Rather, L. J. 1953. Toxicological and pharmacological studies on chlordane. *Arch. Ind. Hyg. Occup. Med.* 7:197–210
6. Anderson, D. W., Hickey, J. J., Ricebrough, R. W., Hughes, D. F., Christensen, R. E. 1969. Significance of chlorinated hydrocarbon residues to breeding pelicans and cormorants. *Can. Field-Nat.* 83:91–112
7. Anderson, R. W., Blakely, R. M., MacGregor, H. I. 1951. The effect of aldrin on growing turkeys. *Poultry Sci.* 30:905
8. Anderson, J. M. 1968. Effect of sublethal DDT on the lateral line of brook trout, *Salvelinus fontinalis*. *J. Fish Res. Bd. Can.* 25:2677–82
9. Anderson, J. M., Peterson, M. R. 1969. DDT: sublethal effects on brook trout nervous system. *Science* 164:440–41
10. Anderson, R. B., Fenderson, O. C. 1970. An analysis of variation of insecticide residues in the landlocked Atlantic salmon (*Salmo salar*). *J. Fish. Res. Bd. Can.* 27:1–11
11. Andrews, A. K., Van Valin, C. C., Stebbings, B. E. 1966. Some effects of heptachlor on bluegills (*Lepomis macrochirus*). *Trans. Am. Fish. Soc.* 95:297–309
12. Arterberry, J. D. Bonifaci, R. W., Nash, E. W., Quinby, G. E. 1962. Potentiation of phosphorus insecticides by phenothiazine derivatives.

J. Am. Med. Assoc. 182:848–50
13. Ash, J. S., Taylor, A. 1964. Further trials on the effects of gamma BHC seed dressings on breeding pheasants. *Game Res. Assoc., 4th Annu. Rep.* 14–20
14. Azevedo, J. A., Jr., Hunt, E. G., Woods, L. A., Jr. 1965. Physiological effects of DDT on pheasants. *Calif. Fish Game* 51:276–93
15. Ball, W. L., Kay, K., Sinclair, J. W. 1953. Observations on the toxicity of aldrin. 1. Growth and estrus in rats. *Arch. Ind. Hyg. Occup. Med.* 7:292–300
16. Bardach, J. E., Fujiya, M., Holl, A. 1965. Detergents: effects on the chemical senses of the fish *Ictalurus natalis* (Le Seuer). *Science* 148:1605–7
17. Barker, P. S., Morrison, F. O. 1966. The basis of DDT tolerance in the laboratory mouse. *Can. J. Zool.* 44:879–87
18. Barnes, J. M. 1966. Carcinogenic hazards from pesticide residues. *Residue Rev.* 13:69–82
19. Barthel, W. F., Murphy, R. T., Mitchell, W. G., Corley, C. 1960. The fate of heptachlor in the soil following granular application to the surface. *J. Agr. Food Chem.* 8:445–47
20. Bernard, R. F., Gaertner, R. A. 1964. Some effects of DDT on reproduction in mice. *J. Mammal.* 45:272–76
21. Bitman, J., Cecil, H. C., Harris, S. J., Fries, G. F. 1968. Estrogenic activity of o,p'-DDT in the mammalian uterus and avian oviduct. *Science* 162:371–72
22. Bitman, J., Cecil, H. C., Fries, G. F. 1970. DDT-induced inhibition of avian shell gland carbonic anhydrase: A mechanism for thin eggshells. *Science* 168:594–96
23. Boyd, C. E., Vinson, S. B., Ferguson, D. E. 1963. Possible DDT resistance in two species of frogs. *Copeia* 1963:426–29
24. Brungs, W. A., Mount, D. I. 1966. Lethal endrin concentration in the blood of gizzard shad. *J. Fish. Res. Bd. Can.* 24:429–31
25. Buhler, D. R., Benville, P. 1969.

Effect of feeding and of DDT on the activity of hepatic glucose 6-phosphate dehydrogenase in two salmonids. *J. Fish. Res. Bd. Can.* 26:3209–16

26. Buhler, D. R., Shanks, W. E. 1970. Influence of body weight on chronic oral DDT toxicity in coho salmon. *J. Fish. Res. Bd. Can.* 27:347–58

27. Burdick, G. E., Harris, E. J., Dean, H. J., Walker, T. M., Skea, J., Colby, D. 1964. The accumulation of DDT in lake trout and the effect on reproduction. *Trans. Am. Fish. Soc.* 93:127–36

28. Butler, P. A. 1964. Commercial fishery investigations, 5–28. In *Pesticide-Wildlife Studies, 1963. Fish Wildl. Serv., U.S., Cir. 199* 130 pp.

29. Butler, P. A. 1965. Commercial fishery investigations, 65–77. In The Effects of Pesticides on Fish and Wildlife, 1964. *Fish Wildl. Serv., U.S., Cir. 226* 77 pp.

30. Butler, P. A. 1966. Pesticides in the marine environment. *J. Appl. Ecol.* 3 (Suppl.):253–60

31. Butler, P. A. 1969. Monitoring pesticide pollution. *BioSci.* 19:889–91

32. Cain, S. A. 1965. Pesticides in the environment, with special attention to aquatic biology resources. *Rep. on U.S.-Japan Planning Meet. Pesticide Res., Honolulu, 1965* 12–18

33. Cairns, J., Jr., Foster, N. R., Loos, J. J. 1967. Effects of sublethal concentrations of dieldrin on laboratory populations of guppies, *Poecillia reticulata* Peters. *Proc. Acad. Nat. Sci. Philadelphia* 119:75–91

34. Cairns, J., Jr., Scheier, A. 1964. The effect upon the pumpkinseed sunfish, *Lepomis gibbosus* (Linn.), of chronic exposure to lethal and sublethal concentrations of dieldrin. *Notulae Naturae. Acad. Nat. Sci. Philadelphia* 370. 10 pp.

35. Chadwick, G. G., Brocksen, R. W. 1969. Accumulation of dieldrin by fish and selected fish-food organisms. *J. Wildl. Manage.* 33:693–700

36. Cope, O. B. 1961. Effects of DDT spraying for spruce budworm on fish in the Yellowstone River sys-

tem. *Trans. Am. Fish. Soc.* 90:239–51

37. Cope, O. B. 1964. Sport fishery investigations, 29–43. In Pesticide-Wildlife Studies, 1963. *Fish Wildl. Serv. U.S., Cir. 199* 130 pp.

38. Cope, O. B. 1965. Sport fishery investigations, 51–64. In The Effects of Pesticides on Fish and Wildlife, 1964. *Fish Wildl. Serv. U.S., Cir. 226* 77 pp.

39. Cope, O. B. 1965. Agricultural chemicals and fresh-water ecological systems. In *Research in pesticides,* 115–28. New York:Academic, 380 pp.

40. Cope, O. B. 1966. Contamination of the fresh-water ecosystem by pesticides. *J. Appl. Ecol.* 3 (Suppl.):33–44

41. Cope, O. B. 1969. Fish-Pesticide Research Laboratory. In Progress in Sport Fishery Research, 1968. *Bur. Sport Fish. Wildl. U.S., Res. Publ. 77* 92–113

42. Cope, O. B., Gjullin, C. M., Storm, A. 1949. Effects of some insecticides on trout and salmon in Alaska, with reference to black fly control. *Trans. Am. Fish. Soc.* 77:160–77

43. Cope, O. B., McCraren, J. P., Eller, L. 1969. Effects of dichlobenil on two fishpond environments. *Weed Sci.* 17:158–65

44. Cope, O. B., Park, B. C. 1957. Effects of forest insect spraying on trout and aquatic insects in some Montana streams. *U.S. Forest Serv.,* 53

45. Cope, O. B., Wood, E. M., Wallen, G. H. 1970. Some chronic effects of 2,4-D on the bluegill (*Lepomis macrochirus*). *Trans. Am. Fish. Soc.* 99:1–12

46. Courtney, K. D., Gaylor, D. W., Hogan, M. D., Falk, H. L., Bates, R. R., Mitchell, I. 1970. Teratogenic evaluation of 2,4,5-T. *Science* 168:864–66

47. Crabtree, D. G. 1965. Wildlife Studies, Denver Wildlife Research Center, 31–48. In The Effects of Pesticides on Fish and Wildlife. *Fish Wildl. Serv. U.S., Cir. 226* 77 pp.

48. Cuerrier, J. P., Keith, J. A., Stone, E. 1967. Problems with DDT in fish culture operations. *Natur. Can.* 94:315–20

49. Cueto, C., Jr., Hayes, W. J., Jr. 1967. Effect of repeated administration of phenobarbital on the metabolism of dieldrin. *Ind. Med. Surg.* 546–51

50. Davis, H. C., Hidu, H. 1969. Effects of pesticides on embryonic development of clams and oysters and on survival and growth of the larvae. *Fish Wildl. Serv. U.S., Fishery Bull.* 67:393–404

51. Decker, G. C. 1966. Significance of pesticide residues: practical factors in persistence. *Ill. Natur. Hist. Surv. Div. Biol. Notes No. 56.* 8 pp.

52. DeWitt, J. B. 1955. Effects of chlorinated hydrocarbon insecticides upon quail and pheasants. *Agr. Food Chem.* 3:672–76

53. DeWitt, J. B., Crabtree, D. G., Finley, R. B., George, J. L. 1962. Effects on wildlife, 4–15. In Effects of Pesticides on Fish and Wildlife. A review of investigations during 1960. *Bur. Sport Fish. Wildl. U.S., Cir. 143* 52 pp.

54. DeWitt, J. B., Stickel, W. H., Springer, P. F. 1963. Wildlife studies, Patuxent Wildlife Research Center, 1961–1962, 74–100. In Pesticide-Wildlife Studies: A Review of Fish and Wildlife Service Investigations During 1961 and 1962. *Fish Wildl. Serv., U.S., Cir. 167,* 109 pp.

55. Dimond, J. B. 1969. DDT in Maine forests. *Maine Agr. Exp. Sta. Misc. Rep. 125* 1–13

56. Durham, W. F. 1967. The interaction of pesticides with other factors. *Residue Rev.* 18:21–103

57. Edwards, C. A. 1966. Insecticide residues in soils. *Residue Rev.* 13:83–132

58. Eisler, R. 1967. Tissue changes in puffers exposed to methoxychlor and methyl parathion. *Bur. Sport Fish. Wildl., U.S. Tech. Paper 17,* 15 pp.

59. Eisler, R., Edmunds, P. H. 1966. Effects of endrin on blood and tissue chemistry of a marine fish. *Trans. Am. Fish. Soc.* 95:153–59

60. Eisler, R., Weinstein, M. P. 1967. Changes in metal composition of the quahog clam, *Mercenaria mercenaria,* after exposure to insecticides. *Chesapeake Sci.* 8:253–58

61. Eller, L. L. 1969. Pathology in redear sunfish exposed to Hydrothol 191. *Trans. Am. Fish. Soc.* 98:52–59

62. Elson, F. P. 1967. Effects on wild young salmon of spraying DDT over New Brunswick forests. *J. Fish. Res. Bd. Can.* 24:731–67

63. Ferguson, D. E. 1967. The ecological consequences of pesticide resistance in fishes. *Trans. 32d N. A. Wildl. Conf.* 103–7

64. Ferguson, D. E., Culley, D. D., Cotton, W. D., Dodds, R. P. 1965. Resistance to chlorinated hydrocarbon insecticides in three species of freshwater fish. *BioSci.* 14:43–44

65. Finley, R. B., Jr. 1965. Adverse effects on birds of phosphamidon applied to a Montana forest. *J. Wildl. Manage.* 29:580–91

66. Gakstatter, J. H., Weiss, C. M. 1967. The elimination of DDT-C^{14}, dieldrin-C^{14} and lindane-C^{14} from fish following a single sublethal exposure in aquaria. *Trans. Am. Fish. Soc.* 96:301–7

67. Genelly, R. E., Rudd, R. L. 1956. Effects of DDT, toxaphene, and dieldrin on pheasant reproduction. *Auk.* 73:529–39

68. Gilderhus, P. A. 1966. Some effects of sublethal concentrations of sodium arsenite on bluegills and the aquatic environment. *Trans. Am. Fish. Soc.* 95:289–96

69. Gough, B. J., Escuriex, L. A., Shellenberger, T. E. 1967. A comparative toxicologic study of a phosphorodithioate in Japanese and bobwhite quail. *Toxicol. Appl. Pharmacol.* 10:12–19

70. Graber, R. R., Wunderle, S. L., Bruce, W. N. 1965. Effects of a low-level dieldrin application on a red-winged blackbird population. *Wilson Bull.* 77:168–74

71. Grant, B. F., Mehrle, P. 1969. Pesticide effects on fish endocrine functions, 97–100. In Progress in Sport Fishery Research, 1968. *Bur. Sport Fish. and Wildl., U.S. Res. Publ. 77,* 259 pp.

72. Greaves, J. H., Ayres, P. 1967. Heritable resistance to warfarin in rats. *Nature* 215:877–78

73. Gunther, F. A., Westlake, W. E., Jaglan, P. S. 1968. Reported solubilities of 738 pesticide chem-

icals in water. *Residue Rev.* 20: 1–148.

74. Hamacker, J. W. 1966. Mathematical prediction of cumulative levels of pesticides in soil. In Organic Pesticides in the Environment. *Advan. Chem. Ser.* 60:122–31

75. Hamelink, J. L., Waybrant, R. C., Ball, R. C. A proposal: Exchange equilibria control the degree chlorinated hydrocarbons are biologically magnified in lentic environments. *Trans. Am. Fish. Soc.* In press

76. Harris, C. R., Lichtenstein, E. P. 1961. Factors affecting volatilization of insecticidal residues from soils. *J. Econ. Entomol.* 54:1038–45

77. Hayes, W. J., Jr. 1965. Review of the metabolism of chlorinated hydrocarbon insecticides especially in mammals. *Ann. Rev. Pharmacol.* 5:27–52

78. Heath, R. G., Spann, J. W., Kreitzer, J. F. 1969. Marked DDE impairment of mallard reproduction in controlled studies. *Nature* 224:47–48

79. Henderson, C., Pickering, Q. H. 1958. Toxicity of organic-phosphorus insecticides to fish. *Trans. Am. Fish. Soc.* 87:39–51

80. Hickey, J. J., Anderson, D. W. 1968. Chlorinated hydrocarbons and eggshell changes in raptorial and fish-eating birds. *Science* 162:271–73

81. Hickey, J. J., Keith, J. A., Coon, F. B. 1966. An exploration of pesticides in a Lake Michigan ecosystem. *J. Appl. Ecol.* 3 (Suppl.): 141–54

82. Holden, A. V. 1962. A study of the absorption of ^{14}C-labeled DDT from water by fish. *Ann. Appl. Biol.* 50:467–77

83. Holland, H. T., III. 1967. Artificial selection of fish. In Report of the Bureau of Commercial Fisheries Biological Laboratory, Gulf Breeze, Florida. *Bur. Comm. Fish., U.S. Cir. 260,* 12

84. Hopkins, C. L., Solly, S. R. B., Ritchie, A. R. 1969. DDT in trout and its possible effect on reproductive potential. *N.Z. J. Marine Freshwater Res.* 3:220–29

85. Houser, A. 1963. Loss of weight of sunfish following aquatic vegetation control using the herbicide Silvex. *Proc. Okla. Acad. Sci.* 43: 232–37

86. Hughes, J. S., Davis, J. T. 1963. Variations in toxicity to bluegill sunfish of phenoxy herbicides. *Weeds* 11:50–53

87. Hunt, E. G. 1966. Biological magnification of pesticides, 257–62. Symposium on the Scientific Aspects of Pest Control, *Nat. Acad. Sci. Nat. Res. Counc. Publ. 1402* 470 pp.

88. Hunt, E. G., Bischoff, A. I. 1960. Inimical effects on wildlife of periodic DDD applications to Clear Lake. *Calif. Fish Game* 46: 91–106

89. Hunt, E. G., Keith, J. A. 1962. Pesticide-wildlife investigations in California—1962. In The *Use of Agricultural Chemicals in California—A summary of the problems and progress in solving them.* 27 pp. Univ. of California, Davis

90. Ivie, G. W., Casida, J. E. 1970. Enhancement of photoalteration of cyclodiene insecticide chemical residues by rotenone. *Science* 167: 1620–22

91. Jefferies, D. J. 1967. The delay in ovulation produced by p,p-DDT and its possible significance in the field. *Ibis* 109:266–72

92. Jefferies, D. J., French, M. C. 1969. Avian thyroid: effect of p,p′-DDT on size and activity. *Science* 166: 1278–80

93. Johnson, D. W. 1968. Pesticides and fishes—a review of selected literature. *Trans. Am. Fish. Soc.* 97: 398–424

94. Kallman, B. J., Cope, O. B., Navarre, R. J. 1962. Distribution and detoxification of toxaphene in Clayton Lake, New Mexico. *Trans. Am. Fish. Soc.* 91:14–22

95. Kay, K. 1966. Effect of pesticides on enzyme systems in mammals. In Organic Pesticides in the Environment. *Advan. Chem. Ser.* 60:54–66

96. Keith, J. A. 1966. Reproduction in a population of herring gulls (*Larus argentatus*) contaminated by DDT. *J. Appl. Ecol.* 3(Suppl.): 57–70

97. Kerswill, C. J., Edwards, H. E. 1967. Fish losses after forest sprayings with insecticides in New Brunswick, 1952–62, as shown by caged

specimens and other observations. *J. Fish. Res. Bd. Can.* 24:709–29

98. Kinne, O. 1963. The effects of temperature and salinity on marine and brackish water animals. I. Temperature. In *Oceanography and Marine Biology: An Annual Review,* I:301–40. London: George Allen and Unwin Ltd. 478 pp.

99. Kitselman, C. H. 1953. Long term studies on dogs fed aldrin and dieldrin in sublethal dosages, with reference to the histopathological findings and reproduction. *J. Am. Vet. Med. Assoc.* 123:28–30

100. Kleinert, S. J., Degurse, P. E. 1968. Survival of walleye eggs and fry of known DDT residue levels from ten Wisconsin waters in 1967. *Wis. Dept. Nat. Res. Res. Rep.* 37 30 pp.

101. Kleinert, S. J., Degurse, P. E., Wirth, T. L. 1968. Occurrence and significance of DDT and dieldrin in Wisconsin fish. *Wis. Dept. Nat. Res. Tech. Bull.* 41 43 pp.

102. Konar, S. K. 1969. Effects of heptachlor and nicotine on the barbels of a catfish (*Heteropneustes fossilis*). *Progr. Fish-Cult.* 31:62–63

103. Korschgen, L. J., Murphy, D. A. 1967. Pesticide-wildlife relationships: reproduction, growth, and physiology of deer fed dieldrin contaminated diets. *Mo. Fed. Aid Project No. 13-R-21* 24 pp.

104. Lemke, A. E., Mount, D. I. 1963. Some effects of alkyl benzene sulfonate on the bluegill, *Lepomis macrochirus. Trans. Am. Fish. Soc.* 92:372–78

105. Lichtenstein, E. P. 1958. Movement of insecticides in soils under leaching and non-leaching conditions. *J. Econ. Entomol.* 51:380–83

106. Locke, L. N., Chura, N. J., Stewart, P. A. 1966. Spermatogenesis in bald eagles experimentally fed a diet containing DDT. *Condor* 68:497–502

107. Luckman, W. H., Decker, G. C. 1960. A 5-year report of observations in the Japanese beetle control area at Sheldon, Illinois. *J. Econ. Entomol.* 53:821–27

108. Macek, K. J. 1968. Reproduction in the brook trout. (*Salvelinus fontinalis*) fed sublethal concentra-

tions of DDT. *J. Fish. Res. Bd. Can.* 25:1787–96

109. Macek, K. J. 1968. Growth and resistance to stress in brook trout fed sublethal levels of DDT. *J. Fish. Res. Bd. Can.* 25:2443–51

110. Macek, K. J. 1969. Screening of pesticides against fish. p. 92. In *Progress in Sport Fishery Research, 1968. Bur. Sport Fish. Wildl., U.S. Res. Publ.* 77 259 pp.

111. Macek, K. J., Hutchinson, C., Cope, O. B. 1969. The effects of temperature on the susceptibility of bluegills and rainbow trout to selected pesticides. *Bull. Environ. Contam. Toxicol.* 4:174–83

112. Macek, K. J., McAllister, W. A. 1970. Insecticide susceptibility of some common fish family representatives. *Trans. Am. Fish. Soc.* 99:20–27

113. Marking, L. L., Hogan, J. W. 1967. The toxicity of Bayer 73 to fish. *Invest. Fish Control* 19, 13 pp. Bur. Sport Fish. Wildl., U.S.

114. Matsumura, F., Patil, K. C. 1969. Adenosine triphosphatase sensitive to DDT in synapses of rat brain. *Science* 166:121–22

115. McCraren, J. P., Cope, O. B., Eller, L. 1969. Some chronic effects of diuron on bluegills. *Weed Sci.* 17:497–504

116. Menzel, D. W., Anderson, J., Randtke, A. 1970. Marine phytoplankton vary in their response to chlorinated hydrocarbons. *Science* 167:1724–26

117. Mestitzova, M. 1966. On reproduction studies and the occurrence of cataracts in rats after long-term feeding of the insecticide heptachlor. *Experientia* 23:42–43

118. Metcalf, R. L. 1955. *Organic Insecticides.* New York:Interscience. 392 pp.

119. Meyer, F. P. 1966. The effect of formulation differences on the toxicity of benzene hexachloride to golden shiners. *Proc. Ann. Conf., SE Assoc. of Game and Fish Comm, 17th* 186–90

120. Mount, D. I. 1962. Chronic effects of endrin on bluntnose minnows and guppies. *Fish and Wildl. Serv., U.S., Res. Rep. 58.* 38 pp.

121. Mount, D. I., Boyle, H. W. 1969. Parathion—use of blood concentration to diagnose mortality of

fish. *Environ. Sci. Toxic.* 3 :1183–85

122. Mount, D. I., Vigor, L. W., Schafer, M. L. 1966. Endrin : Use of concentration in blood to diagnose acute toxicity to fish. *Science* 152 : 1388–90

123. Murphy, S. D., Lauwerys, R. R., Cheever, K. L. 1968. Comparative anticholinesterase action of organophosphorus insecticides in vertebrates. *Toxicol. Appl. Pharmacol.* 12 :22–35

124. Naber, E. C., Ware, G. W. 1965. Effect of Kepone and mirex on reproductive performance in the laying hen. *Poultry Sci.* 44 :875–80

124a. Newsom, L. D. 1967. Consequences of insecticide use on nontarget organisms. *Ann. Rev. Entomol.* 12 :257–86

125. Oborn, E. 1969. Effects of endrin on tissue cations in cutthroat trout, 103. In Progress in Sport Fishery Research. *Bur. Sport Fish. Wildl., U.S. Res. Publ.* 77 259 pp.

126. O'Brien, R. D. 1967. Synergism, antagonism, and other interactions. In *Insecticides. Action and Metabolism,* Chap. 14, 209–30. New York :Academic, 332 pp.

127. Ogilvie, O. M., Anderson, J. M. 1965. Effect of DDT on temperature selection by young Atlantic salmon, *Salmo salar. J. Fish. Res. Bd. Can.* 22 :503–12

128. Peakall, D. B. 1970. p,p'-DDT : Effect on calcium metabolism and concentration of estradiol in the blood. *Science* 168 :592–94

129. Porter, R. D., Wiemeyer, S. N. 1969. Dieldrin and DDT : effects on sparrow hawks and reproduction. *Science* 165 :199–200

130. Reinert, R. E. 1969. Insecticides and the Great Lakes. *Limnos* 2 :3–9

131. Reinert, R. E. 1969. The relationship between fat and insecticide concentration in fish. In Abstr. *Conf. on Great Lakes Research, 12th,* Ann Arbor 61 pp.

132. Rezvin, A. M. 1966. Effects of endrin on telencephalic function in the pigeon. *Toxicol. Appl. Pharmacol.* 9 :75–83

133. Rodgers, C. A. 1970. Uptake and elimination of simazine by green sunfish (*Lepomis cyanellus* Raf.). *Weed Sci.* 18 :134–36

134. Rowe, V. K., Hymas, T. A. 1945. Summary of toxicological information on 2,4-D and 2,4,5-T type herbicides and an evaluation of the hazards to livestock associated with their use. *Am. J. Vet. Res.* 15 :622–29

135. Rubin, M., Bird, H. R., Green, N., Carter, R. H. 1947. Toxicity of DDT to laying hens. *Poultry Sci.* 26 :410–13

136. Rudd, R. L., Genelly, R. E. 1956. Pesticides : their use and toxicity in relation to wildlife. *Calif. Fish Game Bull.* 7, 309 pp.

137. Sanders, H. O. 1970. Pesticide toxicities to tadpoles of the Western chorus frog *Pseudacris triseriata* and Fowler's toad *Bufo woodhouseii. Copeia,* No. 2 :246–51

138. Schoettger, R. A. 1970. Toxicology of Thiodan in several fish and aquatic invertebrates. *Invest. Fish Control No. 35,* 31 pp. Bur. Sport Fish. and Wildl., U.S.

139. Shellenberger, T. E., Newell, G. W., Adams, R. F., Barballia, J. 1966. Cholinesterase inhibition and toxicologic evaluation of two organophosphate pesticides to Japanese quail. *Toxicol. Appl. Pharmacol.* 8 :22–28

140. Shellenberger, T. E., Skinner, W. A., Lee, J. M. 1967. Effect of organic compounds on reproductive processes. IV. Response of Japanese quail to alkylating agents. *Toxicol. Appl. Pharmacol.* 10 :69–78

141. Sherman, M., Herrick, R. B. 1966. Acute and subacute toxicity of apholate to the chick and Japanese quail. *Toxicol. Appl. Pharmacol.* 9 :279–292

142. Stickel, L. 1964. Wildlife studies, Patuxent Wildlife Research Center, 77–115. In Pesticide-Wildlife Studies, 1963. *Fish Wildl. Serv., U.S. Cir.* 199, 130 pp.

143. Stickel, L. F. 1968. Organochlorine pesticides in the environment. *Bur. Sport Fish. Wildl., U.S. Spec. Sci. Rep.—Wildl. No.* 119 32 pp.

144. Stickel, L. F., Heath, R. G. 1965. Wildlife studies, Patuxent Wildlife Research Center, 3–30. In The Effects of Pesticides on Fish and Wildlife, 1964. *Fish Wildl. Serv. U.S. Cir.* 226 77 pp.

145. Stickel, L. F., Stickel, W. H., Christensen, R. 1966. Residues of DDT

in brains and bodies of birds that died on dosage and in survivors. *Science* 151:1549–51

146. Stickel, L., Stickel, W. 1969. Distribution of DDT residues in tissues of birds in relation to mortality, body condition, and time. *Ind. Med.* 38:44–53

147. Stickel, W. H., Hayne, D. W., Stickel, L. F. 1965. Effects of heptachlor-contaminated earthworms on woodcocks. *J. Wildl. Manage.* 29:132–46

148. Stickel, W. H., Stickel, L. F., Spann, J. W. 1969. Tissue residues of dieldrin in relation to mortality in birds and mammals, 174–204. In *Chemical Fallout: Current Research on Persistent Pesticides.* Proceedings of the First Rochester Conference on Toxicity. Springfield, Ill.: Charles C Thomas, 531 pp.

149. Stock, J. N., Cope, O. B. 1969. Some effects of TEPA, an insect chemosterilant, on the guppy, *Poecilia reticulata. Trans. Am. Fish. Soc.* 98:280–87

150. Street, J. C. 1964. DDT antagonism to dieldrin storage in adipose tissue of rats. *Science* 146:1580

151. Surber, E. W., Pickering, Q. H. 1962. Acute toxicity of Endothal, Diquat, Hyamine, Dalapon, and Silvex to fish. *Progr. Fish Cult.* 24:164–71

152. Terriere, L. C., Kligemagi, U., Gerlach, A. R., Borovicka, R. L. 1966. The persistence of toxaphene in lake water and its uptake by aquatic plants and animals. *J. Agr. Food Chem.* 14:66 69

153. Treon, J. F., Cleveland, F. P., Cappel, J. 1955. Toxicity of endrin for laboratory animals. *J. Agr. Food Chem.* 3:842–48

154. Triolo, A. J., Coon, J. M. 1966. Toxicologic interactions of chlorinated hydrocarbon and organophosphate insecticides. *J. Agr. Food Chem.* 14:549–55

155. Tucker, R. K., Crabtree, D. G. 1970. Handbook of toxicity of pesticides to wildlife. *Bur. Sport Fish. Wildl., U.S. Res. Publ. 84* 131 pp.

156. U. S. Bureau of Commercial Fisheries. 1970. Surface slicks have 10,000 more pesticide than encircling water. *Commerc. Fish. Rev.* 32:7

157. U. S. Department of Agriculture.

158. Van Valin, C. C., Andrews, A. K., Eller, L. L. 1968. Some effects of mirex on two warm-water fishes. *Trans. Am. Fish. Soc.* 97:185–96

159. Wade, R. A. 1969. Ecology of juvenile tarpon and effects of dieldrin on two associated species. *Bur. of Sport Fish. Wildl., U.S. Tech. Paper 41* 85 pp.

160. Walker, C. R. 1963. Endothal derivatives as aquatic herbicides in fishery habitats. *Weeds* 11:226–32

161. Walker, C. R. 1964. Toxicological effects of herbicides on the fish environment. II. *Water Sewage Works* 111:173–75

162. Walker, C. R., Lennon, R. E., Berger, B. L. 1964. Preliminary observations on the toxicity of antimycin A to fish and other aquatic animals. *Bur. of Sport Fish. Wildl., U.S. Cir. 186* 18 pp.

163. Ware, G. W., Good, E. E. 1967. Effects of insecticides on reproduction in the laboratory mouse. *Toxicol. Appl. Pharmacol.* 10:54–61

164. Warner, R. E., Peterson, K. K., Bargman, L. 1966. Behavioral pathology in fish: a quantitative study of sublethal pesticide toxicity. *J. Appl. Ecol.* 3 (Suppl.):223–47

165. Webb, R. E., Horsfall, F., Jr. 1967. Endrin resistance in the pine mouse. *Science* 156:1762

166. Weiss, C. M. 1958. The determination of cholinesterase in the brain tissue of three species of freshwater fish and its inactivation *in vivo. Ecology* 39:194–99

167. Weiss, C. M. 1959. Response of fish to sub-lethal exposures of organic insecticides. *Sewage Ind. Wastes* 31:580–93

168. Weiss, C. M. 1961. Physiological effect of organic phosphorus insecticides on several species of fish. *Trans. Am. Fish. Soc.* 90:143–52

169. Woodwell, G. M. 1967. Toxic substances and ecological cycles. *Sci. Am.* 216:24–31

170. Woodwell, G. M., Wurster, C. F.,

Jr., Isaacson, P. A. 1967. DDT residues in an East Coast estuary: a case of biological concentration of a persistent insecticide. *Science* 156:821–24

171. Wright, B. S. 1965. Some effects of heptachlor and DDT on New Brunswick woodcocks. *J. Wildl. Manage.* 29:172–85

172. Young, L. A., Nicholson, H. P. 1951. Stream pollution resulting from the use of organic insecticides. *Progr. Fish Cult.* 13:193–98

REGULATION OF FEEDING[1]   6013

ALAN GELPERIN

*Department of Biology, Princeton University,
Princeton, New Jersey*

The feeding behavior of insects can be studied from a variety of viewpoints. One may be interested in the reasons why an insect selects a particular food for consumption and the efficiency with which it converts that food to insect tissue (2, 10, 34, 55, 60, 66). Feeding may be viewed as one means of introducing a lethal substance into an insect or as the rate-determining step in a biochemical pathway (33, 40). Food intake is an essential component of egg maturation and reproduction (24, 48). Feeding behavior may also be studied as the output of a complex neural circuit whose elements can be studied individually. The latter view will be the prevailing one in this review.

The regulation of food intake is part of a larger mechanism for metabolic homeostasis. By metabolic homeostasis I refer to the regulation of metabolic energy flow into and within the animal. Feeding behavior is involved in introducing energy stores into the animal, gut activity determines the rate of delivery of these stores to the blood, and a third set of controls operates to control delivery of energy stores from blood to tissues. If information is available at all of these levels of analysis, one should be able to trace the causal sequences of events between cellular energy expenditure and the behavior of energy ingestion.

Some insects can regulate their food intake (17, 58). Regulation implies the existence of a feedback loop in the neural circuitry controlling feeding. The feedback loop commonly consists of an internal sensing element which monitors some variable influenced by the ingestion of food. The parameter measured by the internal sensor usually is not directly related to the nutritional value of the food. This poses interesting questions as to how the animal uses such a mechanism to cope with foods of varying nutritional value.

The focus of this review is on physiological mechanisms of feeding regulation. For only a limited number of insect species is there data suggesting that food intake is regulated (17, 58). For only one insect, *Phormia regina,* is there information relating to the neural mechanism underlying that regulation (18). For this reason, I will first present an overview of the mecha-

[1] Supported by National Science Foundation Grant 68 7766.

nism operative in *Phormia* and then deal with the generality of this mechanism by considering data from other insects.

Phormia AS A MODEL SYSTEM

Feeding in the blow fly, *Phormia regina,* is initiated by stimulation of contact chemoreceptors on the tarsi. Proboscis extension, the initial motor act of feeding, results in stimulation of chemoreceptors on the labellum which leads to sucking. The amount of food ingested is determined by the duration of sucking, not its intensity. Feeding is terminated due to receptor adaptation and to increased negative feedback from internal receptors.

The gut of the blow fly is typical of the higher Diptera in the elaboration of the crop as a diverticulum of the foregut. The crop duct arises from the foregut immediately anterior to the proventriculus. It runs posteriorly from thorax to abdomen where it ends in a blind sac, the crop. A valve is present in the crop duct at its junction with the foregut.

The crop is filled with fluid at the time of feeding and emptied over the next several hours (27, 50). After feeding has ceased, slugs of sugar aperiodically move anteriorly up the crop duct, through the crop duct valve and into the foregut. The crop duct valve now closes, the proventriculus opens, and the slug of solution is propelled into the midgut.

There are two sets of internal receptors which provide negative feedback to feeding behavior. One set has been identified as stretch receptors located in a branch of the recurrent nerve which connects with the foregut (29). These cells are activated by expansion of the foregut lumen such as occurs during a wave of peristalsis. The axons of these cells proceed to the brain in the recurrent nerve. If the recurrent nerve is cut between the receptors and the brain, hyperphagia results (20, 22). It is possible to cut the recurrent nerve posterior to the stretch receptors and observe no hyperphagia (31).

A second set of receptors is located in the abdomen. Cutting the median abdominal nerve (MAN) which innervates the abdominal segments from the thoracico-abdominal ganglion (TAG) results in hyperphagia (22). The cells responsible for this effect appear to be nerve cord stretch receptors responsive to tension in the abdominal nerves which are suspended over and stretched by the crop (30). Electrical recording from these cells in a minimally dissected preparation indicates that their short-term activity can be modulated by expansions and contractions of the crop and their average level of activity is dependent on crop volume (30).

The pattern of stimulation of the two sets of internal receptors is directly dependent on gut movements which occur as the crop empties. Emptying can occur in a completely isolated gut and appears to be controlled by the osmotic pressure of the blood (27, 50). The blood sugar, trehalose, accounts for a significant fraction of the total blood osmolarity. As the animal withdraws trehalose from the blood for storage or utilization, the gut responds to the lowered blood osmotic pressure by emptying faster. Conversely, if blood trehalose is increased by digestion of concentrated sugar or injection of sugar into the blood, crop emptying slows (27).

The model which attempts to integrate the total body of information available on *Phormia* feeding behavior views feeding as the resultant of excitation from the external chemoreceptors and inhibition from the internal stretch receptors. Under steady-state conditions when the fly has been allowed ad lib access to a given concentration of sugar for several days, a relatively constant daily food intake and crop volume are maintained (28, 32). The average frequency of stretch receptor input to the brain is just sufficient to counteract the excitatory input from external chemoreceptors. If a small amount of stored fluid is removed from the crop and digested, the negative feedback lessens slightly and upon contacting the food, a small meal is taken. The prediction of infrequent small meals under steady-state conditions has been observed (32).

Evidence for Regulation

The blow fly regulates its daily food intake. This is indicated by the relative constancy of daily food intake under conditions in which it can be quantitatively measured (32). Regulation of intake is further indicated by the fact that flies fed sucrose solutions in the concentration range $0.1-2.0$ M decrease their daily intake as concentration is increased. This indicates that intake is not simply determined by gut capacity or by level of sensory stimulation but rather that some consequence of feeding exerted inside the animal reduces the probability of further feeding.

The regulatory mechanism in *Phormia* does not operate to maintain a constant caloric intake. The intake of sugars such as fucose, which are stimulating to chemoreceptors but metabolically useless, is regulated until death ensues. Similarly, the reduced intake of concentrated sucrose solutions does not result in a constant caloric intake (32).

The praying mantis, *Hierodula crassa,* consumes a relatively constant number of house flies per day when given ad lib access to a surplus of flies (43). Females eat about 34 house flies per day, males about 5 per day. If larger flies are being eaten, a smaller number are consumed daily (31). Approximately 12 *Phormia* or 4 *Calliphora* are eaten per day when *Paratenodera sinesis* is given ad lib access.

Some lepidopterous larvae may regulate their food intake. Larvae of *Celerio euphorbiae* will feed on artificial diets. House compared the intake by larvae feeding on a standard artificial diet with intake of diluted diets containing 85, 70 or 50 percent of the nutrient content of the standard diet (45). Intake increased as the food was diluted. The compensatory increase in food intake as the food was diluted did not result in a constant nutrient intake. If nutrient intake on the standard diet is given as 100 percent, then nutrient intakes on the diluted diets were 96, 83, and 65 percent.

Adult male German cockroaches fed solid carbohydrates such as sucrose, glucose, fructose, maltose, trehalose, or raffinose exhibit a relatively constant daily intake over periods of 30 to 40 days (35). Intake of sucrose is 20 to 25 mg/g per day. If the sucrose is diluted 50 percent with cellulose, intake rises to 40 mg/g per day. Further dilution to 75 percent cellulose increases intake to 60 mg/g per day. A compensatory increase in daily food intake also occurred when the other five sugars listed above were diluted 50 percent with cellulose.

An increase in intake when nutrient content is decreased does not indicate that caloric value is the regulated parameter. It does eliminate the possibility that simply the mass of food ingested is being measured. As the *Phormia* data indicate, it is possible for the regulatory mechanism to cause increased intake of calorically dilute foods without directly measuring the caloric value of the food.

BEHAVIORAL CHANGES WITH DEPRIVATION

Changes in a variety of behavioral parameters occur as the blow fly is deprived of food. The intensity of chemosensory stimulation needed to elicit proboscis extension decreases with deprivation time (19). Threshold may be less than $0.001 M$ in a starved fly and greater than $2 M$ immediately after feeding. The amount of a standard sugar solution which is ingested increases with deprivation time (28). Other changes occur which affect the probability of finding food. Spontaneous locomotor activity increases with deprivation time and drops sharply immediately after feeding (3, 36, 37). Green's experiments indicate that a blood-borne factor controls the level of spontaneous activity. The intensity and duration of the searching movements which occur after brief stimulation with sugar also increase with deprivation time (16).

Holling has studied in a systematic way the changes in behavior shown by the praying mantis as a function of deprivation time (43). The weight of food required to return the animal to an initial level of satiation increases up to a maximum at approximately 40 hr of deprivation for *H. crassa* and 18 hr for a smaller species, *Mantis religiosa*. The maximum distance at which *H. crassa* will react to prey by stalking or striking also increased with deprivation; however, no response is elicited for 8 hr after satiation. The reactive field was determined by measuring the maximum distance at which the mantis visually fixated prey at several points around the periphery of the visual field. The reactive field becomes larger as the mantis is deprived. Hunger level did not affect strike success or rate of eating.

SOURCES OF NEGATIVE FEEDBACK

The two groups of stretch receptors which have been identified as supplying negative feedback to feeding in the blow fly represent the only case in which the cells involved have been identified (29, 30).

The pattern of output of the foregut receptors has been studied while the fly was digesting either a concentrated sucrose solution (1.0 M) or a dilute sucrose solution (0.1 M) (53). The dilute solution empties from the crop more rapidly (27). We therefore expected greater activity in the stretch receptors activated by peristalsis in the foregut when 0.1 M sucrose was in the gut. The opposite result was observed. The foregut receptors produce shorter bursts and longer interburst intervals when 0.1 M sucrose is being digested than when 1.0 M sucrose is being digested (53). The probable explanation is that slugs of fluid can be transferred back and forth between foregut and crop duct without being moved to the midgut for digestion. If this oscillation of fluid between foregut and crop duct occurred vigorously with 1.0 M sucrose in the gut, the observed output from the foregut receptors would result.

The abdominal sense cells in *Phormia* have been provisionally identified as neurons situated in the abdominal peripheral nerves. The cells are activated by stretch applied to the nerves themselves (30). Behavioral operations indicate that sensory neurons located in the first two pairs of lateral branches arising from the MAN are critical for feeding regulation. The cells maintain a steady rate of firing when completely isolated from the CNS and periphery or when the crop is quiescent. If electrical activity is monitored while the fly is fed, the cells are shown to increase their average level of output in response to increased crop volume (30). Since the receptors are also activiated by short-term expansions and contractions of the crop, their in vivo output pattern would be an aperiodic fluctuation around an average value determined by crop volume.

Larvae of *Rhodnius* typically take a single large blood meal during each instar which provides nutrients and triggers brain hormone secretion leading to a molt (67). A bug can drink up to six times its weight in blood (6). If the abdomen is denervated by cutting all nerves posterior to the mesothoracic ganglionic mass, hyperphagia results (51). Hyperphagia can also be produced by an operation which allows ingested blood to leak out of the animal without stretching the abdomen. The abdominal stretch receptors which have been implicated in triggering a molt (65) presumably also act to terminate feeding.

The feeding pattern of the female mosquito, *Aedes aegypti,* is also characterized by large blood meals. Segmental stretch receptors in the abdomen have been implicated in controlling meal size in this insect (39). If the abdominal nerve cord is cut anterior to the second abdominal ganglion, intake is increased by a factor of four. As the transection is made more posteriorly, the hyperphagic effect decreases. Hyperphagia produced by transecting the abdominal cord anterior to ganglion 2 was also demonstrated for *A. taeniorhynchus, A. triseriatus, Armigeres subalbatus, Culex pipiens fatigens,* and *Anopheles quadrimaculatus.*

NEURAL INTEGRATION

Ultimately one would like to obtain the detailed wiring diagram for the interneurons mediating the interactions between the peripheral sensory receptors which drive feeding, the internal receptors which inhibit feeding, and the motor neurons whose activity determines the form of the behavior. Neurophysiological information is not as yet available; however, several types of behavioral experiments indicate what kinds of circuits must be present.

On the sensory side of the mechanism in *Phormia*, Dethier (15) has shown that stimuli applied to separate legs simultaneously summate in the CNS. The sucrose threshold is lower if two legs are stimulated than if only one is stimulated. 2 *M* propanol applied to one leg will decrease the responsiveness to sucrose applied to another leg. It was also found that rejection is greater when opposing compounds stimulate the same leg than when they are applied to opposite legs.

Spatial summation within the population of receptors has also been demonstrated for the labellar hairs (1). The threshold for proboscis extension is lower if two adjacent hairs are stimulated rather than one. Threshold decreased as three and four hairs were stimulated. The threshold values for three, four, and all hairs were not significantly different.

The blow fly possesses neural circuits which can maintain a state of heightened excitation for periods up to 120 sec after brief sensory input (21). If a water-satiated fly is stimulated by the application of water to a single labellar hair no response occurs. However, if another labellar hair is stimulated with sucrose just prior to the water stimulation, a response occurs. The magnitude and probability of a response to water are affected by the strength and timing of the preceding sucrose stimulation, deprivation time, and by which receptor population, labellar or tarsal, is utilized for the sucrose-priming stimulus. The central excitatory state was found to decay slowly in time and to be discharged by the occurrence of proboscis extension. The existence of a central inhibitory state was inferred from the effectiveness of certain aversive stimuli in reducing the central excitatory state set up by sucrose (23).

The integration and summation of foregut and abdominal stretch receptor input have not been extensively studied; however, some preliminary data indicate that the effects of RN and MAN section are additive (53). Intakes and feeding patterns were determined for flies given the two operations singly or in combination. The single operations led to hyperphagia and a feeding pattern characterized by a greatly increased initial drink and more frequent small drinks following it (22, 53). The double operation resulted in continuous drinking until bursting occurred in approximately 85 percent of the flies so treated. Simply measuring the amount of food ingested in the singly operated versus double-operated flies did not reveal the true extent of the behavioral effect because both groups were ingesting nearly maximal amounts of sucrose.

Gut Emptying

The pattern and rate of gut emptying have direct effects on body or gut wall stretch receptors which inhibit feeding. To understand how feeding behavior responds to metabolic energy expenditure, one must discover how gut emptying is controlled and linked to energy expenditure.

Crop emptying in the blow fly was studied using X-ray photography of flies fed sugar solutions containing a radiopaque substance, diatrizoate sodium (27). Increasing the osmotic pressure of the solution in the crop slows crop emptying. This effect does not depend on the nutritive value, stimulating power, or viscosity of the solution. Injections which increase the osmotic pressure of the blood greatly slow crop emptying. The effect of increasing blood osmotic pressure in slowing crop emptying occurs in the absence of the brain, thoracico-abdominal ganglion, or hypocerebral ganglion (27). This agrees with Knight's observation that the gut continues to function in complete isolation (50).

Initial studies on crop emptying in the cockroach (63) indicated that rate of crop emptying is linearly related to the concentration of ingested sugar. Increased concentration results in decreased rate of emptying. Feeding a variety of substances at the same osmotic concentration results in a constant rate of emptying indicating that, as in the blow fly, osmotic pressure of the crop contents is an important parameter.

The physiology of crop function was detailed in a series of papers by Davey & Treherne (11–13). Ingestion of fluid causes little increase in crop volume, rather, the ingested fluid displaces air present in the empty crop. The swallowing of air also contributes to the process of emptying (11). Changes in viscosity of the crop contents have very little effect on rate of emptying. As meal size increases, rate of crop emptying increases so that a constant proportion of the meal is emptied in a given time (11).

The rate-limiting step appears to be the frequency of operation of the proventricular valve. If the stomatogastric system is separated from the brain, valve function continues. However, if the proventricular ganglion is separated from the frontal ganglion by section of the esophageal or ingluvial nerves, the proventriculus no longer operates (12). Cutting N5, a small nerve connecting the frontal ganglion with the dorsal part of the pharynx, greatly slows crop emptying. Histological sections of the pharynx revealed a complex sensory structure which was hypothesized to be an osmoreceptor. As Davey & Treherne indicate, the operation of the system cannot be totally explained by a simple reflex circuit from pharyngeal osmoreceptor to preventriculus because crop emptying can compensate for changes in viscosity or meal size.

Engelmann has investigated the control of crop emptying in *Leucophaea maderae* (25). As in *Periplaneta*, a constant proportion of the meal is emptied from the crop within 24 hr, independent of meal size. Meals of starch paste were removed from the crop more rapidly than meals of solid rat food pellets indicating that consistency of the food can have an effect. The neural control mechanism in *Leucophaea* is strikingly different from *Periplaneta* as shown by the effect of cutting all nerves to the frontal ganglion except the recurrent nerve. This operation has no effect in *Leucophaea*. Cutting the recurrent nerve or osophageal nerve immediately after feeding reduced crop emptying by 50 percent. Engelmann made the very interesting observation that if the same nerves were cut 1 to 12 days before feeding, crop emptying proceeded normally. This suggests a re-examination of the data on *Periplaneta*.

The process of crop emptying in *Schistocerca* is also under nervous control. The removal of the frontal ganglion in females results in the accumulation of very large amounts of food in the foregut while the midgut contains very little food (41). Frontal ganglionectomy obliterates the normal period of intense feeding which occurs 8 to 10 days after ecdysis and causes a constant low level of feeding to be exhibited for at least 14 days following the operation. Some function of the proventricular valve must have occurred as feces production continued at a low level following the operation.

Gut movements in *Locusta* similarly are mediated by the stomatogastric system. If the ventricular ganglion is disconnected from the gut, movements cease "for some time" (7). The role of the frontal ganglion is not yet clear. Clarke & Langley (8) reported that after frontal ganglionectomy, locusts fed normally and their guts contained normal amounts of food. Another report indicated that feces production decreases to a low level following frontal ganglionectomy (42). It may be that the gut was kept full of food but movement along the gut was greatly retarded by the operation. In any event, it is clear that in both *Schistocerca* and *Locusta,* the isolation of the gut and stomatograstric system from the CNS did not lead to hyperphagia.

The mosquito possesses a mechanism for routing blood to the midgut and carbohydrates into the esophageal diverticula (14, 44, 64). Chemoreceptors on the labella, labrum, and within the cibarial pump can apparently determine the distribution of fluids within the alimentary canal (44). It is possible to study gut emptying by radiographic means in this animal (38).

METABOLIC HOMEOSTASIS

Gut emptying in the blow fly is controlled by the osmotic pressure of the blood and therefore in large measure by the concentration of trehalose in the blood. An understanding of the factors which control blood trehalose level should yield some insight into the linkage between metabolic events and feeding behavior.

Fat body is the major site of trehalose synthesis in *Phormia* (9, 26). Isolated fat body can mobilize endogenous reserves to liberate trehalose at a mean rate of 252 μg/mg dry wt fat body/hr (9). This tissue can also utilize exogenous hexoses to synthesize trehalose. The conversion of glucose to trehalose occurs very rapdily. Within 5 min 85 percent of an aliquot of injected glucose is converted to trehalose (9). Gut, blood, and flight muscle do not contribute significantly to the synthesis of blood trehalose.

The trehalose-synthesizing system in both blow fly and silk moth fat body exhibits end product inhibition (26, 56). Increased concentration of trehalose in the blood causes a shift in incorporation of exogenous glucose from trehalose to glycogen. The synthetic system in both fly and cockroach is sensitive to a factor liberated by the corpus cardiacum which causes blood trehalose to be elevated (26, 61). In *Periplaneta* the hyperglycemic hormone apparently activates fat body glycogen phosphorylase (62). In *Phormia* the hormone has no affect while the gut is able to deliver dietary sugar to the blood.

The level of blood trehalose is probably regulated by feedback inhibition of its own synthesis. Rapid withdrawals of trehalose from the blood are met by increased synthesis from exogenous hexoses, increased delivery of gut stores, and perhaps increased glycogen breakdown under the action of the hyperglycemic hormone. Hudson has shown that delivery of gut stores is greatly speeded by increased energy expenditure, e.g., flight (47). This mechanism suggests that blood trehalose could be maintained at a relatively constant level. Measurements of blood sugar level after a single feeding on glucose or mannose indicate the maintenance of a relative constancy for 50 hr after an initial 10-hr period of equilibration. The regulation breaks down when the crop becomes empty. Measurement of blood trehalose level in *Bombyx mori* after injections of saline or trehalose also indicates the existence of a homeostasis in blood sugar (59).

The homeostatic mechanism governing blood sugar level and daily food intake are linked by the gut. Gut responsiveness to blood sugar level is an integral part of the control system for blood trehalose regulation. Gut movements directly influence feeding behavior by their action on foregut and abdominal stretch receptors. In the blow fly, as in no other invertebrate, one can indeed see a general mechanism for metabolic homeostasis at work. This mechanism is diagramatically presented in Figure 1.

For only one other organism in the animal kingdom do data of comparable completeness exist. This organism is the white rat. A comparison of fly and rat as to the mechanism each has evolved to control food intake is instructive. Both the fly and the rat maintain a relatively constant food intake when tested for several days at a single level of nutrient density. When the food is diluted, both creatures increase their intake, the rat maintaining more precisely than the fly a constant nutrient intake. If the fly's behavior is examined over the narrow range of food dilutions commonly used for the rat, the regulation of food intake by rat and fly is not greatly different.

In contrast to the fly, the rat appears to measure some characteristic of its food closely related to metabolic value. According to the glucostatic theory (52), receptors in the rat hypothalamus monitor the level of glucose utilization and stimulate or suppress feeding accordingly. The thermostatic theory (5) postulates that higher animals eat to keep warm. The extra heat released in the assimilation of ingested food, called the "specific dynamic action," is sensed by the hypothalamus and results in the observed regula-

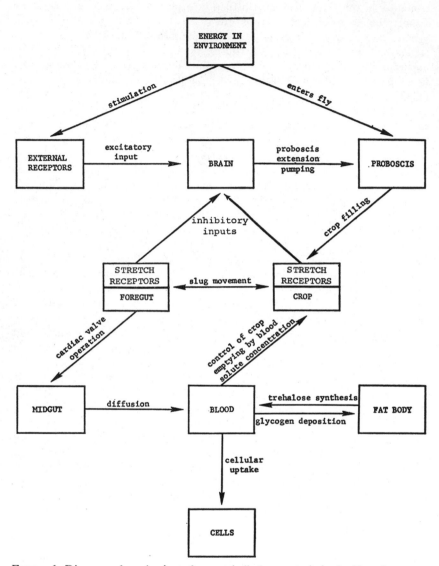

FIGURE 1. Diagram of mechanisms for metabolic homeostasis in the blow fly.

tion of food intake. The ability of food to contribute to either blood glucose or bodily heat is closely related to metabolic value.

Concluding Remarks

Some insects clearly do not regulate their food intake. Adults of some insects do not feed. The feeding rate of the aphid *Tuberolachnus* is influenced by the turgor pressure of the host plant (54). Many insects have a regulatory mechanism containing a sensory component which excites feeding and most probably some type of internal receptor which antagonizes feeding. The nature of the internal receptor is unknown in all but three insects.

The neural control systems which have been characterized are well worked out on the receptor side. The as yet impenetrable interneuron jungle still contains the vast majority of the neural circuitry. Application of intracellular micro-electrode techniques to insect central neurons (4, 46, 49, 57) should produce badly needed data on the wiring diagram and response patterns of the interneurons connecting internal and external sensory receptors with feeding motorneurons.

LITERATURE CITED

1. Arab, Y. M. 1959. Some chemosensory mechanisms in the blowfly. *Bull. Coll. Sci.* Univ. Baghdad 4:77–85
2. Auclair, J. L. 1963. Aphid feeding and nutrition. *Ann. Rev. Entomol.* 8: 439–90
3. Barton-Browne, L., Evans, D. R. 1960. Locomotor activity of the blowfly as a function of feeding and starvation. *J. Insect Physiol.* 4:27–37
4. Bentley, D. R. 1969. Intracellular activity in cricket neurons during the generation of behavior patterns. *J. Insect Physiol.* 15:677–99
5. Brobeck, J. R. 1960. Food and temperature. *Recent Progr. Hormone Res.* 16:439–66
6. Buxton, P. A. 1930. The biology of a blood-sucking bug, *Rhodnius prolixus. Trans. Roy. Entomol. Soc. London,* 78:227–36
7. Clarke, K. U., Grenville, H. 1960. Nervous control of movements in the foregut in *Schistocerca gregaria* Forsk. *Nature* 186:98–99
8. Clarke, K. J., Langley, P. A. 1963. Studies on the initiation of growth and moulting in *Lucusta migratoria migratoroides.* III. The role of the frontal ganglion. *J. Insect Physiol.* 9:411–21
9. Clegg, J. S., Evans, D. R. 1961. The physiology of blood trehalose and its function during flight in the blowfly. *J. Exp. Biol.* 38:771–92
10. Dadd, R. H. 1963. Feeding behavior and nutrition in grasshoppers and locusts. *Advan. Insect Physiol.* 1: 47–109
11. Davey, K. G., Treherne, J. E. 1963. Studies on crop function in the cockroach. I. The mechanism of crop-emptying. *J. Exp. Biol.* 40: 763–73
12. Davey, K. G., Treherne, J. E. 1963. Studies on crop function in the cockroach. II. The nervous control of crop emptying. *J. Exp. Biol.* 40:775–80
13. Davey, K. G., Treherne, J. E. 1964. Studies on crop function in the cockroach. III. Pressure changes during feeding and crop-emptying. *J. Exp. Biol.* 41:513–24
14. Day, M. F. 1954. The mechanism of food distribution to midgut or diverticulum in the mosquito. *Aust. J. Biol. Sci.* 7:515–24
15. Dethier, V. G. 1953. Summation and inhibition following contralateral stimulation of tarsal chemoreceptors of the blowfly. *Biol. Bull.* 105:257–68
16. Dethier, V. G. 1957. Communication by insects: Physiology of dancing. *Science* 125:331–36
17. Dethier, V. G. 1967. Feeding and drinking behavior of invertebrates. In *Handbook of Physiology,* Section 6, *Alimentary Canal,* Vol. 1, Chapter 6:79–96. Washington: Am. Physiol. Soc. 459 pp.
18. Dethier, V. G. 1969. Feeding behavior of the blowfly. *Advan. Study Behavior* 2:112–266
19. Dethier, V. G., Chadwick, L. E. 1948. Chemoreception in insects. *Physiol. Rev.* 28:220–54
20. Dethier, V. G., Bodenstein, D. 1958. Hunger in the blowfly. *Z. Tierpsychol.* 15:129–40
21. Dethier, V. G., Solomon, R. L., Turner, L. H. 1965. Sensory input and central excitation and inhibition in the blowfly. *J. Comp. Physiol. Psychol.* 60:303–13
22. Dethier, V. G., Gelperin, A. 1967. Hyperphagia in the blowfly. *J. Exp. Biol.* 47:191–200
23. Dethier, V. G., Solomon, R. L., Turner, L. H. 1968. Central inhibition in the blowfly. *J. Comp. Physiol. Psychol.* 66:144–50
24. Engelmann, F. 1968. Endocrine control of reproduction in insects. *Ann. Rev. Entomol.* 13:1–26
25. Engelmann, F. 1968. Feeding and crop emptying in the cockroach *Leucophaea maderae. J. Insect Physiol.* 14:1525–31
26. Friedman, S. 1967. The control of trehalose synthesis in the blowfly, *Phormia regina. J. Insect Physiol.* 13:397–405
27. Gelperin, A. 1966. Control of crop emptying in the blowfly. *J. Insect Physiol.* 12:331–45
28. Gelperin, A. 1966. Investigations of a foregut receptor essential to taste threshold regulation in the blowfly. *J. Insect. Physiol.* 12: 829–41
29. Gelperin, A. 1967. Stretch receptors in the foregut of the blowfly. *Science* 157:208–10
30. Gelperin, A. 1970. Abdominal sensory

neurons providing negative feedback to the feeding behavior of the blowfly. In manuscript

31. Gelperin, A. 1970. Unpublished observations

32. Gelperin, A., Dethier, V. G. 1967. Long-term regulation of sugar intake by the blowfly. *Physiol. Zool.* 40 :218–28

33. Gordon, H. T. 1961. Nutritional factors in insect resistance to chemicals. *Ann. Rev. Entomol.* 6 :27–54

34. Gordon, H. T. 1968. Quantitative aspects of insect nutrition. *Am. Zool.* 8 :131–38

35. Gordon, H. T. 1968. Intake rates of various solid carbohydrates by male German cockroaches. *J. Insect Physiol.* 14 :41–52

36. Green, G. W. 1964. The control of spontaneous locomotor activity in *Phormia regina* Meig. I. Locomotor activity patterns of intact flies. *J. Insect Physiol.* 10 :711–26

37. Green, G. W. 1964. The control of spontaneous locomotor activity in *Phormia regina* Meig. II. Experiments to determine the mechanism involved. *J. Insect Physiol.* 10 : 727–52

38. Guptavanij, P., Vanard, C. E. 1965. Radiographic study of oesophageal diverticula and stomach of *Aedes aegypti*. *Mosquito News* 25 :288–93

39. Gwadz, R. W. 1969. Regulation of blood meal size in the mosquito. *J. Insect Physiol.* 15 :2039–44

40. Harvey, W. R., Haskell, J. A. 1966. Metabolic control mechanisms in insects. *Advan. Insect Physiol.* 3 : 133–205

41. Highnam, K. C., Hill, L., Mordue, W. 1966. The endocrine system and oöcyte growth in *Schistocerca* in relation to starvation and frontal ganglionectomy. *J. Insect Physiol.* 12 :977–94

42. Hill, L., Strong, L. 1966. Unpublished observations

43. Holling, C. S. 1966. The functional response of invertebrate predators to prey density. *Mem. Entomol. Soc. Can.*, No. 48

44. Hosoi, T. 1954. Mechanism enabling the mosquito to ingest blood into the stomach and sugary fluids into the esophageal diverticula. *Annot. Zool. Jap.* 27 :82–90

45. House, H. L. 1965. Effects of low levels of the nutrient content of a food and of nutrient imbalance on the feeding and the nutrition of a phytophagous larva, *Celerio euphorbiae*. *Can. Entomol.* 97 :62–68

46. Hoyle, G., Burrows, M. 1970. Intracellular studies on identified neurons of insects. *Fed. Proc.* 29 :589

47. Hudson, A. 1958. The effect of flight on the taste threshold and carbohydrate utilization of *Phormia regina* Meig. *J. Insect Physiol.* 1 :293–304

48. Johansson, A. S. 1964. Feeding and nutrition in reproductive processes in insects. *Insect Reproduction*, ed. K. C. Highnam, *Symp. Roy. Entomol. Soc., 2nd, London* 43–55

49. Kerkut, G. A., Pitman, R. M., Walker, R. J. 1969. Iontophoretic application of acetylcholine and GABA onto insect central neurones. *Comp. Biochem. Physiol.* 31 :611–33

50. Knight, M. R. 1962. Rhythmic activities of the alimentary canal of the black blowfly, *Phormia regina*. *Ann. Entomol. Soc. Am.* 55 :380–82

51. Maddrell, S. H. P. 1963. Control of ingestion in *Rhodnius prolixus*. *Nature* 198 :210

52. Mayer, J. 1955. Regulation of energy intake and body weight. *Ann. N.Y. Acad. Sci.* 63 :15–43

53. McCutchan, M. C., Gelperin, A. 1969. Unpublished observations

54. Mittler, T. E. 1957. Studies on the feeding and nutrition of *Tuberolachnus salignus*. I. The uptake of phloem sap. *J. Exp. Biol.* 34 :334–41

55. Mulkern, G. B. 1967. Food selection by grasshoppers. *Ann. Rev. Entomol.* 12 :59–78

56. Murphy, T. A., Wyatt, G. R. 1965. The enzymes of glycogen and trehalose synthesis in silk moth fat body. *J. Biol. Chem.* 240 :1500–8

57. Rowe, E. C. 1969. Microelectrode records from a cockroach thoracic ganglion : synaptic potentials and temporal patterns of spike activity. *Comp. Biochem. Physiol.* 30 :529–39

58. Rozin, P. 1964. Comparative biology of feeding patterns and mechanisms. *Fed. Proc.* 23 :60–65

59. Saito, G. 1963. Trehalose in the body fluid of the silkworm, *Bombyx mori*. *J. Insect Physiol.* 9 :509–19

60. Schoonhoven, L. M. 1968. Chemo-

sensory bases of host plant selection. *Ann. Rev. Entomol.* 13 :115–36

61. Steele, J. E. 1961. Occurrence of a hyperglycaemic factor in the corpus cardiacum of an insect. *Nature* 192 :680–81

62. Steele, J. E. 1963. The site of action of insect hyperglycaemic hormone. *Gen. Comp. Endocrinol.* 3 :46–52

63. Treherne, J. E. 1957. Glucose absorption in the cockroach. *J. Exp. Biol.* 34 :478–85

64. Trembley, H. L. 1952. The distribution of certain liquids in the oesopha-geal diverticula and stomach of the mosquito. *Am. J. Trop. Med. Hyg.* 1 :693–710

65. Van der Kloot, W. G. 1961. Insect metamorphosis and its endocrine control. *Am. Zool.* 1 :3–9

66. Waldbauer, G. P. 1968. The consumption and utilization of food by insects. *Advan. Insect Physiol.* 5 : 229–88

67. Wigglesworth, V. B. 1934. The physiology of ecdysis in *Rhodnius prolixus.* II. Factors controlling moulting and metamorphosis. *Quart. J. Microsc. Sci.* 77 :191–222

COMPARATIVE MORPHOLOGY OF INSECT GENITALIA[1]

6014

G. G. E. SCUDDER

Department of Zoology, University of British Columbia, Vancouver, British Columbia

INTRODUCTION

This review attempts to look at the comparative morphology of insect genitalia from two aspects: 1. the homology of the parts, and 2. the functional aspects of these organs. In the past, it is the former that has been the traditional aspect to study.

The principal origins ascribed for the genital appendages have been (*a*) of appendicular origin, being derived from segmental appendages; (*b*) composite structures evolved from a combination of appendicular parts plus adjacent segmental papillae or ampullae; (*c*) as *a* or *b* above, or both, but the structures in the male not necessarily homologous with those of the female; (*d*) new organs arising as outgrowths of the sterna, but probably homologous between the sexes (198).

Homologous features in two or more organisms may be defined as those that can be traced back phylogenetically to the same feature in the immediate common ancestor of these organisms (23); this definition does not include any reference to resemblance among features. General resemblance, nevertheless, is one of the basic methods of establishing homology (23), although embryonic development is often regarded as the most reliable operational indicator of homology (70). However, ontogenetic studies may be no better an indicator of homology than comparative morphology, for embryonic processes are subject to selection and modification in evolution just as are the final structures. The genital segments in the higher insects have undergone considerable evolutionary changes, and these changes have probably involved, not only the final adult morphology, but also the ontogeny.

There exists a fair knowledge of the fossil history of the major groups of insects (47) but there is no definite phylogeny, just a series of concepts (169, 197). The interrelationships among the Annelida, Onychophora and

---

[1] No attempt has been made to undertake a comprehensive review of the very early literature for this has been done a number of times (131, 192, 198). The review was written while in receipt of a grant from the National Research Council of Canada.

379

Arthropoda, is a matter on which there is plenty of theoretical opinion
(208). Since the phylogeny of the insects is not known, the homology of the
component parts of these animals is obviously uncertain, and the terminology of the parts, based on homology, most likely will not be agreed on by
all.

The Insect Segmental Appendages

If the genitalia are considered to be modified segmental appendages,
then it is important to know the segmentation of such structures. The monophyletic evolutionary theory of the Arthropoda views the Crustacea as ancestral to the Myriapoda and Insecta, and thus the crustacean appendage
can be considered as ancestral to the insectan appendage (see 192). There is
some debate on the basic structure of the crustacean limb but most authors
(191, 208, 211) agree that it is composed of seven segments, although Sharov (192) would add an extra basal precoxopodite. Thus, Sharov considers
the insect coxa to be homologous with the crustacean basipodite and must
then search in insects for the two segments proximal to the coxa. Such precoxal segments, termed the subcoxa (=coxopodite) and pleuron (=pre-
coxopodite), are said to be present in Chilopoda, Diplopoda, Protura,
Collembola, and Diplura, as well as in the Monura and Thysanura. In the
higher insects, these segments in the thorax are said to be modified such that
the subcoxa forms the trochantin, and the pleuron forms the episternum and
epimeron. A similar concept, generally known as the subcoxal theory, has
been proposed (93, 201, 241) to explain the derivation of these same sclerites by the division of a single subcoxal segment and recently this idea has
been developed in some detail by Matsuda (132, 133).

While this review is not concerned with the evolution of the insect thorax, the above mentioned subcoxal theory clearly is important in supposedly
documenting that the insect appendage has at least one additional segment
proximal to the coxa. Such a subcoxa presumably would be present in all
segmental appendages, and so any modified into genital appendages would
be expected to have a homologous component. Indeed, such subcoxal sclerites have been supposedly recognized in the female genitalia (20, 171, 198,
199).

In contrast, to the preceding, adherants of the polyphyletic origin of the
arthropods (11–14, 86, 125–129, 228) see no connection between insects and
Crustacea and thus do not assume that the basic insect appendage had
seven or more segments. Indeed, they see the Onychophora-Myriapoda-
Hexapod assemblage as having legs composed of only five segments at the
most: the coxa is the basal segment, and no precoxal or subcoxal segments
are recognized (127). Thus, there can be no subcoxa on thoracic legs to be
modified into pleural sclerites (127), and no subcoxa in any appendage
modified into genitalia. Authors accepting this polyphyletic concept cannot
include a subcoxa in their theories, and so must derive parts from alternative structures (81, 183, 185–187).

TERMINOLOGY

Michener (143) has considered the terms that are most appropriate for parts of the male and female genitalia. He notes that the terminology adopted by morphologists is usually composed of names suggested by supposed homologies. Taxonomists, on the other hand, prefer to use more descriptive terms, often without regard for homology. In comparative studies in morphology, a uniform set of terms is necessary and Michener (143) considers that the the morphologically meaningful terminology of Snodgrass (202–204, 206) is most acceptable and this has been the terminology used by most morphologists until recently.

However, the terms are not always easy to apply or do not indicate homologies, and so alternative terms have been used by many authors. Tuxen (230) has given a full glossary of the terms applied in the literature. Most authors in recent years have attempted to avoid coining additional and new names but there are a few exceptions (198, 199). In the present article, for convenience, the terminology of Scudder (183, 185, 186) supplemented by that of Snodgrass (203) is used for the female, and that of Snodgrass (210, 212) is used for the male.

CONCEPTS AND FACTS FROM EMBRYONIC DEVELOPMENT

Either an appendicular or a sternal origin is proposed for the genitalia in insects, and evidence for these ideas is usually sought in the development, with the concept that the ontogeny will recapitulate the phylogeny. Of particular interest is (*a*) the fate of the embryonic abdominal appendages, and (*b*) the composition of the insect sternum.

The insect embryo in its development passes from a protopod stage to what is termed a polypod form in which all the abdominal segments, as well as segments cephalad, possess a pair of limb buds (101): the limb buds on all segments are strictly homologous. From the polypod stage then develops a post-polypod or oligopod type of embryo in most insects, in which most of the abdominal limb buds disappear. It is at this stage that eclosion occurs.

In insects, the abdominal appendages usually do not differentiate (187). Hinton (94) and Menees (138) have stated that the abdominal appendages are lost or disappear in the late embryo of insects and thus homologous structures cannot develop in succeeding instars. However, it is probably impossible to prove that the abdominal appendages have disappeared from the ontogeny (149) and it can be argued that the limbs are not lost (187); the latter concept, in fact, being supported by certain mutations in *Blattella* (170, 171).

At the present time, the fate of the embryonic limb tissue has not been adequately studied, but a number of authors assume that it becomes incorporated into the insect abdominal sternum as the lateral areas (see 131). Experimental studies are clearly needed (187) and, to date, certainly no one has described the limbs as being cast off from the embryo although the pleuropodia may in part be discarded (175). Furthermore, ontogenetic cell death, as it occurs in other animals (75), has not been documented in the insect abdominal limb tissue.

It can be argued that the abdominal limb tissue in the embryo attempts to develop, is unable to do so and so regresses (187). This tissue, however, is not lost but remains and retains the competence to develop at a later time, and thus has provided a reservoir of prior organ tissue which, in the evolution of the Insecta, has been available for new organ formation. Such reactivation proposed by Scudder (187) is thought unlikely by other authors (138).

The most anterior pair of abdominal limb buds has been retained in many insects as pleuropodia and does not have a locomotory function but one of producing a hatching enzyme that weakens the chorion prior to eclosion (144, 194, 196). This example alone shows the origin of a new organ from the segmental abdominal appendages and indicates that these limb buds have evolved along new evolutionary pathways (187). Indeed, embryonic limb tissue develops into the pseudoplacenta in some viviparous insects (83), and the development of the prolegs from this tissue has been reported in Lepidoptera (60, 67) and Hymenoptera (62, 189). Wheeler (242), Else (61), and Roonwall (168) derive the genitalia directly from the posterior abdominal appendages in the Acrididae.

ABDOMINAL STRUCTURE IN THE THYSANURA

The Thysanura are regarded by many as the key to the understanding and interpretation of the genitalia in insects (81, 142, 147, 164, 183, 185–187, 198, 199, 202–204, 206, 239). This is because the abdomen retains in many forms what are regarded as segmental appendages, and the genital segments have similar appendages which are serially homologous with the pregenital structures.

It is the Machilidae that are usually used as a starting point for such comparative studies but Scudder (183, 185–187) has stressed the importance of the Lepismatidae in the interpretation of the female genitalia. Present in adult *Neomachilis,* on each of the first seven abdominal segments, are paired lateral plates, each bearing a stylus and an eversible vesicle. In *Petrobius,* segments II to VII bear paired lateral plates, and styli and two eversible vesicles, the sterna being subdivided into a median and two laterosternites (81). Smith (198) considers the condition in *Petrobius* as being the basic plan on which most insects are built, but other workers (183, 185–187, 202–204, 206) regard the *Neomachilis* condition with a single pair of eversible vesicles as the basic type. This latter view is perhaps correct, since it has been shown that the second pair of vesicles in *Petrobius* arises in post-embryonic development at the side of the initial pair, which is present at eclosion (54, 236). Smith (198) would seem to stress the *Petrobius* condition as primitive because it has sclerites that can be interpreted as subcoxae.

It is generally accepted that the appendages of the machilid abdomen are serially homologous with the thoracic appendages. Since the meso and meta-thoracic coxae in Machilidae usually bear styli, there is an obvious similarity between the thoracic leg coxae and the lateral sclerites on the abdomen. Such lateral sclerites are thus often regarded as the base of abdominal legs, but there is no agreement as to what segments are represented. Said to be involved are coxae (81, 183, 185–187, 208, 237), the subcoxa (195), coxa plus subcoxa (20, 26, 202, 223), precoxa plus subcoxa (192), or several segments of the primitive appendage (170). It is seen that any terminology based on homology will vary according to the author, and so Snodgrass (202) suggested that the noncommital term, limb base or basal plate be used.

If this sclerite on the abdomen represents a limb base, then the rest of the leg (telopodite) is often searched for on these abdominal sclerites: there are many ideas on this telopodite homology. Fortunately, only two structures are present for consideration, namely, the styli and the eversible vesicles. Many workers (see 192) have regarded the abdominal styli as the rudimentary telopodites, the eversible vesicles and the thoracic styli then being considered as special endites and excites, or part of the coxa. However, an almost equal number (see 192) have regarded the abdominal styli as homologous with the thoracic styli. These authors then differ in their interpretation of the telopodite and eversible vesicles. Some regard the eversible vesicles as homologous with the telopodite (20, 21, 124) while others (187) regard the telopodite as absent altogether. Scudder (187) has argued that the telopodite need not be represented on abdominal segments, and thus suggested that the coxae represent all that is able to differentiate in post-embryonic development in this delayed differentiating system. The eversible sacs are regarded as homologous with coxal glands (82, 234, 235), coxal endites (201, 210) or, more recently, as having been derived from metameric genital ampullae (81, 183, 185–187, 223).

External Genitalia in Thysanura

The eighth and ninth segments in both male and female in Microcoryphia are basically alike and according to Smith (198) these were the original genital segments in both sexes. The eighth segment can thus be termed the first genital segment and the ninth segment may be called the second genital segment. The parts of these segments are strictly homologous in the two sexes, and are also homologous with the pregenital structures.

The limb base on segment VIII may be termed the first gonocoxa (= first valvifer), and this bears the first gonostylus. Segment IX, then, carries the second gonocoxa (= second valvifer) and the second gonostylus. These gonocoxae in the female also carry elongate outgrowths, termed gonapophyses, in a position comparable to the eversible vesicle of pregenital segments. Such gonapophyses are also borne on both the first and second gonocoxae in male *Machilis polypoda* (see 239), but most male Microcoryphia lack them on segment VIII. In male *Tricholepidion* (Lepidotrichidae), Wygodzinsky (244) has shown that the first gonocoxa bears a nonfunctional eversible vesicle, while the second gonocoxa bears a regular gonapophysis. This, together with the position of the gonapophyses and eversible sacs, has suggested that the two are homologous (81, 170, 183, 185–187, 195). This homology of the gonapophyses is, however, not the one accepted in the earlier literature, for here they are regarded as telopodites, coxal endites, or special structures of sternal origin (see 131).

The only structure wherein the two sexes of Microcoryphia differ is in the presence of a penis in the male. This structure is situated between the second gonocoxae, but from developmental studies (121) it has been concluded that it is derived from a pair of primary phallic lobes which pertain to segment X and represent secondary ectodermal outgrowths around the opening of the genital ducts (210). Gustafson (81) and Smith (198), on the other hand, think that the penis rudiments arise on segment IX, the former considering them as homologous with an inner pair of eversible vesicles, the latter thinking that they represent papillae which originally bore an accessory gonopore.

Regardless of the correct homology, it is from this thysanuran structure that most authors begin their consideration of the comparative morphology of the male and female genitalia. Most regard the female structure to be composed of appendicular components from the eighth and ninth segments. The male they regard as being constituted from these components plus a pair of primary phallic lobes. The male and female systems are regarded as fundamentally homologous (131, 142, 198, 204). However, this concept is doubted by Snodgrass (210) who regards the male genitalia of the Hemipteroids and Exopterygota as of nonappendicular origin. It should also be noted that Smith (198) considers the structure in Machilidae to be a condition derived from a more archaic structure.

STRUCTURE OF THE MALE GENITALIA IN INSECTA

It is quite inappropriate and undesirable to describe the male genitalia of all the insect orders and then discuss each in turn. A description of all orders, with considerable discussion, is to be found in Tuxen (230). However, in this book, a comparative morphological account has not been attempted. Thus, it is appropriate here to look at the comparative morphology and the ideas that have been expressed (142, 148, 198, 204, 206, 210).

In most insects, the external male genitalia include primary organs for the delivery of sperm to the female, and structures for seizing and holding her during mating. These latter structures are not uniform in insects, and may involve periphallic claspers developed from the ninth segmental appendages (Grylloblattodea, Ephemeroptera), claspers developed from cerci (Dermaptera, Odonata, many Orthoptera) or the paraprocts (Zygoptera, some Tridactylitidae [Orthoptera]) (239).

Within the insects, six general groups[2] can be recognized from published descriptions (44, 155, 206, 210, 230, 239) as follows:

1. Thysanura: single median penis present, situated posterior to, but between, the bases of stylus-bearing limb bases of segment IX; the latter limb bases may bear a gonapophysis in addition to a stylus, and a similar structure may be present on segment VIII.

2. Ephemeroptera: paired penes usually present (sometimes united to form a single organ in which the ejaculatory ducts remain separate), situated caudad of stylus-bearing plates (claspers) and sternum IX. In the Dermaptera the primitive forms have paired penes, but in the higher earwigs the two penes are united basally, but apically may be somewhat divided.

3. Grylloblattodea: have paired phallic lobes, between which at the base is situated the gonopore: these may represent paired penes from which the ducts have been withdrawn (212). In Grylloblattodea, a distinct sternum IX is present, together with asymmetrical limb bases bearing styli (238); these are united to form a composite sclerite with styli in the Blattaria (44, 239), or the structure may lack styli as in the Acrididae. The phallic lobes in most Orthopteroids are divided to form several lobes which either remain separate (Blattaria, Mantodea, Grylloblattodea) or become united to form a single primary phallic organ (Gryllidae, Tettigoniidae, Acrididae).

4. In Hemiptera, there is no sign of appendage plates bearing styli. The ninth segment may be in the form of a genital capsule. Paired claspers (parameres) are usually present, together with an aedeagus which is somewhat variable in structure. In Coleoptera and Lepidoptera, the plan is much the same.

5. In Mecoptera and Megaloptera, the plan is similar to the Hemiptera, but the claspers (parameres) are two-segmented.

6. In Diptera, Trichoptera, and Hymenoptera, there is a median aedeagus, one or two jointed claspers (parameres), and in addition, usually a pair of structures (paraphyses) lateral to the aedeagus. These paraphyses are termed claspettes in Diptera, volsella in Hymenoptera, and titillators in Trichoptera and are attached usually to a basal plate.

Comparative Morphology of the Male Genitalia

The recent general studies of the male genitalia (142, 148, 198) concede, as did those of earlier workers (71, 155, 239), that the eighth segmental limb bases, styli, and gonapophyses present in Machilidae, are not represented in the pterygote insects, except perhaps in situations of anomalous development (97). However, they agree that the ninth segmental components have been retained in virtually all taxa. Thus, they view the male genitalia as being, at least in part, appendicular in origin. In general, they agree that the ninth appendicular parts are to be seen as the outer claspers (parameres). Where these claspers consist of two segments, they are said to

[2] Accessory genitalia of Odonata excluded.

TABLE 1. Homologies of Insect Genitalia Suggested by
Comparative Morphology

Segment	Machilidae	Pterygote male	Pterygote female
VIII	gonapophysis	—	first gonapophysis (=valvula)
VIII	limb base	—	first gonocoxa (=valvifer)
VIII	stylus	—	
IX	gonapophysis	phallus (in part)	second gonapophysis (=valvula)
IX	limb base		second gonocoxa (=valvifer)
		} clasper	+gonangulum+gonoplac (=third valvula)
IX	stylus		second gonostylus
IX or X	primary phallic lobe (male only)	phallus (in part)	

represent the gonocoxae and the gonostylus of segment IX (46), and hence are similar to the condition in Grylloblattodea (Table 1). Where there is but a one segmented outer clasper, it could represent the gonocoxa alone, gonocoxa + gonostylus, or gonostylus alone (45), and is so interpreted according to particular fancy. The inner claspers (paraphyses) present in some forms are usually regarded as subdivisions of the gonocoxa, as is the basal plate.

Significant differences exist in the search for a homologue of the gonapophysis of segment IX and the penis of the Thysanura. Both Michener (142) and Nielsen (148) consider the pterygote aedeagus to be a composite structure made up of the "machilid" penis and the gonapophyses of segment IX (= penis valves of Michener). This, according to Nielsen (148, who believes that the penis came from segment X), has been made possible by the reduction in the venter of segment X.

Smith (198), like George (71) and Metcalf (140) on the other hand, claim that the aedeagus of most pterygotes represents the united gonapophyses alone, the "machilid" penis according to Smith (198) being a special structure not present in other insects. He thus believes that the machilid condition is not ancestral but was derived independently. Nielsen (148) also believes the copulatory organs of the male have been evolved independently several times and further states that those lacking such organs have probably lost them.

The ideas of Acker (1) are somewhat different from the above authors in that he stressed the fact that the musculature of the male arises on segment IX and not VIII as might be expected if the genialia were in part derived from segment IX. He thus has derived the whole male genitalia in Neuroptera from the tenth segmental appendages and sternum X, and Randell (163) believes that the stylus present on the epiphallus of *Turanogryllus* is a clear indication of the appendicular origin from segment X. Brinck (28,

29), from a study of the segmental nerves and muscles in Plecoptera and Ephemeroptera, also concludes that the male gonopore was originally located on segment X.

If one relies on comparative morphology, and accepts that (a) in all insects, the male genitalia have a common structural plan; (b) that they are developed from parts originally belonging to segments IX or X, or both; and (c) that the development has of necessity been modified in the course of evolution, then any of the above interpretations are theoretically possible. On present evidence, there is little to suggest which is most likely to be correct.

COMPARATIVE ONTOGENY OF THE MALE GENITALIA

Snodgrass (210) has pointed out that the great structural diversity of the male genitalia make it difficult to understand the fundamental nature and the homologies of the parts. He states that when homologies cannot be determined by comparative anatomy alone, recourse must be made to ontogeny. He thus has gathered together the scattered accounts of the male genitalia and has attempted to determine this fundamental structure and homology. He believes that these studies show a monophyletic origin of the phallic organs and that this is clearly evident in the ontogeny where, in all cases, the primary organs are derived from a pair of primary phallic lobes. He assigns these primary phallic lobes to segment X, thinks that they are secondary ectodermal outgrowths, and notes that there is a tendency for them to move forward and onto segment IX in the higher insects.

A number of early studies specifically maintain that the primary phallic lobes arise on segment X and represent the segmental appendages (58, 61, 159, 160, 165, 190). Roonwall (168) states that the persistent embryonic limb buds on segment X move forward to fuse with those of segment IX prior to forming the phallus, and Tiegs (227) reports in *Nasonia* that the tenth segmental appendages become closely applied to sternum IX, the claspers then developing from sternum IX and the aedeagus from the tenth segmental component. However, a number of recent studies (138) have emphatically claimed that the primary phallic lobes arise on segment XI and not X.

An analysis of the post-embryonic developmental data shows that the ontogenies fall into two categories: 1. primary phallic lobes giving simple phallic organs (penis or phallomeres); periphallic organs present and forming from other structures (includes Thysanura, Ephemeroptera, Dermaptera, orthopteroids). 2. Primary phallic lobes giving rise to whole male genitalia, i.e., aedeagus and parameres; periphallic organs usually absent (includes hemipteroids and Endopterygota).

Snodgrass (210) regards the primary phallic lobes as strictly homologous throughout, and so the phallic complex in the hemipteroids and Endopterygota is homologous with the penis of the Thysanura. The claspers (parameres) of the higher insects, are not then homologous with the peri-

phallic appendicular parts of segment IX seen in Thysanura, Grylloblattodea, and Ephemeroptera.

Earlier studies (71, 140, 155, 160) on the development of the Hemiptera
have suggested that the limb bases of segment IX form the subgenital plate,
while the gonapophyses of this segment form the aedeagus and parameres.
In the Endopterygota, the limb bases are said to be present as subgenital
plates, with an origin independent of the aedeagus (137, 153, 190).

The paraphyses arise as subdivisions of the parameres and thus have no
simple homologue in the Exopterygota. Christophers (41) showed this origin for the claspettes in the mosquito, and a similar origin for the volsella
has been reported in Hymenoptera (151, 176).

STERNAL ORIGIN OF MALE GENITALIA

Heymons (90–92), from similar studies of the development, maintained
as have other workers (67, 82, 85, 114, 131) that the genitalia are new
structures of sternal origin. Matsuda (131) thinks that the male and female
genitalia are likely to be of similar origin, and he notes that there is no
evidence for the genitalia developing from the appendage anlagen occurring
in the embryo, although the thoracic appendage buds always give the thoracic legs. He emphasizes the studies that show that many endopterygote
larvae have abdominal legs on segment X (60, 157), and thus the appendages of this segment cannot possibly have formed all or part of the male
genitalia.

Since this concept is applicable to both male and female structures, further discussion of this idea is delayed until consideration of the female genitalia in a later section.

STRUCTURE AND FUNCTION IN THE MALE GENITALIA

It is evident that the genitalic structures and their evolution cannot be
understood without a knowledge of their function. The understanding of
the function of these organs is rudimentary (10), with only a few studied in
detail (25, 28, 29, 102, 103, 109, 116, 166, 215, 216, 218, 219, 232).

It is the task of the male genital system to produce and deliver to the
female sufficient spermatozoa to fertilize the eggs (50). It is assumed that
the supposed ancestor of insects had spermatophores, and thus the presence
of spermatophores is thought to be a primitive feature in insects (49, 72, 73,
95, 107).

In the Apterygota, there is no external male genitalia in Collembola, only
a genital papilla in Diplura and a very simple penis in Thysanura (230).
These very simple features are clearly correlated with the indirect sperm
transfer in these insects (18, 177–180, 214, 220–222, 224, 225). The penis in
Thysanura is thus not an intromittant organ. (See also the review by Schaller
on indirect sperm transfer, pp. 407–46 of this volume.)

In virtually all other insects except Odonata, mating takes place directly,
with the male passing the sexual products to the female directly. However,

there is a vast diversity in the structure of the organs for doing this. This diversity is related to the various functions, which are 1. the mode of transfer of germ cells to the female; 2. clasping of the female in mating; 3. mechanical protection of the genitalic parts; and 4. sensory function of the parts (139).

Mode of transfer of germ cells to the female.—In the Pterygota there are five major ways of handling the germ cells (72) which can be arranged in a phylogenetic series as follows:

I. Spermatophore molded in the male prior to copulation.

II. Spermatophore molded in the male organ after aedeagus enters the female bursa copulatrix.

III. Male accessory gland material ejected in definite sequence into bursa copulatrix or vagina after insects *in copula* but before sperm transfer, the sperm being encapsulated on entry.

IV. Male accessory gland material ejected into the female, often after sperm, the material not encapsulating sperm but forming a mating plug.

V. Direct transfer of free sperm into the spermatheca or a site in vagina or bursa copulatrix, near the spermathecal opening.

In many spermatophore-producing types (Blattaria, Tettigoniidae, Gryllidae) the male organ has a spermatophore sac and is not an intromittant organ (7, 19, 163), but serves to transfer the spermatophore to the female genitalia only. In the Gryllidae, although the virga enclosing the spermatophore tube is threaded into the female for anchorage, there is still no real intromittant organ since the sperm-containing ampulla remains outside the female. It is in the Acrididae, that a true intromittant organ is formed for placing the spermatophore directly in the female ducts (63, 78).

There has been little correlation of the structure of the intromittant organ with the mode of germ cell transfer. It is generally thought that those insects which produce spermatophores have a rather sac-like organ, while those that do not possess a more or less elongate aedeagus (49, 145). Certainly, an elongate aedeagus delivering free sperm has been reported in Heteroptera (25, 123), Diptera (56, 107), and Neuroptera (49) while, in the Orthoptera that deliver spermatophores, there is in general a sac-like structure (207). However, in most orders, the long aedeagus is correlated with the delivery of free sperm to a spermatheca, rather than simply the delivery of free sperm (107).

The elongate aedeagus in many forms has been produced by the eversion of the gonopore through the phallotreme with the resulting production of an endophallus (206). It is evident that spermatophores have been lost and free sperm delivery evolved independently many times in the advanced insects (95). Structural features associated with this evolution also must then have been independently evolved a number of times.

Clasping of female in mating.—This task has probably been responsible for most changes in the male genitalia. In many Plecoptera during copulation, the partners are very loosely fitted together by the everted genital cavity of the male, there being no claspers (28). Similar maintenance of the copulatory position, accomplished by the erect phallus pushing against the wall of the female bursa or vagina, often aided by processes on the intromittant organ, is reported in many orders (25, 28, 48, 119, 165, 167). In *Apis,* the aedeagus is so firmly applied to the female vagina that the male genitalia is torn away from the male insect at the end of copulation (209).

However, in most insects, claspers in the male are used to hold the female. There is variation, not only in the organs used for clasping, but also the part or parts clasped. In the Notonemourinae and Nemourinae (Plecoptera) the abbreviated and hooked cerci serve to clasp the subgenital plate and assist in copulation (28). The cerci also serve as claspers in the Dermaptera, Odonata, and Acrididae (63, 205, 239). In Ephemeroptera, they serve to fix loosely the female body for mating, but here it is the stylus-bearing plates of segment IX that hold the eighth and ninth abdominal segments of the female (29). In Gryllidae, the phallomers may serve as claspers, clasping a sclerotic ring round the spermathecal duct, as well as supporting the virga (163).

The parameres have been reported to grasp the ovipositor valves in Hemiptera (25, 165) and in Mecoptera (43), to grip the subgenital plate in Hymenoptera (151), while in Lepidoptera they hold the extremity of the female abdomen (218, 219, 232) and in *Aedes aeqypti* they grasp the female cerci (102, 103, 215).

The organs used for initial seizing of the genitalia may be different from the structures used for final holding of the female apparatus (219). Thus, in the Frenate Lepidoptera, initial seizing is done by the uncus, while final holding is accomplished by the claspers (219). Similarly, in *Periplaneta americana* the hooked left phallomere serves to seize the apical lobes of sternum VII prior to the male moving out and assuming the end-to-end position, while the "clam-shell" sclerites on the right phallomere grip the first gonapophyses for final security after the male movement (79). The clasping structures in insects thus have a different specific function and hence different detailed structure in the various taxa.

While the clasping of the female might seem to be a simple matter of just holding on, this is far from the true situation. Not only must the female be clasped in exactly the correct manner, but also certain movements of the mating insects must be permitted at times. It is the need for this maneuverability that has brought about many of the special structural features of male genitalia. It is in the orthopteroids with extended mating that such movements first seem to be an important behavioral feature, and it is these insects that first show these structural adaptations.

The males of Tettigoniidae, Gryllidae, Gryllacrididae, and Blattaria may possess dorsal glands (7, 8, 24, 69, 80, 172–174) which serve to attract and then position the female on the dorsum of the male, preparatory to mat ing: such a position is thought to be primitive in the pterygotes (7, 10) and is retained for the duration of copulation, only if mating is short. If there is prolonged mating, such as in the Blattaria for the production of spermatophore, or in the Gryllidae for the deposition of the complex spermatophore, then alternative positions are assumed while *in copula.*

The movement from a female above position has taken place by vertical or lateral turning (Gryllidae, Tettigoniidae, Gryllacrididae, Blattaria), or there has been the gradual shifting of the mating act such that no turning is necessary (Acridoidea, Phasmodea, Mantodea). The lateral turning is most prevalent and involves the rotation of the male genitalia on the female structures, or rotation of the male or female genitalia on the abdomen.

Vertical turning.—Male movement vertically under the female has been reported in Tettigoniidae and Gryllidae (7, 10). However, the morphological consequences of this movement have not been studied in detail to date.

Rotation of male genitalia on female structure.—Such movement has been reported in the Gryllidae (10) and is correlated with structural modifications in both male and female. In these forms there is a bifid phallic organ in the male and a flattened copulatory papilla in the female, turning being accomplished by the male complex rotating on the papillae of the female (10).

When the genitalia themselves twist during mating, it is evident that the genitalic parts are often asymmetrical. Thus, in *Blattaria* the left and right phallomeres have different functions and are asymmetrical (79), the left which clasps the female subgenital plate evidently must be twisted or uncoiled as the male turns to the end-to-end position. The aedeagus of many Coleoptera is rotated around its axis during mating, and this movement is permitted by the normally twisted structure when at rest (120, 193). It is likely that the coiling of the terminal parts of the aedeagus in some Heteroptera might serve the same function, since in this case the aedeagus is known to uncoil during erection at a time when the male insect moves relative to the female (25). However, in the Heteroptera, there is rotation of the whole genital capsule on the abdomen also at each copulation (25, 48).

Rotation of the genitalia on the abdomen.—Rotation of the whole male genitalia relative to the body axis in order to accommodate the female position, occurs in many orders. It may involve movement that takes place only at the time of each copulation, as in Heteroptera (25, 48), or may be a permanent rotation, of varying degree, as in the Diptera (40, 66, 74, 87, 105, 106, 135, 181). Such rotation seems not to involve the genitalia itself and so need not be considered further here. Details of the mating process in such Diptera is to be found in the literature (84, 115, 167).

Shifted mating positions.—In many insects, positions are taken that differ from those discussed so far. In one such pose, the male mounts the female and then has to curve the abdomen round and under the female, or in another, the male reaches down one side of the female for genital contact; positions similar to the latter also exist in many side-to-side matings. Such male-above positions, which were once thought to be primitive (167) occur in many taxa (7) and have resulted in structural modification of the genitalia, particularly toward asymmetry (167). Particularly good examples of this are seen in some orthopteroids (239), in the aquatic Heteroptera (116), and in Miridae (109). However, it is possible that the asymmetry in some insects is also a structural consequence of needing to accommodate and protect an excessively large intromittant organ (167).

Mechanical protection of parts.—As has been noted by Pruthi (156), such delicate organs as the aedeagus must be protected against external injury, and many features of the genitalia may have developed to serve this purpose. Thus, in the Heteroptera, segment IX which forms the genital capsule, not only encompasses the aedeagus, but also is retracted into the preceding segment. In the orthopteroids and many Homoptera (Membracidae, Cercopidae, Cicadellidae) the genitalia are protected by a distinct subgenital plate, while in the Cicadidae, the aedeagus is lodged in a groove of tergum X and then protected from below by an enlarged sternum VII (156). The parameres of Fulgoridae are leaf-like and are said to serve not for clasping but for protection of the aedeagus (156).

Many Endopterygota have the genital segments withdrawn into the pregenital segment VII for protection, and in the Diptera, Kessel & Maggioncalda (106) have noted that circumversion may have mechanical protective features in that (*a*) the post-abdomen is often folded under the pre-abdomen for protection, and (*b*) the membranous stalk of the hypopygium may serve as a coiled spring connecting partners in copulation, and thus may provide flexibility and maneuverability without fracture.

The asymmetry in some insects may be associated with protection of an excessively long aedeagus as noted by Richards (167), and species with these elongate structures often have the terminal parts retractable into a basal sclerotized phallotheca for protection.

Many sclerotized areas in the genitalia are probably present to protect approximated parts mechanically, and the membranous areas are certainly there for flexibility. However, our understanding of these aspects is fragmentary.

Sensory functions.—Little is known of the sensory functions in the genitalia, but they obviously exist and are important. In *Oecanthus,* the cerci of the male are reported to be placed on either side of the ovipositor in copulation and there serve to correctly position the spermatophore in the female orifice (68), while in *Gryllus* the upturned cerci serve to position the female on the male for mating (6). Also, in *Culex pipiens* aedeagus eversion appears to be stimulated exclusively through the contact of the male terminalia with those of the female. It is suggested that the prominant curved sensillae on the basal segment of the claspers in the male are the sense organs involved, receiving stimuli on contact with the female postgenital plate (216). Similarly, in many Hemiptera, it is highly probable that the hairs on the distal extremity of the parameres are sensory in function (156), and the same function is noted for the parameres in Carabidae (5).

There can be little doubt that many of the structures in the genitalia have a sensory function, but not until ultrastructural studies are carried out in conjunction with electrophysiological observations, will this become adequately documented.

SPECIAL MODIFICATIONS IN SOME MALE INSECTS

In the bedbug genus, *Cimex,* a special adaptation is found associated with the hemocoelic method of insemination. In *Cimex lectularius* the left clasper is modified to act as a piercing organ and the right is lost (42, 53). This left clasper is used to penetrate Berlese's organ (Ribaga's organ), and in so doing, the aedeagus, which lies in a groove of the clasper, is thrust into this organ (53). Similar modes of insemination are reported in other Cimicidae (32, 39, 104) and in Anthocoridae (36, 37). The beginnings of this habit would seem to be in other cimicoid forms in which the aedeagus is well fortified and pierces the genital chamber (33–35) or anywhere on the abdomen (38).

STRUCTURE OF THE FEMALE GENITALIA

The female ovipositor is an organ for oviposition and can be of two types (212):[3] I, an extension of the abdomen itself, or II, a ventral process from segments VIII and IX.

The ovipositor of the first type (Type I) (= oviscapt) is present in some Thysanoptera, Mecoptera, Lepidoptera, Coleoptera, Diptera, and Trichoptera. The distal segments of the abdomen are tapered and are capable of being retracted, telescope-like (3, 99, 141, 200). Such an ovipositor shows little variation from the comparative morphology point of view, and so the structure will not be considered further.

The second type of ovipositor (Type II) is more widely distributed, being found in most taxa (Table 2).

COMPARATIVE MORPHOLOGY OF THE TYPE II OVIPOSITOR

The basic plan in all taxa is similar to that found in the Thysanura (Table 1). Thus, the first and second gonapophyses in Thysanura are usually

---

[3] An ovipositor is wanting in Protura, Collembola, Diplura, Ephemeroptera, most Isoptera, Plecoptera, Embioptera, most Dermaptera, some Coleoptera, most Trichoptera, Siphonaptera, Strepsiptera, Mallophaga, Siphunculata, and some Hemiptera (Coccidae, Aphididae) (230)

considered homologous with the first and second gonapophyses in the pterygotes (235, 237). Not all early workers however, accepted this homology (146).

In Lepismatidae and Lepidotrichidae, there is a sclerite in the ovipositor complex that is not distinct and part of the machilid organ (183, 185–187, 244). This sclerite, the gonangulum, is triangular in shape and articulates antero-dorsally with tergum IX, postero-ventrally with the second gonocoxa (= second valvifer), and antero-ventrally is attached to the base of the first gonapophyses (= first valvulae). The gonangulum has been homologized with the antero-dorsal corner of the second gonocoxa (111, 183, 185–187) with a subcoxa (192) or is considered as part of the intersegmental membrane between segments VIII and IX (223) or the ventral portion of

TABLE 2. Parts of Ovipositer (Type II) Reported in Various Taxa

	1st gonapophysis	1st gonocoxa	1st gonostylus	2nd gonapophysis	2nd gonocoxa	Gonoplac	Gonangulum	2nd gonostylus
Thysanura: Microcoryphia	X	X	X	X	X			X
Zygentoma	X	X	X	X	X		X	X
Odonata	X	X		X	X	X	X	X
Grylloblattodea	X	X		X	X	X	X	L
Orthoptera: Tettigoniidae	X	X		X	X	X	X	L
Gryllidae	X	R		R	X	X	X	
Acrididae	X	R		R	X	X	X	
Phasmida	X	X		X	X	X	X	
Dermaptera (Echinosoma)	X	X			R	X	X	
Dictyoptera: Blattaria	X	X		X	X	X	X	L
Mantodea	X	X		X	X	X	X	
Isoptera (Mastotermes)	X	?		X	X	X	X	
Psocoptera	X	X		X	X	X	X	
Hemiptera	X	X		X	X	X	X	
Thysanoptera	X	X		X	X		X	
Neuroptera	X	R		R		X		
Mecoptera						X		
Lepidoptera				X				
Trichoptera				X				
Diptera: Tipulidae	X			X	X			
Hymenoptera	X	?		X	X	X	X	
Coleoptera: Adephaga				X	X		X	

X = present; R = reduced; L = present in immatures only; ? = possibly present.

laterotergite IX (122) ; it was said to be the antero-lateral part of the sternum IX by Walker (237). Scudder (183, 185–187) claims that this sclerite
is present as a basic component of the Type II ovipositor in the pterygotes
and always retains its original articulations. This concept has been accepted
by a number of authors, but not all (see 230).

An additional component of the Type II ovipositor, present in many taxa
but absent in Thysanura, is the gonoplacs (= third valvulae). These structures are not serially homologous with the first and second gonapophyses,
although Qadri (159) did accept them as homologous with the first gonapophyses in orthopteroids, and Wheeler (242) thought that they represented
the tenth segmental appendages that have moved onto segment IX. Some authors believe that these gonoplacs are homologous with the second gonostylus of Thysanura (142), but developmental studies have shown that they
arise as outgrowths of the second gonocoxa (162, 165, 170, 176, 203). Further, studies on Blattaria, Odonata, and Grylloblattodea, show the gonoplacs to be present with the gonostyli borne on the ends. Such gonostyli,
although present in immatures, may be absent in the adult, and although
most authors claim that such styli atrophy (152) or are cast off and lost
(158, 159, 170, 171, 203, 233), Ramsay (162) considers in Henicidae (Grylla-
cridoidea) that they are incorporated into the gonoplacs during development. The gonoplacs thus may be partly homologous with the gonostyli in
the latter and other similar forms.

The majority of morphologists accept the thysanuran structure as the
basis of the pterygote Type II ovipositor, and so consider this apparatus of
appendicular origin, involving the appendages of abdominal segments VIII
and IX. This view, however, is not universally acceptable and Matsuda
(131) has reviewed an alternative concept, namely, that the gonapophyses,
like the male genitalia, are special structures of sternal origin.

STERNAL ORIGIN OF THE GONAPOPHYSES

Of particular importance in this concept is the structure of the pterygote
sternum. From the time of Haase (82) and Heymons (92) it has been acknowledged that the abdominal sternum in ptergotes is made up of cells
from three primary embryonic fields, a median true sternal and two lateral
parts derived from the segmental appendage rudiments, although this has
not been demonstrated experimentally. Matsuda (131) has stressed that
much of the mid-ventral area of this supposed composite coxosternum is of
true sternal origin, the appendage fields being located at the side and not
meeting in the mid-line posteriorly as they do in many living Thysanura
(76) : the middle area is said to have no appendage remnants.

Since the gonapophyses, when they occur, usually arise in the median
area of the sternum on either side of the mid-line, it is argued that they
develop from the true sternal field and not from the more lateral appendage
fields. Hence, the gonapophyses must be of sternal origin.

However, in *Gryllus* the second instar has sternum IX divided into three parts, a median and two lateral ones (161): in the third instar a pair of genitalic buds appear at the posterior of sternum VIII, and a pair appear on sternum IX at the postero-inner corner of the lateral areas (161). In *Gryllus*, at least, the second gonapophyses and gonoplacs arise laterally, evidently in the appendage areas of the sternum.

It should also be noted that the median ventral areas of segments VIII and IX are usually concerned with the formation of the bursa copulatrix or vagina and sometimes with the ectodermal ducts of the female (51, 55, 71, 138, 140, 146, 154, 159). A similar process may also occur on segment IX in the male (137). The origin of these structures could well utilize the true sternal field of the coxosternum. The invaginations that occur can have a pronounced effect on the position of the growing gonapophyses. For example, in female Gryllidae where the first gonapophyses arise far apart on sternum VIII, they are dragged inward during ontogeny by the development of the spermathecal invagination that follows. However, in the Blattaria and Acrididae in which the first gonapophyses arise near the mid-line, the spermathecal invagination takes place before the formation of the gonapophyses (187). Indeed, in many forms studied to date, the spermathecal invagination appears prior to the differentiation of the first gonapophyses (159), and so the latter may not in fact arise from the true sternal field.

Matsuda (131) does note that it might be suspected that embryonic limb rudiments might, after the lapse of a latent stage, reappear on the segments as genital rudiments, but he believes that this is most unlikely, that it is too complicated a hypothesis, and certainly is impossible to demonstrate. He considers that it is most improbable that the embryonic appendage rudiment should disappear and then reappear as the genitalia tissue. Not only do they have to become reactivated, but they also have to migrate to the sternum.

In contrast to this, Scudder (187) believes that this is exactly what seems to happen, at least in the female. One of the reasons for so thinking is the belief that new organs in evolution do not arise de novo, but rather arise by modification of pre-existing structures (65, 142, 187, 231). If one believes that the ontogeny is a clear and precise indicator of homology, then the sternal origin idea would appear to be proven. However, ontogeny can change in evolution, and until there is some experimental data to support the claim, this concept is no better and no worse than the comparative morphology data based on the adult and the concept of the genitalia being derived from segmental appendages.

Structure and Function in the Female Genitalia

Forms that lack an ovipositor simply deposit the eggs on the surface, be this in the terrestrial or aquatic habitat, or they may give birth to live young. In general, the insects with an ovipositor are concerned with placing the eggs less obviously in the environment. The species with an ovipositor of Type I usually use this organ as a probe and place the eggs in crevices or similar situations without creating a special oviposition hole; a few like the Trypetidae and some Cecidomyidae (Diptera) use the structure as a spike for ovipositing in soft plant tissue.

Insects with a Type II ovipositor borne on segment VIII and IX usually are concerned with oviposition in firm substrates and tissues, where they have to make a special oviposition hole by movement of the gonapophyses. It is thus to be expected that this type of ovipositor will be rigid, well articulated, and functionally related to the characteristics of the substrate and the tasks to be performed. In addition, this structure must at the same time allow for the requirements of copulation, extension and retraction where necessary, and finally it must be mechanically protected and have some sensory functions.

Ovipositor structure and oviposition.—If one considers the ovipositor of Machilidae, it is evident that two problems exist in moving the eggs any distance between the two sets of gonapophyses (198) : (*a*) it is difficult to force the eggs right to the end of such unconnected parts, and (*b*) it is difficult to coordinate the movements of the two sets of gonapophyses. These mechanical requirements of the ovipositor have recently been considered by Smith (198), although the mechanics of selected types have been studied previously a number of times (31, 88, 89, 203, 209). Smith (198) finds that the oviposition problems were overcome by developing ridges or teeth along the length of the gonapophyses, and by developing an interlocking mechanism for the gonapophyses. This interlocking, according to Smith (198), has been accomplished by the second gonapophyses undergoing a 180° turn, this phylogenetic event being repeated in the ontogeny of all present-day forms. The second gonapophyses then are said to have fused along the new dorsal surface, and finally to have developed a groove along their new ventral edge; into this groove then fits a ridge developed on the dorsal edge of the first gonapophyses. A tongue and groove structure is thus formed which prevents the independent movement of the parts and at the same time prevents buckling of the ovipositor when this is pressed against a firm substrate.

For the effective movement of the interlocking gonapophyses, the terga of segments VIII and IX, which provide the fulcra for the gonocoxae, must then remain stationary in relation to one another. This immobility can be accomplished by fusion of terga VIII and IX, but seems to have been brought about in most insects by the development of a cross-connection between the two segments. This connection is provided by the sclerite called the gonangulum by Scudder (183, 185, 186) and the first gonocoxa by Smith (198, 199). Scudder (183, 185–187) believes that this cross-connection is a basic feature of the Type II ovipositor, but Smith (198, 199) considers that it is a characteristic found only in the Hymenoptera.

Smith (198) sees a four-stage evolutionary series in the various pterygotes, the theme of the series being the increase in the relative movement of

the first gonapophyses relative to the dorsal ovipositor valves (second gonapophyses or gonoplacs, or both). While this series of models no doubt will
explain different degrees of movement of the parts, the structural descriptions presently available in the literature do not exactly fit the models proposed. For example, in most orders there is a connection between the second
gonocoxa and the cross-connecting gonangulum (183, 185–187). Since
Smith (198) indicates an intention to publish a series of papers on the comparative external morphology of the insect genitalia, this apparent discrepancy may be cleared up in the future.

Nevertheless, Smith (198), like Scudder (185, 186), has stressed the importance of the movement of the ventral ovipositor valves against the dorsal ones, either independently or together. Functionally, when the first gonapophyses move independently they produce a sawing action, but when they
move together, they bring about the penetration of the ovipositor (52). In
forms that must undertake much sawing action, it is evident that the first
gonapophyses cannot be united. The efficiency of such ovipositors is dependent on the degree of sclerotization, the rasps and saws, and the interlocking device that unifies the parts, yet permits both longitudinal, and to some
extent, lateral movements (17).

That there is an intimate relationship between oviposition habit and ovipositor structure is clearly shown in the Dytiscidae (27) and Heteroptera
(59, 184). In the latter, the forms that oviposit in tissue have a laciniate ovipositor, while those that deposit eggs on a surface have a "plate-shaped
type" of ovipositor. Not only are the ovipositor valves reduced in the sur-
face-ovipositing types, but the interlocking mechanism of these structures is
redundant and has been reduced or lost (4, 110, 112, 184).

The eggs normally appear to travel down the whole length of the ovipositor valves, as in Gryllidae, where the ends are specially adapted for holding the egg for firm insertion into the oviposition site (203). This movement down the shaft of the ovipositor is said to be aided by ridges and
grooves in the valves (198), but such features are not always evident. In
certain insects, such as the stinging Hymenoptera, the eggs obviously must
issue from the base of the sting, and the organ must be modified accordingly
(203).

Ovipositor structure and copulation.—The female gonopore is primi-
tively located between segments VII and VIII (29, 159, 242), but effectively
usually opens more posteriorly between the bases of the ovipositor valves.
In most forms, the ventral ovipositor valves are not joined along their
length, and so in these the aedeagus can be inserted into the female ducts
between the valves near the base. In such forms, the ovipositor is usually
tilted upward as seen in Cicadellidae (113) and Gryllidae (10). In forms
in which the first gonapophyses are joined along their length mid-ventrally,
such an insertion is not possible. Here, either the aedeagus is inserted down
the length of the ovipositor, as in some Lygaeidae (16, 25) or the aedeagus

is inserted into an opening of the female ducts cephalad of the ovipositor, as in Cicadidae (203).

Mechanical protection of the ovipositor.—A permanently exserted ovipositor is a feature seen in many groups (Tettigoniidae, Gryllidae, Rhaphidiidae, Ichneumonidae) and in these the structure is heavily sclerotized. However, in many other taxa the ovipositor may be protected by the gonaplacs that form an ovipositor sheath, as in many Hymenoptera (31, 88, 89, 198, 199, 203, 204, 209), or it may be protected by pregenital sterna, as in many Hemiptera. In many chalcidoid Hymenoptera, the ovipositor has been rotated proximally into the abdomen, for both protection and function, and has resulted in considerable modification of the pregenital segments (229). All forms with a retracted ovipositor must have the structure appropriately modified for exsertion and retraction, but the mechanical features and movements have been studied in very few forms (52, 113, 203).

Special Modified Female Genitalia

Special structural modifications, associated with peculiar functional requirements, are present in certain groups. Three striking examples exist in the Blattaria, Acrididae, and in the aculeate Hymenoptera.

Many early pterygotes, such as Megasecoptera, had an exserted ovipositor (108). In Blattaria at least some fossil forms had an exserted ovipositor (22, 118), but living representatives, have a vestibulum and a reduced ovipositor (130, 136). This modification is associated with the formation and retention of an ootheca, and the way of handling the spermatophore (77). In these insects a vestibulum has been formed for production of an ootheca, and the ovipositor has been reduced to occupy a position dorsally. Although this ovipositor is usually rather fixed and pointed posteriorly (130, 136), in the forms that rotate and withdraw the ootheca into the vestibulum the valves are reversible so as to point anteriorly. This movement is accomplished by turning or partial twisting of the second gonocoxae, something that is made possible by the reduction of various basal sclerites (130, 136). These same basal sclerites also are important for spermatophore attachment, and their reduction also correlates with the removal of this function and the placing of the spermatophore directly in a bursa copulatrix (77).

In the Acrididae, the females usually oviposit deeply in soil and in such forms the valves are modified to act as an excavating device that not only digs a hole but also draws the abdomen down into the ground (63, 205, 237). The basal sclerites on which the muscles attach are peculiarly modified, with long internal apodemes as a dominant feature.

In the Hymenoptera, the aculeates have the ovipositor acicular and modified into a stinging device. This structure has been described many times (203, 209), its use being facilitated by the waist-like constriction that is characteristic of these insects and their Parasitica relatives.

Sex Anomalies and Homology

The evidence derived from gynandromorphs, intersexes, and mutants of various types, has been used from time to time in attempting to establish the fundamental homologies of the genitalia, and the homologies between male and female parts. While there are gynandromorph examples which appear to demonstrate that the male claspers are homologous with the second gonocoxae (= second valvifer) and gonoplacs (= third valvula) of the female, other gynandromorphs seem to show that the second gonocoxae and gonoplacs correspond to the subgenital plate of the male, while the second gonapophyses correspond to the penis valves. Further, there are still other gynandromorphs which show male claspers and second gonocoxae plus gonoplacs or various combinations of male and female parts (2, 182, 243) present at the same time. Thus, it would seem that there is no conclusive evidence to be obtained from gynandromorphs (142). Intersexes, according to Michener (142), are more reliable, and he describes such an individual in the Hymenoptera which is said to show clearly that the male claspers are homologous with the second gonocoxae and gonoplacs.

Recent work on intersexes (96, 117, 150, 226), temperature-induced sex anomalies (15, 98), and genetic mutants (213) has focused on the higher insect orders with specialized or degenerate structures. Since these insects are lacking many of the important structures required for consideration, no fundamental conceptual facts seem to emerge from those studied to date, other than the point that parts of the two sexes do correspond.

Insect Genitalia and Speciation

The idea that the male and female genitalia of insects form a lock-and-key mechanism is long-standing (57) and there is no doubt that in certain circumstances there is a mechanical isolation between species (240). The concept of intraspecific compatibility and interspecific incompatibility in genitalia is an attractive one, and the fact that there is considerable specific variation in genital structures in most groups is intriguing. However, there is little documentation favoring mechanical isolation in most taxa (134). Indeed, examples of interspecific matings are recorded in the literature, and we know that usually there are many premating isolation mechanisms that come into play prior to insect contact (30, 217). It might also be noted that isolation depending on genitalic incompatibility is inefficient (7), and so is unlikely to occur.

As far as the Fulgoroidea are concerned, Fennah (64) has concluded that the variation in phallic structures raises no mechanical barriers to copulation between different species: the female organs are often so much alike that it is difficult to imagine how any slight modification in the male could bring about a change in genital functions. This conclusion would equally fit many other taxa, as taxonomists will readily admit (109, 113). It has been shown experimentally that some Lepidoptera can manage without parts of the genitalia (188).

Just how such specific differences in genitalia occur and what they are for, is not clear. It is assumed that divergence in genitalia occurs only if copulation is sometimes initiated between species; the differences are assumed to arise in isolation, perhaps as pleiotropic effects (134), as the result of selection other than mechanical isolation. Genitalic differences arise as a result of isolation, and never function to cause it (7, 9).

This aspect of the comparative morphology of insect genitalia is perhaps the most important one from a general point of view, but we are a long way at present from understanding the meaning of this specific variability.

CONCLUSION

Some thirty years ago, Imms (100) wrote "During recent years much attention has been given to the genitalia of Insecta, and more especially to the male organs, in view of their importance in taxonomy. The whole subject has become greatly involved owing to the lack of any convenient, uniform terminology, and to uncertainty which exists with respect to the homology of the parts concerned in different orders. The greater part of the literature deals with the completed organs and especially those of the male. Ontogenetic studies have not been pursued with the same enthusiasm. They are dependent upon obtaining supplies of material in proper succession stages which are often difficult to provide. This fact, coupled with the obscurity which so often attends the initial growth phases of the genitalia and the difficulties of interpretation that are involved, has resulted in a large field, so far, being inadequately explored." More than three decades later, this still clearly expresses the state of our knowledge.

LITERATURE CITED

1. Acker, T. S. 1960. *Microentomology* 24:25–84
2. Agacino, E. M. 1957. *Proc. Roy. Entomol. Soc. London, Ser. A* 32: 169–70
3. Ageyeva, A. G. 1968. *Entomol. Obozr.* 47:285–90 [s/t translation]
4. Ahmad, I., Southwood, T. R. E. 1964. *Tijdschr. Entomol.* 107:365–78
5. Alexander, R. D. 1959. *Ann. Entomol. Soc. Am.* 52:485
6. Alexander, R. D. 1961. *Behaviour* 17:130–223
7. Alexander, R. D. 1964. *Symp. Roy. Entomol. Soc. London* 2:78–94
8. Alexander, R. D., Brown, W. L. 1963. *Occ. Pap. Univ. Michigan Museum Zool.* 628:1–19
9. Alexander, R. D., Moore, T. E. 1962. *Misc. Publ. Univ. Michigan Museum Zool.* 121:1–59
10. Alexander, R. D., Otte, D. 1967. *Misc. Publ. Univ. Michigan Museum Zool.* 133:1–62
11. Anderson, D. T. 1966. *Proc. Linn. Soc. N. S. W.* 91:10–43
12. Anderson, D. T. 1966. *Acta Zool. Stockholm* 47:1–42
13. Anderson, D. T. 1966. *Ann. Mag. Nat. Hist.* (13) 9:445–56
14. Anderson, D. T. 1969. *Phil. Trans. Roy. Soc. London, Ser. B* 256: 183–235
15. Anderson, J. F., Horsfall, W. R. 1963. *J. Exp. Zool.* 154:67–108
16. Ashlock, P. D. 1957. *Ann. Entomol. Soc. Am.* 50:407–26
17. Balduf, W. V. 1933. *Ann. Entomol. Soc. Am.* 26:64–75
18. Bareth, C. 1964. *C. R. Acad. Sci., Paris* 259:1572–75
19. Beier, M. In *Taxonomists Glossary of Genitalia in Insects.* See Ref. 230
20. Bekker, E. G. 1925. *Trud. Nauch. inst. Mosk. Univ.* 1:157–206
21. Bekker, E. G. 1952. *Vestn. Mosk. Univ.* 5:69–83
22. Bekker-Migdisova, Ye. E. 1962. In *Bases of Palaeontology. Arthropods: Tracheates and Chelicerates,* ed. Yu. A. Orlov. *Acad. Nauk SSSR*
23. Bock, W. 1969. *Ann. N.Y. Acad. Sci.* 167:71–73
24. Boldyrev, B. T. 1928. *Eos,* 4:13–56
25. Bonhag, P. F., Wick, J. R. 1953. *J. Morphol.* 93:177–283

26. Börner, C. 1904. *Zool. Anz.* 27:226–43
27. Böving, A. G. 1913. *Int. Rev. Hydrobiol. Hydrogr. Biol. Suppl.* 5:1–28
28. Brinck, P. 1956. *Opusc. Entomol.* 21: 57–127
29. Brinck, P. 1957. *Opusc. Entomol.* 22: 1–37
30. Brown, R. G. B. 1965. *Behaviour* 25: 281–323
31. Callahan, P. S., Blum, M. S., Walker, J. R. 1959. *Ann. Entomol. Soc. Am.* 52:573–90
32. Carayon, J. 1946. *C. R. Acad. Sci., Paris* 226:107–9
33. Carayon, J. 1952. *C. R. Acad. Sci., Paris* 234:751–53
34. Carayon, J. 1952. *C. R. Acad. Sci., Paris* 234:1220–22
35. Carayon, J. 1952. *C. R. Acad. Sci., Paris* 234:1317–19
36. Carayon, J. 1952. *Bull. Mus. Nat. Hist., Paris* 24:89–97
37. Carayon, J. 1953. *C. R. Acad. Sci., Paris* 236:1099–1101
38. Carayon, J. 1954. *C. R. Acad. Sci., Paris* 239:1542–44
39. Carayon, J. 1959. *Rev. Zool. Bot. Afr.* 60:81–104
40. Chillcott, J. C. T. 1958. *Proc. Int. Congr. Entomol., 10th, Montreal, 1958* 1:587–92
41. Christophers, S. R. 1922. *Indian J. Med. Res.* 10:530–71
42. Christophers, S. R., Cragg, F. W. 1922. *Indian J. Med. Res.* 9:445–63
43. Cooper, K. W. 1940. *Am. Midl. Nat.* 23:354–67
44. Crampton, G. C. 1918. *Bull. Brooklyn Entomol. Soc.* 13:49–68
45. Crampton, G. C. 1921. *Can. Entomol.* 53:72
46. Crampton, G. C. 1931. *Psyche* 38: 1–21
47. Crowson, R. A., Rolfe, W. D. I., Smart, J., Waterston, C. D., Willey, E. C., Wootton, R. J. 1967. *The Fossil Record,* 499–534. (Geol. Soc. London)
48. Davey, K. G. 1959. *Quart. J. Microscop. Sci.* 100:221–30
49. Davey, K. G. 1960. *Proc. Roy. Entomol. Soc. London, Ser. A.* 35: 107–13
50. Davey, K. G. 1965. *Reproduction in the Insects.* Edinburgh and London: Oliver and Boyd

51. Davies, R. G. 1961. *Proc. Zool. Soc. London* 136:411–37
52. Davis, N. T. 1955. *Ann. Entomol. Soc. Am.* 48:132–50
53. Davis, N. T. 1956. *Ann. Entomol. Soc. Am.* 49:466–93
54. Delany, M. J. 1959. *Trans. Roy. Soc. Edinburgh* 63:501–33
55. Dodson, M. E. 1935. *Quart. J. Microscop. Sci.* 77:383–403
56. Downes, J. A. 1968. *Can. Entomol.* 100:608–17
57. Dufour, L. 1844. *Ann. Sci. Nat.* 1:244–64
58. Dupuis, C. 1950. *Annee Biol.* (3) 26:21–36
59. Dupuis, C. In *Taxonomists Glossary of Genitalia in Insects.* See Ref. 230
60. Eastham, L. E. S. 1930. *Phil. Trans. Roy. Soc. London, Ser. B* 219:1–50
61. Else, F. L. 1934. *J. Morphol.* 55:577–609
62. Farooqi, M. M. 1963. *Alig. Musl. Univ. Publ. (Zool.) Indian Insect Types* 6:1–68
63. Fedorov, S. M. 1927. *Trans. Entomol. Soc. London* 75:53–61
64. Fennah, R. G. 1946. *Proc. Roy. Entomol. Soc. London, Ser. A* 21:73–80
65. Ferris, G. F., Rees, B. E. 1939. *Microentomology* 4:79–108
66. Feureborn, H. J. 1923. *Zool. Anz.* 55:189–213
67. Friedmann, N. 1934. *Comment. Biol., Helsingfors* 4(10):1–29
68. Fulton, B. B. 1915. *Tech. Bull. N.Y. Agr. Exp. Stat.* 42:3–47
69. Gabbutt, P. D. 1954. *Brit. J. Anim. Behav.* 2:84–88
70. Gans, C. 1969. *Ann. N.Y. Acad. Sci.* 167:506–13
71. George, G. J. 1928. *Quart. J. Microscop. Sci.* 72:447–85
72. Gerber, G. H. 1970. *Can. Entomol.* 102:358–62
73. Ghilarov, M. S. 1958. *Zool. Zh.* 37:707–35
74. Gleichauf, R. 1936. *Z. Wiss. Zool.* 148:1–66
75. Glücksmann, A. 1951. *Biol. Rev. Cambridge Phil. Soc.* 26:59–86
76. Grassi, B. 1889. *Arch. Ital. Biol.* 11(2):1–77
77. Graves, P. N. 1969. *Ann. Entomol. Soc. Am.* 62:595–602
78. Gregory, G. E. 1965. *J. Exp. Biol.* 42:423–35
79. Gupta, P. D. 1947. *Proc. Nat. Inst. Sci. India* 13:65–71
80. Gurney, A. B. 1947. *J. Wash. Acad. Sci.* 37:430–35
81. Gustafson, J. F. 1950. *Microentomology* 15:35–67
82. Haase, E. 1889. *Morphol. Jahrb.* 15:331–435
83. Hagen, H. R. 1951. *Embryology of the Viviparous Insects.* New York: Ronald Press
84. Hardy, G. H. 1944. *Proc. Roy. Entomol. Soc. London, Ser. A* 19:52–65
85. Heberdey, R. 1931. *Z. Morphol. Oekol. Tiere* 22:416–586
86. Hedgpeth, J. W. 1966. *Quart. Rev. Biol.* 42:69
87. Hennig, W. 1936. *Z. Morphol. Oekol. Tiere* 31:328–70
88. Hermann, H. R., Blum, M. S. 1966. *Ann. Entomol. Soc. Am.* 59:397–409
89. Hermann, H. R., Blum, M. S. 1967. *Ann. Entomol. Soc. Am.* 60:1282–91
90. Heymons, R. 1896. *Morphol. Jahrb.* 24:178–204
91. Heymons, R. 1896. *Biol. Zentralbl.* 16:855–64
92. Heymons, R. 1899. *Zool. Zentralbl.* 6:537–56
93. Heymons, R. 1899. *Abh. Kaiser Leopold.-Carolin. Deut. Akad. Naturforsch.* 74:349–456
94. Hinton, H. E. 1955. *Trans. Roy. Entomol. Soc. London* 106:455–556
95. Hinton, H. E. 1964. *Symp. Roy. Entomol. Soc. London* 2:95–107
96. Hollingsworth, M. J. 1960. *J. Exp. Zool.* 143:123–52
97. Horsfall, W. R., Anderson, J. F. 1963. *Science* 141:1183–84
98. Horsfall, W. R., Anderson, J. F. 1964. *J. Exp. Zool.* 156:61–90
99. Huckett, H. C. 1921. *Ann. Entomol. Soc. Am.* 14:290–328
100. Imms, A. D. 1937. *Recent Advances in Entomology, 2nd ed.* London: J. & A. Churchill, Ltd.
101. Johannsen, O. A., Butt, F. H. 1941. *Embryology of Insects and Myriapods.* New York and London: McGraw-Hill
102. Jones, J. C. 1968. *Sci. Am.* 218(4):108–16
103. Jones, J. C., Wheeler, R. E. 1965. *J. Morphol.* 117:401–24
104. Jordan, K. 1922. *Ectoparasites* 1:284–86
105. Kessel, E. L. 1968. *Wasmann J. Biol.* 26:243–53
106. Kessel, E. L., Maggioncalda, E. A.

1968. *Wasmann J. Biol.* 26:33–106

107. Khalifa, A. 1949. *Trans. Roy. Entomol. Soc. London* 100:449–71

108. Kukalova, J. 1960. *Proc. Int. Congr. Entomol., 11th, Vienna, 1960* 1: 292–94

109. Kullenberg, B. 1947. *Zool. Bidrag.* 24:217–48

110. Kumar, R. 1962. *Entomol. Tidskr.* 83:44–88

111. Kumar, R. 1965. *Proc. Roy. Soc. Queensland* 76(3):27–91

112. Kumar, R. 1965. *J. Entomol. Soc. Queensland* 4:41–55

113. Kunze, L. 1959. *Deut. Entomol. Z.* 6:322–87

114. Lacaze-Duthiers, H. 1849. *Ann. Sci. Nat.* (3, *Zool.*) 12:353–75

115. Lamb, C. G. 1922. *Proc. Roy. Soc. London, Ser. B* 94:1–11

116. Larsen, O. 1938. *Opusc. Entomol. Suppl.* 1, 388 pp.

117. Laugé, G. 1968. *Ann. Soc. Entomol. France* 4:481–99

118. Laurentiaux, D. 1951. *Ann. Paleontol.* 37:185–96

119. Leston, D. 1955. *J. Soc. Brit. Entomol.* 5:103–5

120. Lindroth, C. H., Palmen, E. In *Taxonomists Glossary of Genitalia in Insects.* See Ref. 230

121. Lindsay, E. 1939. *Proc. Roy. Soc. Victoria* 52:35–83

122. Livingstone, D. 1967. *J. Anim. Morphol. Physiol.* 14:1–27

123. Ludwig, W. 1926. *Z. Morphol. Oekol. Tiere* 5:291–380

124. Makhotin, A. A. 1929. *Zool. Zh.* 9: 23–74

125. Manton, S. M. 1949. *Phil. Trans. Roy. Soc. London, Ser. B* 233: 483–580

126. Manton, S. M. 1964. *Phil. Trans. Roy. Soc. London, Ser. B* 247:1–183

127. Manton, S. M. 1966. *J. Linn. Soc. London (Zool.)* 46:103–41

128. Manton, S. M. 1966. *Nature* 210: 1303–4

129. Manton, S. M. 1967. *J. Nat. Hist.* 1:1–22

130. Marks, E., Lawson, F. A. 1962. *J. Morphol.* 111:139–72

131. Matsuda, R. 1958. *Ann. Entomol. Soc. Am.* 51:84–94

132. Matsuda, R. 1960. *Ann. Entomol. Soc. Am.* 53:712–31

133. Matsuda, R. 1963. *Ann. Rev. Entomol.* 8:59–76

134. Mayr, E. 1963. *Animals Species and Evolution.* Cambridge, Mass.: Harvard Univ. Press

135. McAlpine, J. F. 1967. *Can. Entomol.* 99:225–36

136. McKittrick, F. A. 1964. *Cornell Univ. Agr. Exp. Sta. Mem.* 389, 197 pp.

137. Mehta, D. R. 1933. *Quart. J. Microscop. Sci.* 76:35–61

138. Menees, J. H. 1963. *Cornell Univ. Agr. Exp. Sta. Mem.* 381, 59 pp.

139. Metcalf, C. L. 1921. *Ann. Entomol. Soc. Am.* 14:169–225

140. Metcalfe, M. E. 1932. *Quart. J. Microscop. Sci.* 75:467–81

141. Metcalfe, M. E. 1933. *Quart. J. Microscop. Sci.* 76:89–105

142. Michener, C. D. 1944. *Ann. Entomol. Soc. Am.* 37:336–55

143. Michener, C. D. 1958. *Proc. Int. Congr. Entomol., 10th, Montreal, 1958* 1:583–86

144. Miller, A. 1940. *Ann. Entomol. Soc. Am.* 33:437–77

145. Muir, F. 1920. *Trans. Entomol. Soc. London, 1919* 404–14

146. Nel, R. I. 1929. *Quart. J. Microscop. Sci.* 73:25–85

147. Newell, A. G. 1918. *Ann. Entomol. Soc. Am.* 11:109–56

148. Nielsen, A. 1957. *Entomol. Medd.* 28:27–57

149. Novak, V. 1961. *Entomol. Obozr.* 40:1–6 [s/t translation]

150. Parker, G. A. 1969. *Trans. Roy. Entomol. Soc. London* 121:305–23

151. Peck, O. 1937. *Can. J. Res.* 15(D): 221–52

152. Peytoureau, A. 1893. *C. R. Acad. Sci., Paris* 117:749–51

153. Pruthi, H. S. 1924. *Proc. Zool. Soc. London, 1924* 857–68

154. Pruthi, H. S. 1924. *Proc. Zool. Soc. London, 1924* 869–83

155. Pruthi, H. S. 1925. *Quart. J. Microscop. Sci.* 69:59–118

156. Pruthi, H. S. 1925. *Trans. Entomol. Soc. London, 1925* 127–267

157. Pryor, M. G. M. 1951. *Quart. J. Microscop. Sci.* 92:351–76

158. Qadri, M. A. H. 1938. *Bull. Entomol. Res.* 29:263–76

159. Qadri, M. A. H. 1940. *Trans. Roy. Entomol. Soc. London* 90:121–75

160. Qadri, M. A. H. 1949. *Proc. Zool. Soc. India* 1:129–43

161. Rakshpal, R. 1961. *Indian J. Entomol.* 23:23–39

162. Ramsay, G. W. 1965. *Proc. Roy. Entomol. Soc. London, Ser. A* 40: 41–50

163. Randell, R. L. 1964. *Can. Entomol.* 96 :1565–1607
164. Rasnitsyn, A. P. 1968. *Entomol. Obozr.* 47 :35–40 [s/t translation]
165. Rawat, B. L. 1939. *Trans. Roy. Entomol. Soc. London* 88 :119–38
166. Rees, D. M., Onishi, K. 1951. *Proc. Entomol. Soc. Wash.* 53 :233–46
167. Richards, O. W. 1927. *Biol. Rev. Cambridge Phil. Soc.* 2 :289–360
168. Roonwall, M. L. 1937. *Phil. Trans. Roy. Soc. London, Ser. B* 227 : 175–244
169. Ross, H. H. 1965. *A Textbook of Entomology,* 3rd ed. New York: Wiley & Sons
170. Ross, M. H. 1966. *Ann. Entomol. Soc. Am.* 59 :473–84
171. Ross, M. H. 1966. *Ann. Entomol. Soc. Am.* 59 :1160–62
172. Roth, L. M. 1969. *Ann. Entomol. Soc. Am.* 62 :176–208
173. Roth, L. M., Willis, E. R. 1952. *Am. Mid. Nat.* 47 :66–129
174. Roth, L. M., Willis, E. R. 1954. *Smithson. Misc. Collect.* 122(12) : 1–49
175. Roth, L. M., Willis, E. R. 1955. *Psyche* 62 :55–67
176. d'Rozario, A. M. 1942. *Trans. Roy. Entomol. Soc. London* 92 :363–415
177. Schaller, F. 1952. *Naturwissenschaften* 39 :48
178. Schaller, F. 1952. *Verh. Deut. Zool. Ges. Suppl.* 17 :184–89
179. Schaller, F. 1954. *Naturwissenschaften* 41 :406–7
180. Schaller, F. 1954. *Forsch. Fortschr. Deut. Wiss.* 28 :321–26
181. Schräder, T. 1927. *Z. Morphol. Oekol. Tiere* 8 :1 44
182. Scudder, G. G. E. 1957. *Entomol. Monthly Mag.* 92 :377–79
183. Scudder, G. G. E. 1957. *Nature* 180 : 340–41
184. Scudder, G. G. E. 1959. *Trans. Roy. Entomol. Soc. London* 111 :405–67
185. Scudder, G. G. E. 1961. *Can. Entomol.* 93 :267–72
186. Scudder, G. G. E. 1961. *Trans. Roy. Entomol. Soc. London* 113 :25–40
187. Scudder, G. G. E. 1964. *Can. Entomol.* 96 :405–17
188. Sengün, A. 1944. *Rev. Fac. Sci., Istanbul Univ., Seri B* 9 :239–53
189. Shafig, S. A. 1954. *Quart. J. Microscop. Sci.* 95 :93–144
190. Sharif, M. 1937. *Phil. Trans. Roy. Soc. London, Ser. B* 227 :465–538
191. Sharov, A. G. 1959. *Trud. Inst. Morfol. Zhivotn. Akad. Nauk SSR* 27 :175–86
192. Sharov, A. G. 1966. *Basic Arthropodan Stock with Special Reference to Insects.* Oxford : Pergamon
193. Sharp, D., Muir, F. 1912. *Trans. Entomol. Soc. London, 1912* 477–642
194. Shutts, J. H. 1952. *Proc. S. Dak. Acad. Sci.* 31 :158–63
195. Silvestri, F. 1905. *Zool. Jahrb., Suppl.* 6 :773–806
196. Slifer, E. H. 1937. *Quart. J. Microscop. Sci.* 79 :493–506
197. Smart, J. 1963. *Proc. Linn. Soc. London* 172 :125–26
198. Smith, E. L. 1969. *Ann. Entomol. Soc. Am.* 62 :1051–79
199. Smith, E. L. 1970. *Ann. Entomol. Soc. Am.* 63 :1–27
200. Smith, G. S. S. 1938. *Parasitology* 30 :441–76
201. Snodgrass, R. E. 1927. *Smithson. Misc. Collect.* 80(1) :1–108
202. Snodgrass, R. E. 1931. *Smithson. Misc. Collect.* 85(6) :1–128
203. Snodgrass, R. E. 1933. *Smithson. Misc. Collect.* 89(8) :1–148
204. Snodgrass, R. E. 1935. *Principles of Insect Morphology.* New York and London : McGraw-Hill
205. Snodgrass, R. E. 1935. *Smithson. Misc. Collect.* 94(6) :1–89
206. Snodgrass, R. E. 1936. *Smithson. Misc. Collect.* 95(14) :1–96
207. Snodgrass, R. E. 1937. *Smithson. Misc. Collect.* 96(5) :1–107
208. Snodgrass, R. E. 1952. *A Textbook of Arthropod Anatomy.* Ithaca, N.Y.: Comstock Publ. Assoc.
209. Snodgrass, R. E. 1956. *Anatomy of the Honey Bee.* Ithaca, N.Y.: Comstock Publ. Assoc.
210. Snodgrass, R. E. 1957. *Smithson. Misc. Collect.* 135(6) :1–60
211. Snodgrass, R. E. 1958. *Smithson. Misc. Collect.* 138(2) :1–77
212. Snodgrass, R. E. 1963. *Smithson. Misc. Collect.* 146(2) :1–48
213. Sokoloff, A., Hoy, M. A. 1968. *Ann. Entomol. Soc. Am.* 61 :550–53
214. Spencer, G. J. 1930. *Can. Entomol.* 62 :1–2
215. Spielman, A. 1964. *Biol. Bull.* 127 : 324–44
216. Spielman, A. 1966. *Ann. Entomol. Soc. Am.* 59 :309–14
217. Spieth, H. T. 1949. *Evolution* 3 :67–81
218. Stekol'nikov, A. A. 1965. *Entomol. Obozr.* 44 :258–71 [s/t translation]

219. Stekol'nikov, A. A. 1967. *Entomol. Obozr.* 46 :400–9 [s/t translation]
220. Stürm, H. 1952. *Naturwissenschaften* 39 :308
221. Stürm, H. 1955. *Z. Tierpsychol.* 12 : 337–63
222. Stürm, H. 1956. *Z. Tierpsychol.* 13 : 1–12
223. Stys, P. 1959. *Acta Univ. Carolinae (Biol.)* 1 :75–85
224. Sweetman, H. L. 1934. *Bull. Brooklyn Entomol. Soc.* 29 :158–61
225. Sweetman, H. L. 1938. *Ecol. Monogr.* 8 :285–311
226. Thomas, H. T. 1950. *Proc. Zool. Soc. London* 120 :155–63
227. Tiegs, O. W. 1922. *Trans. Proc. Roy. Soc. S. Australia* 46 :319–527
228. Tiegs, O. W., Manton, S. M. 1958. *Biol. Rev. Cambridge Phil. Soc.* 33 :255–337
229. Tryapitsyn, V. A. 1968. *Entomol. Obozr.* 47 :277–84 [s/t translation]
230. Tuxen, S. L., Ed. 1970. *Taxonomists Glossary of Genitalia in Insects,* 2nd ed. Copenhagen : Munksgaard
231. Tuxen, S. L. 1970. *Bull. Soc. Entomol. Italy* In press

232. van Goethem, J. 1967. *Natuurwetensch. Tijdschr. Ghent* 48 :163–201
233. van Wyk, L. E. 1952. *J. Entomol. Soc. S. Afr.* 15 :3–62
234. Verhoeff, K. W. 1896. *Zool. Anz.* 19 :378–88
235. Verhoeff, K. W. 1902. *Zool. Anz.* 26 :60–77
236. Verhoeff, K. W. 1910. *Zool. Anz.* 36 :385–99
237. Walker, E. M. 1919. *Ann. Entomol. Soc. Am.* 12 :267–316
238. Walker, E. M. 1919. *Can. Entomol.* 51 :131–39
239. Walker, E. M. 1922. *Ann. Entomol. Soc. Am.* 15 :1–76
240. Watson, J. A. L. 1966. *Proc. Roy. Entomol. Soc. London, Ser. B* 41 : 171–74
241. Weber, H. 1928. *Z. Wiss. Zool.* 131 : 181–254
242. Wheeler, W. M. 1893. *J. Morphol.* 8 :1–160
243. Willis, E. R., Roth, L. M. 1959. *Ann. Entomol. Soc. Am.* 52 :420–29
244. Wygodzinsky, P. 1961. *Ann. Entomol. Soc. Am.* 54 :621–27

INDIRECT SPERM TRANSFER BY SOIL ARTHROPODS

6015

F. Schaller

1. Zoological Institute of the University of Vienna, Vienna, Austria

The sperm cells of primitive terrestrial arthropods and vertebrates (Amphibia) are often transferred in special containers from the males to the females. Such sperm containers or sperm carriers are called spermatophores. They appear in different aquatic animals, e.g., Crustacea in which they facilitate more direct and efficient fertilization of the eggs in the body of the female. However, most of the aquatic animals simply empty their sperm cells into the water (e.g., many annelids, echinoderms, and fish).

During the transition from aquatic to terrestrial life, the formation of spermatophores was particularly necessary mainly for animals which had not developed special copulatory organs. Sperm transfer in these animals is often very complicated and long-lasting so that the very delicate sperms are in danger of drying up.

Some annelid ancestors of arthropods developed spermatophores and appropriate methods of transfer in the preliminary stages of transition to terrestrial life. Some small psammobiotic polychaetes, semiterrestrial oligochaetes, and leeches are examples. The 2-mm-long polychaete *Hesionides arenaria* produces double spermatophores (13, 110, 111). The males, which have the genital openings on the head, attach these 60–175 μ long tubes anywhere on the integument of the females. The sperm penetrates the skin of the female in a day and passes actively through the body cavity to the eggs. The related, also psammobiotic species, *Microphthalmus aberrans* has a similar sperm transfer (111). It is significant that the females do not show any copulation reaction and always react defensively to the males, so that the latter have to act rapidly. The males apparently cannot recognize with certainty their females because they attach the spermatophores also to other males of the species.

The copulation of the hermaphrodite earthworms and leeches is well known. These semiterrestrial annelids produce simple spermatophores. The earthworms transfer simple spermballs in temporarily formed skin grooves from one partner to the other. Many Rhynchobdellidae among the Hirudineae produce spermatophores which they attach to the skin of their partners with a histolytic secretion. The double tubes filled with sperm cells contract and inject their contents through the skin into the body cavity, where the sperm reaches the eggs by chemotaxis. However, many sperm cells are engulfed by phagocytes in the process. This is apparently the reason that these Hirudineae copulate repeatedly, attaching reciprocally as many as 10 spermatophores (45).

The spermatophore transfer takes place in the same way in the strictly terrestrial proarthropods. The males of the bisexual Onychophora (*Peripatus*) form sperm packets and attach them, frequently in large numbers, anywhere to the integument of the females. After about seven days, amoebocytes dissolve the integument at the places of attachment, and the sperm can penetrate and fertilize the eggs in the ovaries.

The females of the Onychophora are also completely passive and the males "copulate" quite unselectively; they attach the spermatophores also to males and immature individuals (58).

Different mechanisms and methods of sperm transfer have evolved in the lower terrestrial arthropods. Many species which have emancipated themselves completely from water and from habitats with high relative humidity, evolved copulatory organs for internal insemination. Other species produce spermatophores which they place directly into or at least on the female genital opening (e.g., grasshoppers). Many others transfer the sperm indirectly with secondary genital appendages or gonopods (e.g., spiders). The more primitive forms, which are usually ecologically restricted to humid habitats, have complicated indirect methods of sperm transfer in which the sperm remains for a long time outside the body of males and females. These methods of sperm transfer apparently evolved independently in the lower Tracheata (Myriapoda and Apterygota), and in the Arachnida.

DIPLOPODA

Many Diplopoda have gonopods with which they transfer the sperm directly. However, the African pill-millipede *Sphaerotherium dorsale* (33, 34) which has no gonopods, uses its normal walking legs for the transfer of sperm. The males clutch with their pincer-like terminal legs the vulvae of the females. The male and female genital openings are nearly 2 cm apart in this inverse position. The male then secretes an approximately 1-mm large sperm drop from his genital pores which are situated anteriorly, and passes it posteriorly with the tarsi from one pair of legs to the next, to the female genital opening (Fig. 1). This is probably the most primitive case of sperm transfer by means of normal, i.e., unspecialized extremities in terrestrial arthropods, that is known. This is in accordance with the long-standing view of systematists that sphaerotherians are the least evolved Diplopoda.

Recent observations of the sexual behavior of the related Glomeridae (35) are of the greatest interest. The males of *Glomeris marginata* and *G. transalpina*, and of *Loboglomeris pyrenaica* and *L. rugifera* clutch the vulvae of the females with their telopods; they then roll up to clean their terminal

FIGURE 1. Sperm transport in the direction of arrow by male *Sphaerotherium dorsale.*

FIGURE 2. Pairing of *Loboglomeris pyrenaica.*

FIGURE 3. Initiation of sperm transport by *Glomeris marginata.* The male in the middle has grasped a soil particle.

legs with the mandibles (Fig. 2), grasp a particle of substrate (usually a frass pellet) with their forelegs, roll it on the ground and gnaw it until it is round (Fig. 3). They then apply this particle for 10–20 sec to their own genital pores on the second pair of legs, and put a drop of sperm on it. Finally they transport the particle with the drop of sperm along their legs posteriorly to the telopods, where it is deposited at the syncoxite, so that the sperm drop is directly opposite the genital opening of the female. The paired projections of the syncoxite then introduce the sperm synchronically and rhythmically into the vulvae. The particle used as sperm carrier is then dropped to the side. This complicated and risky manipulation may be almost considered as a case of use of a tool for sperm transfer.

The sperm transfer of the Pselaphognatha (*Polyxenus lagurus*, 86–88), is entirely indirect. The males of these usually bark-inhabiting diplopods (Fig. 4a) spin a double zig-zag thread with both their "penis appendages" above a small depression, and place on it two drops of sperm (Fig. 4b,d). They then turn around and, by means of glands located on the seventh segment, spin two thick, parallel threads as they retreat (Fig. 4b,c). These thick threads apparently contain pheromones which attract sexually mature specimens. Mature females run along them, search for the sperm drops and take them up with their vulvae. Males and females of the Pselsphognatha have no contact whatever in this extremely indirect "sperm transfer." The signal threads also stimulate mature males to directed searching. They eat the sperm drops of their predecessors, renew or repeat the threads in the same place, and place on it fresh sperm drops. This behavior is biologically meaningful, because it constantly provides fresh sperm for the females.

CHILOPODA

Typical indirect methods of sperm transfer occur in many Chilopoda. After a tactile pre-mating behavior ("courtship"), the males of the southern European *Scolopendra cingulata* (49, 52) spin a web and deposit in it a bean-shaped spermatophore which the closely following female takes up. The web is not only a sperm support but also a signal for the receptive female.

The sexual partners of the euedaphic Geophilidae (50) behave similarly. Here, too, courtship and sperm transfer occur in narrow galleries. The male *Geophilus longicornis* uses his genital appendage to attach an irregular web within a gallery and deposits a drop of sperm on it (Fig. 5). The female waits behind the male, mostly without contact with him, and generally proceeds some hours later. As she crawls forward, the posterior part of her body sways to and fro. When the web is reached, she "feels" her way carefully through it and attempts to wipe the sperm drops onto her terminal segments.

The mating behavior and the transfer of the spermatophores in the central European centipede *Lithobius forficatus* (51) is performed essentially in the same way as in scolopenders, but with substantial improvement in the con-

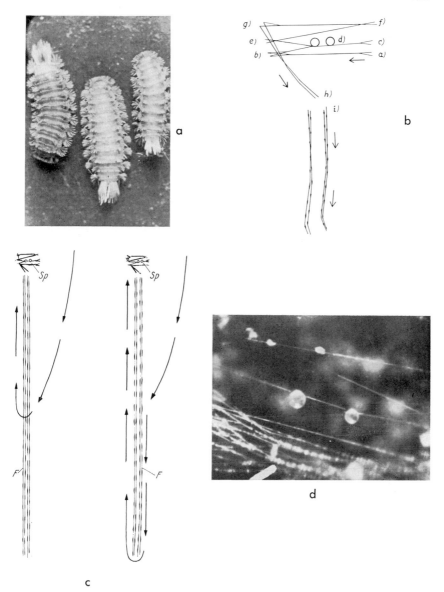

FIGURE 4. a. *Polyxenus lagurus*. b. Sperm web of *P. lagurus* (foreshortened below; d, sperm drops). c. As in b but complete. Length 1.5 cm. The arrows indicate direction of movement of a searching female. d. Sperm drops of *P. lagurus*.

tact of the partners. After spinning the web and depositing the spermato-phore, the male turns the anterior part of his body around and with his antennae gives a tactile signal to the female to proceed. In addition, the web of the male *Lithobius* always contains some conspicuously broad thread bands transversely across the longitudinal axis of the animals (Fig. 6). These threads, without doubt, provide special stimuli for the female.

The males of Scutigeridae take an even more active part in the sperm transfer (48, 52, 53). After a long prelude, the male of *Scutigera coleop-trata* places a spermatophore on the ground and then pushes the waiting female with antennae and legs towards the sperm packet (Fig. 7). He finally transfers the spermatophore actively with his mouth into the genital opening of the female. This has been closely observed in the southern Asiatic *Thereu-poda decipiens cavernicola* (53).

PAUROPODA

We have known for some years that the males of Pauropoda also spin primitive webs into which they deposit two sperm drops (55). Although the uptake of sperm by the females of *Stylopauropus pedunculatus* has not been observed, it is probable that the two conspicuous V-shaped double threads with their sperm drops (Fig. 8) have signaling significance. It is certain that the males of Pauropoda and Pselaphognatha spin sperm webs and produce spermatophores in the absence of females.

SYMPHYLA

The sexual biology of the Symphyla is significant for phylogenetic reasons. These small, pale, blind soil-inhabiting Myriapoda are considered to be the group most closely related to the Apterygota. The males of *Scutigerella im-maculata* and *S. silvestrii* (38–43) deposit simple sperm drops on 1.5-mm-long stalks (Fig. 9a) also in the absence of females. They make peculiar move-ments and twistings with the posterior part of the body and then walk away. They can produce up to 24 spermatophores in a few days. Fertile females of Symphyla regularly bite off the "heads" of these spermatophores. They do not swallow the sperm but collect it in special gnathal sperm pockets. They can collect and store in these pockets up to 18 sperm drops in a day. The eggs are fertilized only when laid. With her mouth parts, the female pulls one egg after another from the anteriorly situated genital atrium and deposits it in a suitable place (Fig. 9b). Each egg is then worked over with the mouth parts for 1.5–5 min, and some small, sperm-containing drops are smeared onto the egg. The fertilization of the eggs of the Symphyla is thus performed quite externally and in a special indirect manner, by the exclusive activity of the females.

APTERYGOTA
COLLEMBOLA

Collembola produce a simple spermatophore as do the Symphyla. Males

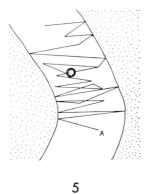

5

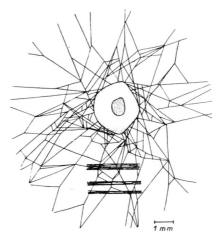

6

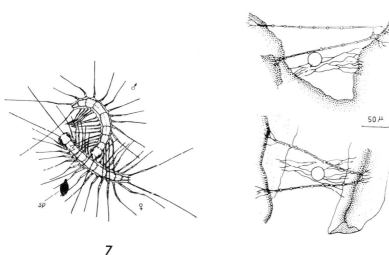

7

8

FIGURE 5. Sperm web and sperm drop of *Geophilus*.
FIGURE 6. Sperm web of *Lithobius* with spermatophore.
FIGURE 7. *Scutigera*. Pair with a spermatophore.
FIGURE 8. Web of *Stylopauropus* with sperm drop viewed laterally and dorsally.

California Baptist College
Riverside, California

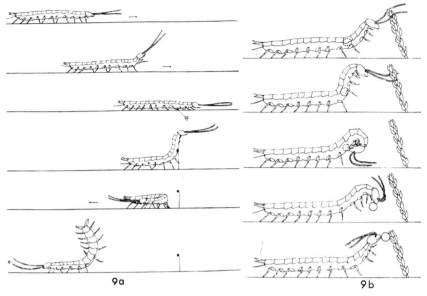

FIGURE 9a.   Spermatophore deposition by a male *Scutigerella*.
FIGURE 9b.   Egg deposition and fertilization by a female *Scutigerella*.

of *Orchesella, Tomocerus,* and other genera deposit bare sperm drops on
stalks (Fig. 10a), even in the absence of females (73). Receptive females must find these spermatophores, brush over them actively with the
vulvae and take them up (Fig. 10b). This is facilitated by the fact that Collembola have a high population density, and that each male can produce more
than 100 spermatophores in his reproductive life span. They also always provide fresh sperm, similar to the males of *Polyxenus,* by regularly eating the
old spermatophores (their own as well as those of other individuals), and
replace them immediately with new ones. They examine the sperm drops with
their antennae (61, 73).

Some species of Collembola evolved a behavior pattern which has resulted
in at least one-sided contact between males and females, ensuring sperm
transfer. *Podura aquatica* (84, 85) which lives on the surface of water, is
an example. Only the male searches actively for a mature female carrying
ripe eggs as a partner. He places a fence of spermatophores near her and
pushes her to it (Fig. 11b), so that she can easily strip off a sperm drop (Fig.
11a). The female is quite passive to all attempts at contact by the male.

These contacts are especially frequent and intensive in the Symphy-
pleona (Sminthuridae) (22, 26, 60, 61). The males of the forest-inhabiting
species *Dicyrtomina minuta* search actively for females, which they first see
and then examine olfactorily. A fence of spermatophores is placed
around the female, so that she must come into contact with a sperm drop

10a   11a

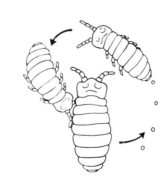

10b

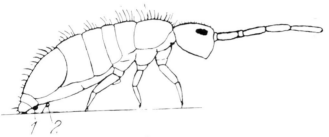

11b

FIGURE 10a. Spermatophore of a Collembola.
FIGURE 10b. Uptake of a spermatophore 2 by a female *Orchesella*.
FIGURE 11a. Spermatophore of *Podura aquatica*.
FIGURE 11b. Pairing behavior of *P. aquatica*.

when moving. Here also the females practically take no notice of their partners, who defend "their" female against other males by "jostling."

Pairing behavior involving only loose contact but activity of both partners is known to occur among the symphypleonid *Bourletiella (Deuterosminthurus) bicincta* and *B. repanda* which live on herbs and bushes. The male circles around the mature female and palpates her until both are standing head to head. In this position, both of them perform lengthy coordinating and synchronizing movements. Finally the male turns 180° and places a

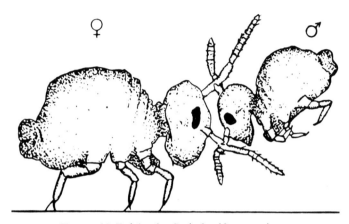

FIGURE 12. Pairing by *Sminthurides aquaticus.*

stalked spermatophore immediately before the female who steps slowly over it and takes up the sperm drop with the genital opening, while the male turns around and assists the uptake with his mouth (22).

The behavior of *Sminthurides aquaticus* is considered the most advanced stage of such close contact and pairing. Like *Podura,* this species lives on the surface of water and is distinguished by its conspicuous sexual dimorphism. The males are much smaller than the females and have specialized prehensile antennae. With these they clasp the antennae of the females and let themselves be carried about for days (Fig. 12). However, the male can lead actively and, by a lengthy play of movement, determines whether the female is ready to accept his attentions without resistance. If this is the case, he places a spermatophore on the water and draws his partner over it, stepping carefully backwards, so that she can easily take up the sperm with her genital opening. The male can also lead his partner to the spermatophore by directing her in a half-circle.

Among the springtails, one can therefore recognize all transitions from an indirect spermatophore transfer without association between partners to specific pairing and bodily contact (75, 78–83).

DIPLURA

The Diplura also produce simple stalked spermatophores, as proved for different Campodeidae (74, 14–18). The stalks are usually relatively short (50–100μ), and thick (7–12μ). They may be minimally 30μ, maximally 300μ long in extreme cases. The round sperm drops have a diameter of 50–100μ (min 35μ, max 150μ) and contain conspicuous spiral filaments which con-

TABLE 1.

	Sperm droplet		Stalk
	No. of sperm bundles	Diameter of sperm bundle	Length
Campodea rhopalota	1	55–60 μ	30–65 μ
C. kervillei	1	50–55 μ	60–70 μ
C. plusiochaeta	1	45–50 μ	40–85 μ
Lepidocampa juradoi afra	1	40–55 μ	170–300 μ
Plusiocampa vandeli	1	50–55 μ	60 μ
Plusiocampa humilis	1 ?	50 μ	—
C. sensillifera	1–2	80–100 μ	90–170 μ
C. remyi	1–4	40–50 μ	50–100 μ
C. pieltaini	2–4 (3)	70–80 μ	100–180 μ
C. fragilis	2–5 (3)	35 μ	65–80 μ
C. lubbocki	3–5 (4)	30 μ	50 μ
C. (Paurocampa) suensoni	3–7 (5)	28–30 μ	45–50 μ
C. (Paurocampa) ribauti	5–9 (7)	20–25 μ	100–140 μ
C. chardardi	5–12	25–30 μ	50–140 μ
Litocampa (C.) solomonis	—	9–10 μ	—
C. staphylinus	—	42 μ	—
C. lankesteri	—	57 μ	—
P. (Stygiocampa) remyi	—	50 μ	—
Hystrichocampa pelletieri	—	57 μ	—

sist of bundles of large sperms. Their number varies from one bundle in *Campodea rhopalota, C. plusiochaeta,* or *Plusiocampa vandeli,* to 12 bundles in *Campodea chardardi.* Their form and size depend on the number of spiral turns (cf Table 1 and Fig. 13). The uptake of sperm by the female was observed only once; it takes place in the absence of the male.

It is very probable that closer contact between the sex partners exists among the Japygidae. "Pairing preludes" have been observed among them. However, it is unknown whether they produce spermatophores and how these are ventually transferred (32, 64, 65).

THYSANURA

Thysanura exhibit a diversity of indirect sperm transfer methods. Because

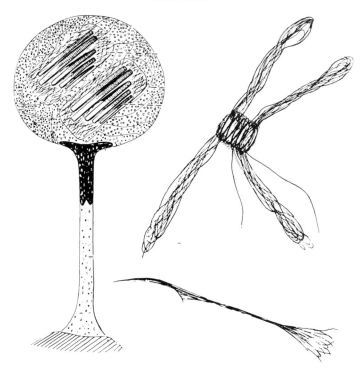

FIGURE 13. Spermatophore and sperm bundles of *Campodea remyi.*

they have external genital appendages it was generally assumed that true copulation occurs among them. But they have not evolved adequate pairing behavior. The males of Thysanura use their so-called organs of copulation only as spinning organs.

The male of *Machilis germanica* and *Lepismachilis y-signata* (97, 99) performs a lengthy prelude with a willing partner in order to coordinate and synchronize her readiness with his own behavior. He then spins a thread one end of which is fastened to the ground, while the other is held with the "penis," so that this taut thread extends obliquely upwards. Some sperm drops are suspended on this thread. While spinning the thread, the male forces the female with his antennae, palps, and legs to the thread in such a manner that she can collect the sperm drops from the thread with her genital appendages (ovipositor) (Fig. 14).

No less intimate is the tactile contact between the sexes in *Lepisma saccharina* (98, 100) (Fig. 15). After a complicated play of run-and-touch, the male *Lepisma* rapidly spins several threads across the running direction of the female. The threads are attached by one end to the ground and with the other slightly elevated, so that the running partner is caught by her raised

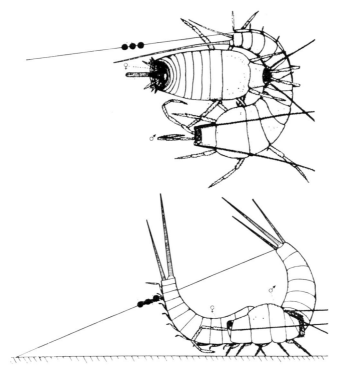

FIGURE 14. Pairing position of *Machilis* during sperm transfer.

tail appendages. As soon as she feels this touch, she stops suddenly, and begins to make careful searching movements. The male has placed a spermatophore on the ground exactly where he spun the threads, and it can now be taken up and received easily by the female. It has been demonstrated experimentally that these restraining threads are essential to stop the female and to stimulate her searching behavior at the right place.

ARACHNIDA

The sexual behavior of the various arachnids is as diverse as their morphology.

SCORPIONIDA

As we know from Fabre, scorpions make a long "promenade à deux," in which the male holds the female by the chelae. During this prelude, the male of *Euscorpius italicus* stings its partner repeatedly (Fig. 16a) in the joint proximal to the chela (9, 10, 12). The spermatophores are formed, deposited and received only after this promenade (1–3, 45, 69, 92, 94, 95). The male presses his genital opening against the ground to which he attaches the

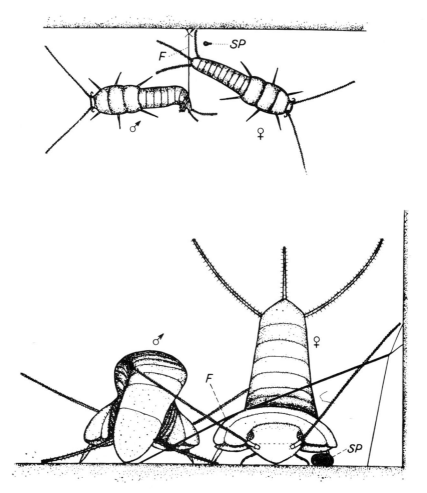

FIGURE 15. Pairing of *Lepisma* viewed dorsally and frontally.
F = restraining thread, SP = spermatophore.

stalk of the spermatophore, and pulls out the spermatophore by raising his body. His pectines remain in firm contact with the ground. Moving backwards, the male then draws his partner to the spermatophore (Figs. 16b, 17a). She feels for it with her pectines and orientates herself so that the paired sperm containers are situated exactly before her genital opening. She then opens the spermatophore cover with a sudden movement and injects the sperm into the genital atrium. The male immediately releases his partner, and subsequently she consumes the empty spermatophore (Fig. 17b).

The spermatophores of scorpions are complicated formations, which are produced in special gland pockets ("paraxial organs") of the males as two halves which are mirror images of each other (11, 12) (Fig. 33).

The pairing behavior of the Pedipalpi is much more diverse than that of the scorpions.

Uropygida.—In the Uropygida, the pairing of the schizopeltid *Trithyreus sturmi* was observed closely (101). First, the male runs after the female and touches her with vibrating movements. She then turns around and he does the same, so that she now stands behind him. In this position, she hooks her

16 a

16 b

17a   17 b

FIGURE 16a. "Promenade à deux" of *Euscorpius;* the male stings the female.
FIGURE 16b. The male pulls the female toward the spermatophore.
FIGURE 17a. Intact spermatophore of *Euscorpius italicus.*
FIGURE 17b. Opened spermatophore of *E. italicus.*

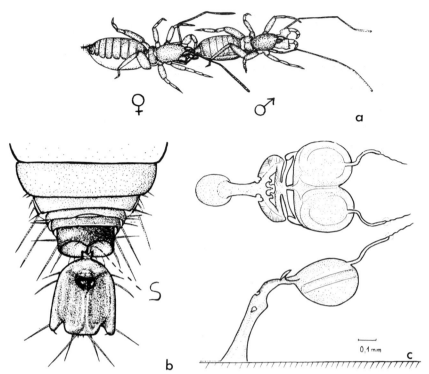

FIGURE 18a. Pairing by *Trithyreus sturmi*.
FIGURE 18b. Metasoma of *Trithyreus;* S = point at which female clutches male.
FIGURE 18c. Spermatophore of *Trithyreus* viewed dorsally and laterally.

chelicerae into the base of his peculiarly knob-shaped metasoma (Fig. 18a,b) and allowed herself to be carried forward. The male then deposits a spermatophore which contains two sperm balls as in the case of scorpions (Fig. 18c). Repeatedly jerking his posterior part, he finally pulls the female to the spermatophore, and she takes up the sperm with her vulva.

Holopeltida.—The southern Asiatic species *Telyphonus caudatus* has been examined (54). The male touches the female with vibrating movements and grasps her finally with pedipalps and chelicerae on her crossed, long, first pair of legs (Fig. 19a). He then pulls and pushes the female here and there, often for days. After repeated touching of the female genital opening, the male turns slowly around and passes over the prosoma of the female. At the same time, he releases the tactile forelegs of the female from his pedipalps, and holds her only with the chelicerae. The male, standing with his posterior end before the female, then deposits a spermatophore and advances with the

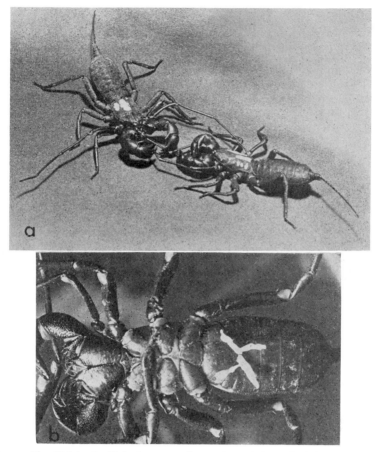

FIGURE 19a Pairing by *Telyphonus caudalus;* the male is on the left.
FIGURE 19b. Female of *Telyphonus* with spermatophore at the genital opening.

female until she can take up the sperm (Fig. 19b). He then strokes her for about 20 minutes with his flagellum before they separate.

Amblypygi.—The sexual behavior of several species of Amblypygi has been studied (4, 5, 19, 54, 117). The sex partners of the Malayan species *Sarax sarawakensis* perform long palpating movements before deposition and uptake of the spermatophore. Unlike the Uropygida, the partners do not make firm bodily contact. After several hours of premating activity the male turns around and deposits a spermatophore while maintaining contact with his partner with his posteriorly directed first pair of legs. After this he turns again, and signals her by vibrating his tactile forelegs. The female

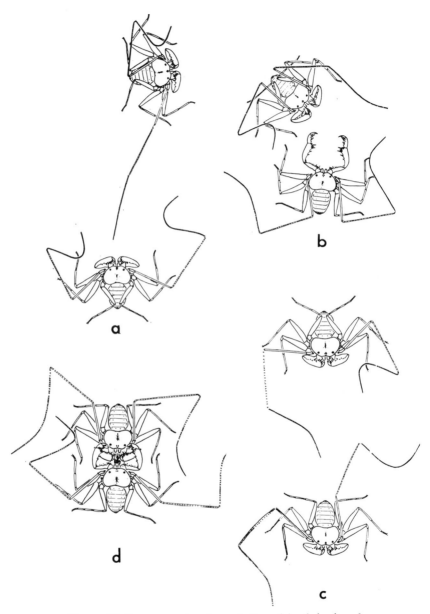

FIGURE 20. Four successive phases in the pairing behavior of
Tarantula marginemaculata.

then moves forward while the male moves slowly backwards, leading her exactly to the spermatophore.

The pairing and indirect sperm transfer of other species of Amblypygi studies (*Ademetus barbadensis, A. pumilio,* and *Damon variegatus*) proceed in essentially the same way. However, the male of *Tarantula marginemaculata* takes a more active part in the sperm transfer (Fig. 20). First he performs intensive movements before the female which resemble mock attacks, he then turns around and deposits the spermatophore. Then he steps over it rubbing it with the anterior part of his body. The female moves repeatedly over the spermatophore, making high swinging and circulating movements above it. The male finally seizes the female with his palps and pulls her over the spermatophore. She tears the sperm packets rapidly out with the claw-shaped sclerites of her genital operculum and moves away. She usually eats the empty spermatophore afterwards (117).

Araneida

The sexual biology of the Araneae is generally well known. The males do not produce spermatophores, but place a sperm drop on a web, and then put the sperm into specially differentiated tarsal appendages of the pedipalps, with which they finally transfer the sperm into the vulva of the female.

Pseudoscorpionida

The pairing behavior and indirect methods of sperm transfer of the pseudoscorpions (46, 47, 108, 112–116) have been known for a long time. Many species exhibit very simple behavior; the males produce stalked spermatophores without the presence of females, and leave their uptake entirely to the females (Fig. 21). They belong to the primitive families Chthoniidae, Neobisiidae, Garypidae, and Cheridiidae. Only species of the families Chernctidae and Cheliferidae form pairs and show more complicated behavior patterns and transfer methods.

There are variations among the Cheiridiidae between the extremes of pairing and nonpairing behavior. The male of *Cheiridium museorum* often makes characteristic vibrating movements with his palps when he meets a female. She does not react at all, but later searches independently (possibly chemotactically) for the abandoned spermatophores, which the male generally produces in great numbers. This may be called "semi" or "one-sided" pairing, similar to that of some Collembola. The males of these pseudoscorpions regularly demolish the older spermatophores then replace them with new ones.

A species of *Serianus* inhabiting sand dunes in North Carolina resembles the Pselaphognatha in its sexual behavior (114). However, the male produces the spermatophore and the signaling threads only after it has had tactile contact with a female whether mature or not. He is not interested in the female while he deposits the spermatophores. The sexually stimulated male searches for a convenient small crack and deposits a stalked sperm drop (Fig. 22, "1" and "3"). He then spins some threads on the ground and a double row of vertical threads converging towards the spermatophore (Fig. 22, "3") with the opisthosomal spinning glands which open at the anus. Thus, a narrowing, fenced way to the spermatophore is made. Mature fe-

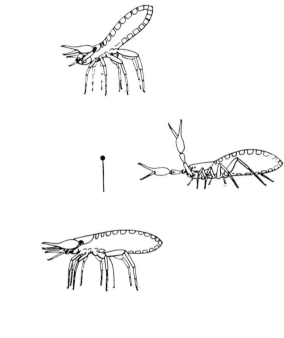

FIGURE 21. Spermatophore deposition and uptake by *Neobisium muscorum.*

males of *Serianus* use this way to reach the spermatophore. Since its stalk always has a bend, the uptake of the sperm drop appears possible from only one direction (Fig. 22, "3" & "4"). When males find such a web with a spermatophore, they destroy it and push the spermatophore over regardless of how old it is. But if a female is in the vicinity, they immediately produce a fresh web and spermatophore.

Species of *Chelifer* and *Chernes* perform intriguing pairing dances before the production and transfer of a spermatophore (Fig. 23) (108, 116). The pairing of *Chelifer cancroides, Dactylochelifer latreillii,* and *Dendrochernes*

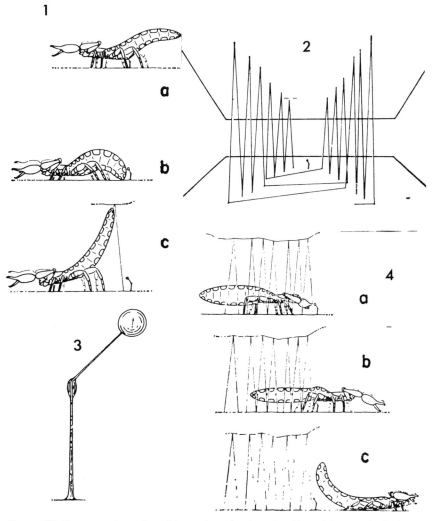

FIGURE 22. Spermatophore deposition and production of guide web by a male *Serianus* (1a,b,c, 2, 3), and uptake of spermatophore by the female (4a,b,c).

cyrneus has been known for a long time. The *Chelifer* male performs a dance before the more or less inactive female, in the course of which a spermatophore is deposited. During this action he evaginates two long glandular tubes from his genital opening, which may have a chemotactically alluring and stimulating effect on the female. Only shortly before the uptake of the sperm drop, the male grasps his partner and brings her into contact with it by strong shaking movements.

The male of *Dactylochelifer latreillii* grasps the female on both pedi-

palps and also evaginates both its glandular tubes during the following movements. Before deposition of the spermatophore, he releases his partner who
moves alone toward the spermatophore. When the female has nearly
reached the spermatophore, the male again rushes forward, hooks the long

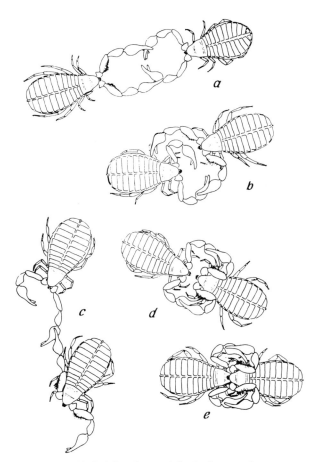

FIGURE 23. Pairing dance of *Lasiochernes pilosus*.

claws of his forelegs into the genital opening of the female, and pulls her to
the spermatophore.

The male of *Dendrochernes* grasps one chela of the female and does not
release it during the whole pairing dance and the following sperm transfer.
The pairing ceremony of this species may be repeated several times over
many hours. The pairing of *Dinocheirus tumidus* (115) is similar. The male
of this species, which lives in Florida, holds the female with one chela,
stroking her back and pedipalp with the other. In the meantime, he deposits

the spermatophore and pulls the female toward it. After the transfer, he shakes the partner energetically with his free chela.

The *Withius subruber* male grasps the partner first by the palps, later by the prosoma, and pulls her toward him several times, vibrating with the third pair of legs. He then deposits a spermatophore, pulls the female over it, and presses her hard on it. This species produces an especially complicated spermatophore with a lever-like opening mechanism (118).

SOLIFUGIDA

The apparently brutal pairing behavior of the males of *Solifuga* is generally well known. Their nearly direct method of sperm transfer closely resembles the methods of Araneae.

ACARINA

Among the Acarina a number of means of indirect sperm transfer has apparently been independently evolved. Extremities are often used as gonopods, to place spermatophores or simple sperm balls on or into the genital opening. For instance, many gamasids transfer the sperm with their chelicerae, which are suitably modified. Many hydrachnellids use the third pair of legs for sperm transfer. But many of them also produce stalked spermatophores and transfer them in close contact (20). The formation of free-standing spermatophores and behavioral transitions between close pairing and complete independence of the sex partners takes place mainly in the edaphic oribatid mites.

The long (maximally 0.9 mm) stalked sperm of the oribatids (Fig. 24) (66, 67, 91, 103, 119) have been known for a long time. The males of *Belba geniculosa, B. clavipes, B. gracilipes,* and *Euzetes seminulum* and of other species, deposit many spermatophores on the ground during the period of reproduction also in the absence of females. The high population density and large numbers of spermatophores ensure that the females always find sperm in time, even without signals and without the help of the males. Males and females always examine all sperm drops which they find, and apparently destroy those which are no longer fresh. The males then deposit new ones.

Closer relations between sex partners apparently exist in many oribatids (89). In the marked sexually dimorphous species *Collohmannia gigantea,* the male places his forelegs on the flanks of the female, follows her for an hour (Fig. 25), and pushes occasionally with his rostrum so violently under the posterior part of the female that she may fall over. When she finally sits quietly, he releases her, goes forward, stands with his posterior part steeply erect immediately before her, and extends his hind legs toward her, on which he has previously smeared a fluid (sperm?) with his penis. Although the female gnaws on this fluid, it is not certain that this represents the actual means of sperm transfer.

Tropical Galumnidae have also been fleetingly observed during pairing marches (31).

FIGURE 24. Spermatophore of *Belba geniculosa*.

The male *Allothrombium fuliginosum* is known to deposit his spermatophores only after having circled around a female and having stroked her with his legs by beating up and down (8). Stalked sperm drops similar to those produced by Oribatida were also found in Trombiculidae (109) and Labidostomidae (56, 90). It was clearly shown that the males of *Labidostoma cornuta* are stimulated to intensified spermatophore production by the presence of females. The same is probably true also for other mites in which sperm transfer occurs without "personal" contact.

Stalked spermatophores occur also in the Hydrachnellidae (20, 21). But the transfer takes place during pairing with close bodily contact. The strongly sexually dimorphous species *Arrenurus globator* has been studied in detail. The males are immobilized by contact with the females and assume the so-called preparatory position. The male extends his posterior appendix to the female. If the female is ready for pairing, she moves toward the male and is immediately fixed to the appendix with a sticky secretion (Fig. 26). After a lengthy pairing movement, the male deposits a spermatophore, and with an exact forward movement brings the genital opening of the attached female directly to the sperm ball. The male assists the transfer of the sperm by cautious rubbing movements.

The sperm transfer of other water mites is less indirect. They are clasped together in pairs and transfer the sperm with their gonopods (third pair of legs) or directly from genital opening to genital opening [e.g., *Piona nodata* (Fig. 27) and *Eylais infundibulifera*].

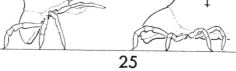

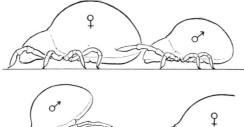

25

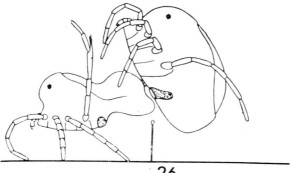

26

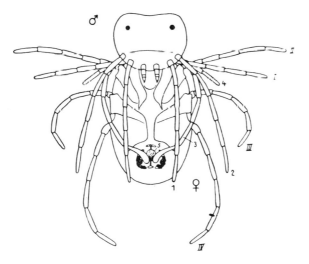

27

FIGURE 25. Pairing behavior of *Collohmannia.*
FIGURE 26. Pairing of *Arrenurus globutor.*
FIGURE 27. Pairing of *Piona nodata.*

GENERAL CONSIDERATIONS

If one reviews the variety of phenomena of indirect sperm transfer described here among the lower arthropods, the following four aspects clearly emerge:

Ecophysiological aspect.—Nearly all species in which indirect sperm transfer occurs live in very humid habitats (mainly in the soil). Many of them are only incompletely adapted to an existence in dry air. Only those which show close contact between the sexes are able to settle in drier habitats, because their sperm transfer takes place so rapidly that there is no danger of the sperm drying up.

Population-ecological aspect.—The greater the population density of a species, the less intimate the contact between the sex partners. The extremes are exhibited by the vegetarian pselaphognathids, Collembola, and oribatids on the one hand, and the predaceous chilopods, scorpions, and pseudoscorpions on the other.

Phylogenetic aspect.—Nearly all species exhibiting spermatophore production and indirect sperm transfer are considered to be primitive and of a low systematic position compared to related groups. Sexually they have not achieved full emancipation from water.

Systematic aspect.—Although the lower terrestrial arthropods are often closely related systematically, it is unlikely that all the different morphological and behavioral aspects of their sexual biology are differentiations of homologous phenomena. It is clear in some cases that striking similarities are only adaptive analogies, e.g., the simple spermatophores of Collembola and Oribatida and their primitive indirect methods of sperm transfer (7, 27–30, 73, 75–83). The well-studied pseudoscorpions provide the best opportunity for comparison (116). They provide a beautiful series of ascending structural and functional differentiations of the mechanisms of sperm transfer.

1. Members of families with completely indirect, contact-free transfer of spermatophores (Chthoniidae and Neobisiidae) invariably withdraw upon contact with specimens of the same species. However, the Cheiridiidae occasionally permit one animal to take the other by the chela. Finally, tactile contacts are common in the Chernetidae and Cheliferidae, which frequently result in violent but harmless fights between males. The males of Chernetidae attack any specimen of the same species and either a fight with a rival or a pairing dance with a female results. The males of Cheliferidae estab-

lish separate territories and perform epigamic behavior only in the presence of females.

2. Chthoniidae, Neobisiidae, Garypidae, and Cheiridiidae do not form pairs. The males in isolation produce spermatophores and step forward over them. Chernetidae and Cheliferidae form pairs. The males put down spermatophores only after a pairing dance and retreat over them.

3. Pseudoscorpionida that do not pair always produce numerous spermatophores. The males of Chthoniidae also always provide fresh sperm. The males of Cheiridiidae deposit spermatophores only in the presence of females. These are very attractive to all members of the species, so that the removal of old spermatophores is always assured.

4. The spermatophores of Pseudoscorpionida are of different structure, according to the method of sperm transfer. The simpler droplet-spermatophores which are deposited without pairing are placed upright and are accessible from all directions. However, the spermatophores formed during pairing are inclined and are distinctly bilaterally symmetrical so that the females have access from one side only, i.e., from the front.

5. A cover for the sperm balls is absent only in the relatively primitive genus *Apocheiridium,* and in the Chthoniidae it is at least partially covered. The emptying of the covered sperm packets takes place by swelling in simple cases. The females press them mechanically only in the case of the especially large spermotophores of the Cheliferidae.

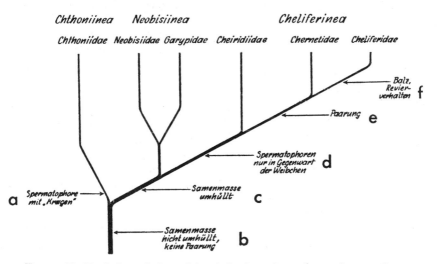

FIGURE 28. Evolution of the pairing behavior of pseudoscorpions. a, Spermatophore with "collar"; b, sperm not enclosed, no pair formation; c, sperm enclosed; d, spermatophores formed only in presence of female; e, pair formation; f, territories established, mating.

6. Chemical stimulants (pheromones), associated with the spermatophores, enhance the transfer especially clearly in the Neobisiidae and
Cheiridiidae. The effect of such sex stimulants is very probable in the Cheliferidae, the males of which evaginate long tubes from the genital atrium
during pairing, and extend them to the females. They also establish territories during the period of reproduction and perform very intensive pairing
dances. They keep very close contact with the female during sperm transfer.

The described sexual-biological phenomena of behavior of the pseudoscorpions can be used in systematic classification, as shown in Figure 28
(116). As with the pseudoscorpions, comparative considerations and evaluations could be made also for other lower terrestrial arthropods with indirect
sperm transfer. The Collembola provide the best material for comparison.
Further examples of differentiation in the methods of pairing and sperm
transfer will probably be found, mainly in the group of the Symphypleonae.
A taxonomic evaluation of the phenomena is therefore better postponed.

Finally, we will attempt to describe comparatively the morphological aspects of the formation and structure of the stalked spermatophores.

Stalked spermatophores have been found in the following groups:

I. Antennata
   1. Symphyla
   2. Diplura
   3. Collembola

II. Chelicerata
   1. Scorpionida
   2. Pedipalpida
   3. Pseudoscorpionida
   4. Acarina
      a. Trombiculida
      b. Labidostomida
      c. Oribatida
      d. Hydrachnellidae:
         Arrenurida

The three small, usually euedaphic groups of trignathous Antennata,
produce the simplest type of spermatophore (Fig. 29). They stick a rapidly
hardening secretion to the ground with their genital opening, and pull it
into a tapering stalk by raising the body. In the same operation, a sperm
droplet is placed at the upper end of the stalk, which becomes spherical because of its viscosity, and remains attached the the upper end of the stalk.
The stalks of these spermatophores are either straight (the short stalks of
Diplura or sometimes also the longer stalks of the Collembola), or usually
slightly curved at the end. This curvature (most distinct in the Symphyla)
results from the movements of the male as he draws out the secretion for
the stalk. The end of the stalk may be simply drawn out smoothly (among
many Collembola), or it forms a knob (as in Symphyla) or, more rarely, it
is branched (among some Collembola).

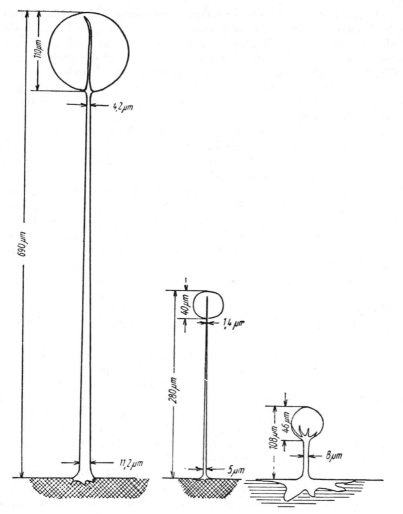

FIGURE 29. Various sizes of spermatophore of Collembola; left, *Sminthurus fuscus;* middle, *Orchesella cincta;* right, *Podura aquatica.*

The sperm drops have no cover, but in the air form only a condensation membrane which may shrink or swell hygroscopically and which burst when in contact with water. The females which strip off the spermatophores, therefore, probably always exude a drop of fluid from the genital opening. The sperm within the intact sperm drop apparently is normally inactive.

The peculiar spiral bundles of sperm of the Campodeidae (17, 74) pose a special question. Their extremely long spermatozoa (up to 1000μ in *Campodea sensillifera*) are inactive before transfer (Fig. 13). Similar, but nonspiral shaped bundles of spermatozoa are found in the Pselaphognatha (87). The function of these formations requires further study.

A special base plate at the lower end of the stalk of the spermatophore (Figs. 11a, 29) is formed only in *Podura aquatica* which lives on the surface of water (84, 85).

The spermatophores of mites (Arrenurida, Trombiculida, Labidostomida, and Orbatida) are identical in principle with the spermatophores of Symphyla, Diplura, and Collembola in formation and structure. The spermatophores of Oribatida (66, 67, 85, 91, 103, 119) have been studied in most detail. They, too, consist only of a stalk ending in a little knob. The sperm drops are also probably bare in this case, that is, they form a condensation membrane which is apparently reinforced below like a cup. The stalks of the spermatophores of the Oribatida are often undulate in the upper quarter (Figs. 24, 30c) which is probably due to the movements of the male during deposition.

The problem of where the secretion for the stalk is formed, and how it is secreted through the same genital opening neatly isolated from the sperm fluid, and drawn out to a stalk has been clarified with certainty only in the Collembola (85). The sperm and the secretion for the stalk are produced at different times (phase of rest and phase of deposition) but are mixed inside the testes. Sperm and stalk secretion are separated and stored in individual layers in the ductus ejaculatorius by the action of a special gland secretion —the sperm is stored dorsally, the stalk substance, ventrally. The males of Collembola activate the different muscles of the ductus ejaculatorius during deposition of the spermatophores in such a way that first the stalk secretion and then the sperm is pressed out. During the periods of sexual activity (61, 68), there is always so much material stored in the ductus that several spermatophores can be produced successively.

The production method of the Oribatida is probably the same in principle (85), although there are other interpretations (119). It is not very likely that formation of stalk and sperm proceeds differently in Diplura, i.e., through different glands and pores.

The stalks of the spermatophores of the arthropod groups described in the preceding section are certainly not produced in the sexual tubes of the males and then deposited, as it has been assumed on occasion (11, 65, 93). On the other hand, the complicated structures of the larger spermatophores of the other groups of Chelicerata (Scorpionida, Pedipalpida, and Pseudoscorpionida) can only be explained by the assumption that they are prepared in matrix-like structures of the genital organs of the males. This has been proved without doubt by anatomical studies of the scorpions (11, 12). The

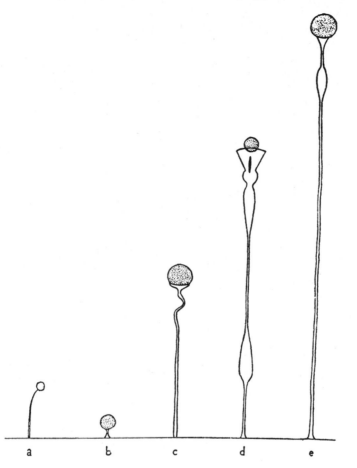

FIGURE 30. Relative sizes of spermatophores of a, Collembola (*Orchesella*), b, Diplura (*Campodea*), c, Oribatida (*Belba*), and d-e, Pseudoscorpionida (*Chthonius* and *Obisium*).

so-called paraxial organs of male scorpions (Fig. 33) are ideal forms for casting the spermatophores. In *Euscorpius* males, the corresponding halves of the spermatophores are formed in these organs within a four-day period. At the time of their deposition they are glued together. Only shortly before this event do the seminal vesicles form the sperm balls which fill the paired sperm supports on the stalk (Fig. 34). The wing-shaped appendage of the spermatophore is a lever which permits the opening of the sperm container. The female activates it during pairing.

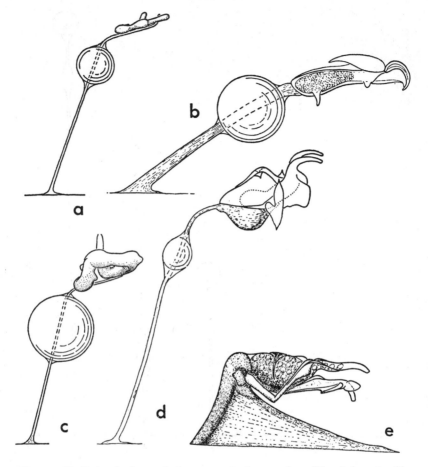

FIGURE 31. Lateral views of the spermatophores of a, *Dinocheirus tumidus;* b, *Dactylochelifer latreillei;* c, *Chernes cimicoides;* d, *Chelifer cancroides;* e, *Withius subruber.*

The different spermatophores of the Pedipalpida are not less compli-cated. A most exact study of its structure has been made for *Tarantula marginemaculata* (117). Figure 32 shows that the sloping stalk consists of two transparent stripes with a V-shaped cross section. The sperm container is strongly sculptured and forms an isosceles triangle, which is situated with one point on the stalk. There are arm-like appendages at both distal ends, which form slightly concave guiding rails. Two small Y-shaped, tube-like sperm balls are situated inside; they are attached to re-movable sperm plates. The spiral-shaped, inactivated sperm are also situ-

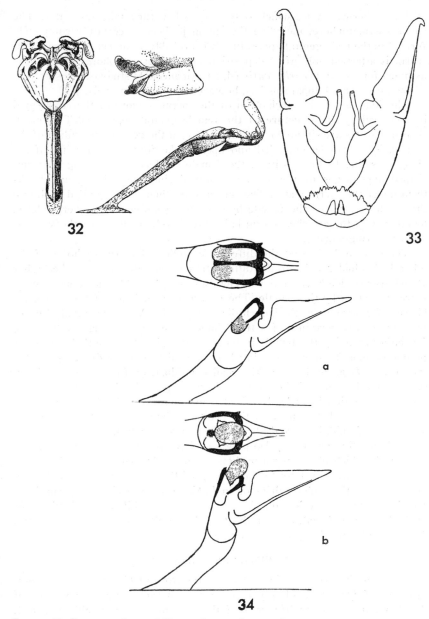

FIGURE 32. Spermatophore of *Tarantula marginemaculata*.

FIGURE 33. Paraxial organs of *Euscorpius italicus* in which the two halves of the spermatophore are formed.

FIGURE 34. Spermatophore of *E. italicus*; a, intact; b, opened.

ated in spherical cysts, which only burst when they take up water. The whole spermatophore, including the sperm packets, is certainly completely formed in the male genital apparatus. It is probably incorrect to suppose that sperm is attached only after deposition of the spermatophore, as has been assumed for *Admetus barbadensis* (4, 5). An adequate study of the structure of the male genital apparatus has, however, not yet been made.

The function of the mechanism of the spermatophore of the Amblypygi is explained by the structure of the female genital apparatus. Two claw-shaped sclerites are situated on a soft cushion at the genital operculum of the female. They can be extended and opened like forceps by raising of the blood pressure. In this position, the female conducts them along the arm-shaped appendages of the spermatophore during the pairing, and hooks them under the sperm plates. The sclerites are then rapidly pulled back into the genital atrium of the female by two strong muscles. The sperm plates, together with the attached sperm balls, are thereby torn out of the sperma-tophore during pairing.

There is no doubt that the different complicated spermatophores of the other Pedipalpida, and also the differentiated spermatophores of the higher Pseudoscorpionida have similar mechanisms, which enable and support active sperm reception by the female. The different appendages of the spermato-phores of Scorpionida, Schizopeltidia, Uropygida, Amblypygi, and Pseudo-scorpionida, are certainly of marked importance for ensuring sperm trans-fer, either assisting the female in tactile orientation (e.g., probably in some pseudoscorpions), or representing hook-shaped levers for drawing out the sperm balls (e.g., probably in the Schizopeltidia), or functioning as direct opening mechanisms of the sperm containers (e.g., in the scorpions). The actual fertilization is probably initiated in all cases by the fact that the re-ceived sperm packets, on or in the genital atrium of the female, swell by absorption of fluid, burst and liberate the activated sperm.

Such swelling processes also play a decisive role in the lower terrestrial arthropods which place sperm packets or sperm drops on webs. The sperm packets are stuck normally to the outside of the genital opening of the fe-male (e.g., in the scolopendrids or in *Lepisma*) or the viscose sperm drops remain attached there by adhesion (e.g., in the Machilidae, Collembola, or Oribatida). The females assist the transfer in these cases, by secreting a drop of fluid from their genital opening.

FINAL REMARKS

Since the discovery of the stalked sperm drops of Collembola (Schaller, 71) and of Oribatida (Pauly 66), of the characteristic sperm webs of the Thysanura (Sturm 97), and of the Pselaphognatha (Schömann 86), and of the complicated spermatophores of the scorpions (Angermann 9), the sexual biology of the lower terrestrial Arthropods has become an unexpectedly rich chapter of comparative research on behavior, with many interesting morphological, ecological, and phylogenetical aspects. But, in spite of all the progress in our knowledge of the phenomenon of indirect sperm transfer, further research will certainly discover many surprising and fascinating facts.

ACKNOWLEDGMENT

Translation of this article was aided by a grant from the National Science Foundation.

LITERATURE CITED

1. Alexander, A. J. 1955. Mating in scorpions. *Nature* 178:867–68
2. Alexander, A. J. 1957. The courtship and mating of the scorpion, *Opisthophthalmus latimanus. Proc. Zool. Soc. London* 128:529–44
3. Alexander, A. J. 1959. Courtship and mating in the Buthid scorpions. *Proc. Zool. Soc. London* 133:145–69
4. Alexander, A. J. 1962. Courtship and mating in amblypygids. *Proc. Zool. Soc. London* 138:379-83
5. Alexander, A. J. 1962. Biology and behaviour of *Damon variegatus* Perty of South Africa and *Admetus barbadensis* Pocock of Trinidad, W. J. *Zoologica, New York* 47:25–37
6. Alexander, A. J., Ewer, D. W. 1957. On the origin of mating behavior in spiders. *Am. Natur.* 91:311–17
7. Alexander, R. D. 1964. The evolution of mating behavior in arthropods.. In Insect Reproduction, ed. K. C. Highman. *Roy. Entomol. Soc. London, Symp.* No. 2:78–94
8. André, M. 1953. Observation sur la fecundation chez *Allothrombium fuliginosum* Herm. *Bull. Mus. Hist. Natur. Paris* 25:383–86
9. Angermann, H. 1955. Indirekte Spermatophorenübertragung bei *Euscorpius italicus* Hbst. *Naturwissenschaften* 42:303
10. Angermann, H., Schaller, F. 1955. Die Spermatophore von *Euscorpius italicus* und ihre Übertragung. *Verh. Deut. Zool. Ges. Erlangen* (= *Zool. Anz. Suppl. 19*) 459–62
11. Angermann, H., Schaller, F. 1956. Spermatophorenbau und -bildung bei Arthropoden mit indirekter Spermatophorenübertragung. *Ber. 100. Jahrb. Deut. Entomol. Ges. Berlin* 228-37
12. Angermann, H. 1957. Über Verhalten, Spermatophorenbildung und Sinnesphysiologie von *Euscorpius italicus* Hbst. und verwandten Arten. *Z. Tierpsychol.* 14:276–302
13. Ax, P. 1968. Populationsdynamik, Lebenszyklen u. Fortpflanzungsbiologie der Mikrofauna des Meeressandes. *Verh. Deut. Zool. Ges. Innsbruck* (= *Zool. Anz. Suppl. 32*) 66–113
14. Bareth, C. 1964. Structure et dépôt des spermatophores chez *Campodea remyi. C. R. Acad. Sci. Paris* 259:1572–75
15. Bareth, C. 1965. Le spermatophore de *Lepidocampa* (Diploures Campodéidés). *C. R. Acad. Sci. Paris* 260:3755–57
16. Bareth, C. 1966. Régression temporaire d'un caractère sexuel secondaire chez *Campodea remyi* Denis. *C. R. Acad. Sci. Paris* (D) 262:165–67
17. Bareth, C. 1966. Études comparatives des spermatophores chez les Campodéidés. *C. R. Acad. Sci. Paris* 262:2055-58
18. Bareth, C. 1968. Biologie sexuelle et formations endocrines de *Campodea remyi* Denis (Diploures, Campodéidés). *Rev. Ecol. Biol. Sol* 5:303–426
19. Beck, L. 1968. Aus den Regenwäldern am Amazonas II. *Natur Mus.* 98:71–80
20. Böttger, K. 1962. Zur Biologie und Ethologie der einheimischen Wassermilben *Arrenurus* (*Megaluracarus*) *globator* (Müll.) 1776, *Piona nodata nodata* (Müll.) 1776, und *Eylais infundibulifera meridionalis* (Thon) 1899. *Zool. Jahrb.* (*Abt. System.*) 89:501–84
21. Böttger, K. 1965. Ökologie und Fortpflanzungsbiologie von *Arrenurus valdiviensis* K. O. Viets. 1964. *Z. Morphol. Oekol. Tiere* 55:115–41
22. Bretfeld, G. 1969. Neuer Paarbildungstyp bei der indirekten Spermatophoren-Übertragung der Collembolen. *Naturwissenschaften* 56:425
23. Davey, K. G. 1960. The evolution of spermatophores in insects. *Proc. Roy. Entomol. Soc. London* 35:107–13
24. Demange, J. M. 1956. Contribution á l'étude de la biologie, en captivité, de *Lithobius piceus gracilitarsis* Bröl. *Bull. Mus. Natur. Hist. Natur. Ser. 2* 28:388–93
25. Demange, J. M. 1959. l'Accouplement chez *Graphidostreptus tumuliporus* (Karsch) avec quelques remarques sur la morphologie des gonopodes et leur fonctionnement. *Bull. Soc. Entomol. Fr.* 64:198–207
26. Falkenhan, H. 1932. Biologische Beobachtungen an *Sminthurides aqua-*

ticus. Z. Wiss. Zool. 141 :525–80
27. Ghilarov, M. S. 1958. Evolution of insemination character in terrestrial arthropods. *Zool. J. Moskau* 37 :707–35 (In Russian)
28. Ghilarov, M. S. 1959. Evolution of insemination type in insects as the result of the transition from aquatic to terrestrial life in the course of the phylogenesis. In The Ontogeny of Insects. *Acta Symp. Evolutione Insectorium, Prague* 50–55 (In Russian with Engl. sum.)
29. Ghilarov, M. S. 1962 Vergleichend stammesgeschichtliche Analyse der Besamungsweisen bei den Arthropoden. Vorlesungen zum Andenken an Nikolai Alexandrowitsch Cholodkowski, 37–76. Acad. Sci. USSR (In Russian)
30. Ghilarov, M. S. 1961. Evolution des modes d'insémination chez les insectes au cours de leur phylogénèse. *Scientia* 55 :1–6
31. Grandjean, F. 1956. Observations sur les Columnidae. Ser. 1 (Acariens, Oribates). *Rev. Fr. Entomol.* 23 : 137–46
32. Gyger, H. 1960. Untersuchungen zur postembryonalen Entwicklung von *Diplojapyx humberti* (Grassi). *Verh. Naturforsch. Ges. Basel* 71 : 29–95
33. Haacker, U. 1968. Sperma-Transport beim Kugeltausendfüssler (*Sphaerotherium*) *Naturwissenschaften* 55 :89
34. Haacker, U. 1968. Das Sexualverhalten von *Sphaerotherium dorsale. Verh. Deut. Zool. Ges. Innsbruck* (= *Zool. Anz. Suppl. 32*) 454–63
35. Haacker, U. 1969. Spermaübertragung von *Glomeris. Naturwissenschaften* 56 :467
36. Hale, W. G. 1965. Observations on the breeding biology of Collembola *Pedobiologia* 5 :146–52, 161–77
37. Jouin, C. 1968. Sexualité et biologie de la reproduction chez *Mesonerilla* Remane et *Meganerilla* Boaden. *Cah. Biol. Mar.* 9 :31–52
38. Juberthie-Jupeau, L. 1956. Existence de spermatophores chez les Symphyles. *C. R. Acad. Sci. Paris* 243 :1164–66
39. Juberthie-Jupeau, L. 1959. Données sur les phénomènes externes de l'émission des spermatophores chez les Symphyles. *C. R. Acad. Sci. Paris* 248 :469–72
40. Juberthie-Jupeau, L., 1959. Sur une modalité nouvelle de prise des spermatophores et sur l'existence des poches spermatiques gnathales chez les Scutigerellidae. *C. R. Acad. Sci. Paris* 248 :862–65
41. Juberthie-Jupeau, L. 1959. Étude de la ponte chez les Symphyles avec mise an évidence d'une fécondation externe des oeufs par la femelle. *C. R. Acad. Sci. Paris* 249 :1821–23
42. Juberthie-Jupeau, L. 1960. Cycle d'émission des spermatophores et évolution des testicules et des vésicules séminales au cours de l'intermue chez *Scutigerella pagesi* Jupeau. *C. R. Acad. Sci. Paris* 250 : 2285–87
43. Juberthie-Jupeau, L. 1963. Recherches sur la réproduction et la mue chez les Symphyles. *Arch. Zool. Exp. Gen.* 102 :1–172
44. Junqua, C. 1966. Recherches biologiques et histophysiologiques sur un solifuge saharien *Othoes saharae* Panouse. *Tes. Fac. Sci. Univ. Paris, Ser. A* No. 4689 (5537) 1–124
45. Kaestner, A. 1965. *Lehrbuch der Speziellen Zoologie,* Vol. 1, Pt. 1. Jena : Fischer
46. Kew, H. W. 1912. On the pairing of pseudoscorpions. *Proc. Zool. Soc. London* 25 :376–90
47. Kew, H. W. 1930. On the spermatophores of the pseudoscorpions *Chthonius* and *Obisium. Proc. Zool. Soc. London* 253–56
48. Klingel, H. 1956. Indirekte Spermatophorenübertragung bei Chilopoden, beobachtet bei der "Spinnenassel" *Scutigera coleoptrata* Latz. *Naturwissenschaften* 43 :311
49. Klingel, H. 1957. Indirekte Spermatophorenübertragung beim Skolopender (*Scolopendra cingulata* Latreille). *Naturwissenschaften* 44 :338
50. Klingel, H. 1959. Indirekte Spermatophorenübertragung bei Geophiliden. *Naturwissenschaften* 46 : 632–33
51. Klingel, H. 1959. Die Paarung des *Lithobius forficatus* L. *Verh. Deut. Zool. Ges. Münster* (= *Zool. Anz. Suppl. 23*) 326–32
52. Klingel, H. 1960. Vergleichende Ver-

444 SCHALLER

haltensbiologie der Chilopoden *Scutigera coleoptrata* L. ("Spinnenassel") und *Scolopendra cingulata* Latreille (Skolopender). *Z. Tierpsychol.* 17 :11–30

53. Klingel, H. 1962. Das Paarungsverhalten des malaischen Höhlentausendfüssers *Thereuopoda decipiens cavernicola* Verhoeff. *Zool. Anz.* 169 :458–60

54. Klingel, H. 1962. Paarungsverhalten bei Pedipalpen (*Telyphonus caudatus* L., *Holopeltidia, Uropygi* und *Sarax sarawakensis* Simon) *Verh. Deut. Zool. Ges. Wien* (= *Zool. Anz. Suppl. 26*) 452–59

55. Laviale, M. L. 1964. Présence de spermatophores chéz *Stylopauropus pedunculatus* (Lubb.) *C. R. Acad. Sci.* 259 :652–54

56. Lipowsky, L. J., Byers, G. W., Kardos, E. H. 1957. Spermatophores—the mode of insemination of chiggers (Acarina : Trombiculidae). *J. Parasitol.* 43 :256–62

57. Lundblad, O. 1929. Über den Begattungsvorgang bei einigen *Arrenurus*-Arten. *Z. Morphol. Oekol. Tiere* 15 :705–22

58. Manton, S. M. 1938. Studies on the onychophora IV. The passage of spermatozoa into the ovary in *Peripatopsis* and the early development of the ova. *Phil. Trans. Roy. Soc. London, Ser. B* 228 :421–42

59. Matthiesen, F. A. 1960. Sobre o acasalamento de *Tityus bahiensis* (Perty, 1834). *Rev. Agr.* 35 :341–46

60. Mayer, H. 1956. Fortpflanzungsbiologie symphypleoner Collembolen. *Naturwissenschaften* 43 :137–38

61. Mayer, H. 1957. Zur Biologie und Ethologie einheimischer Collembolen. *Zool. Jb. System* 85 :501–71

62. Muma, M. H. 1966. Mating behavior in the solpugid genus *Eremobates* Banks. *Anim. Behav.* 14 :346–50

63. Orelli, M. von. 1956. Untersuchungen zur postembryonalen Entwicklung von *Campodea. Verh. Naturforsch. Ges. Basel* 67 :501–74

64. Pagés, J. 1967. Données sur la biologie de *Diplojapyx humberti* (Grassi). *Rev. Ecol. Biol. Sol* 4 : 187–281

65. Paget, J. 1963. Observations biologiques sur les Diploures Japygidés. *Proc. Int. Congr. Zool., 16th, Wash.* 300

66. Pauly, F. 1952. Die "Copula" der

Onbatiden. *Naturwissenschaften* 39 :572–73

67. Pauly, F. 1956. Zur Biologie einiger Belbiden u. zur Funktion ihrer pseudostigmatischen Organe. *Zool. Jb. System* 84 :275–328

68. Poggendorf, D. 1956. Über rhythmische sexuelle Aktivität und ihre Beziehung zur Häutung und Haarbildung bei arthropleonen Collembolen. *Naturwissenschaften* 43 :45

69. Rosin, R., Shulov, A., 1963. Studies on the scorpion *Nebo hierodonticus. Proc. Zool. Soc. London* 140 : 547–76

70. Sahrhage, D., 1953. Ökologische Untersuchungen an *Thermobia domestica* Packard und *Lepisma saccharina* L. *Z. Wiss. Zool.* 157 : 77–168

71. Schaller, F. 1952. Die "Copula" der Collembolen. *Naturwissenschaften* 39 :48

72. Schaller, F. 1952. Das Fortpflanzungsverhalten apterygoter Insekten (Collembolen und Machiliden). *Verh. Deut. Zool. Ges. Freiburg* (= *Zool. Anz. Suppl. 16*) 184–89

73. Schaller, F. 1953. Untersuchungen zur Fortpflanzungsbiologie arthropleoner Collembolen. *Z. Morphol. Oekol. Tiere* 41 :265–77

74. Schaller, F., 1954. Indirekte Spermatophorenübertragung bei *Campodea. Naturwissenschaften* 41 : 406–7

75. Schaller, F. 1954. Die indirekte Spermatophorenübertragung und ihre Probleme. *Forsch. Fortschr.* 28 : 321–26

76. Schaller, F. 1955. Indirekte Samenübertragung im Tierreich. *Umschau* 55 :7-9

77. Schaller, F. 1955. Zwei weitere Fälle indirekter Samenübertragung: Skorpione u. Silberfischchen. *Forsch. Fortschr.* 29 :261–63

78. Schaller, F. 1956. Deskriptive und analytische Instinktforschung bei Spinnentieren und Insekten. *Forsch. Fortschr.* 30 :225–31

79. Schaller, F. 1958. Weitere Beiträge zum Problem der indirekten Spermatophorenübertragung u. Versuch eines Systems der Verhaltensphänomene. *Forsch. Fortschr.* 32 : 200–4

80. Schaller, F. 1962. Dei Unterwelt des Tierreichs (*Verständliche Wissenschaft*, Vol. 78). Berlin, Göttingen,

Heidelberg :Springer

81. Schaller, F. 1964. Das Paarungsverhalten der Bodentiere. Bodenbewohnende Arthropoden (= Gliederfüsser) in ökologischer und stammesgeschichtlicher Sicht. *Naturwiss. Rundsch.* 17 :384–91

82. Schaller, F. 1968. *Soil Animals.* Univ. of Michigan Press, Ann Arbor Sci. Library

83. Schaller, F. 1964. Mating behaviour of lower terrestrial arthropods from the phylogenetic point of view. *Proc. Int. Congr. Entomol. London, 12th* 297–98. Sect. 5, Behaviour

84. Schliwa, W., Schaller, F. 1963. Die Paarbildung des Springschwanzes *Podura aquatica* (Apterygota, Collembola). *Naturwissenschaften* 50 : 698

85. Schliwa, W. 1965. Vergleichend anatomisch-histologische Untersuchungen über die Spermatophorenbildung bei Collembolen (mit Berücksichtigung der Dipluren und Oribatiden). *Zool. Jb. Anat.* 82 : 445–520

86. Schömann, K. 1954. Das "Paarungs"-Verhalten von *Polyxenus lagurus* L. *Naturwissenschaften* 41 :13

87. Schömann, K. 1956. Zur Biologie von *Polyxenus lagurus* (L. 1758). *Zool. Jb.* 84 :195–256

88. Schömann,, K., Schaller, F. 1964. Das Paarungsverhalten von *Polyxenus lagurus* L. *Verh. Deut. Zool. Ges. Tübingen* (= *Zool. Anz. Suppl. 18*) 342–46

89. Schuster, R. 1962. Nachweis eines Paarungszeremoniells bei den Hornmilben. *Naturwissenschaften* 49 :502

90. Schuster, R., Schuster, I. G. 1969. Gestielte Spermatophoren bei Labidostomiden. *Naturwissenschaften* 56 :145

91. Sengbusch, H. 1958. Zuchtversuche mit Oribatiden. *Naturwissenschaften* 45 :498–99

92. Southcott, R. V. 1955. Some observations on the biology, including mating and other behaviour, of the Australian scorpion *Urodacus abruptus* Pocock. *Trans. Roy. Soc. Austr.* 78 :145–54

93. Sharma, G. D., Kevan, D. K. McE. 1963. Observations on *Pseudosinella petterseni* and *Pseudosinella alba* in eastern Canada. *Pedobio-*

logia 3 :62–74

94. Shulov, A., Amitai, P. 1959. On the mating habits of two species of scorpions, *Leiurus quinquestriatus* H. et E. and *Buthotus judaicus* E. S. *Bull. Res. Counc. Isr. Sect. B* 8 :41–42

95. Shulov, A., Amitai, P. 1958. On mating habits of three scorpions, *Leiurus quinquestriatus* H. et E., *Buthotus judaicus* E. Sim, and *Nebo hierochonticus* E. Sim. *Arch. Inst. Pasteur Alger.* 36 :351–69

96. Spencer, G. J. 1930. The firebrat, *Thermobia domestica*, in Canada. *Can. Entomol.* 62 :1–11

97. Sturm, H. 1952. Die Paarung bei *Machilis* (Felsenspringer). *Naturwissenschaften* 39 :308

98. Sturm, H. 1955. Die Paarung von *Lepisma saccharina* L. (Silberfischchen). *Verh. Deut. Zool. Ges. Erlangen* (= *Zool. Anz. Suppl. 19*) 463–66

99. Sturm, H. 1955. Beiträge zur Ethologie einiger mitteldeutscher Machiliden. *Z. Tierpsychol.* 12 :337–63

100. Sturm, H. 1956. Die Paarung beim Silberfischchen *Lepisma saccharina.* *Z. Tierpsychol.* 13 :1–12

101. Sturm, H., 1958. Indirekte Spermatophorenübertragung bei dem Geisselskorpion *Trithyreus sturmi* Kraus. *Naturwissenschaften* 45 : 142–43

102. Sweetman, H. L. 1938. Physical ecology of the firebrat *Thermobia domestica* Packard. *Ecol. Monogr.* 8 :285–311

103. Taberly, G. 1957. Observations sur les spermatophores et leur transfert chéz les Oribates. *Bull. Soc. Zool. Fr.,* 82 :139–45

104. Tuzet, O., Manier, J. F. 1951. La spermiogenèse du *Lithobius calcaratus* C. Koch. *C. R. Acad. Sci. Paris* 232 :882–84

105. Tuzet, O., Manier, J. F. 1953. Les spermatozoides de quelques Myriapodes Chilopodes et leur transformation dans le réceptacle séminal de la femelle. *Ann. Sci. Nat. Zool.* 15 :221–30

106. Tuzet, O., Manier, J. F. 1956. Contribution à l'étude de la spermatogenèse des Aptérygotes Entotrophes : *Orchesella villosa* L., *Entomobrya* du groupe *nivalis* L., *Entomobrya* du groupe *nigrocincta* Denis, *Sminthurus viridis* L. Lubb.

(Collembola) et *Campodea mon-spessulana* Condé (1953) (Dip-loures). *Ann. Sci. Nat. Zool.* 18 : 15–32

107. Tuzet, O., Manier, J. F. 1957. La spermatogenèse de *Polyxenus lucidus* Chalande et de *Polyxenus lagurus* Latr. *Ann. Sci. Nat. Zool.* 19 :1–14

108. Vachon, M., 1938. Récherches anatomiques et biologiques sur la réproduction et le développement des Pseudoscorpiones. *Ann. Sci. Nat. Zool.* (11) 1 :1–207 (= Thès. Fac. Sci. Univ. Paris, Sér. A, No. 1779 [2645] 1–207)

109. Wen, T. W. 1958. Observations on the mating process of *Acomatacarus josanoi* Fukuzumi a. Obata 1953, with the discovery of its spermatophores. *Acta Zool. Sinica (Peking)* 10 :213–22

110. Westheide, W., Ax, P. 1964. Bildung und Übertragung von Spermatophoren bei Polychaeten (Untersuchungen an *Hesionides arenarius* Friedrich). *Verh. Deut. Zool. Ges. Kiel* (= *Zool. Anz. Suppl. 28*) 196–203

111. Westheide, W. 1967. Monographie der Gattungen *Hesionides* Friedrich und *Microphthalmus* Mecznikow. Ein Beitrag zur Organisation u. Biologie psammobionter Polychaeten. *Z. Morphol. Oekol. Tiere* 61 :1–159

112. Weygoldt, P. 1965. Mechanisms der Spermienübertragung bei einem Pseudoskorpion. *Naturwissenschaften* 52 :218

113. Weygoldt, P. 1965. Das Fortpflanzungsverhalten der Pseudoskorpione. *Naturwissenschaften* 52 :436

114. Weygoldt, P. 1966. Spermatophore web formation in a Pseudoscorpion. *Science* 153 :1647–49

115. Weygoldt, P. 1966. Mating behaviour and spermatophore morphology in the pseudoscorpion *Dinocheirus tumidus* Banks. *Biol. Bull.* 130 : 462–67

116. Weygoldt, P. 1966. Vergleichende Untersuchungen zur Fortpflanzungsbiologie der Pseudoskorpione. Beobachtungen über das Verhalten, die Samenübertragungsweisen u. die Spermatophoren einiger einheimischer Arten. *Z. Morphol. Oekol. Tiere* 56 :39–92

117. Weygoldt, P. 1969. Beobachtungen zur Fortpflanzungsbiologie u. zum Verhalten der Geisselspinne *Tarantula marginemaculata* C. L. Koch. *Z. Morphol. Oekol. Tiere* 64 :338–60

118. Weygoldt, P., 1969. Paarungsverhalten und Samenübertragung beim Pseudoscorpion *Withius subruber* Simon. *Z. Tierpsychol.* 26 :230–35

119. Woodring, J. P., Cook, E. F. 1962. The internal anatomy, reproductive physiology and molting process of *Ceratozetes cisalpinus*. *Ann Entomol. Soc. Am.* 55 :164–81

120. Zolessi, L. C. de. 1956. Observationes sobre el comportiamiento sexual de *Bothriurus bonariensis* (Koch). *Bol. Fac. Agron. Montevideo* 35 : 1–10

AUTHOR INDEX

SUBJECT INDEX

A

Abate, 328
Abdominal appendages, 381-82
Abdominal limb buds, 382
Abiskomyia, 216
Ablabesmyia cingulata, 222
Abrasive dusts, 148
Abrasive index, 135
Acarina
  bioecology of, 249-88
  blood-sucking, 3, 5, 17
  sperm transfer in, 429, 434
Acceptance stimuli, 172
Acertagallia fuscoscripta, 193
Acetylcholine, 351-52
Acleris variana, 308
Acrididae
  copulation by, 390-91
  genitalia of, 382, 385, 389, 394, 396
  oviposition by, 399
  ovipositor of, 399
  spermatophores of, 389
Activation of clays, 142-43
Aculeata
  silk glands of, 88
Acute toxicity of pesticides, 325, 334
Adaptation
  host-dependent, 162
  in tissue culture, 32-34, 37-38, 46-47
Ademetus barbadensis, 425, 440
Ademetus pumilio, 425
Adenosine 5-phosphate, 5
Adenosine triphosphatase, 352
Adenosine triphosphate, 237
Adephaga
  genitalia of, 394
Adult chironomids, 221-25
Aedeagus
  of leafhoppers, 174
  structure of, 385-93, 398
Aedes, 7, 35, 44
Aedes aegypti
  autogeny of, 231-32, 235-37, 241, 369, 390

blood-sucking behavior of, 1, 5, 11-12
  eyes of, 4
  and tissue culture, 29-30, 33-34, 36-37, 41, 45
Aedes albonotatus, 241
Aedes albopictus, 29, 33, 36
Aedes atropalpus, 5, 233, 241-42
Aedes a. atropalpus, 241
Aedes canadensis, 9
Aedes communis, 235
Aedes dorsalis, 238
Aedes impiger, 232
Aedes nigripes, 232
Aedes rempeli, 235
Aedes stimulans, 5, 12
Aedes taeniorhynchus, 233, 242, 369
Aedes togoi, 242
Aedes triseriatus, 369
Aestivation, 15
Agallia brachyptera, 196
Agallia constricta, 29, 31
Agallia quadrinotata, 180
Agalliidae, 179
Agallinae, 193
Agasicles, 167-68
Age structure of populations, 274
Aggregate glands, 95
Aggregations
  and bioluminescence, 104-5
  in blood-feeding, 16
  of cells in tissue culture, 35
  of Collembola, 272-73
  of mites, 271-72
Aggressive mimicry, 99
Aglaostigma, 83-84
Agrilus bilineatus, 311
Agrogorytes and leafhoppers, 195
Aldrin, 328, 334, 345, 352
Aliesterase, 352
Algicides and wildlife, 348
Allantinae, 83
Allantus, 83
Allatectomy, 236
Allochironomus crassiforceps, 224
Allodermanyssus sanguineus, 147
Allothrombium fuliginosum, 430
Altica caerulea, 166-67

Altica carduorum, 165-66
Alumina dusts, 126
Amabilis fir
  defoliation of, 309
Amblypygi
  sperm transfer in, 423-25
American cockroach, 57, 63, 124-25, 132, 146
Amino acids
  of collagens, 82
  requirements of, 42
  and silk, 73, 75-77, 81-82, 85-87
  and tissue culture, 41
Ammalo, 161
Ammalo insulata, 161, 166
Ammonium fluosilicate, 141
Ammonium silicofluoride, 140
Amphibians and pesticides, 327, 330
Amphotericin B, 45
Amylase, 185-86
Anagasta kühniella, 129
Anaitis plagiata, 168
Anemotaxis, 8
Anopheles, 8, 44, 231, 236
Anopheles elutus, 232
Anopheles gambiae, 232
Anopheles quadrimaculatus, 369
Anopheles stephensi, 29, 33
Anoplura
  blood-sucking, 2-3
Antagonism
  of pesticides, 333-34
Antennata, 433
Anteoninae and leafhoppers, 193
Antheraea, 39-40
Antheraea eucalypti, 27, 29-30, 34, 37, 42-47
Antheraea pernyi, 61, 76
Anthocoridae, 3
  copulation by, 393
Anthonomus grandis, 55
Antibiosis, 197-98
Antibiotics and tissue culture, 45
Antigenic markers, 29
Antimycin, 329, 332
Ants
  and dusts, 135
  and leafhoppers, 194
Anurida maritima, 252

461

CUMULATIVE INDEXES

VOLUMES 7-16

INDEX OF CONTRIBUTING AUTHORS

INDEX OF CHAPTER TITLES

VOLUMES 7-16